Statistical Mechanics

Part B: Time-Dependent Processes

MODERN THEORETICAL CHEMISTRY

Editors: ***William H. Miller****, University of California, Berkeley*
Henry F. Schaefer III*, University of California, Berkeley*
Bruce J. Berne*, Columbia University, New York*
Gerald A. Segal*, University of Southern California, Los Angeles*

Volume 1 ***DYNAMICS OF MOLECULAR COLLISIONS***
Part A
Edited by William H. Miller

Volume 2 ***DYNAMICS OF MOLECULAR COLLISIONS***
Part B
Edited by William H. Miller

Volume 3 ***METHODS OF ELECTRONIC STRUCTURE THEORY***
Edited by Henry F. Schaefer III

Volume 4 ***APPLICATIONS OF ELECTRONIC STRUCTURE THEORY***
Edited by Henry F. Schaefer III

Volume 5 ***STATISTICAL MECHANICS***
Part A: Equilibrium Techniques
Edited by Bruce J. Berne

Volume 6 ***STATISTICAL MECHANICS***
Part B: Time-Dependent Processes
Edited by Bruce J. Berne

Volume 7 ***SEMIEMPIRICAL METHODS OF ELECTRONIC STRUCTURE CALCULATION***
Part A: Techniques
Edited by Gerald A. Segal

Volume 8 ***SEMIEMPIRICAL METHODS OF ELECTRONIC STRUCTURE CALCULATION***
Part B: Applications
Edited by Gerald A. Segal

Statistical Mechanics

Part B: Time-Dependent Processes

Edited by
Bruce J. Berne
Columbia University, New York

SPRINGER SCIENCE+BUSINESS MEDIA, LLC

Library of Congress Cataloging in Publication Data

Main entry under title:

Statistical mechanics.

(Modern theoretical chemistry; v. 5-6)
Includes bibliographical references and indexes.
CONTENTS: pt. A. Equilibrium techniques. – pt. B. Time-dependent processes.
1. Statistical mechanics. 2. Chemistry, physical and theoretical. I. Berne, Bruce J., 1940- II. Series.
QC174.8S7 530.1'3 76-46977
ISBN 978-1-4615-7908-3 ISBN 978-1-4615-7906-9 (eBook)
DOI 10.1007/978-1-4615-7906-9

Originally published by Plenum Press, New York in 1977
Softcover reprint of the hardcover 1st edition 1977

Contributors

Bruce J. Berne, Department of Chemistry, Columbia University, New York, New York

J. R. Dorfman, Institute for Physical Science and Technology, and Department of Physics and Astronomy, University of Maryland, College Park, Maryland

Jerome J. Erpenbeck, Theoretical Division, Los Alamos Scientific Laboratory, University of California, Los Alamos, New Mexico

Leon Glass, Department of Physiology, McGill University, Montreal, Quebec, Canada

T. Keyes, Department of Chemistry, Yale University, New Haven, Connecticut

J. Kushick, The James Franck Institute, University of Chicago, Chicago, Illinois

Gene F. Mazenko, The James Franck Institute and Department of Physics, The University of Chicago, Chicago, Illinois

H. van Beijeren, Institute for Physical Science and Technology, and Department of Physics and Astronomy, University of Maryland, College Park, Maryland

William W. Wood, Theoretical Division, Los Alamos Scientific Laboratory, University of California, Los Alamos, New Mexico

Sidney Yip, Department of Nuclear Engineering, Massachusetts Institute of Technology, Cambridge, Massachusetts

Preface

The last decade has been marked by a rapid growth in statistical mechanics, especially in connection with the physics and chemistry of the fluid state. Our understanding in these areas has been considerably advanced and enriched by the discovery of new techniques and the sharpening of old techniques, ranging all the way from computer simulations to mode–mode coupling theories.

Statistical mechanics brings together under one roof a broad spectrum of mathematical methods. The aim of these volumes is to provide a didactic treatment of those topics that are most useful for the study of problems of current interest to theoretical chemists. The emphasis throughout is on the techniques themselves and not on reviewing the enormous literature in statistical mechanics. Each author was charged with the following task. Given N pages, (a) pose the problem, (b) present those aspects of the particular technique that clearly illustrate its internal workings, (c) apply the technique to the solution of several illustrative examples, and (d) write the chapter so that it will enable the reader to approach key citations to the literature intelligently.

These volumes are designed for graduate students and research workers in statistical mechanics. Nevertheless, they should be useful in other areas as well.

The choice of topics was dictated not only by the taste and interests of the editor, but also by the proviso that there did not already exist a didactic treatment of a topic in the literature. The topics fall rather neatly into two categories: equilibrium and nonequilibrium properties of fluids. Thus, Volume 5 is devoted to equilibrium techniques and Volume 6 to nonequilibrium techniques.

Volume 6 begins with two chapters on the computer simulation of molecular dynamics in fluids. The first of these chapters is concerned with fluids consisting of hard elastic particles, whereas the second of these chapters is concerned with particles that interact via a continuous pair potential. These techniques have led to a renaissance in the theory of fluids by providing an accurate picture of fluids with known force laws. The chapters on molecular dynamics are followed by two chapters on the kinetic theory of fluids. The first of the chapters covers many new topics in the kinetic theory of gases including the role of correlated collisions in producing long time-persistent effects and long time tails in time correlation functions, whereas the second of these involves the kinetic theory of dense gases and liquids and aims at a renormalized theory. This is followed by a treatment of projection operator techniques and symmetry relations mainly to provide background for a subsequent chapter on molecular hydrodynamics and mode–mode coupling theory. The book closes with a chapter on nonlinear processes in chemistry. This last chapter does not entirely fit with the rest of the book, but was judged to be of such importance to chemistry that its absence would be sorely missed.

Bruce J. Berne

Contents of Volume 6

Chapter 6. Principles of Mode–Mode Coupling Theory

T. Keyes

Chapter 7. Global Analysis of Nonlinear Chemical Kinetics

Leon Glass

Contents of Volume 5

1

Molecular Dynamics Techniques for Hard-Core Systems

Jerome J. Erpenbeck
and
William W. Wood

1. Introduction

Computational studies of equilibrium and transport properties of simple models of molecular systems have become an important part of statistical-mechanical research. Such studies include Monte Carlo calculations, which are described by Valleau in Volume 5 and the molecular dynamics calculations described in this chapter and in Chapter 2 by Kushick and Berne. Here we consider the molecular dynamics (MD) method for systems of hard-core particles. Because of the simplicity of the intermolecular interaction, the integration of the classical equations of motion is trivial and the methods used for the study of various material properties are frequently different from those for soft potentials.

Our objective here is to present a didactic exposition of the theoretical foundations and methodology of molecular dynamics rather than a review of the various results that have been obtained by these techniques. For the latter type of survey, the reader is referred to the recent review by Wood.[1]

The term "molecular dynamics method" is used here to refer to any study of the properties of matter that requires the explicit integration of the dynamical equations of motion for a many-body system. Thus our discussion

---

Jerome J. Erpenbeck and William W. Wood • Theoretical Division, Los Alamos Scientific Laboratory, University of California, Los Alamos, New Mexico. Work performed under the auspices of the U.S. Energy Research and Development Administration

includes methods that are appropriate to both equilibrium and nonequilibrium properties. Nevertheless, we introduce our discussion (Section 2) with a review of statistical-mechanical equilibrium averages, including both ensemble methods and time-averaging methods.

In the next four sections, we discuss the four principal types of application of molecular dynamics. Section 3 very briefly describes the problem of the approach to equilibrium. Section 4 deals with the evaluation of equilibrium thermodynamic functions through a discussion of the dynamical equation of state. In Section 5, we consider the evaluation of equilibrium time correlation functions, detailing the application of the combined Monte Carlo–molecular-dynamics method to the time correlation functions for self-diffusion. Section 6 deals with nonequilibrium molecular dynamics and in particular with a calculation for self-diffusion.

In the final two sections, we discuss some of the numerical considerations that are important in the implementation of a molecular dynamics calculation. Section 7 concerns the question of the accuracy of the dynamical trajectory. In Section 8 some detailed programming techniques are discussed.

2. Statistical-Mechanical Background

2.1. Canonical-Ensemble Averages

Consider a system of N identical particles, each of mass m, contained in the d-dimensional "volume" V, at temperature T. The restriction to identical particles is made principally for notational convenience and does not, of course, imply any restriction on the MD method. Let particle i have position $\mathbf{r}_i$ and velocity $\mathbf{v}_i$. We designate a point in $2dN$-dimensional phase space by

$$\mathbf{x}^N = [\mathbf{r}^N, \mathbf{v}^N]$$

$$\mathbf{r}^N = [\mathbf{r}_1, \mathbf{r}_2, \ldots, \mathbf{r}_N], \qquad \mathbf{v}^N = [\mathbf{v}_1, \mathbf{v}_2, \ldots, \mathbf{v}_N]$$

For particles interacting through the potential $U(\mathbf{r}^N)$, the canonical-ensemble probability density in phase space is

$$\begin{aligned} \rho(\mathbf{x}^N) &= Z(N, V, T)^{-1} \exp[-\beta H(\mathbf{x}^N)] \\ H(\mathbf{x}^N) &= \frac{1}{2} m \sum_{i=1}^{N} v_i^2 + U(\mathbf{r}^N) \\ Z(N, V, T) &= \int d\mathbf{x}^N \exp[-\beta H(\mathbf{x}^N)] \\ \beta &= 1/kT \end{aligned} \tag{1}$$

where k is the Boltzmann constant and the phase integral is defined by

$$\int d\mathbf{x}^N \equiv \int_V d\mathbf{r}^N \int d\mathbf{v}^N$$

$$\int_V d\mathbf{r}^N \equiv \int_V d\mathbf{r}_1 \int_V d\mathbf{r}_2 \cdots \int_V d\mathbf{r}_N, \qquad \int d\mathbf{v}^N \equiv \int d\mathbf{v}_1 \int d\mathbf{v}_2 \cdots \int d\mathbf{v}_N$$

The velocity integrals in (1) can be evaluated explicitly to yield

$$\begin{aligned} Z(N, V, T) &= (2\pi/m\beta)^{dN/2} Q(N, V, T) \\ Q(N, V, T) &= \int_V d\mathbf{r}^N \exp[-\beta U(\mathbf{r}^N)] \end{aligned} \tag{2}$$

where $Q(N, V, T)$ is called the configurational integral.

The ensemble average of a phase function $f(\mathbf{x}^N)$ is then

$$\langle f(\mathbf{x}^N)\rangle \equiv \int d\mathbf{x}^N \rho(\mathbf{x}^N) f(\mathbf{x}^N) \tag{3}$$

which reduces for functions of position to

$$\langle f(\mathbf{r}^N)\rangle = Q(N, V, T)^{-1} \int_V d\mathbf{r}^N \exp[-\beta U(\mathbf{r}^N)] f(\mathbf{r}^N) \tag{4}$$

The equilibrium thermodynamic functions of the system follow from the Helmholtz free energy $A(N, V, T)$,

$$\begin{aligned} \exp[-\beta A(N, V, T)] &= (m/h)^{dN}(N!)^{-1} Z(N, V, T) = \lambda^{-dN}(N!)^{-1} Q(N, V, T) \\ \lambda &= (\beta h^2/2\pi m)^{1/2} \end{aligned} \tag{5}$$

where h is Planck's constant. For particles interacting through the pairwise-additive potential $u(r_{ij})$,

$$\begin{aligned} U(\mathbf{r}^N) &= \sum_{i<j} u(r_{ij}) \\ \sum_{i<j} &\equiv \sum_{i=1}^{N-1} \sum_{j=i+1}^{N}, \qquad r_{ij} \equiv |\mathbf{r}_i - \mathbf{r}_j| \end{aligned} \tag{6}$$

one obtains the internal energy

$$E(N, V, T) = (\partial \beta A/\partial \beta)_{N,V} = \tfrac{1}{2} dNkT + \langle U(\mathbf{r}^N)\rangle \tag{7}$$

and the pressure

$$\begin{aligned} p &= -(\partial A/\partial V)_{N,T} \\ pV/NkT &= 1 - 2\beta\langle W(\mathbf{r}^N)\rangle/dN, \qquad W(\mathbf{r}^N) = \frac{1}{2}\sum_{i<j} r_{ij}\, du(r_{ij})/dr_{ij} \end{aligned} \tag{8}$$

The function W is the virial and (8) is the so-called virial equation of state.

The Monte Carlo method is the principal numerical method for the evaluation of expressions of the form (4), as typified by the averages appearing in (7) and (8). In the following we shall assume the reader to be familiar with the elements of this technique.

2.2. Molecular Dynamics Averages

While the calculation of the properties of matter by ensemble methods has attained preeminence in statistical mechanics, nonetheless the method of time averages along a dynamical trajectory is no less fundamental a technique to determine such properties. We define then the molecular dynamics average of a phase function $f(\mathbf{x}^N)$ as

$$\bar{f} = \lim_{t\to\infty} \bar{f}(t), \qquad \bar{f}(t) = \frac{1}{t}\int_0^t dt'\, f[\mathbf{x}^N(t')] \tag{9}$$

Here $\mathbf{x}^N(t)$ denotes the phase space trajectory determined by the classical equations of motion

$$\frac{d\mathbf{r}_i}{dt} = \mathbf{v}_i, \qquad \frac{d\mathbf{v}_i}{dt} = -\frac{1}{m}\frac{\partial}{\partial \mathbf{r}_i} U(\mathbf{r}^N) \tag{10}$$

together with certain boundary conditions (to be discussed below) and initial data $\mathbf{x}^N(0) \equiv \mathbf{x}^N$. While dynamical methods can evidently have application beyond the field of equilibrium averages, for the most part we shall be concerned with the latter. The relationship between ensemble averages and time averages then introduces the field of ergodic theory.

The quasi-ergodic hypothesis asserts the equivalence of the MD time average (9) with an ensemble average in an ensemble characterized by the constants of the dynamical motion, viz., the volume V, the number of particles N, and at least for most numerical applications, the energy $H(\mathbf{x}^N)$. (For an up-to-date discussion of the ergodic problem, see the lectures by Ford.[2]) The existence of other constants of motion depends on the boundary conditions; for periodic boundary conditions, discussed below and used in most MD studies, linear momentum

$$\mathbf{M}(\mathbf{v}^N) = m \sum_{i=1}^{N} \mathbf{v}_i \tag{11}$$

is also conserved. We refer to an ensemble characterized by specified values of N, V, energy E, and linear momentum $\mathbf{M}$ as the "molecular dynamics ensemble" and write averages therein as

$$\begin{aligned}
\langle f(\mathbf{x}^N)\rangle_{NVEM} &= \int d\mathbf{x}^N\, \rho_{NVEM}(\mathbf{x}^N) f(\mathbf{x}^N) \\
\rho_{NVEM}(\mathbf{x}^N) &= Z(N, V, E, \mathbf{M})^{-1}\, \Delta(\mathbf{x}^N; E, \mathbf{M}) \\
\Delta(x^N; E, \mathbf{M}) &= \delta[E - H(\mathbf{x}^N)]\, \delta[\mathbf{M} - \mathbf{M}(\mathbf{v}^N)] \\
Z(N, V, E, \mathbf{M}) &= \int d\mathbf{x}^N\, \Delta(\mathbf{x}^N; E, \mathbf{M})
\end{aligned} \tag{12}$$

where the δ function for a d-vector argument $\mathbf{y}$ is defined by

$$\delta(\mathbf{y}) = \prod_{j=1}^{d} \delta(y_j)$$

The time average $\bar{f}(t)$ in (9) is called the integral time average and is useful in numerical applications normally only when $f(\mathbf{x}^N)$ involves δ functions such that the time average reduces to a sum; the calculation of the equation of state discussed in Section 4 is such an application. Otherwise a summation time average is normally used,

$$\bar{f}_\Delta(S) = \frac{1}{S} \sum_{s=0}^{S-1} f[\mathbf{x}^N(s\Delta)] \tag{13}$$

under a somewhat different quasi-ergodic hypothesis,

$$\langle f(\mathbf{x}^N)\rangle_{NVEM} = \lim_{S\to\infty} \bar{f}_\Delta(S) \tag{14}$$

independent of Δ. If one writes Eq. (9) in a similar form,

$$\bar{f}(S\Delta) = \frac{1}{S} \sum_{s=0}^{S-1} \tilde{f}_\Delta(s\Delta), \qquad \tilde{f}_\Delta(s\Delta) = \frac{1}{\Delta} \int_{s\Delta}^{(s+1)\Delta} dt\, f[\mathbf{x}^N(t)]$$

then we see that the equivalence of the time averages depends upon the difference $f[\mathbf{x}^N(s\Delta)] - \tilde{f}_\Delta(s\Delta)$ having zero mean. While for systems having periodic orbits $\mathbf{x}^N(t+T) = \mathbf{x}^N(t)$ this is evidently not true for $\Delta = T$, for hard spheres or hard disks it seems reasonable but unproven. Apparently summation time averaging introduces an addition parameter Δ, which can be expected to affect the rate of convergence of (14).

It is also possible to combine ensemble averaging with time averaging by using Liouville's theorem to write, for any equilibrium distribution function,

$$\langle f(x^N)\rangle = \int d\mathbf{x}^N\, \rho(\mathbf{x}^N) f[\mathbf{x}^N(t)] \tag{15}$$

for arbitrary t. Multiplying (13) by $\rho(\mathbf{x}^N)$ and integrating over $\mathbf{x}^N$, we obtain

$$\langle f(x^N)\rangle = \langle \bar{f}_\Delta(S)\rangle \tag{16}$$

Numerically this relation can be exploited by using the Monte Carlo method to obtain a sequence $\{\mathbf{r}_p^N; p = 1, 2, \cdots, P\}$ of configurations. Using one of the standard methods (such as the Box–Muller[3]) to sample from the normal distribution so as to obtain a sequence of velocities $\{\mathbf{v}_p^N; p = 1, 2, \ldots, P\}$, then the sequence of phases $\{\mathbf{x}_p^N; p = 1, 2, \ldots, P\}$ represents a Monte Carlo sample from the distribution $\rho(\mathbf{x}^N)$.* Determining the molecular dynamic trajectories with initial phases $\mathbf{x}_p^N$, we obtain the combined Monte Carlo–molecular dynamics average

$$\bar{f}_\Delta(S, P) = \frac{1}{SP} \sum_{p=1}^{P} \sum_{s=0}^{S-1} f(\mathbf{x}_p^N(s\Delta)] \tag{17}$$

We note that this method does not depend upon the validity of (14). Moreover, a similar result using integral time averaging is readily obtained. We observe that the normal Monte Carlo sum is the special case $S = 1$ of (17) for functions independent of velocity.

2.3. Periodic Boundary Conditions

In the application of either Monte Carlo or molecular dynamics, the number of particles that can be treated is small (at most a few thousand) compared to Avogadro's number. In order that such small systems have properties similar to macroscopic systems, we impose periodic boundary conditions. It is then necessary that the volume V of the system have a shape that is d-dimensional space filling under appropriate translations. For notational convenience, we will suppose it to be a d-dimensional cube of edge $L = V^{1/d}$, although in practice in two dimensions V is usually a rectangle rather than a square. Thus we imagine space to be filled by indefinite replications of the volume V and its constituent particles.

If particle i is in primary volume V at $\mathbf{r}_i$, then "image" particles are located at positions $\mathbf{r}_i + \boldsymbol{\nu}L$ for all d-vectors $\boldsymbol{\nu}$ having integer components, $-\infty < \nu_l < \infty$. Under periodic boundary conditions, each particle interacts with all other particles, whence the potential energy and virial become

$$\begin{aligned} U(\mathbf{r}^N) &= \sum_{\boldsymbol{\nu}} \sum_{i<j} u(|\mathbf{r}_{ij} + \boldsymbol{\nu}L|), \qquad \sum_{\boldsymbol{\nu}} \equiv \sum_{\nu_1=-\infty}^{\infty} \sum_{\nu_2=-\infty}^{\infty} \cdots \sum_{\nu_d=-\infty}^{\infty} \\ W(\mathbf{r}^N) &= \frac{1}{2} \sum_{\boldsymbol{\nu}} \sum_{i<j} |\mathbf{r}_{ij} + \boldsymbol{\nu}L| \, du(|\mathbf{r}_{ij} + \boldsymbol{\nu}L|)/d(|\mathbf{r}_{ij} + \boldsymbol{\nu}L|) \end{aligned} \tag{18}$$

For hard-core interactions,

$$u(r) = \begin{cases} 0 & \text{for } r > \sigma \\ \infty & \text{for } r < \sigma \end{cases} \tag{19}$$

*To obtain Monte Carlo averages in the microcanonical ensemble, one can radially project the velocities $\mathbf{v}_p^N$ onto the hypersphere of constant energy.

which we assume hereafter, the $\boldsymbol{\nu}$-sum in (18) reduces to neighbors of the primary volume only, i.e., $-1 \leq \nu_l \leq 1$ provided $\sigma < L/2$.

The properties of such periodic systems are discussed in some detail elsewhere[4] and we state the results of interest.

1. The system has d ideal-gas degrees of freedom corresponding to free translation of the center of mass.

2. The singlet particle density $\rho_1(\mathbf{r}) = \langle \sum_i \delta(\mathbf{r}_i - \mathbf{r}) \rangle$ is uniform, $\rho_1 = n = N/V$.

3. The radial distribution function

$$g(\mathbf{s}_1, \mathbf{s}_2) = n^{-2} \left\langle \sum_{\boldsymbol{\nu}_1} \sum_{\boldsymbol{\nu}_2} \sum_{i<j} \delta(L\boldsymbol{\nu}_1 + \mathbf{r}_i - \mathbf{s}_1)\, \delta(L\boldsymbol{\nu}_2 + \mathbf{r}_j - \mathbf{s}_2) \right\rangle$$

depends only on $\mathbf{s}_{12} = \mathbf{s}_1 - \mathbf{s}_2$. For finite N, g depends both on the magnitude and orientation of $\mathbf{s}_{12}$, with the directional dependence vanishing in the thermodynamic limit for fluids.

4. Except perhaps at very low densities, the volume V should be chosen to be a unit cell for the crystallographic lattice appropriate to the high-density behavior of the system. For three-dimensional systems this will be either the face-centered or hexagonal close-packed lattice. For hard disks, it is the planar triangular lattice for which the possible values of N are of the form $2\xi_1\xi_2$ with ξ_1 and ξ_2 integers and V is a rectangle with sides in the ratio $\sqrt{3}\xi_2/\xi_1$. For hard spheres, V is usually taken to be a cube, with N of the form $4\xi^3$, with ξ an integer. For both hard spheres and disks, we will express the volume V in units of the close-packed volume V_0, viz.,

$$\tau = V/V_0, \qquad V_0 = (3/2d)^{1/2} N\sigma^d, \quad d = 2 \text{ or } 3 \tag{20}$$

While the imposition of periodic boundary conditions is desirable, it should be mentioned particularly in connection with so-called nonequilibrium molecular dynamics that other boundary conditions have been used. This will be discussed in Section 6, but we note that a greater dependence on system size can be expected in such calculations.

3. Approach to Equilibrium

In this and the following three sections we discuss the principal types of application of the molecular dynamics method. We begin with the problem of the approach to equilibrium.

As noted in Section 2, the molecular dynamics ensemble is characterized by fixed values of N, V, E and $\mathbf{M}$. All states $\mathbf{x}^N$ consistent with these values are equiprobable. According to the quasi-ergodic hypothesis, the trajectory $\mathbf{x}^N(t)$ starting from $\mathbf{x}^N$ should, except possibly for a set of initial phases of zero

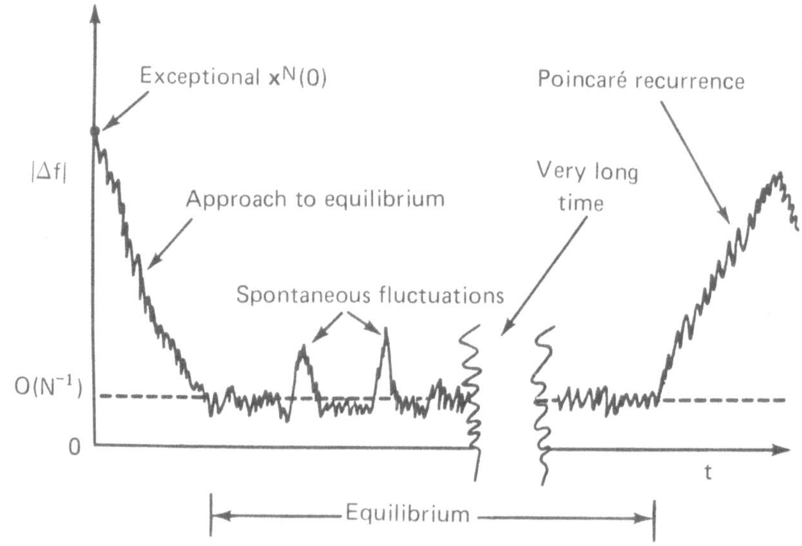

Fig. 1. Time evolution of a dynamical system (reproduced from Wood[1] with permission of the publisher).

measure, spend equal amounts of time in regions of allowed phase space having equal measure. Nonetheless there exist exceptional states $\mathbf{x}^N$ for which the short-time behavior of $\mathbf{x}^N(t)$ is unusual in that the phase function $f(\mathbf{x}^N)$ corresponding to a macroscopic observable departs from its ensemble average significantly. In Fig. 1 we illustrate such a situation by a sketch of $\Delta f(t) = f[\mathbf{x}^N(t)] - \langle f(\mathbf{x}^N)\rangle_{NVEM}$. The initially large values of Δf tend to decrease with time until the fluctuations become of order $1/N$, with only rare fluctuations of larger magnitude.

An example of such an initial state is provided by the situation in which the particles in one-half of V have much higher energy than those in the other half, at least if N is of macroscopic magnitude. Other such exceptional states will not necessarily appear so unusual. Nonetheless we speak of the system as non-equilibrium initially, as approaching equilibrium when Δf is large and decreasing, and as in equilibrium at subsequent times, when Δf is of order $1/N$. While large fluctuations should reappear after the Poincaré recurrence time, this time is enormous for fluid systems, at least if N is not too small.

Despite the evident appropriateness of the MD method to the quantitative study of the approach to equilibrium, the few published studies (see Wood[1] for a discussion of some of these) have provided largely qualitative information.

4. Equilibrium Thermodynamic Functions

The study of the equilibrium properties of hard-core systems through MD was pioneered by Alder and Wainwright.[5] For the most part these equilibrium calculations have been concerned with the equation of state and the question of the existence of a first-order phase transition.

The calculations of Alder and Wainwright use time-averaging exclusively, as described in Section 2. The definition then of a dynamical pressure and its relation to the canonical ensemble pressure are matters of central importance.

It has been customary to base the dynamical pressure on the virial theorem. However, the usual derivation, as given for example by Hirschfelder *et al.*,[6] is based on rigid-wall boundary conditions, rather than the periodic boundary conditions usually employed in the numerical calculations. As we shall show below, for periodic boundary conditions the virial theorem actually leads to a somewhat unusual result. Thus we derive a dynamical pressure based on the momentum flux across an element of area within the fluid, as discussed for example by Chapman and Cowling.[7] As we shall see, the resulting expression for the pressure is identical to the one usually attributed to the virial theorem.

4.1. Dynamical Pressure

Consider then a system of N hard-core particles in volume V subject to periodic boundary conditions. Define the configuration $\mathbf{R}^N(t)$ as a function of time by following the N particles, initially in the primary cell, as they move through the infinite-checkerboard space illustrated in Fig. 2. Let $\mathbf{R}$ denote a point in the fluid, say in the primary cell, with surface element dS (unit normal $\mathbf{n}$). The pressure $\mathbf{p}(\mathbf{n})$ across dS in the direction $\mathbf{n}$ is defined as the average

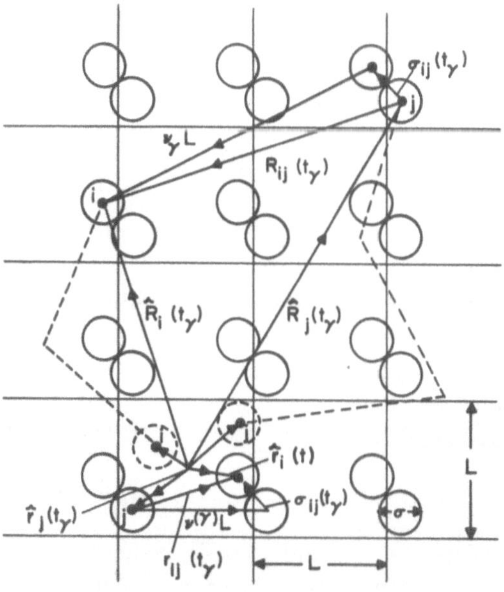

Fig. 2. Dynamics of hard disks for periodic boundary conditions in the center of mass frame of reference. The broken circles show the initial positions of particles i and j, which collide at time t_γ. The broken lines show the particle paths from $t=0$ to $t=t_\gamma$ in the infinite-checkerboard version of periodic boundary conditions for which the particle coordinates are $\hat{\mathbf{R}}_i(t_\gamma)$ and $\hat{\mathbf{R}}_j(t_\gamma)$. The solid circles show the position of the particles at t_γ including the "image" particles. In the single-cell version of periodic boundary conditions, the particles are at $\hat{\mathbf{r}}_i(t_\gamma)$ and $\hat{\mathbf{r}}_j(t_\gamma)$ in the original cell.

momentum flux per unit area of surface. The momentum flux evidently has two contributions, one due to molecular flow and the other from collisional transfer,

$$\mathbf{p}(\mathbf{n}) = \mathbf{p}_{\text{mf}}(\mathbf{n}) + \mathbf{p}_{\text{ct}}(\mathbf{n}) \tag{21}$$

The rate of flow of momentum per unit area of surface at time t from the motion of particles across dS is

$$\mathbf{p}_{\text{mf}}(\mathbf{R}, t, \mathbf{n}) = \sum_{\boldsymbol{\nu}} \sum_{i=1}^{N} \delta[\mathbf{R}_i(t) + \boldsymbol{\nu}L - \mathbf{R}]\mathbf{n} \cdot \hat{\mathbf{v}}_i(t) m \mathbf{v}_i(t) \tag{22}$$

where we have taken the surface element to be moving with center-of-mass velocity $\mathbf{v}_0$,

$$\mathbf{v}_0 = \frac{1}{N} \sum_{i=1}^{N} \mathbf{v}_i(t) = \frac{\mathbf{M}}{Nm} \tag{23}$$

and where

$$\hat{\mathbf{v}}_i(t) = \mathbf{v}_i(t) - \mathbf{v}_0 \tag{24}$$

is the velocity relative to the center of mass. Because of the motion of the surface element the pressure will not include the d ideal-gas degrees of freedom associated with the center-of-mass motion. Averaging $\mathbf{R}$ over the volume V and averaging over time, we obtain

$$\mathbf{p}_{\text{mf}}(\mathbf{n}) = \mathbf{P}_{\text{mf}} \cdot \mathbf{n}, \qquad \mathbf{P}_{\text{mf}} = \frac{m}{V} \sum_{i=1}^{N} \overline{\hat{\mathbf{v}}_i \hat{\mathbf{v}}_i} \tag{25}$$

where $\mathbf{P}$ denotes the pressure tensor.

The collisional transfer contribution arises from collisions across dS. The contribution is impulsive for hard-core particles so that the time integral includes a contribution from each collision,

$$\mathbf{p}_{ct}(\mathbf{R}, t, \mathbf{n}) = \sum_{\gamma=1}^{c(t)} m\, \delta\mathbf{v}_\gamma(\mathbf{R}, \mathbf{n})\, \delta(t - t_\gamma) \tag{26}$$

where the sum is over the $c(t)$ collisions at times $t_1, t_2, \ldots$ up to time t, and where $\delta\mathbf{v}_\gamma(\mathbf{R}, \mathbf{n})$ is the change in velocity per unit area across dS on the γth collision. Denote the particles in the γth collision by $i(\gamma)$ and $j(\gamma)$ at $\mathbf{R}_{i(\gamma)}(t_\gamma)$ and $\mathbf{R}_{j(\gamma)}(t_\gamma)$, respectively, having "images" in contact in each of the periodic cells. Denote the line of centers by $\boldsymbol{\sigma}_{i(\gamma)j(\gamma)}$, the vector from the image of particle j to the image of particle i, with $|\boldsymbol{\sigma}_{ij}| = \sigma$, and define $\boldsymbol{\nu}_\gamma$, as shown in Fig. 2, so that

$$\mathbf{R}_{ij} = \boldsymbol{\sigma}_{ij} + \boldsymbol{\nu}_\gamma L \tag{27}$$

where $i = i(\gamma)$ and $j = j(\gamma)$. Now the velocity change $\delta\mathbf{v}_\gamma$ vanishes unless particle j (or its appropriate image) lies in the cylinder with base dS and axis $\boldsymbol{\sigma}_{ij}$, with volume $|\mathbf{n} \cdot \boldsymbol{\sigma}_{ij}|dS$. Because $\delta \mathbf{v}_\gamma \, dS$ for particle j in this element of volume is $\Delta \mathbf{v}_i = \mathbf{v}_i' - \mathbf{v}_i$ (where the prime denotes the postcollision velocity) if $\mathbf{n} \cdot \boldsymbol{\sigma}_{ij}$ is positive, and $\Delta \mathbf{v}_j$ if $\mathbf{n} \cdot \boldsymbol{\sigma}_{ij}$ is negative, and because

$$\Delta \mathbf{v}_i = -\Delta \mathbf{v}_j$$

(from momentum conservation), the average of $\mathbf{p}_{\text{ct}}$ over $\mathbf{R}$ and t yields

$$\bar{\mathbf{p}}_{\text{ct}}(t, \mathbf{n}) = \frac{m}{Vt} \sum_{\gamma=1}^{c(t)} \mathbf{n} \cdot \boldsymbol{\sigma}_{ij}(t_\gamma) \Delta \mathbf{v}_i(t_\gamma)$$

Thus

$$\bar{\mathbf{p}}_{ct}(\mathbf{n}) = \mathbf{P}_{ct} \cdot \mathbf{n}, \qquad \mathbf{P}_{ct} = \lim_{t \to \infty} \frac{m}{Vt} \sum_{\gamma=1}^{c(t)} \boldsymbol{\sigma}_{ij}(t_\gamma) \, \Delta \mathbf{v}_i(t_\gamma) \tag{28}$$

We determine now the molecular dynamic pressure as $d^{-1} \operatorname{Tr}(\mathbf{P})$ and obtain

$$p_{\text{MD}} = \frac{2}{dV}(\hat{E} - \bar{W})$$

$$\hat{E} = \frac{1}{2} m \sum_{i=1}^{N} \hat{v}_i^2, \qquad \bar{W}(t) = -\frac{m}{2t} \sum_{\gamma=1}^{c(t)} \boldsymbol{\sigma}_{ij}(t_\gamma) \cdot \Delta \mathbf{v}_i(t_\gamma) \tag{29}$$

where $\hat{E}$ is the kinetic energy in the center-of-mass frame of reference and $\bar{W}$ the time-averaged virial of (18), as seen from the form

$$W(\mathbf{r}^N) = -\frac{1}{2} \sum_{\boldsymbol{\nu}} \sum_{i<j} (\mathbf{r}_{ij} + \boldsymbol{\nu} L) \cdot \mathbf{F}_{ij}(\mathbf{r}_{ij} + \boldsymbol{\nu} L) \tag{30}$$

where $\mathbf{F}_{ij}$ is the force on i due to particle j.

4.2. Virial Theorem

While (29) has the form of the usual virial-theorem equation of state, its derivation from the virial theorem appears to fail for periodic boundary conditions. Consider, for example, the virial in the *infinite checkerboard*,

$$Z(t) = -\frac{1}{2} \sum_{i=1}^{N} \mathbf{R}_i(t) \cdot \mathbf{F}_i(t)$$

where $\mathbf{F}_i$ is the force on particle i, $m \, d^2\mathbf{R}_i/dt^2$. Integration by parts yields

$$\overline{Z(t)} = E - \Delta(t)$$

$$E = \frac{1}{2} m \sum_{i=1}^{N} v_i^2, \qquad \Delta(t) = \frac{m}{2t} \sum_{i=1}^{N} [\mathbf{R}_i(t) \cdot \mathbf{v}_i(t) - \mathbf{R}_i(0) \cdot \mathbf{v}_i(0)] \tag{31}$$

Noting that $\Delta(t)$ has a nonvanishing component due to the center-of-mass motion, we define

$$\mathbf{R}_0(t) = \frac{1}{N}\sum \mathbf{R}_i(t), \qquad \hat{\mathbf{R}}_i(t) = \mathbf{R}_i(t) - \mathbf{R}_0(t)$$

with $\mathbf{R}_0(t) = \mathbf{R}_0(0) + \mathbf{v}_0 t$ and obtain

$$\bar{Z}(t) = \hat{E} - \hat{\Delta}(t)$$

$$\hat{\Delta}(t) = \frac{m}{2t}\sum_{i=1}^{N} [\hat{\mathbf{R}}_i(t)\cdot\hat{\mathbf{v}}_i(t) - \hat{\mathbf{R}}_i(0)\cdot\hat{\mathbf{v}}_i(0)]$$

Because $|\hat{\mathbf{v}}_i(t)| \leq (2\hat{E}/m)^{1/2}$,

$$|\hat{\Delta}(t)| \leq \frac{m}{2t}\left(\frac{2\hat{E}}{m}\right)^{1/2}\sum_{i=1}^{N}|\hat{\mathbf{R}}_i(t)| + O(t^{-1})$$

and because $|\hat{\mathbf{R}}_i|$ is expected to grow as $t^{1/2}$, $\hat{\Delta}(t)\to 0$ and

$$\bar{Z} = \hat{E} \tag{32}$$

The usual derivation recomputes $\bar{Z}$ as the sum of a collision part and an external force part, but here there is no external force. The collision calculation is nonetheless of interest, for writing

$$\mathbf{v}_i(t) = \mathbf{v}_i(0) + \sum_{\gamma=1}^{\infty} \Delta\mathbf{v}^{(i)}(t_\gamma)A(t-t_\gamma) \tag{33}$$

where $\Delta\mathbf{v}^{(i)}(t_\gamma)$ is the velocity change of particle i at collision time t_γ and $A(x)$ the unit step function, one readily obtains

$$\bar{Z}(t) = \bar{W}(t) + \bar{Q}(t), \qquad \bar{Q}(t) = -\frac{mL}{2t}\sum_{\gamma=1}^{c(t)} \boldsymbol{\nu}_\gamma\cdot\Delta\mathbf{v}_i(t_\gamma) \tag{34}$$

where $\boldsymbol{\nu}_\gamma$ is the cell translation vector of Fig. 2 as defined in (27). Combining (34), (32), and (29), we obtain the remarkable relation

$$p_{\mathrm{MD}} = 2\bar{Q}/dV \tag{35}$$

which evidently permits a rather different means of computing the equation of state than (29).

Some additional insight into the significance of (35) can be obtained by writing the equivalent relation for a general interaction law. The same arguments as used above for hard spheres lead again to (29), with the previously conserved kinetic energy in the center-of-mass reference frame $\hat{E}$ being now replaced by its time average $\bar{\hat{E}}$, and with $\bar{W}$ being replaced by the time average of the general virial function (30),

$$\bar{W} = -\frac{1}{2}\lim_{t\to\infty}\sum_{\boldsymbol{\nu}}\sum_{i<j}\frac{1}{t}\int_0^t dt'\,[\mathbf{R}_{ij}(t') + \boldsymbol{\nu}L]\cdot\mathbf{F}_{ij}[\mathbf{R}_{ij}(t') + \boldsymbol{\nu}L]$$

Here we use the infinite-checkerboard coordinates $\mathbf{R}^N(t)$ of the particles that initially are in the primary ($\boldsymbol{\nu} = 0$) cell. If we make the decomposition

$$\bar{W} = \bar{W}_1 + \bar{W}_2$$

$$\bar{W}_1 = -\frac{1}{2} \lim_{t\to\infty} \sum_{\boldsymbol{\nu}} \sum_{i<j} \frac{1}{t} \int_0^t dt' \, \mathbf{R}_{ij}(t') \cdot \mathbf{F}_{ij}[\mathbf{R}_{ij}(t') + \boldsymbol{\nu}L]$$

$$\bar{W}_2 = -\frac{L}{2} \lim_{t\to\infty} \sum_{\boldsymbol{\nu}} \sum_{i<j} \frac{1}{t} \int_0^t dt' \, \boldsymbol{\nu} \cdot \mathbf{F}_{ij}[\mathbf{R}_{ij}(t') + \boldsymbol{\nu}L]$$

then by the same arguments used to obtain (32), we find

$$\bar{W}_1 = \bar{\hat{E}}$$

Thus an alternative form of the equation of state for a system with a general interaction law, equivalent to (35) for a hard-core system, is

$$p = -\frac{2}{dV} \bar{W}_2$$

It should be noted that the separation of $\bar{W}$ into $\bar{W}_1$ and $\bar{W}_2$ is dynamical in character, and that there appears to be no analogous statistical ensemble average equivalent to this form of the equation of state.

4.3. Relationship of Dynamical and Canonical-Ensemble Pressures

In order to relate the dynamical pressure (29) to the canonical-ensemble piessure (8), we compute

$$\langle W \rangle = Z(N, V, T)^{-1} \int d\mathbf{x}^N \exp\left[-\frac{1}{2}\beta m \sum_{i=1}^{N} v_i^2 \right] B(\mathbf{r}^N) W(r^N) \tag{36}$$

$$B(\mathbf{r}^N) = \begin{cases} 1 & \text{if } U(\mathbf{r}^N) = 0 \\ 0 & \text{if } U(\mathbf{r}^N) = \infty \end{cases}$$

Introduce on the right-hand side the factor

$$\int_0^\infty dE \int d\mathbf{M} \, \Delta(\mathbf{x}^N; E, \mathbf{M}) = 1$$

with Δ the product of δ functions in (12). Inverting the order of integration and introducing the definition of the molecular dynamics ensemble average (12), we have

$$\langle W \rangle = Z(N, V, T)^{-1} \int_0^\infty dE e^{-\beta E} \int d\mathbf{M} Z(N, V, E, M) \langle W \rangle_{NVEM} \tag{37}$$

By the quasi-ergodic hypothesis we replace $\langle W\rangle_{NVEM}$ by $\bar{W}$ and determine the dependence of $\bar{W}$ on E and $\mathbf{M}$ as follows. From the conservation of momentum, one has

$$\Delta\mathbf{v}_i = \frac{1}{2}\Delta\mathbf{v}_{ij}$$

Thus $\bar{W}(t)$, Eq. (29), depends on relative coordinates and velocities only and therefore for fixed N, V, $\mathbf{r}^N$, and $\hat{\mathbf{v}}^N$ is independent of $\mathbf{M}$. Moreover, the hard-core trajectory scales with the magnitude of the velocity as

$$\mathbf{r}^N(t;\mathbf{r}^N,\alpha\mathbf{v}^N) = \mathbf{r}^N(\alpha t;\mathbf{r}^N,\mathbf{v}^N), \qquad \mathbf{v}^N(t;\mathbf{r}^N,\alpha\mathbf{v}^N) = \alpha\mathbf{v}^N(\alpha t;\mathbf{r}^N,\mathbf{v}^N) \tag{38}$$

where $\mathbf{x}^N(t;\mathbf{x}^N)$ denotes the trajectory with initial data $\mathbf{x}^N$. Thus, using similar notation, we find

$$\bar{W}(t;\mathbf{r}^N,\alpha\mathbf{v}^N) = \alpha^2\bar{W}(\alpha t;\mathbf{r}^N,\mathbf{v}^N)$$

so that we can write

$$\bar{W} = \hat{E}h(n,N) \tag{39}$$

where $n = N/V$ is the number density. Substituting into (37) and integrating over E and $\mathbf{M}$, we find

$$\langle W\rangle = h(n,N)\langle\hat{E}\rangle$$

Evaluating $\langle\hat{E}\rangle = (1-1/N)\langle E\rangle$ and replacing $h(n,N)$ by the dynamical pressure, we find the canonical-ensemble pressure from (8) to be

$$\left(\frac{pV}{NkT}-1\right)_{NVT} = \left(\frac{dpV}{2\hat{E}}-1\right)_{MD}(1-1/N) \tag{40}$$

Hoover and Alder[8] found the Monte Carlo[9] and molecular dynamics results for the equation of state of systems of 12 hard disks to be consistent with (40) to within statistical error.

4.4. Pressure from Collision Rate

Another method for determining the dynamical pressure is from the collision rate.[5,8] In this method, one rewrites (29) as

$$\left(\frac{dpV}{2\hat{E}}-1\right)_{MD} = \frac{m\Lambda\Theta}{2\hat{E}}$$
$$\Lambda = \lim_{t\to\infty}\frac{c(t)}{t}, \qquad \Theta = \lim_{t\to\infty}\frac{1}{c(t)}\sum_{\gamma=1}^{c(t)}\boldsymbol{\sigma}_{ij}\cdot\Delta\mathbf{v}_i(t_\gamma) \tag{41}$$

where Λ is the collision rate and Θ the mean value per collision of $\boldsymbol{\sigma}_{ij}\cdot\Delta\mathbf{v}_i$.

While Θ can be determined in several ways, the simplest is perhaps to observe that it is independent of density and use (40) to introduce the *NVT*-ensemble pressure.

$$\Theta = \frac{2N\hat{E}n}{m(N-1)}\left(\frac{pV}{NkT}-1\right)_{NVT}\Big/\Lambda n$$

Taking the limit as $n \to 0$, we obtain Θ in terms of the second virial coefficient $B_2(N)$ and the low-density collision rate Λ_0. Using the known[10] N dependence of B_2, we obtain the equation of state in the form

$$\left(\frac{dpV}{2\hat{E}}-1\right)_{MD} = \frac{\pi^{d/2}\sigma^d n\Lambda}{d\Gamma(d/2)\Lambda_0} \tag{42}$$

where $\Gamma(x)$ is the gamma function.

For a finite system with periodic boundary conditions, one readily finds the molecular dynamics ensemble average low-density collision rate to be

$$\Lambda_0 = \frac{\pi^{(d-1)/2}N(1-1/N)\sigma^{d-1}n}{(d-1)\Gamma[(d-1)/2]}\langle v_{12}\rangle_{NVEM}$$

The evaluation of $\langle v_{12}\rangle$ requires the *NVEM*-ensemble velocity distribution function for hard-core particles. From Eq. (12) we have

$$\begin{aligned} p(\mathbf{v}^N) &\equiv \int d\mathbf{r}^N \rho_{NVEM}(\mathbf{x}^N) = \Delta(v^N; E, \mathbf{M})/Q_v(N, E, \mathbf{M}) \\ \Delta(\mathbf{v}^N; E, \mathbf{M}) &= \delta\Big[2E/m - \sum_i v_i^2\Big]\,\delta\Big[\mathbf{M}/m - \sum \mathbf{v}_i\Big] \\ Q_v(N, E, \mathbf{M}) &= \int d\mathbf{v}^N\, \Delta(\mathbf{v}^N; E, \mathbf{M}) \end{aligned} \tag{43}$$

The normalization Q_v is calculated by transformation to velocities $\hat{\mathbf{v}}^N$ relative to the center of mass, followed by an orthogonal transformation of the velocities $\mathbf{u}^N = \mathbf{T}\cdot\hat{\mathbf{v}}^N$ such that $\mathbf{u}_1 = N^{-1/2}\sum \hat{\mathbf{v}}_i$. Because the sum of squares $\sum v_i^2$ transforms to $\sum u_i^2$, one obtains

$$Q_v(N, E, \mathbf{M}) = \int d\mathbf{u}_2\, d\mathbf{u}_3 \cdots d\mathbf{u}_N\, \delta\Big[2\hat{E}/m - \sum_{i=2}^{N} u_i^2\Big]$$

which is readily evaluated in terms of the surface area of the $d(N-1)$-hypersphere of radius $(2\hat{E}/m)^{1/2}$. One obtains

$$Q_v(N, E, \mathbf{M}) = \frac{\pi^{(N-1)d/2}}{N^{d/2}\Gamma[(N-1)d/2]}\left(\frac{2\hat{E}}{m}\right)^{[(N-1)d/2]-1} \tag{44}$$

For the calculation of $\langle v_{12}\rangle$ it is convenient to write

$$\begin{aligned} \langle v_{12}\rangle_{NVEM} &= \int d\mathbf{v}^N\, p(\mathbf{v}^N)v_{12} = \int d\mathbf{v}_{12}\, h(\mathbf{v}_{12})v_{12} \\ h(\mathbf{v}) &= \int d\mathbf{v}^N\, p(\mathbf{v}^N)\,\delta(\mathbf{v}-\mathbf{v}_{12}) \end{aligned} \tag{45}$$

Table 1. Molecular Dynamics Equation of State $dpV/2\hat{E}-1$ for 12 Hard Disks at Reduced Volume $\tau=2$

Formula used	$dpV/2\hat{E}-1$	$c(t)$
Eq. (29)	2.53 ± 0.02	13,563
Eq. (35)	2.62 ± 0.17	13,563
Eq. (42)	2.53 ± 0.02	13,563
Eq. (29)[a]	2.54 ± 0.04	4,000
Eq. (40)[b]	2.59 ± 0.07	—

[a] From Hoover and Alder.[8] [b] Monte Carlo data from Wood.[9]

The distribution function $h(\mathbf{v})$ of relative velocity can be evaluated by transformations of the velocities, first to $\hat{\mathbf{v}}^N$ and thence by the orthogonal transformation $\mathbf{w}^N=\mathbf{R}\cdot\hat{\mathbf{v}}^N$ such that $\mathbf{w}_1=\mathbf{v}_{12}/\sqrt{2}$ and $\mathbf{w}_2=N^{-1/2}\sum\hat{\mathbf{v}}_i$. One readily obtains

$$h(\mathbf{v})=\frac{\Gamma[(N-1)d/2]}{\Gamma[(N-2)d/2]}\left(\frac{m}{4\pi\hat{E}}\right)^{d/2}\left(1-\frac{mv^2}{2\hat{E}}\right)^{[(N-2)d-2]/2} \tag{46}$$

from which follows

$$\langle v_{12}\rangle_{NVEM}=\frac{\Gamma[(d+1)/2]\Gamma[(N-1)d/2]}{\Gamma(d/2)\Gamma[((N-1)d+1)/2]}\left(\frac{4\hat{E}}{m}\right)^{1/2} \tag{47}$$

We obtain then the low-density collision rate,*

$$\Lambda_0=\Lambda_0(\infty)\frac{(1-1/N)(Nd/2)^{1/2}\Gamma[(N-1)d/2]}{\Gamma[((N-1)d+1)/2]}$$
$$\Lambda_0(\infty)=\frac{N}{2t_{00}},\qquad t_{00}=\frac{\Gamma(d/2)}{2\pi^{(d-1)/2}\sigma^{d-1}n}\left(\frac{mNd}{2\hat{E}}\right)^{1/2} \tag{48}$$

where $\Lambda_0(\infty)$ and t_{00} are the infinite-system, low-density values of the collision rate and the mean free time. Using (48) in (42) one then obtains a relation by means of which the pressure can be obtained from a molecular dynamics calculation of the collision rate Λ.

It should be noted that the pressure determined from (42) becomes identical with the strictly dynamical pressure (29) or (35) only through the application of the quasi-ergodic hypothesis in proceeding from (41) to (42).

The three formulas (29), (35), and (42) can all be used to determine the molecular dynamics pressure. The resulting values from a molecular dynamics calculation for 12 hard disks at a reduced volume $\tau=2$ are shown in Table 1 along with the value reported by Hoover and Alder[8] from (29) and the value

*This exact result for the low-density collision rate in the molecular dynamics ensemble differs slightly in $O(N^{-1})$ from the approximate expression given by Hoover and Alder.[8]

from (40) using the Monte Carlo estimate[9] of NVT-ensemble pressure. The values are seen to agree within the statistical errors, but we note that the pressure determined from (35) is not nearly so precise as the others.

5. Equilibrium Time Correlation Functions

The statistical-mechanical theory of transport processes has led in recent years to the study of the time correlation function expressions for the transport coefficients,

$$\mu = \lim_{t\to\infty} c_\mu \int_0^t dt' \, \tilde{\rho}_\mu(t') \tag{49}$$

where μ denotes generically a transport coefficient, c_μ is a constant, and $\tilde{\rho}_\mu$ denotes a time correlation function

$$\tilde{\rho}_\mu(t) = \text{tlim}\, \rho_\mu(t), \qquad \rho_\mu(t) = \langle J_\mu(0) J_\mu(t)\rangle \tag{50}$$

Here J_μ is the appropriate microscopic current and tlim denotes the thermodynamic limit. (For reviews of this theory, see Zwanzig[11] and Steele.[12])

We will here focus our attention on the self-diffusion process, both because it is the simplest case from a pedagogical standpoint and because it is the one that has been most extensively studied by means of molecular dynamics calculations. Calculations of the coefficients of viscosity and thermal conductivity and the associated time-correlation functions have been reported by Alder *et al.*[13] for hard spheres.

5.1. Time Correlation Functions for Self-Diffusion

The self-diffusion process is discussed by Dorfman,[14] who writes the self-diffusion current and conservation law as

$$\begin{aligned} \mathbf{j}_1(\mathbf{r}, t) &= -D^{(0)}(t)\nabla n_1(\mathbf{r}, t) - D^{(2)}(t)\nabla\nabla^2 n_1(\mathbf{r}, t) - \cdots \\ \partial n_1(\mathbf{r}, t)/\partial t &= -\nabla \cdot \mathbf{j}_1(\mathbf{r}, t) \end{aligned} \tag{51}$$

where n_1 is the number density of a tagged particle at position $\mathbf{r}$ at time t, ∇ the gradient operator, $D^{(0)}(t)$ the *time-dependent* Fick's law self-diffusion coefficient, and $D^{(2)}(t)$ the *time-dependent* super-Burnett coefficient. No first-order Burnett term is present due to the symmetry of the process.

1. Fick's Law Coefficient

The Fick's law coefficient $D^{(0)}(t)$ is given by

$$D^{(0)}(t) = (\beta m)^{-1} \int_0^t dt' \rho_D(t') \tag{52}$$

in which

$$\rho_D(t) = \beta m \langle v_{1x}(0) v_{1x}(t) \rangle \tag{53}$$

is the familiar normalized velocity autocorrelation function. Alternatively, one can write $D^{(0)}$ in the form

$$D^{(0)}(t) = \langle v_{1x}(0) \Delta x_1(t) \rangle \tag{54}$$

where

$$\Delta x = x_1(t) - x_1(0)$$

is the x component of displacement of particle 1 at time t. Finally, the mean square x component of displacement is related to $D^{(0)}$

$$D^{(0)}(t) = (d/dt) t \Delta^{(0)}(t), \qquad \Delta^{(0)}(t) = \langle \Delta x_1{}^2(t) \rangle / 2t \tag{55}$$

For periodic boundary conditions we will understand the particle coordinates that enter the formulas to be the infinite-checkerboard description, e.g., $\Delta x_1(t) = \Delta \mathbf{R}_1(t) \cdot \hat{\mathbf{e}}_x$, so that the transformation of (52) to (54) holds. In making this identification, one also expects to minimize effects arising from the finite size of the system.

Equations (52), (54), and (55) are all suitable for the calculation of the Fick's law transport coefficient. Indeed, Alder and Wainwright[15,16] studied the velocity autocorrelation function (53) using summation time averaging for both hard disks and hard spheres and discovered the existence of the well-known long-time tails:

$$\rho_D(t) \sim \alpha_D (t/t_0)^{-d/2} \tag{56}$$

2. Super-Burnett Coefficient

The $D^{(2)}(t)$ term in (51) is called the linear super-Burnett term and we have analogous correlation function expressions for it.[14] Corresponding to (55) we have

$$\begin{aligned} D^{(2)}(t) &= (d/dt) t \, \Delta^{(2)}(t) \\ \Delta^{(2)}(t) &= [\langle \Delta x_1^4(t) \rangle - 3 \langle \Delta x_1^2(t) \rangle^2] / 24t \end{aligned} \tag{57}$$

By carrying out the differentiation and using Liouville's theorem, we obtain

$$D^{(2)}(t) = \tfrac{1}{6}\langle v_{1x}(0)\,\Delta x_1^3(t)\rangle - \tfrac{1}{2}\langle \Delta x_1^2(t)\rangle\langle v_{1x}(0)\,\Delta x_1(t)\rangle \tag{58}$$

corresponding to (54), and from another differentiation,

$$D^{(2)}(t) = \int_0^t dt'\, c^{(2)}(t') \tag{59}$$

$$c^{(2)}(t) = \tfrac{1}{2}\langle v_{1x}(0)v_{1x}(t)\,\Delta x_1^2(t)\rangle - \langle v_{1x}(0)\,\Delta x_1(t)\rangle^2 - \tfrac{1}{2}\langle \Delta x^2(t)\rangle\langle v_{1x}(0)v_{1x}(t)\rangle$$

corresponding to (52).

In order to develop some of the considerations involved in molecular dynamics calculations of time correlation functions, we will discuss in some detail the evaluation of the super-Burnett coefficient $D^{(2)}(t)$ from Eq. (58). At the outset we observe that one could also evaluate $\Delta^{(2)}(t)$ from (57) or $c^{(2)}(t)$ from (59). While in theory an exact knowledge of any one of these functions (for all t) permits a determination of the others, from the practical point of view which function one calculates can be a matter of some importance. The principal aim in the study of $D^{(2)}(t)$ is the determination of its long-time behavior and particularly the question of whether $D^{(2)}(t)$ diverges as t becomes large as predicted (with some differences in details) by Keyes and Oppenheim,[17] by Dufty and McLennan,[18] and by de Schepper *et al.*[19] In general the long-time behavior of $D^{(2)}(t)$ is most sensitively reflected in its derivative $c^{(2)}(t)$. If the time dependence of $c^{(2)}(t)$ could be inferred from the numerical calculations, then it would certainly be the most decisive quantity to study. On the other hand, if $c^{(2)}(t)$ cannot be determined with sufficient accuracy, then $D^{(2)}(t)$ itself would be the next candidate. The calculation of $D^{(2)}(t)$ from $\Delta^{(2)}(t)$ requires that the differentiation of (57) be performed numerically. The chief difficulty with the latter arises in attempting to estimate error bounds on the results. In actual fact, the numerical data for $c^{(2)}(t)$ turn out to have relatively large uncertainties, so that $D^{(2)}(t)$ as determined from (59) or even from $\Delta^{(2)}(t)$ appears to be the best quantity to study.

Actually rather than $D^{(2)}(t)$, it is somewhat more appropriate to calculate

$$\hat{D}^{(2)}(t) = \tfrac{1}{6}\langle \Delta\hat{x}_1^3(t)\hat{v}_{1x}\rangle - \tfrac{1}{2}\langle \Delta\hat{x}_1^2(t)\rangle\langle \hat{v}_{1x}\,\Delta\hat{x}(t)\rangle \tag{60}$$

containing only velocities and displacements in the frame of reference of the center of mass,

$$\hat{\mathbf{v}}_i(t) = \mathbf{v}_i(t) - \mathbf{v}_0, \qquad \hat{\mathbf{r}}_i(t) = \mathbf{r}_i(t) - \mathbf{v}_0 t, \qquad \mathbf{v}_0 = \mathbf{M}/Nm \tag{61}$$

(An MD program would normally use these coordinates in order to eliminate the effects of bulk flow.) In the NVT ensemble,

$$\hat{D}_2(t) = D_2(t) \tag{62}$$

as shown in the Appendix. Although such an identity does not hold for the *NVE* ensemble, for which the calculations described below were done, the correction term is an ensemble correction and hence of order $1/N$.

As seen from (60) the determination of $\hat{D}^{(2)}$ requires the correlation functions

$$\hat{D}_0^{(0)}(t) = \langle \hat{v}_{1x}\,\Delta\hat{x}_1(t)\rangle, \qquad \hat{s}_2(t) = \langle \Delta\hat{x}_1^2(t)\rangle, \qquad \hat{d}_0(t) = \langle \Delta\hat{x}_1^3(t)\hat{v}_{1x}\rangle$$
$$\hat{D}_0^{(2)}(t) = \tfrac{1}{6}\hat{d}_0(t) - \tfrac{1}{2}\hat{s}_2(t)\hat{D}_0^{(0)}(t) \tag{63}$$

We have added the subscript 0 to $\hat{D}^{(0)}$, $\hat{d}$, and $\hat{D}^{(2)}$ to distinguish these quantities from the alternative expressions,

$$\hat{D}_f^{(0)}(t) = \langle \hat{v}_{1x}(t)\,\Delta\hat{x}_1(t)\rangle, \qquad \hat{d}_f(t) = \langle \Delta\hat{x}_1^3(t)\hat{v}_{1x}(t)\rangle$$
$$\hat{D}_f^{(2)}(t) = \tfrac{1}{6}\hat{d}_f(t) - \tfrac{1}{2}\hat{s}_2(t)\hat{D}_0^{(0)}(t) \tag{64}$$

It is easy to show that

$$\hat{D}_0^{(0)}(t) = \hat{D}_f^{(0)}(t), \qquad \hat{d}_0(t) = \hat{d}_f(t), \qquad \hat{D}_0^{(2)}(t) = \hat{D}_f^{(2)}(t) \tag{65}$$

Because the quantities being averaged, however, are not identical for a given trajectory, these functions may be expected to have different fluctuation behavior. This distinction will be seen to have considerable effect on the numerical estimates of $\hat{D}^{(2)}(t)$.

The determination of these functions can be effected by any of the three methods discussed in Section 2. Alder and Wainwright used the summation time average method in the calculation of the velocity autocorrelation function[15,16] and the calculation of the super-Burnett coefficient[20] $D^{(2)}(t)$ from $\Delta^{(2)}(t)$. Here we detail the calculation of $D^{(2)}(t)$ using the combined Monte Carlo–molecular dynamics method.

5.2. Computational Procedures and Results for $D^{(2)}(t)$

It is instructive to specialize Eq. (17) to the time correlation function $\hat{D}_0^{(0)}(t)$. First, however, we use spatial symmetry and particle indistinguishability to write

$$\hat{D}_0^{(0)}(t) = \frac{1}{dN}\left\langle \sum_{i=1}^{N} \hat{\mathbf{v}}_i \cdot \Delta\hat{\mathbf{R}}_i(t)\right\rangle \tag{66}$$

In our program the correlation functions [$\hat{D}_0^{(0)}(t)$ in the present example] are "observed" at times $t = h, 2h, \ldots, Mh$, where h is a specified time step (in units of the Boltzmann mean free time) and M an integer determining the maximum time to which the correlation functions will be estimated. Each N-particle trajectory is generated out to time θh for specified integer $\theta \geq M$. Summation time averaging is specified with respect to "time origins" 0, ωh, $2\omega h, \ldots$, $(I_\theta - 1)\omega h$, with ω a specified integer and with the already mentioned integer θ being required to be equal to the integer I_θ times ω (i.e., $\theta = I_\theta \omega$ for integer I_θ).

In addition, we also require $M = I_M\omega$ with integer I_M. Provision is made for retaining up to I_M sets of time origin phases $\mathbf{x}^N$, which imposes a nontrivial mass storage requirement for large values of N. If the argument t of the time correlation function is written in the form

$$t = mh = (I_m\omega + \mu)h$$

for integers $1 \leq m \leq M$, with $1 \leq \mu \leq \omega$ and $0 \leq I_m \leq I_M - 1$, then the combined Monte Carlo and time-averaged correlation function is

$$\hat{D}_0^{(0)}[(I_m\omega + \mu)h] = \frac{1}{dPN(I_\theta - I_m)} \sum_{p=1}^{P} \sum_{i=1}^{N} \sum_{I=I_m}^{I_\theta - 1} \hat{\mathbf{v}}_{ip}[(I - I_m)\omega h] \cdot \{\hat{\mathbf{R}}_{ip}[(I\omega + \mu)h] - \hat{\mathbf{R}}_{ip}[(I - I_m)\omega h]\} \quad (67)$$

with $\hat{\mathbf{R}}_{ip}(t)$ and $\hat{\mathbf{v}}_{ip}(t)$ denoting the position (infinite checkerboard) and velocity (both relative to the center of mass) of particle i on the pth trajectory. Expressions analogous to (67) can be written for $\hat{s}_2(t)$, $\hat{D}_f(t)$, $\hat{d}_0(t)$, and $\hat{d}_f(t)$.

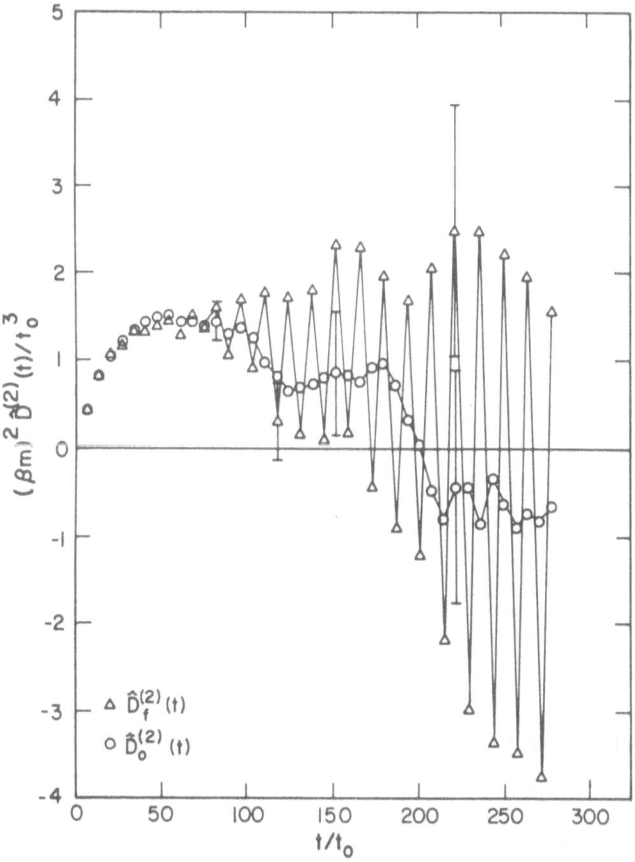

Fig. 3. The reduced super-Burnett self-diffusion coefficient $(\beta m)^2 \hat{D}^{(2)}(t)/t_0^3$ as a function of t/t_0, as computed using Eq. (63) (○) and Eq. (64) (△), for a system of 5822 hard disks at $\tau = 1.5$.

The numerical estimates of $\hat{D}_0^{(0)}$ and $\hat{D}_f^{(0)}$ and of $\hat{d}_0$ and $\hat{d}_f$ will differ, as previously indicated, although they should be consistent within their statistical errors. The effectiveness of using either or both sets of correlation functions in estimating $\hat{D}^{(2)}$ depends upon details of the statistical distributions of the various quantities being averaged.

The effect of these details of the calculation can be seen in Fig. 3 in which the reduced super-Burnett coefficient $(\beta m)^2\hat{D}^{(2)}(t)/t_0^3$ in the microcanonical ensemble is plotted as a function of t for a system of 5822 hard disks at a reduced volume $\tau = 1.5$. The figure includes points labeled $\hat{D}_0^{(2)}$ calculated as in (63) and points labeled $\hat{D}_f^{(2)}$ calculated as in (64). The jumpy character of the $\hat{D}_f^{(2)}$ data is apparently related to the time origin spacing $\omega = 2$ used in the calculation and to a resulting oscillation in the statistical correlation coefficients[21] r_{ij} between the observed values of the time correlation functions at times ih and $(i+j)h$. Table 2 shows a portion of this correlation matrix for both $\hat{d}_0$ and $\hat{d}_f$ in the present calculation, for which $h = 6.96t_0$. We see that the serial correlations for both variables are very strong, but weaker in the case of $\hat{d}_f$ at adjacent points ($j = \pm 1$) as compared with points twice removed ($j = \pm 2$). We believe that this behavior arises from the fact that values of $\hat{v}_i(t)$ at non-time origin points enter the calculation analogous to (67) for $\hat{d}_f$ and $\hat{D}_f^{(0)}$, but not for $\hat{d}_0$ and $\hat{D}_0^{(0)}$. A behavior similar to that shown in Table 2 is also observed for $\hat{D}_0^{(0)}$ and $\hat{D}_f^{(0)}$. It is notable that the displacement between successive $\hat{D}_f^{(2)}$ points in Fig. 3 exceeds three standard deviations. At lower densities such effects are less noticeable, although appreciable statistical correlation is still present.

It is probably worth emphasizing that serial correlations among time correlation functions computed in molecular dynamics calculations are commonly present and should be taken into account when making comparisons with theoretical calculations and with other similar calculational results.

Table 2. Statistical Correlation Coefficients r_{ij} for $\hat{d}_0$ (First Line of Each Row) and $\hat{d}_f$ (Second Line)

	j						
i	−3	−2	−1	0	1	2	3
5	0.70	0.88	0.94	1.00	0.96	0.91	0.89
	0.59	0.92	0.66	1.00	0.76	0.93	0.79
10	0.92	0.97	0.99	1.00	0.98	0.98	0.96
	0.83	0.96	0.83	1.00	0.80	0.96	0.78
20	0.98	0.99	0.99	1.00	0.99	0.99	0.98
	0.81	0.98	0.83	1.00	0.83	0.99	0.84
30	0.99	0.99	1.00	1.00	1.00	0.99	0.98
	0.89	0.99	0.89	1.00	0.89	1.00	0.88
40	0.99	1.00	1.00	1.00			
	0.88	1.00	0.88	1.00			

Space does not permit us to give here a detailed discussion of the effects of finite-system size on molecular dynamic calculations of time correlations functions. We have given elsewhere[1] a discussion of such effects on the velocity autocorrelation function from a hydrodynamic point of view. This reference can also be consulted for a more extensive discussion of results for both the velocity autocorrelation function and the super-Burnett self-diffusion coefficient, including comparisons with theoretical predictions.

6. Nonequilibrium Molecular Dynamics

Recently a new approach to the computer study of transport properties was initiated by Ashurst and Hoover[22] with somewhat similar methods also being introduced by Lees and Edwards[23] and Gosling *et al.*[24] In the Ashurst–Hoover method, the basic notion is to alter the boundary conditions in such a way as to drive the system into a nonequilibrium steady state. From the computer then, one obtains directly observed values of the appropriate gradients and currents and studies the relation between these quantities as the driving forces at the boundary are varied.

Because of the microscopic size of systems that can in practice be studied, an enormous gradient is required to produce an average current differing significantly from zero in the face of the natural fluctuations of the system. These calculations, then, can be expected to involve not only the linear Navier–Stokes coefficients, but also both the linear and nonlinear Burnett coefficients.

In order to illustrate the nonequilibrium MD method, we describe the self-diffusion calculation by Holian *et al.*[25] The calculation treats a system of *labeled* hard spheres or hard disks, with the usual periodic boundary conditions augmented by the following label-changing algorithm. A labeled particle crossing one of the planes $x = \nu L$ (for $\nu = 0, \pm 1, \ldots$) in the direction $\hat{e}_x$ is unlabeled with probability p. An unlabeled particle crossing in the direction $-\hat{e}_x$ is labeled with probability p. Otherwise particle labels remain fixed.

Evidently the effect of the label change will be to increase the number density of labeled particles in the primary cell near the $x = L$ boundary relative to that near the $x = 0$ boundary. In the long time limit, it is expected that the system will approach a one-dimensional steady state, in which a self-diffusion current j_1 of labeled particles will flow in the $-\hat{e}_x$ direction independent of $\mathbf{r}$ and t. The calculation depends on the establishment of this steady state and is to be contrasted with the use of an initial nonequilibrium ensemble in which one might study the number density and current as transients. Here the number density and current are to be evaluated as time averages, beginning at such a time that initial transients have vanished.

It is of interest to observe that the calculation is simply a standard (time-averaging) MD calculation if one ignores the particle labels. Thus, simultaneously with the nonequilibrium self-diffusion calculation, one can determine time averages for the equation of state and for time correlation functions.

6.1. Numerical Details

From the computational point of view, the direct evaluation of number density and current at a given point x is not practical because the microscopic expressions

$$\tilde{n}_1(x, t) = \frac{1}{L^{d-1}} \sum_{i=1}^{N} \delta[x - \hat{x}_i(t)] l_i(t)$$

$$\tilde{j}_1(x, t) = \frac{1}{L^{d-1}} \sum_{i=1}^{N} \delta(x - \hat{x}_i(t)] \hat{v}_{ix}(t) l_i(t) \tag{68}$$

$$l_i(t) = \begin{cases} 1 & \text{if particle } i \text{ is labeled at time } t \\ 0 & \text{otherwise} \end{cases}$$

contain δ functions. The simplest alternative is to compute the spatially averaged number density and current within layers of thickness Δx. That is, one evaluates

$$\tilde{N}_1(x, t) = \frac{1}{\Delta x L^{d-1}} \sum_{i=1}^{N} \{A[x - \tfrac{1}{2}\Delta x - \hat{x}_i(t)] - A[x + \tfrac{1}{2}\Delta x - \hat{x}_i(t)]\} l_i(t)$$

$$\tilde{J}_1(x, t) = \frac{1}{\Delta x L^{d-1}} \sum_{i=1}^{N} \{A[x - \tfrac{1}{2}\Delta x - \hat{x}_i(t)] - A(x + \tfrac{1}{2}\Delta x - \hat{x}_i(t)]\} \hat{v}_{ix}(t) l_i(t) \tag{69}$$

for a series of K layers with centers

$$x_k = (k - \tfrac{1}{2})\Delta x, \qquad k = 1, 2, \ldots, K, \quad \Delta x = L/K$$

and at the sequence of M times

$$t_m = mh, \qquad m = 1, 2, \ldots . M$$

for some specified value of h. The function

$$A(x) = \begin{cases} 0, & x < 0 \\ 1, & x \geq 0 \end{cases}$$

is the familiar unit step function. From these expressions then, one omits as transient the data before some time s_0 and calculates the summation time averages,

$$\bar{N}_1(x; M) = \frac{1}{M - m_0 + 1} \sum_{m=m_0}^{M} \tilde{N}_1(x, t_m)$$
$$\bar{J}_1(x; M) = \frac{1}{M - m_0 + 1} \sum_{m=m_0}^{M} \tilde{J}_1(x, t_m) \tag{70}$$
$$s_0 = m_0 h$$

which approach for large M the steady-state values $N_1(x)$ and j_1.

The label change probability p can be achieved through a standard Monte Carlo procedure, using pseudorandom numbers uniformly distributed on the (0, 1) interval. The generation of such numbers is extensively discussed by Jansson[(3)] and by Dieter and Ahrens.[(26)]

It is perhaps of interest to note that instead of a single label, particles can be assigned a set of labels, $\mathbf{l}_i = (l_{i1}, l_{i2}, \ldots, l_{ic})$ for particle i, with one label for each of a set of probabilities $\mathbf{p} = (p_1, p_2, \ldots, p_c)$ for label change determination. Since each value of p is expected to yield a different steady-state number density gradient, one can, with a single molecular dynamics trajectory, perform a number of self-diffusion experiments. Moreover, a different set of labels $\mathbf{l}_i^{(2)}$ can be used to label particles with respect to crossings of the y-boundaries and, in three dimensions, a third set $\mathbf{l}_i^{(3)}$ can be used with respect to the z-boundaries.

One final note concerns the initial state. In order to reduce the initial transient, we impose an initial labeling having a number density gradient approximating that expected in the steady state. The initial current, however, is not so readily adjusted, so that in the calculation reported below the initial current turns out to be very small, of the order of the fluctuations in the particle velocity.

6.2. Analysis of Data

The analysis of the results depends strongly on the achievement of a steady state. If, moreover, the Burnett effects are unimportant, then from (51) we have the linear profile of labeled-particle number density

$$n_1(x) = n[0.5 + b(x - L/2)] \tag{71}$$

where we have required from symmetry that $n_1(L/2) = \frac{1}{2}n$. By determining the current j_1 and the gradient b, one obtains the self-diffusion constant

$$D^{(0)} = -bn/j_1 \tag{72}$$

The principal tasks are then to assure the validity of the assumptions and to assign error bounds to the final value of $D^{(0)}$.

On the microscopic level, both spatial and temporal fluctuations in number density are to be expected. The decay of these should be rapid compared to the time and distance scales appropriate to the macroscopic self-diffusion phenomenon. Thus to study the presence of a steady state, the M observations of $\tilde{N}_1$ and $\tilde{J}_1$ are pooled ΔM at a time so that the resulting observations appear to be uncorrelated. Typically $(\Delta M)h \approx 200t_0$ seems to be satisfactory for this purpose at a reduced volume $\tau = 3$ for hard disks (but pooling over even longer intervals is frequently used to assure the independence of the observations). While there are rather long-time correlation effects, there seems little doubt that on any macroscopic time scale the "flow" is steady.

The importance of Burnett effects can be ascertained in several ways. With respect to nonlinear Burnett effects, we simply analyze the results for each probability p separately, with the expectation that the resulting values of $D^{(0)}$ should display a trend with p if such effects are present. Effects due to the linear super-Burnett terms in (51) should affect the linearity of the profile (71). The latter is averaged over each layer,

$$N_1(x) = \frac{1}{\Delta x} \int_{x-\frac{1}{2}\Delta x}^{x+\frac{1}{2}\Delta x} dx'\, n_1(x') \tag{73}$$

to obtain

$$N_1(x) = n_1(x) \tag{74}$$

The pooled observations of N_1 are fitted to the linear expression (71) via least squares. No significant departures from linearity are observed, which is consistent with the existence of theoretical reasons[27] to believe that the super-Burnett effects have a much smaller distance scale than the typical layer thicknesses used in these calculations. Similarly the layers adjacent to the boundaries show no atypical behavior, but the layer thicknesses are many mean free paths, while kinetic boundary layers are typically of the order of a mean free path. Thus the absence of effects beyond the linear Fick's law term is essentially a result of the spatial coarse graining used.

In Table 3 we summarize the computed self-diffusion coefficients, reduced by the Enskog dense-gas value

$$D_E^{(0)} = D_B^{(0)}/\chi \tag{75}$$

where $D_B^{(0)}$ is the Boltzmann theory result and χ the pair correlation function at contact (Enskog's χ), all at a reduced volume $\tau = 3$. The quoted standard

Table 3. Self-Diffusion Coefficient from Nonequilibrium Molecular Dynamics

d	N	Mh/t_0	p	b	$D^{(0)}/D_E^{(0)}$
2	168	21,203	1.0	0.924	1.391 ± 0.028
			0.8	0.636	1.430 ± 0.042
	1512	4920	1.0	0.938	1.794 ± 0.045
			0.8	0.643	1.786 ± 0.065
	5822	40,097	1.0	0.970	1.852 ± 0.011
			0.9	0.796	1.881 ± 0.014
3	108	14,980	1.0	0.889	1.166 ± 0.014
			0.8	0.612	1.183 ± 0.022
	500	6670	1.0	0.922	1.231 ± 0.012
			0.8	0.630	1.238 ± 0.019
	4000	7450	1.0	0.955	1.292 ± 0.017
			0.8	0.649	1.276 ± 0.028

deviations are computed from the variance of the gradient b, the variance of the current j_1, and their covariance through the usual propagation of error formula. In each case the d results for each spatial direction have been combined. Results for the smaller values of p have larger standard deviations and are not included. We note that the results are consistent with the hypothesis that $D^{(0)}$ is independent of gradient, so that nonlinear effects are imperceptible for self-diffusion.

The rather large N dependence in the two-dimensional case is qualitatively consistent with the expectation that $D^{(0)}$ increases without bound with N. On the other hand, in the three-dimensional case one expects a finite limit for large N; the data seem consistent with the infinite-system estimate $D^{(0)}/D_E^{(0)} - 1.34$ given by Alder *et al.*[13] To go beyond such qualitative remarks will apparently require some quantitative theoretical estimate for the self-diffusion coefficient for such a steady-state process in a finite system.

7. Accuracy of Molecular Dynamics Trajectory

The trajectory as determined on a computer is not exact because of the finite register length. Here we describe what is known about the growth of these errors.

7.1. Constants of Motion

One measure of the accuracy of a trajectory is provided through the constants of the motion, viz., the kinetic energy and the linear momentum. The change in kinetic energy can be accounted for, approximately, by taking into account that the velocity change calculated for each collision will tend to be smaller in magnitude than the true value,

$$\Delta \mathbf{v}_{12} = -2\sigma^{-2}(\mathbf{v}_{12} \cdot \boldsymbol{\sigma}_{12})\boldsymbol{\sigma}_{12}$$

if the calculation truncates (rather than rounds) the results of the multiplication. If we introduce notation $\mathcal{M}(x)$ to denote the machine value of x, then

$$\mathcal{M}(\Delta \mathbf{v}_{12}) = -2\sigma^{-2}(\mathbf{v}_{12} \cdot \boldsymbol{\sigma}_{12})\boldsymbol{\sigma}_{12}(1-\alpha) \tag{76}$$

where α is a positive number of the order $\frac{1}{2} \times 10^{-m}$ for a calculation using m-digit registers.

Computing the post collision kinetic energy, one finds

$$\mathcal{M}(2E'/m) = 2E/m - 2(\mathbf{v}_{12} \cdot \boldsymbol{\sigma}_{12})^2\sigma^{-2}\alpha + O(\alpha^2)$$

For a series of many collisions, we replace $(\mathbf{v}_{12} \cdot \boldsymbol{\sigma})^2$ by its average value to find that the energy change per mean free time is

$$dE/d(t/t_0) \sim -2\alpha N/\beta m = -4\alpha E/d \tag{77}$$

a result that agrees fairly well with the observed energy loss.

With regard to the linear momentum, for a single collision one has

$$\mathcal{M}(\mathbf{v}_1' + \mathbf{v}_2') \approx \mathcal{M}(\mathbf{v}_1 + \tfrac{1}{2}\Delta \mathbf{v}_{12}) + \mathcal{M}(\mathbf{v}_2 - \tfrac{1}{2}\Delta \mathbf{v}_{12})$$

so that the truncation error in $\Delta \mathbf{v}_{12}$ cancels approximately and only the truncation error in $\mathbf{v}_1 + \mathbf{v}_2$ persists,

$$\mathcal{M}(\mathbf{v}_1 + \mathbf{v}_2) = (\mathbf{v}_1 + \mathbf{v}_2)(1-\alpha)$$

Over many collisions $\mathbf{v}_1 + \mathbf{v}_2$ is on the average zero, so that in first approximation the momentum would be expected to fluctuate slightly, but the error should not grow.

These errors do not seem particularly serious. The energy loss (77) is independent of density and is small except for trajectories of the order $1/\alpha$ collisions per particle, viz., 10^{10} collisions per particle for a 10-digit register length, which is beyond the capability of current machinery.

7.2. Growth of Trajectory Error

A direct measure of the trajectory error is provided by the use of so-called multiple-precision arithmetic. Here one considers a set of P trajectories $\{\mathbf{x}_p^N(t); p = 1, 2, \ldots, P\}$. Suppose, for simplicity, that the initial phase $\{\mathbf{x}_p^N(0)\}$ is precisely represented in the computer by the m-digit registers of the machine.

Then the numerical estimates $\{\mathbf{x}_p^N(t; m); p = 1, 2, \ldots, P\}$ calculated using m-digit precision will gradually move farther from the exact trajectories with time. While it is not possible to measure this directly, because the exact trajectories are not known, one can in fact calculate the growth of the error by using multiple-precision arithmetic to obtain more accurate trajectories, e.g., in double precision $\{\mathbf{x}_p^N(t; 2m); p = 1, 2, \ldots, P\}$. Because the error in the double-precision trajectory should be 10^{-m} of the single precision error, one can compute the error in the single-precision trajectory until the double-precision errors are themselves large.

We define then the reduced root mean square errors $\Delta_r(t, m)$ and $\Delta_v(t, m)$ in position and velocity,

$$\Delta_r^2(t, m) = \frac{1}{N\sigma^2}\left\langle \sum_{i=1}^{N} [\hat{\mathbf{R}}_i(t; m) - \hat{\mathbf{R}}_i(t)]^2 \right\rangle$$

$$\Delta_v^2(t, m) = \frac{\beta m}{dN}\left\langle \sum_{i=1}^{N} [\mathbf{v}_i(t; m) - \mathbf{v}_i(t)]^2 \right\rangle \tag{78}$$

$$\hat{\mathbf{R}}_i(0; m) = \mathcal{M}[\hat{\mathbf{R}}_i(0)], \qquad \mathbf{v}_i(0; m) = \mathcal{M}[\mathbf{v}_i(0)]$$

These can be calculated for t not too large by the Monte Carlo–molecular dynamics averages

$$\Delta_r^2(t, m) = \frac{1}{NP\sigma^2} \sum_{p=1}^{P} \sum_{n=1}^{N} [\hat{\mathbf{R}}_{np}(t; m) - \hat{\mathbf{R}}_{np}(t; 2m)]^2$$

$$\Delta_v^2(t, m) = \frac{\beta m}{dNP} \sum_{p=1}^{P} \sum_{n=1}^{N} [\mathbf{v}_{np}(t; m) - \mathbf{v}_{np}(t; 2m)]^2 \tag{79}$$

where $\hat{\mathbf{R}}_{np}$ and $\mathbf{v}_{np}$ denote the position and velocity of particle n on the pth trajectory. While the detailed behavior of Δ_r and Δ_v is expected to depend somewhat on the details of both the computer program and the computer hardware, intuitively such effects are expected to be small.

In Fig. 4 are shown Δ_r and Δ_v for systems of 12 hard disks at relative volumes τ of 2, 3, and 10, and for 168 hard disks at $\tau = 3$ as functions of t in units of the mean free time t_0. Each point is the average of 9 trajectories generated on the CDC 7600 computer, having $m \approx 14$, using initial phases selected from the microcanonical ensemble, and with the position update time (see Section 8) fixed at $t_u = 0$. The standard deviations (not shown) in Δ_v and Δ_r are in the range of 10 to 50%. We note that the errors grow essentially exponentially with time, until they reach the order of magnitude of the variable itself. Moreover, we observe the dramatic increase of growth rate of the errors with decreasing density, viz., 0.3 digits of accuracy lost per collision at $\tau = 2$, 1.9 digits at $\tau = 3$, and 3.6 digits for $\tau = 10$. The effect of system size is not so dramatic but the calculations do show a slow increase of error with N.

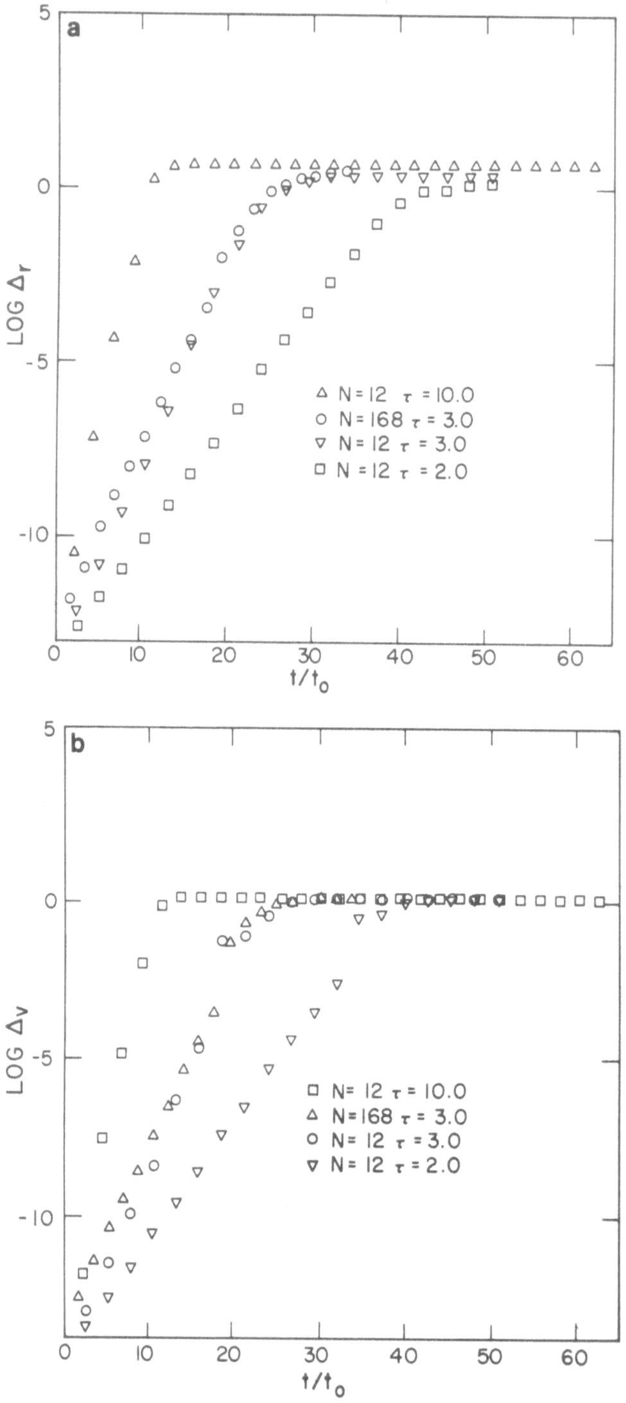

Fig. 4. Growth of the errors (a) in position Δ_r and (b) in velocity Δ_v defined by (79) with increasing time t in units of the mean free time t_0 for several hard-disk systems.

The results in Fig. 4 are, of course, specific to a particular computer program and can be expected to vary somewhat with details of the method. We expect the latter dependence to be small on the basis of another calculation of Δ_r and Δ_v for $\tau = 2$ and $N = 12$ in which position updating (see Section 8) at $\Delta t_u \approx 2t_0$ was used, in contrast to the above results having $\Delta t_u = \infty$. While this change effects a very sizable change in the method of calculation, the errors Δ_r and Δ_v are essentially identical with those in Fig. 4.

It is interesting to note that the long-time values for Δ_v agree reasonably well with the value one obtains assuming that $v_{np}(t; m)$ and $v_{np}(t; 2m)$ are uncorrelated stochastic variables with the usual Maxwellian distribution, viz., $\Delta_v \approx \sqrt{2}$.

The effect of the loss of accuracy on the physical properties one wishes to estimate by molecular dynamics calculations is not readily determined. Evidently MD calculations that employ a large degree of "time averaging" have in fact a sizable *statistical* component in the averaging process, rather than giving true time averages. The agreement found by Hoover and Alder[(8)] between the molecular dynamics and the Monte Carlo equation of state results indicates that the trajectory inaccuracy introduces no bias into the calculation of equilibrium properties. For time correlation function calculations there could well be an effect arising in, say, the velocity autocorrelation function $\rho_D(t)$, for t greater than the time of total loss of accuracy. While the numerical trajectory might well be "just as good" as the exact trajectory for the purposes of the quasi-ergodic hypothesis, the time correlations along the exact trajectory could conceivably be expected to differ from those along the inexact one. A comparison of the velocity autocorrelation function calculated from both single- and double-precision trajectories for 870 hard disks[(1)] shows qualitative agreement beyond the time the single-precision trajectory has completely lost accuracy, but a more complete analysis of the statistical correlations present in the data is needed to make the comparison quantitative.

Another comparison was made by Bishop and Berne,[(28)] who calculated the velocity autocorrelation function of one-dimensional hard rods by a time average molecular dynamics method. Because the system is nonergodic (velocities simply interchange at collision) the comparison of these results with the theoretical results of Jepsen,[(29)] as extended to nonzero rod length by Lebowitz and Percus,[(30)] is not clear. While the numerical data do not display the expected negative t^{-3} long-time tail, the source of this difficulty is probably not accuracy loss in the trajectory. Indeed, Anderson *et al.*[(31)] found that for a system of hard rods with a short-range linear attraction, the (double-precision) trajectory could be reversed (i.e., regenerated from the final phase $\mathbf{x}^N(t)$ by reversing the signs of the velocities) to reasonable precision for times t of the order of $50t_0$ (i.e., 50 collisions per particle), while the negative part of the velocity autocorrelation function is expected at $t \approx 2t_0$ in the hard-rod calculation. Bishop and Berne believe that the negative tail was obscured in their calculation by relatively large statistical fluctuations.

8. Programming Techniques

Here we describe numerical techniques for the MD method that have been mentioned in the literature and others that we have found useful in our own work. Unfortunately the applicability and desirability of many techniques depend upon the computer hardware in question and different techniques are required to take advantage of new types of hardware.

The simplicity of the trajectory calculation for hard-core particles permits an extremely simple (albeit inefficient) MD program. We begin by describing such a program so as to establish a point of departure for discussing more efficient programs.

In the discussion that follows we use the configuration vector $\mathbf{r}^N(t)$ (in which the N particles remain in the primary cell) and the velocity vector $\mathbf{v}^N(t)$. In actual fact it is usually more convenient to use velocities $\hat{\mathbf{v}}^N$ relative to the center of mass. The distinction is of no consequence here. In computing $\mathbf{r}^N(t)$, however, it must be remembered that there are applications in which the "checkerboard" coordinates are required, e.g., the calculation of the mean square displacement $\Delta^{(0)}(t)$, Eq. (55). In such cases, provision must be made to obtain $\mathbf{R}^N(t)$. One could equally well compute $\mathbf{R}^N(t)$, without appreciable change in the numerical methods.

8.1. Simple Molecular Dynamics Program

Consider the problem of advancing the phase space trajectory from an initial phase $\mathbf{x}^N$. Because particle trajectories are straight between collisions, we need only find the first collision. For each pair ij, consider the time of collision t_{ij}, ignoring the presence of other particles. Introduce relative coordinates and velocities

$$\mathbf{r}_{ij} = \mathbf{r}_i - \mathbf{r}_j, \qquad \mathbf{v}_{ij} = \mathbf{v}_i - \mathbf{v}_j \tag{80}$$

and examine the condition that particles i and j are approaching each other, viz., $\mathbf{r}_{ij} \cdot \mathbf{v}_{ij} < 0$. Otherwise, define the collision time to be infinite,

$$t_{ij} = \infty \qquad \text{for } \mathbf{r}_{ij} \cdot \mathbf{v}_{ij} \geq 0 \tag{81}$$

For particles that are approaching, require that they lie in the collision cylinder, viz.,

$$b_{ij} \leq \sigma, \qquad b_{ij}^2 = r_{ij}^2 - (\mathbf{r}_{ij} \cdot \mathbf{v}_{ij}/v_{ij})^2 \tag{82}$$

as shown in Fig. 5. Otherwise, again define t_{ij} to be infinite,

$$t_{ij} = \infty \qquad \text{for } b_{ij} > \sigma \tag{83}$$

Finally, for $b_{ij} < \sigma$ compute the collision time

$$t_{ij} = -\frac{1}{v_{ij}}\left[\frac{\mathbf{r}_{ij} \cdot \mathbf{v}_{ij}}{v_{ij}} + (\sigma^2 - b_{ij}^2)^{1/2}\right] \tag{84}$$

as seen from Fig. 5. From (81), (83), and (84), one determines $N(N-1)/2$ collision times.

For periodic boundary conditions, collisions between a particle i in the primary cell and image particles can also occur. Provided $\sigma < \frac{1}{2}L$, a simultaneous collision of i with j and its image is impossible—we always assume this condition. Under this same condition, collisions are possible only with an image particle in a neighbor of the primary cell, i.e., for particles j at $\mathbf{r}_j + \boldsymbol{\nu}L$ where $\boldsymbol{\nu}$ has components -1, 0, or $+1$. We label such collision times

$$\begin{aligned} t_{ij}^{(\nu)} &= -\frac{1}{v_{ij}}\left[\frac{(\mathbf{r}_{ij} - \boldsymbol{\nu}L) \cdot \mathbf{v}_{ij}}{v_{ij}} + (\sigma^2 - b_{ij}^{(\nu)2})^{1/2}\right] \\ b_{ij}^{(\nu)2} &= (\mathbf{r}_{ij} - \boldsymbol{\nu}L)^2 - \left[\frac{(\mathbf{r}_{ij} - \boldsymbol{\nu}L) \cdot \mathbf{v}_{ij}}{v_{ij}}\right]^2 \end{aligned} \tag{85}$$

Finally each particle has a time $t_i^{(b)}$ at which, in the absence of other particles, it would cross a boundary of the primary cell

$$\begin{aligned} t_i^{(\mathrm{b})} &= \min\{t_{ik}^{(\mathrm{b})};\ k = 1, \ldots, d\} \\ t_{ik}^{(\mathrm{b})} &= -\frac{1}{\mathbf{v}_i \cdot \hat{\mathbf{e}}_k}\left[\mathbf{r}_i \cdot \hat{\mathbf{e}}_k - \frac{L}{2}\left(1 + \frac{\mathbf{v}_i \cdot \hat{\mathbf{e}}_k}{|\mathbf{v}_i \cdot \hat{\mathbf{e}}_k|}\right)\right] \end{aligned} \tag{86}$$

where the $\hat{\mathbf{e}}_k$ are unit vectors along the edges of the periodic cell. At such an event, an image particle also enters the primary cell through the opposite boundary, so that the coordinates of the particle are to be changed appropriately.

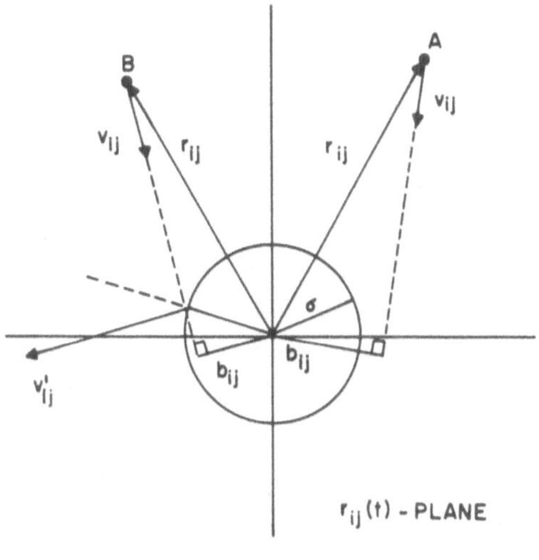

Fig. 5. Collision calculation in the plane of $\mathbf{r}_{ij}$ and $\mathbf{v}_{ij}$ for two initial (relative) phases. In both cases, particles i and j are approaching, but for A the collision coordinate $b_{ij} > \sigma$, while for B, $b_{ij} < \sigma$. For the latter case, $\mathbf{v}'_{ij}$ indicates the postcollision velocity for an elastic collision.

Thus a simple MD program could consist of the following steps. Given phase $\mathbf{x}^N(t)$:

1. For each i, find $t_i^{(b)}$, $\{t_{ij}^{(\nu)}\}$, where the latter set includes $j = 1, 2, \ldots, i-1$ as well as all $\boldsymbol{\nu}$ having $\nu_k \in \{-1, 0, +1\}$. Select the minimum such time, t_m.

2. Advance the phase to $t+t_m$ using the straight-line trajectory, $\mathbf{r}_i(t+t_m) = \mathbf{r}_i(t) + t_m \mathbf{v}_i(t)$ for $i = 1, 2, \ldots, N$.

3a. For t_m given by a $t_i^{(b)}$, translate the appropriate coordinate of particle i by L.

3b. For t_m given by a $t_{ij}^{(\nu)}$, compute the postcollision velocities of particles i and j,

$$\mathbf{v}_i' = \mathbf{v}_i + \Delta\mathbf{v}_i, \qquad \mathbf{v}_j' = \mathbf{v}_j - \Delta\mathbf{v}_i$$
$$\Delta\mathbf{v}_i = -\frac{\mathbf{v}_{ij} \cdot \boldsymbol{\sigma}_{ij}\boldsymbol{\sigma}_{ij}}{\sigma^2}, \qquad \boldsymbol{\sigma}_{ij} = \mathbf{r}_{ij}(t+t_m) - \boldsymbol{\nu}L \tag{87}$$

By iteration of these three stages, the phase can be advanced indefinitely.*

In practice, it is frequently desirable to apply the MD method to as large a system as possible. This need arises particularly in the study of transport coefficients. Thus we assess the efficiency of the method, particularly as a function of the number of particles N. For step 1, we see that there are $N(N-1)3^d/2$ calculations of $t_{ij}^{(\nu)}$. For step 2, there are N calculations $\mathbf{r}_i(t+t_m)$. For step 3, a single calculation is performed. It follows, then, that to advance the phase by a mean free time t_0 (i.e., $N/2$ collisions), the relative "cost" of step 1 increases as N^3, of step 2 as N^2, and of step 3 as N. We consider now methods to reduce the number of computations for the various steps.

8.2. Time Table Method

The inefficiency of step 1 is partly associated with the fact that almost the same set of collision times are recomputed on each iteration. In fact, after a collision of particles i and j, only the collision times between each of these and the other $N-2$ particles are distinct from those on the previous iteration. Alder and Wainwright[33,34] described a method for taking advantage of this; we call it the time table method.

Consider a table $\mathcal{T} = \{T_i, i = 1, 2, \ldots, N\}$ consisting of the collision and boundary-crossing times for each particle,

$$T_i = \{t_i^{(b)}, T_i^{(c)}\} \tag{88}$$

*The molecular dynamics of so-called rough elastic spheres and disks is essentially identical to the above. For this case, however, each particle has an angular velocity ω_i. The calculation of the postcollision velocities and angular momenta for this model is given by Chapman and Cowling.[32]

The list $T_i^{(c)}$ of collision times $t_{ij}^{(\nu)}$ could not very well contain an entry for each j and $\boldsymbol{\nu}$, inasmuch as it would not only be large for large N, but would also consist mostly of values that are infinite. It is more to the point to maintain only the finite values that are less than some preset value t_m^*. A manageable table is obtained by letting

$$T_i^{(c)} = \{(t_{ij}^{(\nu)}, j, \boldsymbol{\nu})\} \tag{89}$$

where the indices j and $\boldsymbol{\nu}$ run over only those values for which $t_{ij}^{(\nu)}$ are less than t_m^*. (The j and $\boldsymbol{\nu}$ are made part of the table so as to identify the partner in the collision.)

The program for this method is as follows:

1. Initialize the $\mathcal{T}$-table.
2. Find the minimum time t_m in $\mathcal{T}$.
3. Advance the positions to time $t + t_m$.

4a. For a cell change, translate the appropriate coordinate of the particle by L.

4b. For a collision, determine the post collision velocities as previously.

5. For each (one or two) particle i involved in the event, eliminate from $\mathcal{T}$ all $t_{ij}^{(\nu)}$ that involve that particle and recompute T_i.

6. Provided $\mathcal{T}$ is nonempty, return to step 2. If empty return to step 1.

Provided t_m^* is at least a mean free time, the advance from $\mathbf{x}^N(t)$ to $\mathbf{x}^N(t+t_0)$ involves, on the average, the following N dependences: Step 1 requires N^2 calculations, but is done only once; step 2 involves N calculations (provided the number of $t_{ij}^{(\nu)}$ is of order N) performed N times; step 3 has N^2 calculations; step 4 has order N; step 5 also has order N^2, with $2N$ calculations of $t_{ij}^{(\nu)}$ for each collision. The computation time then no longer increases as N^3 but as N^2. The improvement is obtained at the cost of maintaining the $\mathcal{T}$-table in steps 1 and 5.

8.3. Cell Table Method

One limitation of the time table method arises from the fact that, in step 5, the $t_{ij}^{(\nu)}$ calculation is done for all $N-1$ particles j. In actual practice, only the particles that are near particle i (say within a mean free path or so) are expected to collide without intermediate collisions with it. The number of "neighbors," in this sense, is a function of density but is independent of system size N. The cell table method is intended to take advantage of this circumstance.

Let the primary volume V be subdivided by ξ planes perpendicular to each coordinate axis into ξ^d cells. Provided $\sigma < L/\xi$, we see that only particles in the same cell or in neighboring cells can collide within the time interval in which each particle remains in its cell. By keeping track of the cell occupied by each particle, the calculation of collision times t_{ij} for particle i can be limited to those particles j occupying any of 3^d cells centered at the cell of i. To keep track of cells, we compute also for each particle the time $t_i^{(c)}$ at which it changes cells. The $t_i^{(b)}$ are no longer especially required but are included in the $t_i^{(c)}$.

The actual structure of the cell tables is a matter permitting considerable variation. Perhaps the simplest structure involves a table $\mathscr{C} = \{C_\alpha; \alpha = 1, 2, \ldots, \xi^d\}$ in which each C_α is a list of particles' indices

$$C_\alpha = \{j_{\alpha 1}, j_{\alpha 2}, \ldots, j_{\alpha n(\alpha)}\} \tag{90}$$

for $n(\alpha)$ particles in cell α. The phase would then be contained in a separate $\mathscr{P}$-table,

$$\mathscr{P} = \{\mathbf{x}_i; i = 1, 2, \ldots, N\} \tag{91}$$

and the collision and cell change times would be contained in a separate $\mathscr{T}$-table analogous to (88) and (89).

For large systems, however, $\mathscr{P}$ and $\mathscr{T}$ tend to be so large that they cannot be kept within high-speed memory of the computer. If this is so, then it is advantageous to maintain the appropriate portions of $\mathscr{P}$ and $\mathscr{T}$ within the cell table itself, i.e.,

$$C_\alpha = \{\mathbf{x}_i, T_i, i = j_{\alpha 1}, j_{\alpha 2}, \ldots, j_{\alpha n(\alpha)}\}, \qquad T_i = \{t_i^{(c)}; [t_{ij}, j]\} \tag{92}$$

where $[t_{ij}, j]$ denotes the set of collision times for particle i with other particles in C_α and its neighbors. With this structure only those cells C_α (and their neighbors) containing the particles involved in a collision need be transferred to high-speed memory from secondary storage in order to complete the calculations for a given event.

It is perhaps worthwhile to note that the time tables are not readily searched for the minimum when made part of $\mathscr{C}$. In this case, the minimum time $t_{m\alpha}$ for each cell can be kept in a table $Q = \{t_{m\alpha}; \alpha = 1, 2, \ldots, \xi^d\}$ in high-speed memory. When the T_i portion of C_α is changed, the $t_{m\alpha}$ is updated.

By the use of cell tables, the calculation of new collision times at each event becomes independent of N so that the step 5 contribution to the computer time to advance the phase by a mean free time is of order N.

Finally we note that the time table limit t_m^* can normally be made large, because large but finite t_{ij} do not occur frequently for neighboring particle pairs.

8.4. Position Updating

While the use of cell tables can reduce the number of calculations in step 5 to order N, we observe that the advance of the particle positions in step 3 also has an order N^2 dependence. This contribution can be readily reduced by the use of different position variables. Define coordinates $\mathbf{q}_i(t)$,

$$\mathbf{q}_i(t) = \mathbf{r}_i(t) - \mathbf{v}_i(t)(t - t_u) \tag{93}$$

in terms of an update time t_u. At time $t = t_u$, the $\mathbf{q}_i$ are identical with the $\mathbf{r}_i$, while at a later time they correspond to the positions obtained by free-streaming of all particles to time t_u. In particular, the $\mathbf{q}_i(t)$ are invariant with t except at a collision, for which the postcollision values are

$$\mathbf{q}_i' = \mathbf{q}_i - (t - t_u)\,\Delta\mathbf{v}_i \tag{94}$$

where t is the time of the collision.

By setting $t_u = 0$ throughout a trajectory, the positions are never updated and step 3 calculation is eliminated while step 4b is expanded to include the calculation (94), but with the cost remaining of order N. By choosing a sequence of update times $0, \Delta t_u, 2\Delta t_u, \ldots$ at which the position of all N particles are to be recalculated, step 3 evidently involves N calculations each time it is done. Provided Δt_u is of the order of the mean free time, the total cost of step 3 is of order N.

Appendix

We prove the equivalence of $D^{(2)}t$, Eq. (58), and $\hat{D}^{(2)}(t)$, Eq. (60) in the canonical ensemble. Substitute v_1 and Δx_1 from Eq. (61) into (58) to obtain

$$\begin{aligned} D^{(2)}(t) = {} & \frac{1}{6}[\langle \Delta\hat{x}_1^3(t)\hat{v}_{1x}\rangle + 3\langle \Delta\hat{x}_1^2(t)v_{0x}^2\rangle t + 3\langle \Delta\hat{x}_1(t)\hat{v}_{1x}v_{0x}^2\rangle t^2 + \langle v_{0x}^4\rangle t^3] \\ & - \frac{1}{2}[\langle \Delta\hat{x}_1^2(t)\rangle + \langle v_{0x}^2\rangle t^2]\cdot[\langle \hat{v}_{1x}\,\Delta\hat{x}_1(t)\rangle + \langle v_{0x}^2\rangle t] \end{aligned} \tag{A.1}$$

where terms containing odd powers of v_{0x} have been set to zero, e.g., $\langle \Delta\hat{x}_1^3(t)v_{0x}\rangle$. Such terms vanish because $\Delta\hat{x}_1(t)$ is independent of v_{0x} for fixed $(\mathbf{r}^N, \hat{\mathbf{v}}^N)$, and because for fixed $(\mathbf{r}^N, \hat{\mathbf{v}}^N)$ the velocities $\mathbf{v}_0$ and $-\mathbf{v}_0$ are equiprobable in the canonical ensemble. (Indeed such terms also vanish in the microcanonical ensemble by this same argument.)

Consider the term

$$\langle \Delta\hat{x}_1^2(t)v_{0x}^2\rangle = \int dx^N \rho(\mathbf{x}^N)\,\Delta\hat{x}_1^2(t)v_{0x}^2 \tag{A.2}$$

For the hard-core potential, write

$$E = \sum_i \frac{1}{2}m\hat{v}_i^2 + \frac{N}{2}mv_0^2 \tag{A.3}$$

and transform

$$\int d\mathbf{x}^N = \int_v d\mathbf{r}^N \int d\hat{\mathbf{v}}_1 \int d\hat{\mathbf{v}}_2 \cdots \int d\hat{\mathbf{v}}_{n-1} \int d\mathbf{v}_0 \tag{A.4}$$

whence

$$\begin{aligned}\langle \Delta \hat{x}_1^2(t) v_{0x}^2 \rangle &= \int d\mathbf{v}_0 \exp\left(-\frac{\beta N m}{2} v_0^2\right) v_{0x}^2 \int_V d\mathbf{r}_N \int d\hat{\mathbf{v}}_1 \cdots \\ &\quad \times \int d\hat{\mathbf{v}}_{N-1} B(\mathbf{r}^N) \exp[-\beta \hat{E}(\hat{\mathbf{v}}^N)] \, \Delta x_2^1(t) \\ &= \langle \Delta \hat{x}_1^2(t) \rangle \langle v_{0x}^2 \rangle \end{aligned} \tag{A.5}$$

where $B(\mathbf{r}^N)$ is the overlap function defined in (36). Similarly

$$\langle \Delta \hat{x}_1(t) \hat{v}_{1x} v_{0x}^2 \rangle = \langle \Delta \hat{x}_1(t) \hat{v}_{1x} \rangle \langle v_{0x}^2 \rangle \tag{A.6}$$

so that (A.1) becomes

$$D^{(2)}(t) = \hat{D}^{(2)}(t) + \frac{t^3}{6}[\langle v_{0x}^4 \rangle - 3\langle v_{0x}^2 \rangle^2] \tag{A.7}$$

The quantity in square brackets in the cumulant of the v_{0x} distribution, which from (A.5) is evidently a normal distribution, with

$$\langle v_{0x}^4 \rangle - 3\langle v_{0x}^2 \rangle^2 = 0 \tag{A.8}$$

whence follows (62).

ACKNOWLEDGMENTS

The authors are grateful to Dr. Berni J. Alder and Dr. Thomas E. Wainwright for discussions of their methods. We are also grateful to Dr. Brad L. Holian for permission to discuss the nonequilibrium self-diffusion experiment and for calling to our attention the question of generalizing (35) to soft interactions. Finally we acknowledge the continued counsel of Professors E. G. D. Cohen and J. R. Dorfman.

This work was performed under the auspices of the United States Energy Research and Development Administration.

References

1. W. W. Wood, *in*: *Fundamental Problems in Statistical Mechanics III* (E. G. D. Cohen, ed.), pp. 331–338, North-Holland Publishing Co., Amsterdam (1975).
2. J. Ford, *in*: *Fundamental Problems in Statistical Mechanics III* (E. G. D. Cohen, ed.), pp. 215–255, North-Holland Publishing Co., Amsterdam (1975).
3. Birger Jansson, *Random Number Generators*, Pettersons, Stockholm (1966).

4. W. W. Wood, *in*: *Physics of Simple Liquids* (H. N. V. Temperley, J. S. Rowlinson, and G. S. Rushbrooke, eds.), pp. 116–230, North-Holland Publishing Co., Amsterdam (1968).
5. B. J. Alder and T. E. Wainwright, Studies in molecular dynamics. II. Behavior of a small number of elastic spheres, *J. Chem. Phys.* **33**, 1439–1451 (1960).
6. J. O. Hirschfelder, C. F. Curtiss, and R. B. Bird, *Molecular Theory of Gases and Liquids*, p. 134, John Wiley and Sons, New York (1954).
7. S. Chapman and T. G. Cowling, *The Mathematical Theory of Non-Uniform Gases*, p. 32, 3rd ed., Cambridge University Press, Cambridge (1970).
8. W. G. Hoover and B. J. Alder, Studies in molecular dynamics. IV. The pressure, collision rate, and their number dependence for hard disks, *J. Chem. Phys.* **46**, 686–691 (1967).
9. W. W. Wood, Monte Carlo Calculations of the Equation of State of Systems of 12 and 48 Hard Circles, Los Alamos Scientific Laboratory Report, LA-2827, July, 1963.
10. J. L. Lebowitz and J. K. Percus, Thermodynamic properties of small systems, *Phys. Rev.* **124**, 1673–1681 (1963).
11. R. Zwanzig, Time-correlation functions and transport coefficients in statistical mechanics, *Am. Rev. Phys. Chem.* **16**, 67–102 (1965).
12. W. A. Steele, *in*: *Transport Phenomena in Fluids* (H. J. M. Hanley, ed.), pp. 209–312, Marcel Dekker, New York (1969).
13. B. J. Alder, D. M. Gass, and T. E. Wainwright, Studies in molecular dynamics. VIII. The transport coefficients for a hard-sphere fluid, *J. Chem. Phys.* **53**, 3813–3826 (1970).
14. J. R. Dorfman, *in*: *Fundamental Problems in Statistical Mechanics III* (E. G. D. Cohen, ed.), pp. 277–330, North-Holland Publishing Co., Amsterdam (1975).
15. B. J. Alder and T. E. Wainwright, Velocity autocorrelations for hard spheres, *Phys. Rev. Lett.* **18**, 988–990 (1967).
16. B. J. Alder and T. E. Wainwright, Decay of the velocity autocorrelation function, *Phys. Rev. A* **1**, 18–21 (1970).
17. T. Keyes and I. Oppenheim, On the moments of the displacement of a tagged particle, *Physica* **70**, 100–104 (1973).
18. J. W. Dufty and J. A. McLennan, Persistent correlations in diffusion, *Phys. Rev. A* **9**, 1266–1272 (1974).
19. I. M. de Schepper, H. van Beijeren, and M. H. Ernst, The non-existence of the linear diffusion equation beyond Fick's law, *Physica* **75**, 1–36 (1974).
20. B. J. Alder and T. E. Wainwright (private communication).
21. A. Hald, *Statistical Theory with Engineering Applications*, p. 624, John Wiley and Sons, Inc., New York (1952).
22. W. T. Ashurst and W. G. Hoover, Argon shear viscosity via a Lennard-Jones potential with equilibrium and non-equilibrium molecular dynamics, *Phys. Rev. Lett.* **31**, 206–207 (1973).
23. A. W. Lees and S. F. Edwards, The computer study of transport processes under extreme conditions, *J. Phys. C: Solid State Phys.* **5**, 1921–1929 (1972).
24. E. M. Gosling, I. R. McDonald, and K. Singer, On the calculation by molecular dynamics of the shear viscosity of a simple fluid, *Mol. Phys.* **26**, 1475–1484 (1973).
25. B. L. Holian, W. W. Wood, and J. J. Erpenbeck, Self-diffusion of hard spheres and disks by non-equilibrium molecular dynamics (in preparation).
26. U. Dieter and J. H. Ahrens, *Uniform Random Numbers*, Institut für Math. Statistik, Graz, Austria (1974).
27. J. R. Dorfman (private communication).
28. M. Bishop and B. J. Berne, Molecular dynamics of one-dimensional hard rods, *J. Chem. Phys.* **60**, 893–897 (1974).
29. D. W. Jepsen, Dynamics of a simple many-body system of hard rods, *J. Math. Phys.* **6**, 405–413 (1965).
30. J. L. Lebowitz and J. K. Percus, Kinetic equations and density expansions: exact solvable one-dimensional system, *Phys. Rev.* **155**, 122–138 (1967).
31. J. L. Anderson, J. K. Percus, and J. K. Steadman, Numerical investigations of a simple model of a one-dimensional fluid, *J. Comp. Phys.* **1**, 68–86 (1966).
32. S. Chapman and T. G. Cowling, *The Mathematical Theory of Non-Uniform Gases*, 3rd ed., p. 218, Cambridge University Press (1970).

33. B. J. Alder and T. E. Wainwright, *Int. Symp. Statist. Mech. Brussels, 1956*, pp. 97–131 (I. Prigogine, ed.), Wiley (Interscience), New York (1958).
34. B. J. Alder and T. E. Wainwright, Studies in molecular dynamics, *J. Chem. Phys.* **31**, 459–466 (1959).

2

Molecular Dynamics Methods: Continuous Potentials

J. Kushick
and
B. J. Berne

1. Introduction

Molecular dynamics techniques have been used with considerable success during the past two decades to probe the often subtle relationships between the motion of individual molecules and the observable properties of matter. Most of the molecular dynamics studies carried out thus far have dealt with systems containing structureless spherical particles. Recently, there has been growing activity in the simulation of assemblies of anisotropic molecules.

The purpose of this chapter is to discuss some of the techniques that are currently being used in molecular dynamics, with special emphasis on those techniques involved in the study of polyatomic fluids. The scientific accomplishments of molecular dynamics research have been reviewed elsewhere,[1] and will not be covered here.

Generally, in molecular dynamics a computer is used to solve numerically the classical equations of motion for an assembly of N (usually several hundred) interacting particles in a box of side L. We shall restrict our attention to those cases where the interaction can be specified by a pairwise-additive, continuous potential function. Because discontinuous potentials require very different techniques, they are discussed in Chapter 1. The output of the

J. Kushick • The James Franck Institute, University of Chicago, Chicago, Illinois. Present address: Department of Chemistry, Amherst College, Amherst, Massachusetts, and *B. J. Berne* • Department of Chemistry, Columbia University, New York, New York

computation is a trajectory in phase space—a complete classical description of the system at discrete times t_n separated by a time increment $\Delta t = t_{n+1} - t_n$.

In order to simulate more closely the behavior of an infinite system, periodic boundary conditions are imposed on the solution to the equations of motion. If a molecule labeled i is located at position (x_i, y_i, z_i) at time t, we imagine that there are 26 additional images of i located at $(x_i \pm L, 0; y_i \pm L, 0; \pm L, 0)$. The particle and its 26 images have the same orientation and velocity. Another molecule j may interact with any i within its interaction range. If molecule i should cross a face of the box, it is reinserted at the opposite face. Constant density is thus maintained. These periodic boundary conditions avoid the strong surface effects that would result from a box with reflecting walls.

Once the interaction potential is specified, the equations of motion may be written. Consider the case of molecules with cylindrical symmetry. The pair potential between molecules i and j is of the form $V(\hat{\mathbf{u}}_i, \hat{\mathbf{u}}_j, (\mathbf{r}_j - \mathbf{r}_i))$, where $\hat{\mathbf{u}}_i$ is a unit vector parallel to the symmetry axis of molecule i, and $\mathbf{r}_i$ is the center of mass position of molecule i. For the translational degrees of freedom we have the Newton equations:

$$\dot{\mathbf{r}}_i = \mathbf{v}_i \tag{1}$$

$$m\dot{\mathbf{v}}_i = \mathbf{F}_i = -\nabla_{\mathbf{r}_i} \sum_{j \neq i} V(\hat{\mathbf{u}}_i, \hat{\mathbf{u}}_j, (\mathbf{r}_j - \mathbf{r}_i)) \tag{2}$$

Here $\mathbf{v}_i$ is the translational velocity of molecule i. The force on molecule i, $\mathbf{F}_i$, is found from the spatial gradient with respect to $\mathbf{r}_i$ of the total interaction energy of molecule i.

The orientation of each molecule may be specified by the three Euler angles α, β, and γ.[2] If we associate with a molecule a body-fixed coordinate system (to be denoted by primes), then the Euler angles locate the body-fixed system with respect to the space-fixed system. If the symmetry axis $\hat{\mathbf{u}}$ is taken to be in the z' direction, then

$$\hat{\mathbf{u}} = \hat{\mathbf{i}} \sin\beta \sin\alpha - \hat{\mathbf{j}} \sin\beta \cos\alpha + \hat{\mathbf{k}} \cos\beta$$

We see immediately that γ is a cyclic coordinate, as required by symmetry.

The partial derivative of the interaction potential with respect to any Euler angle is a torque antiparallel to the axis about which the angle is defined. If

$$V_i = \sum_{j \neq i} V(\hat{\mathbf{u}}_i, \hat{\mathbf{u}}_j, (\mathbf{r}_j - \mathbf{r}_i))$$

then $-\partial V_i/\partial \alpha_i$ is the torque on molecule i parallel to the z axis, $-\partial V_i/\partial \beta_i$ is the torque parallel to the line of nodes, and $-\partial V_i/\partial \gamma_i$ is the torque parallel to the z' axis. The components of the total torque on molecule i expressed in its body-fixed system are then given by linear combinations of these partial derivatives:

$$N_{ix'} = -\cos \gamma_i \frac{\partial V_i}{\partial \beta_i} - \frac{\sin \gamma_i}{\sin \beta_i} \frac{\partial V_i}{\partial \alpha_i}$$

$$N_{iy'} = \sin \gamma_i \frac{\partial V_i}{\partial \beta_i} - \frac{\cos \gamma_i}{\sin \beta_i} \frac{\partial V_i}{\partial \alpha_i} \tag{3}$$

$$N_{iz'} = -\frac{\partial V_i}{\partial \gamma_i} = 0$$

The Euler equations of motion relate the time derivative of the molecular angular velocity $\boldsymbol{\omega}_i$, expressed in the body-fixed (principal) axis system, to the torques given above[2]:

$$I_\perp \dot{\omega}_{ix'} = N_{ix'} + (I_\perp - I_\parallel)\omega_{iy'}\omega_{iz'}$$

$$I_\perp \dot{\omega}_{iy'} = N_{iy'} - (I_\perp - I_\parallel)\omega_{ix'}\omega_{iz'} \tag{4}$$

$$\dot{\omega}_{iz'} = 0$$

where $I_\perp$ is the molecular moment of inertia for rotation about an axis perpendicular to $\hat{\mathbf{u}}$ and $I_\parallel$ is the moment of inertia for rotation parallel to $\hat{\mathbf{u}}$. The primed components of $\boldsymbol{\omega}_i$ are themselves related to the time derivatives of the Euler angles:

$$\dot{\alpha}_i = \frac{1}{\sin \beta_i}(\omega_{ix'} \sin \gamma_i + \omega_{iy'} \cos \gamma_i)$$

$$\dot{\beta}_i = \omega_{ix'} \cos \gamma_i - \omega_{iy'} \sin \gamma_i \tag{5}$$

$$\dot{\gamma}_i = -\frac{1}{\sin \beta_i}(\omega_{ix'} \cos \beta_i \sin \gamma_i + \omega_{iy'} \cos \beta_i \cos \gamma_i - \omega_{iz'} \sin \beta_i)$$

Equations (1), (2), (4), and (5) are the equations of motion for a system of particles with cylindrical symmetry. In the following pages, we shall discuss a number of aspects of the numerical solution of these equations such as the choice of numerical integration algorithm, some shortcuts in the numerical integration, the difficult problem of choosing a potential function for polyatomic molecules, and the calculation of quantities of statistical-mechanical interest from the phase space trajectory.

2. Integration Procedures for the Equations of Motion

The variety of algorithms currently in use by molecular dynamics groups for the numerical integration of the equations of motion may be taken as evidence that no single procedure has been demonstrated to be superior to any other procedure under all conditions. As is generally the case in large-scale computing operations, one must strike a proper balance in the use of the available computational resources. Thus, the choice of algorithm will depend on the particular system being simulated, the hardware configuration and mode of operation of the computer center, and to some degree, personal preference.

The computational parameters of greatest concern in molecular dynamics production runs are execution time and memory storage capacity (region). Other factors, such as input/output (I/O) requirements, do not generally present as significant a problem as these two. In the following paragraphs, we shall briefly discuss several integration algorithms that have been used in molecular dynamics research. These methods will be compared with regard to their use of computer time and region. We do not intend this discussion to be exhaustive. Rather, our aim is to illustrate the considerations involved in the selection of a particular method.

Before actually discussing the various algorithms, a few general observations are in order. Suppose that we know the values of the molecular coordinates and velocities (translational and rotational) at time t_n; that is, we have specified the point Γ_n in phase space. The object of any integration procedure is to determine the phase point Γ_{n+1} representing the state of the system at time $t_{n+1} = t_n + \Delta t$. Γ_{n+1} will be calculated by means of a power series in Δt, which is truncated after a given term, say $(\Delta t)^l$, where l is called the "order" of the method. Now, Δt must be selected in such a way that the integration procedure generates a stable solution to the equations of motion, that is, a solution that does not diverge with time from the "true" solution. The stability of the solution is checked by monitoring the value of a constant of the motion, usually the total energy, and making sure that it does not drift away from its initial value. Too large a value of Δt will result in nonconservation of energy and an unstable phase space trajectory. The range of permissible values for Δt will depend to some degree on the order of the integration algorithm. A high-order method may allow a larger Δt to be used than a low-order method, since errors resulting from series truncation are smaller with a high-order method.

All integration algorithms incorporate a subroutine that calculates, for any given phase point Γ, the forces and torques that appear in the equations of motion. This subroutine is the computational workhorse of the dynamics program; most of the execution time in a production run is spent in the evaluation of forces. Needless to say, careful coding and the use of time-saving devices are extremely important in this subroutine. In the generation of Γ_{n+1} from Γ_n, higher order integration algorithms require knowledge of the forces at intermediate or trial phase points, in addition to Γ_n itself. This means that the force subroutine must be called more than once during each integration step. As a result, the cycle time (the execution time required to advance the state of the system one step in time, from Γ_n to Γ_{n+1}) for a high-order algorithm may be larger than for a low-order procedure, even though the actual value of Δt may be greater.

In some procedures, knowledge of Γ_n alone is not sufficient to generate Γ_{n+1}. In addition, phase points at times $t < t_n$, or equivalently, high-order time derivatives of Γ_n, are required. A procedure where Γ_n by itself suffices is called a self-starting procedure. Methods that are not self-starting may pose additional problems. Region requirements may be greater because of the larger amount of information to be stored. Furthermore, such a procedure must include a separate program segment to generate the first few phase points until enough information has been gathered for the main integration algorithm to take over. This program segment must be referenced every time the integration is started or restarted. Therefore, a method that is not self-starting will generally be more complex than a self-starting method, and may be especially inconvenient during the equilibration phase of a run.

Having made these few remarks, we shall turn to some specific examples. A method especially attractive for its simplicity is due to Verlet.[3] If x_n is the x component of a particular molecule's center of mass at time t_n, and v_n^x is the x component of the molecule's translational velocity, the Verlet algorithm is obtained by addition and subtraction of the Taylor series for $x_{n+1} = x(t_n + \Delta t)$ and $x_{n-1} = x(t_n - \Delta t)$:

$$x_{n+1} = -x_{n-1} + 2x_n + f(\Gamma_n)(\Delta t)^2 \tag{6a}$$

$$v_n^x = (x_{n+1} - x_{n-1})/2\,\Delta t \tag{6b}$$

where $f(\Gamma_n)$ is the total force (in the x direction) on the molecule at time t_n, divided by the molecular mass. This method requires only one evaluation of forces per step and is therefore relatively fast. Even though the method is not self-starting, its region requirements are small compared to the higher order methods. The fact that this is a low-order procedure, however, may force one to select a relatively small Δt to ensure energy conservation.[4] There is an additional problem with Eq. (6b). Here one must find the difference between two quantities of roughly the same magnitude. Such a machine operation results in a loss of numerical precision, and may give rise to a serious round-off error.

This last objection is overcome in a method recently used by Schofield[5]:

$$x_{n+1} = x_n + v_n^x\,\Delta t + \tfrac{1}{6}[4f(\Gamma_n) - f(\Gamma_{n-1})](\Delta t)^2 \tag{7a}$$

$$v_{n+1}^x = v_n^x + \tfrac{1}{6}[2f(\Gamma_{n+1}) + 5f(\Gamma_n) - f(\Gamma_{n-1})]\,\Delta t \tag{7b}$$

This method also requires only one evaluation of forces per integration step, but its storage requirements are greater than Verlet's method. Schofield reports good results with energy conservation for large Δt, even though the method is low order. As in the previous case, this procedure is not self-starting.

Neither of the above procedures is suitable for rotational degrees of freedom. We shall illustrate with Eqs. (6). Suppose we replace x_n by one of the Euler angles α_n. Then v_n^x must be replaced by $\dot{\alpha}_n$. In Eq. (6a), $f(\Gamma_n)$ is now related to the time derivative of the angular velocity at t_n. From Eqs. (4), however, we see that this derivative is a function of the angular velocity itself, in addition to the angular coordinates. Because $\dot{\alpha}_n$ cannot be determined until α_{n+1} is known [see Eq. (6b)], $f(\Gamma_n)$ cannot be calculated, and the method fails. A similar problem arises in Schofield's method, where $f(\Gamma_{n+1})$ in Eq. (7b) cannot be calculated for an angular degree of freedom.

In the simulation of anisotropic particles, therefore, a separate algorithm for the orientational coordinates must be coded if either of the above methods is chosen for the center-of-mass coordinates. In their simulation of liquid nitrogen, Barojas *et al.* used the Verlet method in conjunction with special expansions for the derivatives of the angular coordinates.[6]

So-called predictor–corrector methods have been used in molecular dynamics from its inception.[7,8] In this method, molecular forces at time t_n are used to predict a trial phase point $\bar{\Gamma}_{n+1}$. This prediction requires information about the state of the system at times earlier than t_n, so that the method is not self-starting. The forces at the phase point $\bar{\Gamma}_{n+1}$ are then calculated and are used to refine the prediction and generate the actual phase point Γ_{n+1}. The method can be formulated to high order (fourth- and fifth-order procedures have been used) and is therefore very accurate. Two evaluations of the forces are required per step. The method is relatively fast given its high accuracy. However, the storage requirements are quite large, so that the method can be inconvenient in some applications.

The last algorithm that we shall mention in this section is the Runge–Kutta–Gill method.[9] This is a fourth-order procedure that requires four evaluations of forces per step. It is therefore the slowest of all the methods we have discussed. However, its high accuracy permits the use of a relatively large Δt. Furthermore, it is self-starting and requires significantly less storage than a predictor–corrector method.

It was noted earlier that the choice of integration algorithm will depend partially on the mode of operation of the computer center. We would like now to illustrate the role that this factor might play in a decision between the predictor–corrector and the Runge–Kutta–Gill procedures.

At some installations, a particular user might be assigned the use of the computer for a predetermined period of time. During this time, the user will have available all of the resources of the computer system. In such a situation, there is no incentive to minimize the storage requirements, as long as sufficient storage exists to accommodate the job. The most important consideration is the number of time steps that can be calculated within the assigned time period. Therefore, the relative speed of the predictor–corrector method would make it the method of choice under these conditions.

On the other hand, at a time-sharing installation, the situation is changed. Here, multiple jobs can be active simultaneously, and a support processor dynamically allocates system resources and assigns priorities to jobs on the basis of their system requirements. The requirement of large region would now be a liability, since a large-region job would be assigned a low priority and would be queued until the requested storage becomes available. Thus, a job will have a shorter turnaround time if its region requirement is low. It may then be advantageous to expend additional execution time and use the Runge-Kutta-Gill procedure to improve the throughput of the calculation. It is obvious that local circumstances will dictate the choice of the integration algorithm to be employed.

3. *Time-Saving Techniques*

The careful selection of an integration algorithm is not the only means of conserving computer resources in molecular dynamics. In this section we shall discuss two devices that can save a significant amount of computer time during the generation of a phase space trajectory: equilibration on subsystems and bookkeeping.

At the beginning of a molecular dynamics calculation, one must normally assign initial conditions, that is, the values of the position and velocity for each particle in the system. There are a number of ways of doing this: the molecules may be arranged on a lattice, a configuration from another molecular dynamics calculation may be modified to satisfy the new density condition, or a random configuration may be generated using a Monte Carlo technique. The velocities may be assigned from a Maxwell–Boltzmann distribution at the desired temperature by a rejection technique, or they may even all be assigned an initial value of zero. In any case, the initial state selected will in general not be an "equilibrium state" consistent with the desired temperature and density. As a result, the temperature (which is found from the sum of the squares of the molecular velocities) will be seen to drift away from its initial value as the integration proceeds. When this happens, the integration must be halted and the velocities multiplied by a scale factor to restore the temperature to its desired value. This procedure is then repeated until equilibrium is achieved, that is, until the temperature drift is replaced by fluctuations about the desired value.

In some molecular dynamics work, the process of equilibration has accounted for a substantial fraction of the total computer time devoted to the calculation.[8] If the intermolecular potential has a sufficiently short range, however, it is possible to use only a fraction of the molecular dynamics system for equilibration. One can set aside a subsystem containing, say, an eighth of the total number of molecules, select initial conditions, and allow the subsystem to come to equilibrium. After equilibrium has been reached, the subsystem is replicated eightfold in space to generate the full system. All of the velocities are reassigned from a Maxwell–Boltzmann distribution, and equilibration is resumed. After the replication, equilibrium for the full system is usually established quickly. Since the integration for the subsystem is fast compared to the full system, this equilibration procedure results in a significant saving of computer time.

Bookkeeping procedures take advantage of the finite range of the intermolecular potential. At any given time during a molecular dynamics calculation, most pairs of molecules are separated by a distance much greater than the effective range of the intermolecular potential r_t. It makes little sense then to calculate the distance between each pair of molecules at each time step of the integration. In the "standard" method of bookkeeping, one imagines a sphere of radius slightly larger than r_t centered at each molecule.[3] A list is then kept in computer storage of all the other molecules that are within each molecule's sphere. Of course, if molecule j is listed in molecule i's sphere, it should not be necessary to list molecule i explicitly in molecule j's sphere. Then when the total force and torque exerted on molecule i are to be calculated, only these neighbors need be considered. The list of neighbors must, of course, be updated periodically. In sufficiently dense fluids this is done approximately every ten integration steps. The difference between the radius of the bookkeeping sphere and r_t must be larger than the maximum distance a molecule might reasonably travel in these ten steps, at the given temperature of the system. This method consequently requires that after each ten steps the list of particles within the bookkeeping sphere of each particle must be updated—a procedure requiring the calculation of the distance between every distinct pair of molecules, of which there are $N(N-1)/2$. For an N-molecule system, the computer time required for this updating is roughly proportional to N^2.

An alternative bookkeeping scheme has recently been proposed.[5,10] In this method, the molecular dynamics box is divided into a number of small cells. At the beginning of each time step, the particular molecules in each cell are listed. The cells are then scanned, and relative distances are calculated only for pairs of molecules in neighboring cells. In this method the time required at each step is always proportional to N. Thus, as the number of particles in the system increases, this method should become faster than the older bookkeeping method.

One of the important parameters in this form of bookkeeping is the size of the cell v. There is a trade off between cycle time and accuracy. This procedure only includes interactions between particles in nearest-neighbor cells. If $v^{1/3}$ is less than or equal to the interaction range, then important interactions will be missed. On the other hand, if v is large so that on the average a large fraction of the N particles occupy each cell, the procedure becomes as inefficient as if no bookkeeping were used. The optimum value of v has been located in an empirical study where it was shown that for Lennard-Jones spheres, v is optimum if on the average a cell contains four particles. The optimum v may be different for anisotropic molecules.

The use of bookkeeping procedures will in general yield dramatic reduction in the consumption of computer time in molecular dynamics and can therefore greatly increase the scope of such calculations.

4. Potential Functions for Anisotropic Molecules

Perhaps the most significant difficulty in the computer simulation of polyatomic fluids is the formulation of the intermolecular potential function. Extremely little is known about the details of anisotropic molecular interactions, and the possibilities for modeling are restricted by considerations of practicality for computer applications. In this section we shall discuss several approaches that have been used to model the interactions of anisotropic molecules.

The interactions of diatomic molecules have been treated in two fashions. In one case, orientation-dependent dipolar and quadrupolar interactions are superimposed on a spherically symmetric potential. Computer simulations of N_2 and CO have been carried out using the Stockmayer potential, which is a sum of a center-to-center Lennard-Jones potential and a number of multipole interaction terms.(4) Alternately, the N_2 molecule can be envisioned as two bound force centers, each of which interacts isotropically with force centers on other molecules.(6) The total potential of two nitrogen molecules is thus the sum of four terms.

The force center representation has been the most widely used method for simulating the interactions of polyatomic molecules. A series of computer experiments has recently been conducted on water. Two potentials have been used in these studies: the BNS potential(8) and the ST2 potential.(11) Both of these potentials represent the water molecule as five bound force centers. One center is located at the oxygen atom. This force center interacts with the force centers on other oxygen atoms through a Lennard-Jones (12-6) potential. In addition, each molecule possesses four point charges (two positive and two negative) representing the partially shielded protons and the nonbonding electron pairs. These force centers interact via a Coulomb potential with the point charges on other molecules. The Coulomb potential is modulated by a "switching function" to avoid divergences when two point charges approach each other. Thus, the interaction of two water molecules is composed of seventeen terms, that is, one Lennard-Jones interaction and sixteen Coulomb interactions. These potential functions have proven quite useful in the simulation of the static and dynamic properties of liquid water.

Corner has generalized the force center concept to include all moderately elongated molecules with approximate cylindrical symmetry (e.g., normal butane).[12] In Corner's scheme, each molecule is represented by four force centers arranged on a line. The force centers are equally spaced by a distance determined by the overall length ratio of the molecule. The interaction of two molecules is then taken to be the sum of sixteen Lennard-Jones terms between the force centers on distinct molecules. In order to simplify the complicated superposition of potentials, Corner numerically fit the sum of Lennard-Jones terms to a *single* Lennard-Jones potential for the two molecules. The Lennard-Jones parameters σ and ε that appear in the resultant potential are no longer constant, but are functions of the relative orientation of the two molecules.

The representation of molecular interactions by means of force centers is quite useful for relatively small molecules of moderate anisotropy. However, if one whishes to study phenomena such as long-range orientational order or strongly hindered rotation, one must be able to simulate the interactions of highly anisotropic molecules. A highly elongated molecule would require a large number of force centers to maintain its impenetrability. The very large number of terms comprising the interaction of two molecules would then make the model unwieldy. Furthermore, as Corner noted, as the degree of elongation becomes substantial, it becomes increasingly difficult to obtain a satisfactory fit of the sum of force center terms to a single Lennard-Jones interaction. What is needed here is a method for incorporating the essential geometric characteristics of a strongly anisotropic interaction, while maintaining computational simplicity. Berne and Pechukas have suggested a method of modeling the interactions of molecules with cylindrical symmetry.[13] In their model, each molecule is visualized as an ellipsoidal rigid rotor. In the following paragraphs, we shall summarize the development of the model.

An ellipsoidal molecule can be located in space by specifying the coordinates of its center of mass and a unit vector $\hat{\mathbf{u}}$, parallel to its symmetry axis. The force centers in this molecule are imagined to be distributed in space according to an ellipsoidal Gaussian distribution centered on the molecule's center of mass, and characterized by a half-width $\sigma_{\parallel}$ parallel to the molecular axis $\hat{\mathbf{u}}$ and $\sigma_{\perp}$ perpendicular to $\hat{\mathbf{u}}$. It is then assumed that the repulsive potential energy of two interacting molecules is measured by the mathematical volume overlap of the two respective Gaussian distributions of force centers. This overlap integral I depends on the relative center-of-mass distance and on the relative orientation angles. The overlap integral I may be cast in the following form:

$$I = \varepsilon(\hat{\mathbf{u}}_1, \mathbf{u}_2) \exp[-r^2/\sigma^2(\hat{\mathbf{u}}_1, \hat{\mathbf{u}}_2, \hat{\mathbf{r}})] \tag{8}$$

$$\varepsilon(\hat{\mathbf{u}}_1, \hat{\mathbf{u}}_2) = \varepsilon_0[1-\chi^2(\hat{\mathbf{u}}_1 \cdot \hat{\mathbf{u}}_2)^2]^{-1/2} \tag{9}$$

$$\sigma(\hat{\mathbf{u}}_1, \hat{\mathbf{u}}_2, \hat{\mathbf{r}}) = \sigma_0\left\{1-\frac{1}{2}\chi\left[\frac{(\hat{\mathbf{r}}\cdot\hat{\mathbf{u}}_1+\hat{\mathbf{r}}\cdot\hat{\mathbf{u}}_2)^2}{1+\chi(\hat{\mathbf{u}}_1\cdot\hat{\mathbf{u}}_2)}+\frac{(\hat{\mathbf{r}}\cdot\hat{\mathbf{u}}_1-\hat{\mathbf{r}}\cdot\hat{\mathbf{u}}_2)^2}{1+\chi(\hat{\mathbf{u}}_1\cdot\hat{\mathbf{u}}_2)}\right]\right\}^{-1/2} \tag{10}$$

Here, χ is determined by the anisotropy of the ellipsoids. Given the axial ratio $a = \sigma_{\|}/\sigma_{\perp}$, then

$$\chi = \frac{a^2 - 1}{a^2 + 1} \tag{11}$$

$\hat{\mathbf{u}}_1$ and $\hat{\mathbf{u}}_2$ are the orientation vectors of the two molecules; $\mathbf{r}$ is a vector joining the two centers, with magnitude r, and $\hat{\mathbf{r}} = \mathbf{r}/r$.

Thus, the Gaussian overlap model generates a strength parameter ε and a range parameter σ that are determined by the relative orientation of the two molecules. These two parameters may now be used in any of a variety of two-parameter potentials that have been proposed to describe atomic interactions. For example, we may use $\varepsilon(\hat{\mathbf{u}}_1, \hat{\mathbf{u}}_2)$ and $\sigma(\hat{\mathbf{u}}_1, \hat{\mathbf{u}}_2, \hat{\mathbf{r}})$ in the Lennard-Jones (12-6) potential

$$V(\hat{\mathbf{u}}_1, \hat{\mathbf{u}}_2, \mathbf{r}) = 4\varepsilon(\hat{\mathbf{u}}_1, \hat{\mathbf{u}}_2)\left\{\left[\frac{\sigma(\hat{\mathbf{u}}_1, \hat{\mathbf{u}}_2, \hat{\mathbf{r}})}{r}\right]^{12} - \left[\frac{\sigma(\hat{\mathbf{u}}_1, \hat{\mathbf{u}}_2, \hat{\mathbf{r}})}{r}\right]^{6}\right\} \tag{12}$$

By this means, we construct a differentiable potential energy function that depends realistically on the distance r between the centers and whose dependence on molecular orientation is intuitively reasonable. Furthermore, through the parameter χ we may handle molecules of any eccentricity with equal ease.

This potential has been studied in connection with two prolate molecules, CO_2 and N_2, and one oblate molecule, benzene. The potential parameters were chosen to optimize agreement with the temperature dependence of measured second virial coefficients.(14) These potentials were then used to compute the cohesive energy of the respective solids with some success.(14)

The ellipsoidal Lennard-Jones potential has recently been applied in molecular dynamics studies of the angular velocity relaxation of hindered rotors, collective reorientation, and the stability of nematic-like orientational ordering.(15–17) The model was found to be a convenient and flexible representation of the interactions of polyatomic molecules.

In order to limit the total number of interactions experienced by each molecule in a computer simulation, the potential energy is usually truncated so that a molecule's interaction range is finite. For example, the ordinary (spherical) Lennard-Jones potential is truncated at about 2.5σ; the interactions between all molecules separated by more than this distance are weak enough to be neglected. In order to maintain conservation of energy, an anisotropic potential function should be truncated at an equipotential surface. However, if the potential is not too anisotropic, truncation at a fixed distance leads to only minor effects on energy conservation.(8)

The combination of a truncated potential function and periodic boundary conditions leads to an important programming simplification. Let L be the length of the molecular dynamics box and let r_t be a molecule's interaction range. (In the case of an anisotropic potential, r_t is the maximum distance of the truncation equipotential surface from the center of the molecule.) In principle, a molecule labeled i might interact simultaneously with 27 periodic images of another molecule j. It may readily be seen, however, that if $L \geq 2r_t$, molecule i will interact with *at most* one image of molecule j. If $L \geq r_t$, there will be no more than eight interactions between any pair of molecules. Because a significant fraction of the computer time used in a molecular dynamics calculation is spent testing pairs of molecules for possible interactions, this information can be put to good use in the dynamics program.

When we consider molecules of substantial anisotropy, the interaction range r_t for the orientation-dependent Lennard-Jones potential is quite large. For example, if the truncation equipotential surface corresponds to an energy of $-0.01\varepsilon_0$, then for $a = 3.5$ we find $r_t \sim 10\sigma_0$. As a result, a very large number of pairs of molecules must be examined for possible interaction.

A recent study examined the effect on molecular dynamics of neglecting the long-range attractive part of the Lennard-Jones potential (spherical and ellipsoidal).[18] This study was motivated by the successful static perturbation theory of simple Lennard-Jones fluids developed by Weeks *et al.*[19] The latter theory decomposed the spherical Lennard-Jones potential

$$V(r) = 4\varepsilon((\sigma/r)^{12} - (\sigma/r)^6) \tag{13}$$

into a sum of a strong short-range purely repulsive part,

$$V_R(r) = \begin{cases} V(r) + \varepsilon, & r/\sigma < 2^{1/6} \\ 0, & r/\sigma > 2^{1/6} \end{cases} \tag{14}$$

and a weak longer-range attractive part,

$$V_A(r) = \begin{cases} -\varepsilon, & r/\sigma < 2^{1/6} \\ V(r), & r/\sigma > 2^{1/6} \end{cases} \tag{15}$$

It was found that, especially at high densities, the structure of Lennard-Jones fluids is determined almost entirely by $V_R(r)$ alone.

This finding has significant consequences for molecular dynamics. If the dynamic behavior of sufficiently dense Lennard-Jones fluids is dominated by $V_R(r)$, as was shown for static properties, then one can realize an enormous saving of computer time by using the purely repulsive part of the spherical or ellipsoidal Lennard-Jones potential. Since the computer time used in molecular dynamics depends partially on the number of interactions experienced per molecule, the savings factor is of the order $(2^{1/6}/2.5)^3 = 0.09$.

In the molecular dynamics comparison of the full and the purely repulsive Lennard-Jones potentials, the velocity correlation function was calculated using both potentials in a variety of systems. However, the extremely close agreement shown in dense systems for static properties was not exhibited for the velocity correlation function, possibly because of coherence effects in the mutual interactions of large groups of molecules. It was concluded that for accurate numerical simulation intended for the computation of particular properties, the attractive part of the Lennard-Jones potential cannot be neglected in molecular dynamics, but that the qualitative physical attributes of a Lennard-Jones system are not disturbed when only the repulsive part of the potential is used.

The preceding has illustrated some of the considerations involved in the representation of intermolecular interactions. Needless to say, the models used so far in molecular dynamics are empirical and are dictated more by convenience than by detailed experimental knowledge. However, a good deal of information has been gained from the use of these models, and it is to be hoped that molecular dynamics will soon reap the benefits of current research into the nature of intermolecular potential surfaces.

5. The Calculation of Time Correlation Functions and Static Properties

The preceding sections were devoted to the generation of the phase space trajectory. The goal of a molecular dynamics calculation is the evaluation of static and dynamical properties of statistical-mechanical interest. Because these evaluations often turn out to be formidable computations, we would like to make a few observations on this phase of molecular dynamics.

In the study of dynamical properties, one usually wishes to evaluate a variety of time correlation functions. If $A(t)$ is a real dynamical property of the system, then the time correlation function $C(t)$ is defined by an ensemble average:

$$C(t) = \langle A(0)A(t) \rangle \tag{16}$$

For ergodic systems, the ensemble average may be replaced by an infinite-limit time integral over the phase space trajectory of a single system:

$$C(t) = \lim_{T\to\infty} \frac{1}{T} \int_0^T d\tau\, A(\tau)A(t+\tau) \tag{17}$$

In molecular dynamics, one calculates $C(t)$ by a time average over discrete phase points:

$$C(t_n) = \frac{1}{M} \sum_{m=1}^{M} A(t_m)A(t_{m+n}) \tag{18}$$

The sum over m need not include each consecutive time step. As shall be seen, it may be advantageous from the viewpoint of error analysis to skip some time steps and average instead over a longer time span.

If $A(t)$ is a property of a single molecule, $C(t)$ is called a single-particle correlation function, whereas if $A(t)$ is a collective property of the entire system, $C(t)$ is called a collective correlation function. In general, the calculation of a single-particle correlation function is more complicated than a collective correlation function, because in the former case, contributions from distinct molecules must be handled separately. There are two methods in use for the calculation of time correlation functions. One of these is the straightforward application of Eq. (18). This procedure is quite convenient for collective correlation functions, since each time step in the trajectory is represented by one number and, at least in some cases, all of the relevant data may be contained in computer memory. In the case of single-particle correlation functions, the amount of data is much greater, and efficient procedures must be devised for rereading the data corresponding to a single time step many times for use in forming different correlation products. In these procedures, data are not always read in the order corresponding to increasing values of the time. It is therefore of great advantage to have the relevant data stored as a direct-access dataset during the computation of the correlation function.[17]

Futrelle and McGinty have described an alternate method of calculating time correlation functions in molecular dynamics.[20] In their method, $A(t)$ as calculated from the phase space trajectory is Fourier transformed to give $\tilde{A}(\omega)$. It follows from the Wiener–Khintchine theorem that the Fourier transform of $|\tilde{A}(\omega)|^2$ gives $C(t)$. Futrelle and McGinty claim that their method is much faster than the "standard" method [Eq. (18)]. Another advantage of the Fourier transform method is that it is possible to calculate a time correlation function dynamically, that is, while the phase space trajectory is being generated. This feature can eliminate a great many of the input–output operations. However, the region requirements of the Fourier transform method may be considerable, especially in the case of single-particle correlation functions and when a fine grid in the frequency domain is needed.

Zwanzig and Ailawadi have studied the error arising in $C(t)$ as a result of finite time averaging.[21] Suppose $C(t)$ is normalized $[C(0)=1]$ and is characterized by a relaxation time τ_A. Zwanzig and Ailwadi estimate that if $C(t)$ is calculated by a time average over the time $T=(M-1)\,\Delta t$, then the error in the calculated function is

$$\varepsilon(t) = \pm(2\tau_A/T)^{1/2}(1-C(t)) \tag{19}$$

If $C(t)$ is a time correlation function of a single-particle property, then in addition to time averaging, $C(t)$ may be averaged over the N particle labels in the molecular dynamics system. This additional averaging decreases the error by the factor $(1/N)^{1/2}$. Even if $C(t)$ is the time correlation function of a collective property, the error may in some cases be less than the above expression indicates. If the volume of the system were sufficiently large so that many of the molecules that contribute to the collective property are separated by more than some specified correlation length, we intuitively feel that we are getting a better sampling of the collective property than if the system volume were small. More precisely, if the correlation length of the system is L_c and the side of the molecular dynamics box is L, we might expect the error to be decreased by the factor $(L_c/L)^{3/2}$.

Erpenbeck and Wood have recently made an interesting observation with regard to the effect of trajectory round-off errors on the calculated time correlation functions.[22] They performed parallel dynamics simulations with hard-sphere systems using single-and double-precision arithmetic. It was found that the single-precision trajectory, which is more subject to serious accumulation of machine round-off error, slowly diverged from the double-precision trajectory during the course of the calculation. However, when the velocity correlation function was calculated for the two trajectories, the two functions fluctuated around each other for a surprisingly long time. The time-averaging process thus seems to "wash out" the inaccuracy in the single-precision trajectory.

Time correlation functions can be used in conjunction with the Green–Kubo relations to calculate the various transport coefficients in the system. For example, the self-diffusion coefficient D is related to the time integral of the velocity correlation function:

$$D = \frac{1}{3}\int_0^\infty dt\, \langle \mathbf{v}(0) \cdot \mathbf{v}(t) \rangle \tag{20}$$

Alternately, D may be found from an Einstein relation:

$$D = \lim_{t\to\infty} \frac{\langle (\Delta r(t))^2 \rangle}{6t} \tag{21}$$

where $\langle (\Delta r(t))^2 \rangle$ is the mean square displacement. In the latter case, the mean square displacement is calculated as a function of time until its variation in time is linear. The diffusion coefficient is then found from the slope of the mean square displacement.

Helfand[23] has shown that for each transport coefficient it is possible to define a dynamical variable, such that the transport coefficient can be determined from the mean square displacement of this dynamical variable in a manner completely analogous to Eq. (21). This approach has certain definite advantages over the use of the Green–Kubo relation and has been used recently to determine the transport coefficients in hard-sphere fluids.[24]

Other methods have been developed for calculating transport coefficients and time correlation functions. The viscosity of argon has been computed[25,26] by determining the steady flux of momentum that arises in a system in which a shearing force is applied at the boundaries. The appropriate ratio of flux to force then gives the viscosity. This is an example of nonequilibrium molecular dynamics. It is not difficult to imagine other variations of this procedure (see Chapter 1 of this volume).

Recently a very important method for bypassing the direct evaluation of the time correlation function has been exploited.(27) This method is based on linear response theory.(28) In this method all the molecular dynamics phase space trajectories $\Gamma(t)$ are subdivided into "segments" lasting typically 60 time steps. The trajectories in each segment are computed twice starting from the same initial conditions. In one case the calculation proceeds in the normal way, whereas in the other case a small external force is applied. The perturbed system evolves differently with time than the unperturbed system with a concomitant difference $\Delta B_i(t)$ in the value of certain dynamical variables B_i. By calculating the *difference* in B_i in the perturbed and unperturbed trajectories, and averaging this difference over all segments comprising the run, the "ensemble" averaged difference $\overline{\Delta B_i(t)}$ is calculated. By a standard result of linear response theory,(28) the mean induced change in B_i can be related to a time correlation function. Specifically, if $\Delta B_i(t)$ is the change induced in B_i by the application of an external force $F_j(\tau)$ that is conjugate to the property B_j, then in the limit of sufficiently small $F_j(\tau)$,

$$\overline{\Delta B_i(t)} = \beta \int_{-\infty}^{t} d\tau \, \langle B_i(\tau)\dot{B}_j(0)\rangle F_j(\tau - t) \tag{22}$$

where $\dot{B}_j$ is the rate of change of B_j, $\beta \equiv (k_B T)^{-1}$, and the force $F_j(\tau)$ has a prescribed time dependence. In the event that $F_j(\tau) = \delta(\tau)$, the calculated response is

$$\overline{\Delta B_i(t)} = -\beta \frac{d}{dt}\langle B_i(t)B_j(0)\rangle \tag{23}$$

so that the time correlation function is immediately obtainable from the computed response. In molecular dynamics one cannot generate a true delta function force. In fact, because finite time intervals are used in the numerical integrations, the force is applied as a unit square pulse of width Δt. Given the statistical accuracy of the calculations this is usually not an important distinction. If on the other hand the force is given by

$$F_j(\tau) = F_j \theta(\tau) \tag{24}$$

where $\theta(t)$ is the step function

$$\theta(t) = \begin{cases} 0, & t < 0 \\ 1, & t \geq 0 \end{cases} \tag{25}$$

the response will be

$$\overline{\Delta B_i}(t) = +\beta\left(\int_0^t d\tau\, \langle B_i(\tau)\dot{B}_j(0)\rangle\right)F_j \tag{26}$$

These formulas have been applied with great success to the study of fused salts and electrolyte solutions. If B_i is identified as the electric current along the x direction $J_Q^x(t)$ and the applied force is the electrical force acting on all the ions due to the imposition of a step function electrical field, it follows that $\dot{B}_j = J_Q^x$ and Eq. (26) becomes

$$\overline{J_Q^x}(t) = \left[\beta \int_0^t d\tau\, \langle J_Q^x(\tau) J_Q^x(0)\rangle\right]E \tag{27}$$

The induced electrical current is proportional to the applied field. For short times $\overline{J_Q^x}(t)$ should increase linearly with time. For sufficiently long times, that is, times long compared to the correlation time of the current, the integral reaches a plateau value and the term in the square brackets can be used to determine the specific conductivity σ,

$$\sigma = \frac{\overline{J_Q^x}(\infty)}{VE} \tag{28}$$

where V is the volume of the system.

Kubo has shown[28] that all transport coefficients are related to time integrals of appropriate currents. To obtain transport coefficients by the above method it is necessary to generate a large number of trajectory segments such that each is sufficiently long compared to the correlation time of the appropriate time correlation function. Unfortunately there is an inherent limitation on the length of a segment. Let us consider a given equilibrium and corresponding perturbed segments. The cartesian distance $S(t)$ between these two segments can be defined as $S(t) = |\boldsymbol{\Gamma}'(t) - \boldsymbol{\Gamma}(t)|$, that is, the cartesian distance between the state of the perturbed system (primed variable) and the unperturbed system (unprimed variable). Since both systems have the same state at the beginning of the segment, $S(0) = 0$. As time evolves, $S(t)$ grows exponentially with time, and the values of any observable in the two systems become statistically uncorrelated. This should result from any source of noise. Wood and Erpenbeck (see Chapter 1) have shown that if two trajectories are generated on a CDC computer, one in single precision (60 bit word) and one in double precision (120 bit word), starting from the same state, the velocity correlation functions in the two systems fluctuate around each other for a long time, but the single-precision result cannot be used to compute long-time behavior. Likewise here, this linear response method cannot be used to study long-time phenomena. This method is useful only if it is possible to generate segments longer than the appropriate correlation times but still short enough that the noise does not become an obstacle.

This method has been applied to the electrical properties of charged systems and to the calculation of the viscosity in fluids. It offers certain advantages over the brute-force calculation of time correlation functions. Although the two methods must agree if an infinite time average is performed, it is not clear which of the two methods is more accurate for finite time averages. Gicotti and Jacucci feel that this new method has certain advantages from the point of view of minimizing statistical noise.[27]

Molecular dynamics trajectories are also used to determine a variety of static properties. Such properties can, of course, also be found from Monte Carlo calculations. In molecular dynamics, a given property is evaluated at given time intervals and then averaged.

The temperature is, of course, found from the mean square velocity, and the pressure p is conveniently calculated from the virial theorem:

$$pV = NkT + \frac{1}{3}\left\langle \sum_{i=1}^{N} \mathbf{r}_i \cdot \mathbf{F}_i \right\rangle \tag{29}$$

where $\mathbf{r}_i$ is the center-of-mass position of the ith molecule and $\mathbf{F}_i$ is the total force on that molecule due to the other molecules. In a fixed energy and volume system both the temperature and the pressure fluctuate with time. Lebowitz *et al.* have shown that in a microcanonical ensemble the mean square fluctuation in the temperature, for example, may be used to calculate the specific heat C_v.[29] Accurate determinations of C_V are very difficult to achieve.

The microscopic static property that is usually of primary concern is the pair correlation function $g(r)$. To calculate $g(r)$, each molecule in turn is imagined to be at the center of a series of concentric spheres. The number of molecules in each spherical shell is divided by the volume of that shell, with attention being paid to periodic image locations of molecules outside the box. The results are averaged over all the molecules, and then over many time steps. The pair correlation function is usually calculated in this way only for distances less than the range of the potential r_t. Verlet has provided a method for extending $g(r)$ beyond r_t using the direct correlation function determined from the Percus–Yevick equation.[30]

6. Systems with Long-Range Forces

Molecular dynamics methods have been used to study systems with long-range forces. Early work was done on gravitationally interacting mass points. The purpose of this work was to study galactic evolution starting from different initial configurations. The methods developed in this application have also been applied to point particles with a Coulomb interaction. More recent applications have been made to the study of electrolyte solutions and fused salts. Here it is necessary to include the short-range forces between the ion cores as well as the long-range Coulomb interactions. Very important in this regard are the interactions that arise because of the polarizabilities of the ions. It is clear that these systems introduce a series of subtle problems. In the following we merely mention some of these problems and indicate the progress that has been made to resolve them. The reader interested in more details should consult the literature.*

Electric charges within a cavity in a dielectric medium polarize the material outside the cavity. This polarization in turn makes a contribution, called the reaction field, to the electric field within the cavity. In systems containing polar molecules or ions, or both, the reaction field plays an important role in the theory of the dielectric constant[31,32] ε and cannot be disregarded in the study of molecular dynamics. The usual periodic boundary conditions do not give a realistic representation of the actual reaction field for a system of N polar molecules, confined in a volume V and interacting with the dielectric medium outside it. In an attempt to remedy this inadequacy, Barker and Watts have used periodic boundary conditions supplemented by a uniform approximation to the reaction field.[33]

Periodic boundary conditions seem especially inappropriate when the MC or MD method is applied to solutions in polar solvents. Ionic or uncharged solute particles are expected to organize their neighboring solvent molecules. Periodic boundary conditions force this order to be repeated from cell to cell of the periodic system, even when the direct interaction of solute molecules in different cells is suppressed. Thus for reasonable N the calculation introduces unrealistic long-range correlations in the solvent configuration. The Barker–Watts procedure[33] reduces these correlations but does this in an approximate fashion.

Friedman[34] has recently proposed that the usual periodic boundary conditions, which give an unrealistic representation of the reaction field, should be replaced by "image boundary conditions." These new boundary conditions are based on an image charge approximation to the reaction field that is highly accurate when the dielectric constant of the medium is large. These boundary conditions have to our knowledge not yet been applied in molecular dynamics despite their apparent simplicity. It remains to be seen whether this approximation is accurate; nevertheless it offers an attractive alternative to the uniform approximation.[32] In addition, it might well be more efficient than using periodic boundary conditions on a system with sufficiently large N to give the same accuracy. Friedman gives the following argument to support this conjecture. There are $N(N-1)/2$ pair forces to compute in the usual periodic boundary conditions, whereas there are $\frac{1}{2}N(3N-1)$ pair forces to computed

*A very fruitful source of information is the Report of the C.E.C.A.M. Workshop on Ionic Liquids, Orsay, France, 1974.

with the more accurate image boundary conditions. To get comparable accuracy with periodic boundary conditions might require a doubling or tripling of the system size with a corresponding enormous increase in cycle time. Unfortunately this estimate does not take into account various bookkeeping devices.

Those workers who use periodic boundary conditions[34,35] must contend with the calculation of the Coulomb energy, which because of its infinitely long range must be summed over all pair interactions in the primary cell and all interactions between a particle in the cell and the infinite number of image particles. The procedure adopted by these workers is to evaluate the electrostatic energy using a Ewald summation technique.[37]

Hockney has developed what seems to be a very fast method for the simulation of systems with both short- and long-range forces.[38] In this method the system is divided into cells of a certain mesh. The particles in each cell are counted, and the number of particles (occupation number) of each component in each cell are listed. The occupation numbers of the charged particles when substituted into the Poisson equation give an equation for the electrostatic potential. This equation is solved subject to periodic boundary using fast Fourier transform techniques. The electrostatic potential in each cell due to all primary charges and all their periodic images is thus computed. The electrostatic force is then computed by differencing this electrostatic potential. The short-range force on each particle is computed by summing over particles in nearest-neighbor cells as was described in Section 3. The total force on a particle is then obtained by summing both the short-range forces and the long-range mesh forces exerted on that particle, and the equations of motion are integrated for one step Δt as in conventional molecular dynamics. This gives rise to a new set of occupation numbers and the above procedure is repeated.

The great advantage of this procedure is that the cycle time is proportional to N rather than N^2. A very large number of particles can be described in this fashion.[39]

7. Brownian Molecular Dynamics

There are many problems that would require so much computer time that their study by the previous method would not be possible. For example polyelectrolyte solutions, or motions of particles in membranes would not be susceptible to study because of wide separations in the time scales for different dynamic processes characterizing solute and solvent or because the property of interest evolves so slowly that an excessively long trajectory would be required. The study of these systems requires a different approach. A beginning was made many years ago by Simon,[40] who studied the melting of DNA by solving the coupled set of stochastic Langevin equations on a computer. This required an assumption about the statistical distribution of random forces. The precise values of the forces were then sampled from this distribution.

The molecular theory of fluctuations and Brownian motion offers a generalization of the Langevin equations. These equations provide a set of equations of motion that are stochastic in nature and that can be modeled on the basis of phenomenology. For example the velocity of the jth Brownian particle in solution is described by the equation of motion

$$M_j \frac{d\mathbf{V}_j}{dt} = \sum_i \mathbf{F}_{ji}(|\mathbf{R}_j - \mathbf{R}_i|) - \int_0^t d\tau\, K_j(\tau)\mathbf{V}_j(t-\tau) + \mathbf{F}_j + \mathbf{F}^{\text{ext}} \tag{30}$$

where $\mathbf{F}_{ji}$ is the force on particle j due to Brownian particle i, $\mathbf{F}_j$ is the fluctuating force on j, $\mathbf{F}^{\text{ext}}$ is an external force, and $K_j(\tau)$ is the memory function representing the frictional effect of the solvent on j. By the second fluctuation dissipation theorem,

$$K_j(t) = \langle \mathbf{F}_j(t)\mathbf{F}_j(0)\rangle / \langle V_j^2 \rangle \tag{31}$$

We will not mention all the assumptions that are required to establish Eq. (30).

Equation (30) can be solved if $\mathbf{F}_j$ is sampled at each time step in some specified way. If $\mathbf{F}_j$ is assumed to be a Gaussian Markov process it follows from Doob's theorem that $K_j(\tau)$ is an exponential function of time. Then only two parameters need be specified before $\mathbf{F}_j$ can be sampled from the Gaussian two-time probability distribution and these are the mean square value $\langle F_j^2 \rangle$ and the correlation time of $\mathbf{F}_j$, say τ_j. Equation (30) then forms a set of coupled stochastic differential equations that can be solved by methods similar to those already mentioned.

A modification of this very convenient procedure has recently been adopted by Lantelme *et al.*[41] They applied it to the study of a small system, but the method looks very promising and should open up new horizons in computer simulations.

8. Conclusion

Continual advances in computer technology and the development of new algorithms have greatly broadened the range, in size and complexity, of systems that can be simulated by molecular dynamics. Applications of the techniques discussed in this chapter have already yielded many important results, and we expect molecular dynamics to remain a valuable tool in the furthering of our understanding of fundamental dynamical processes.

References

1. B. J. Berne and D. Forster, *in*: *Annual Review of Physical Chemistry* (H. Eyring, ed.), Vol. 22, pp. 563–596, Annual Reviews Inc., Palo Alto, California (1971).
2. H. Goldstein, *Classical Mechanics*, Addison-Wesley, Reading, Massachusetts (1950).

3. L. Verlet, Computer "experiments" on classical fluids I, *Phys. Rev.* **159**, 98–103 (1967).
4. B. J. Berne and G. D. Harp, *in*: *Advances in Chemical Physics* (I. Prigogine and S. A. Rice, eds.), Vol. 17, pp. 63–227, Wiley (Interscience), New York (1970).
5. P. Schofield, Computer simulation studies of the liquid state, *Comput. Phys. Comm.* **5**, 17–23, (1973).
6. J. Barojas, D. Levesque, and B. Quentrec, Simulation of diatomic homonuclear liquids, *Phys. Rev. A* **7**, 1092–1105 (1973).
7. A. Rahman, Correlations in the motion of atoms in liquid argon, *Phys. Rev.* **136**, A405–A411(1964).
8. A. Rahman and F. H. Stillinger, Molecular dynamics study of liquid water, *J. Chem. Phys.* **55**, 3336–3359 (1971).
9. A. Ralston and H. Wilf, *Mathematical Methods for Digital Computers*, Wiley, New York (1966).
10. B. Quentrec and C. Brot, New method for searching for neighbors in molecular dynamics computations, *J. Comput. Phys.* **13**, 430–432 (1973).
11. F. H. Stillinger and A. Rahman, Improved simulation of liquid water by molecular dynamics, *J. Chem. Phys.* **60**, 1545–1557 (1974).
12. J. Corner, The second virial coefficient of a gas of nonspherical molecules, *Proc. Roy. Soc.* (*London*) **A192**, 275–292 (194).
13. B. J. Berne and P. Pechukas, Gaussian model potentials for molecular interactions, *J. Chem. Phys.* **56**, 4213–4216 (1972).
14. T. B. MacRury, W. A. Steele, and B. J. Berne, Intermolecular potential models for anisotropic molecules, with applications to N_2, CO_2, and benzene, *J. Chem. Phys.* **64**, 1288 (1976).
15. J. Kushick and B. J. Berne, Methods for experimentally determining the angular velocity relaxation in liquids, *J. Chem. Phys.* **59**, 4486–4490 (1973).
16. J. Kushick and B. J. Berne, Computer simulation of anisotropic molecular fluids, *J. Chem. Phys.* **64**, 1362 (1976).
17. J. Kushick, The Dynamics of Anisotropic Liquids, Ph.D. Thesis, Columbia University, New York (1974).
18. J. Kushick and B. J. Berne, Role of attractive forces in self-diffusion in dense Lennard-Jones fluids, *J. Chem. Phys.* **59**, 3732–3736 (1973).
19. J. D. Weeks, D. Chandler, and H. C. Andersen, Role of repulsive forces in determining the equilibrium structure of simple liquids, *J. Chem. Phys.* **54**, 5237–5247 (1971).
20. R. P. Futrelle and D. J. McGinty, Calculation of spectra and correlation functions from molecular dynamics data using the fast Fourier transform, *Chem. Phys. Lett.* **12**, 285–287 (1971).
21. R. Zwanzig and N. K. Ailawadi, Statistical error due to finite time averaging in computer experiments, *Phys. Rev.* **182**, 280–283 (1969).
22. W. W. Wood, private communication.
23. E. Helfand, *Phys. Rev.* **119**, 1 (1960).
24. B. J. Alder, D. M. Gass, and T. E. Wainwright, *J. Chem. Phys.* **53**, 3813 (1970).
25. W. T. Ashurst and W. G. Hoover, *Phys. Rev. Lett.* **31**, 206 (1973).
26. E. M. Gosling, I. R. Singer, and K. Singer, *Mol. Phys.* **26**, 1475 (1973).
27. G. Gicotti and G. Jacucci, *Phys. Rev. Lett.* **35**, 789 (1975).
28. R. Kubo, *Rep. Progr. Phys.* **29**, 255 (1966).
29. J. L. Lebowitz, J. K. Percus, and L. Verlet, Ensemble dependence of fluctuations with application to machine computations, *Phys. Rev.* **153**, 250–254 (1967).
30. L. Verlet, Computer "experiments" on classical fluids II, *Phys. Rev.* **165**, 201–214 (1968).
31. C. J. F. Boettcher, *Theory of Electric Polarization*, 2nd ed., Elsevier, Amsterdam, (1973). [Additional relevant results are in the first edition.]
32. H. Frolich, *Theory of Dielectrics*, Clarendon Press, Oxford (1959).
33. J. A. Barker and R. O. Watts, *Mol. Phys.* **26**, 789 (1973).
34. H. Friedman, *in*: *Rep. C.E.C.A.M. Workshop on Ionic Liquids*, Orsay, France (1974).
35. J. P. Hansen and I. R. MacDonald, *J. Phys. C* **7**, L384 (1974).
36. J. P. Hansen, I. R. MacDonald, and E. L. Pollack, *Phys. Rev. A* **11**, 1025 (1975).

37. S. G. Brush, H. L. Sahlin, and E. Teller, *J. Chem. Phys.* **45**, 2102 (1966).
38. R. W. Hockney, *Chem. Phys. Lett.* **21**, 589 (1973).
39. R. W. Hockney, *Methods Comput. Phys.* **9**, 135 (1970).
40. E. Simon and B. Zimm, *J. Chem. Phys.* **51**, 4937 (1969).
41. F. Lantelme, P. Turq, and H. L. Friedman, *in*: *C.E.C.A.M. Workshop on Ionic Liquids*, Orsay, France (1974).

3

The Kinetic Theory of Gases*

J. R. Dorfman
and
H. van Beijeren

1. Introduction

The kinetic theory of gases attempts to explain the macroscopic nonequilibrium properties of gases in terms of the microscopic properties of the individual gas molecules and the forces between them. A central aim of this theory is to provide a microscopic explanation for the fact that a wide variety of gas flows can be described by the Navier–Stokes hydrodynamic equations and to provide expressions for the transport coefficients appearing in these equations, such as the coefficients of shear viscosity and thermal conductivity, in terms of the microscopic properties of the molecules. We devote most of our attention in this article to this problem.

The kinetic theory of gases has had a long and rich history.† In its modern form, kinetic theory begins with the work of D. Bernoulli, Clausius, and most importantly Maxwell, who first used statistical methods to compute the properties of gases, recognizing that the random motions of the gas molecules could be best described by a distribution function.‡ Besides giving the form of this distribution function for a gas at equilibrium, Maxwell derived equations for the transport of mass, momentum, and energy in a dilute gas. For a gas composed of so-called Maxwell molecules, which interact with repulsive forces

*Work supported in part by the National Science Foundation, Grant No. CHE-73-08856 A 03.
†For a detailed description of the history, the reader should consult the monographs of Brush.[1]
‡A number of important papers of historical and scientific interest can be found in Brush[1] and Maxwell.[2]

J. R. Dorfman and H. van Beijeren • Institute for Physical Science and Technology, and Department of Physics and Astronomy, University of Maryland, College Park, Maryland

varying as the inverse fifth power of the distance between their centers, he derived microscopic expressions for the transport coefficients and showed that the coefficients of shear viscosity and thermal conductivity are independent of the gas density.

The next important advance in the theory, and the one that provided the foundation for all later work in this field, was made by Boltzmann, who in 1872 derived an equation for the time rate of change of the distribution function for a dilute gas that is not in equilibrium—the Boltzmann transport equation. (See Boltzmann[3] and also Klein.[4]) Boltzmann's equation gives a microscopic description of nonequilibrium processes in the dilute gas, and of the approach of the gas to an equilibrium state. Using the Boltzmann equation, Chapman and Enskog derived the Navier–Stokes equations and obtained expressions for the transport coefficients for a dilute gas of particles that interact with pairwise, short-range forces.* Even now, more than 100 years after the derivation of the Boltzmann equation, the kinetic theory of dilute gases is largely a study of special solutions of that equation for various initial and boundary conditions and various compositions of the gas.†

Although Boltzmann may have supposed at one time that his derivation of the transport equation was based on purely mechanical arguments, it quickly became clear that this was not so, and that he had used in his derivation an assumption that is not in accord with mechanics. This assumption, the *Stosszahlansatz*, accounted for the monotonic approach of the gas to equilibrium predicted by the Boltzmann equation.[14] It could be justified only by supposing that the Boltzmann equation itself is the result of a more fundamental statistical-mechanical theory of gases, based on the Liouville equation.

Several attempts were made to derive the Boltzmann equation from the Liouville equation, but it was not until the work of Bogoliubov in 1947[15] and the later work of Cohen[16–19] and of Green[20,21] that a satisfactory derivation was given. The approach to the Boltzmann equation from the Liouville equation has proved to be important for two reasons. First, it makes it possible to replace the *Stosszahlansatz* by a more fundamental assumption about the statistical ensemble that is sampled when an experiment is performed. Second, it allows one to generalize the Boltzmann equation to dense gases.

The kinetic theory of dense gases began with the work of Enskog, who in 1922, generalized Boltzmann's derivation of the transport equation to apply it to a dense gas of hard spheres.[1] Enskog showed that for dense gases there is a mechanism for the transport of momentum and energy by means of the intermolecular potential, which is not taken into account by the Boltzmann equation at low densities, and he derived expressions for the transport coeffi-

*The original papers are reprinted in an English translation in Brush,[1] Vol. 3.

†The basic references on the Boltzmann equation and its applications are Chapman and Cowling,[5] Ferziger and Kaper,[6] Cercignani,[7] Hirschfelder *et al.*,[8] Kogan,[9] Grad,[10] Harris,[11] Liboff,[12] and Waldmann.[13]

cients of a dense gas of identical hard spheres that reduced to the results of the Boltzmann equation at low densities. However, Enskog's theory is not completely satisfactory for dense gases since it applies only to hard-sphere molecules and, as became clear from the later work on dense gases,[20–23] even for hard spheres it does not consider a large class of dynamical processes that play an important role in determining the nonequilibrium properties of the gas.

The difficulties inherent in the Enskog theory of dense gases were resolved with the derivation of the Boltzmann equation from the Liouville equation by Bogoliubov, Cohen, and Green. For the first time, a systematic kinetic theory for dense gases whose particles interact with short-range forces was developed. Yet even this systematic theory is not free of difficulties, and their resolution remains one of the central problems in the kinetic theory of dense gases.[23]*

At the same time as the generalized Boltzmann equation was being worked out, a different approach to the microscopic theory for hydrodynamic equations was developed. The time correlation function method derives the hydrodynamic equations directly from the Liouville equation, for certain types of initial ensembles, and it provides expressions for transport coefficients for a wide class of fluid systems, including simple liquids as well as dilute and dense gases.† The explicit evaluation of these expressions for dilute or dense gases, however, requires precisely the same methods that are used to derive the Boltzmann equation and its generalization to higher densities from the Liouville equation, and it is the similarity in method that is exploited when one attempts to show that the time correlation function method and the generalized Boltzmann equation method lead to identical expressions for the transport coefficients in a dense gas.[29,30]

The principal advantage of the time correlation function method is that it provides a new set of microscopic functions for a fluid, the time correlation functions, which can be studied directly by experimental observations of the fluid‡ or by computer-simulated molecular dynamics.§ The time correlation functions depend even more sensitively on the microscopic properties of the fluid molecules than the transport coefficients, which are expressed as time integrals of the correlation functions. Thus, a further test of kinetic theory has been found: it must not only lead to expressions for the transport coefficients for dilute and dense gases that are in agreement with experiment, but also describe the dependence of the time correlation functions on both time and the density of the gas. One of the principal successes of kinetic theory is that it provides a quantitatively correct description of the short- and long-time

*For a list of references on these difficulties, see Ernst *et al.*,[24] and for a more recent discussion, see Pomeau and Gervois,[25] and Gervois *et al.*[26]

†For a review and references see Zwanzig[27] and Steele.[28]

‡For a discussion of experimental studies on time correlation functions, see Chu[31] and Berne and Harp.[32]

§For a discussion of computer studies on time correlation functions, see Erpenbeck and Wood, Chapter 1 of this volume and Wood.[33]

behavior of the correlation functions over a wide range of densities.[34-37] The kinetic theory for the intermediate time region is still being developed.

Our plan of presentation of the kinetic theory of gases is as follows. In Section 2.1 we present Boltzmann's derivation of the transport equation for a dilute gas. We generalize his equation somewhat to take into account the interactions of the particles with the boundaries of the container. In Section 2.2 we derive the H-theorem for a dilute gas, which provides a microscopic theory for the Clausius form of the second law of thermodynamics, $dS \geq dQ/T$, and under certain restrictions on the interactions of the particles with the boundaries, predicts a monotonic approach of the gas to equilibrium. Special solutions of the Boltzmann equation are discussed in Section 2.3, including the Chapman–Enskog solution, which leads to a derivation of the hydrodynamic equations; the resulting expressions for the transport coefficients are compared with experiment. In Section 3 we discuss the kinetic theory of dense gases, first the Enskog theory for dense hard-sphere gases, and then the systematic kinetic theory of dense gases as derived from the Liouville equation. We conclude this discussion with a description of the divergence difficulties that appear in the density expansion of the generalized Boltzmann equation, and of how a "renormalized" theory is obtained. In Section 4 we discuss the time correlation functions and show how the results of Section 3 can be applied to describe their time and density behavior. The results of kinetic theory for the long-time behavior of the correlation functions are compared with the computer results of Alder and Wainwright[38-42] and of Wood and Erpenbeck (Chapter 1 of this volume), and some of the implications of these results for the theory of transport processes in dense gases are discussed. Finally, in Section 5 we discuss a number of open problems, which are of fundamental interest for the kinetic theory of gases.

We will use classical mechanics throughout, since most of the topics we deal with are more easily presented in terms of classical, rather than quantum, mechanics. Those topics for which a quantum formulation is essential, such as inelastic collisions of polyatomic molecules, are not discussed here, but appropriate literature references will be given. In addition, we assume that some of the elementary concepts of the kinetic theory of gases, such as the mean free path, are already familiar to the reader.

2. *The Boltzmann Equation*

2.1. Boltzmann's Derivation

2.1.1. *One-Component, Monatomic Gas*

We consider a collection of N identical monatomic molecules contained in a volume V. We suppose that the energy of the system can be expressed as the

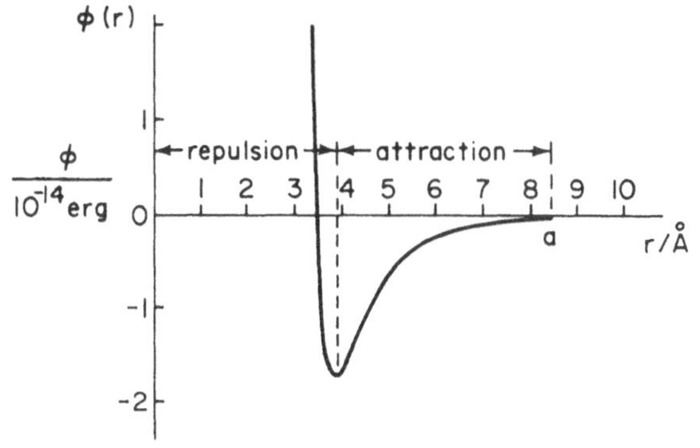

Fig. 1. Qualitative sketch of a typical interatomic or intermolecular potential.

sum of (a) the kinetic energies of the molecules, (b) the potential energies of interaction between pairs of particles in the system, and (c) the potential energy of interaction between the particles and the walls of the container. That is, the Hamiltonian function for the system is

$$H_N = \sum_{i=1}^{N} \frac{m\mathbf{v}_i^2}{2} + \sum_{1\le i<j\le N} \phi(r_{ij}) + \sum_{i=1}^{N} \Phi_w(\mathbf{r}_i) \tag{1}$$

where m is the mass of a particle; $\mathbf{r}_i$, $\mathbf{v}_i$ are the position and velocity of particle i, $r_{ij} = |\mathbf{r}_i - \mathbf{r}_j|$; $\phi(r_{ij})$ is the potential energy of interaction between particle i at $\mathbf{r}_i$ and particle j at $\mathbf{r}_j$, and $\Phi_w(\mathbf{r}_i)$ is the potential energy of interaction between a particle at $\mathbf{r}_i$ and the walls of the system. We assume that $\phi(r)$ has the general shape illustrated in Fig. 1. It is repulsive at short distances, may be attractive at larger distances, and vanishes for $r > a$. For our purposes, the length a characterizes the diameter of a molecule. The purpose of the wall potential is to restrict the particle to remain inside the container. Therefore $\Phi_w(\mathbf{r}_i)$ is taken to be infinite whenever all or part of molecule i is outside the container, and $\Phi_w(\mathbf{r}_i)$ is zero otherwise. In our later discussions we will consider collisions of the particles with the walls of the container, and to describe the results of these collisions we will sometimes use a model for the molecule–wall interaction that cannot be described in terms of a potential energy of interaction.

To specify the state of the system, we will assume that the system is a dilute gas such that the average free volume per molecule $v = V/N$ is much larger than the volume of a molecule, a^3, i.e.,

$$a^3 \ll v \qquad \text{or} \qquad n_0 a^3 \ll 1 \tag{2}$$

where $n_0 = 1/v$ is the number density of the gas.* Under these circumstances, one can picture the gas molecules as moving freely most of the time and only infrequently making collisions with other gas molecules. Since the particles are mechanically identical and since they move freely most of the time, we suppose

*For air at STP, $n_0 \sim 3\times10^{19}$ particles per cm^3, $a \sim 2\times10^{-8}$ cm, $n_0 a^3 \sim 3\times10^{-4}$.

that the entire state of the gas at any time t can be adequately specified by determining the number of particles in every small region of the vessel, together with their distribution as a function of velocity. That is, we construct the distribution function $f(\mathbf{r}_1, \mathbf{v}_1, t)$,* where $f(\mathbf{r}_1, \mathbf{v}_1, t)\delta\mathbf{r}_1\, \delta\mathbf{v}$, is the number of particles† in $\delta\mathbf{r}_1$ about $\mathbf{r}_1$ with velocity in the range $\delta\mathbf{v}_1$ about $\mathbf{v}_1$ at time t, and assume that this function is sufficient to characterize all processes taking place in the gas.‡

Following Boltzmann,[(3-13)] we derive an equation for the change in $f(\mathbf{r}_1, \mathbf{v}_1, t)$ with time by considering the various processes taking place in the gas. We focus our attention on a small region in $(\mathbf{r}_1, \mathbf{v}_1)$ space of volume $\delta\mathbf{r}_1\, \delta\mathbf{v}_1$ about the point $\mathbf{r}_1, \mathbf{v}_1$. Then the number of particles $f(\mathbf{r}_1, \mathbf{v}_1, t)\, \delta\mathbf{r}_1\, \delta\mathbf{v}_1$ in this region may change in time through four processes:

(a) A particle may move into or out of this region through its free motion with velocity $\mathbf{v}_1$,

(b) a particle with velocity $\mathbf{v}_1$ inside this region may suffer collisions with other molecules, changing its velocity from $\mathbf{v}_1$ to $\mathbf{v}_1' \neq \mathbf{v}_1$,

(c) a particle with velocity $\mathbf{v}_1'' \neq \mathbf{v}_1$ but in the region $\delta\mathbf{r}_1$ about $\mathbf{r}_1$ may suffer collisions with other molecules such that after the collision the particle's velocity is equal to $\mathbf{v}_1$,

(d) in the case that the region $\delta\mathbf{r}_1$ is near a wall of the container, particles with velocity $\mathbf{v}_1$ may change their velocity by colliding with a wall, or particles may acquire velocity $\mathbf{v}_1$ after colliding with a wall.

To determine $(\partial/\partial t)f(\mathbf{r}_1, \mathbf{v}_1, t)$ we consider a small time interval δt, and use the results of the four processes to compute

$$f(\mathbf{r}_1, \mathbf{v}_1, t+\delta t)\, \delta\mathbf{r}_1\, \delta\mathbf{v}_1 - f(\mathbf{r}_1, \mathbf{v}_1, t)\, \delta\mathbf{r}_1\, \delta\mathbf{v}_1 = \frac{\partial f(\mathbf{r}_1, \mathbf{v}_1, t)}{\partial t}\, \delta t\, \delta\mathbf{r}_1\, \delta\mathbf{v}_1 + \cdots \tag{3}$$

We assume that:

(a) Since the gas is dilute, only binary collisions need to be taken into account,

(b) δt can be chosen to be large compared to the average duration of a binary collision, but small compared to the average time it takes for a particle to cross the volume $\delta\mathbf{r}_1$,

(c) the volume $\delta\mathbf{r}_1$ can be chosen to be large enough to contain several molecules at any given time, but is small enough that $f(\mathbf{r}_1, \mathbf{v}_1, t)$ does not vary

*For later convenience we will indicate the position of a particular point in space by $\mathbf{r}_1$ and the velocity of a representative particle in the vicinity of this point, by $\mathbf{v}_1$.

†We will see farther on that $\delta\mathbf{r}_1$ must satisfy $a^3 \ll \delta\mathbf{r}_1 \ll l^3$, where l is the mean free path between collisions for a typical molecule of the gas.

‡Here assumption of dilution is crucial. The average potential energy in the gas, for example, cannot be expressed in terms of $f(\mathbf{r}_1, \mathbf{v}_1, t)$, but it is assumed to be small compared to the kinetic energy.

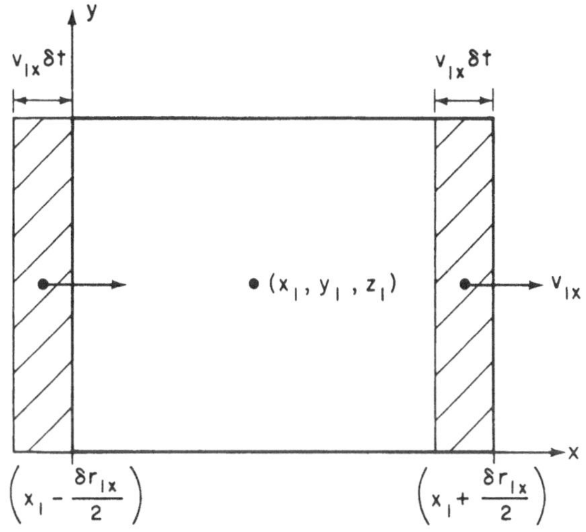

Fig. 2. Projection of the volume $\delta\mathbf{r}_1\,\delta\mathbf{v}_1$ onto the xy plane. The hatched areas represent regions from which particles enter and leave $\delta\mathbf{r}_1\,\delta\mathbf{v}_1$ in time δt.

much over distances of the order $(\delta\mathbf{r}_1)^{1/3}$. In fact, we assume that on the average $\delta\mathbf{r}_1$ contains a large number of molecules.

Under these circumstances* we can regard the contributions of each of the four processes to $\partial f/\partial t$ as being independent of the other three processes. That is, since only molecules close to the boundaries† will enter or leave $\delta\mathbf{r}_1$ in time δt and since these are a small fraction of the molecules in $\delta\mathbf{r}_1$, we can neglect the effect of collisions that prevent molecules close to the boundaries from entering (or leaving) $\delta\mathbf{r}_1\,\delta\mathbf{v}_1$ in time δt. Similarly, we suppose that no molecule suffers more than one collision in δt, so that once having left or entered $\delta\mathbf{r}_1\,\delta\mathbf{v}_1$ by a collision a molecule does not return or leave by a second collision. In short, we assume that we can neglect any contributions to $\delta f/\partial t$ coming from processes where two or more events of the four types listed above happen to the same particle in time δt.

We therefore write

$$[f(\mathbf{r}_1, \mathbf{v}_1, t+\delta t)-f(\mathbf{r}_1, \mathbf{v}_1, t)]\,\delta\mathbf{r}_1\,\delta\mathbf{v}_1=\Gamma_f+\Gamma_+-\Gamma_-+\Gamma_w \tag{4}$$

where Γ_f denotes the change in $f\,\delta\mathbf{r}_1\,\delta\mathbf{v}_1$ through the free motion of particles entering or leaving the region, Γ_+ the increase in $f\,\delta\mathbf{r}_1\,\delta\mathbf{v}_1$ due to collisions, Γ_- the decrease in $f\,\delta\mathbf{r}_1\,\delta\mathbf{v}_1$ due to collisions, and Γ_w the change in $f\,\delta\mathbf{r}_1\,\delta\mathbf{v}_1$ due to collisions with the walls.

Γ_f can be easily computed by calculating the rate at which particles enter and leave the region $\delta\mathbf{r}_1$ if they are traveling with velocity $\mathbf{v}_1$. A glance at Fig. 2

*It is clear that these assumptions cannot be true for all $\mathbf{r}_1$, $\mathbf{v}_1$. It is sufficient for our purposes if we can neglect the contributions to quantities of interest where the assumptions break down.
†We mean here the *boundaries* of $\delta\mathbf{r}_1$ and not the boundaries of the container.

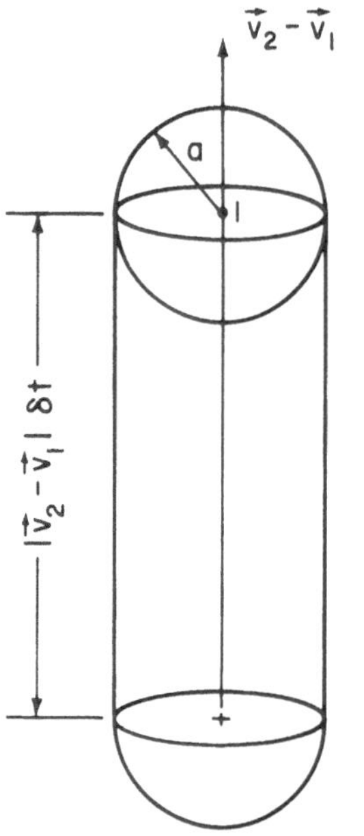

Fig. 3. The collision cylinder for a $(\mathbf{v}_1, \mathbf{v}_2)$ collision taking place in time δt. The sphere of radius a about particle 1 is the action sphere.

shows that

$$\begin{aligned}\Gamma_f = \delta \mathbf{v}_1 \{ & v_{1x}\, \delta t\, \delta r_{1y}\, \delta r_{1z} [f(x_1 - \tfrac{1}{2}\delta r_{1x}, y_1, z_1, \mathbf{v}_1, t) - f(x_1 + \tfrac{1}{2}\delta r_{1x}, y_1, z_1, \mathbf{v}_1, t)] \\ & + v_{1y}\, \delta t\, \delta r_{1x}\, \delta r_{1y} [f(x_1, y_1 - \tfrac{1}{2}\delta r_{1y}, z_1, \mathbf{v}_1, t) - f(x_1, y_1 + \tfrac{1}{2}\delta r_{1y}, z_1, \mathbf{v}_1, t)] \\ & + v_{1z}\, \delta t\, \delta r_{1x}\, \delta r_{1y} [f(x_1, y_1, z_1 - \tfrac{1}{2}\delta r_{1z}, \mathbf{v}_1, t) - f(x_1, y_1, z_1 + \tfrac{1}{2}\delta r_{1z}, \mathbf{v}_1, t)]\} \end{aligned} \tag{5}$$

or

$$\Gamma_f = -\delta \mathbf{r}_1\, \delta \mathbf{v}_1\, \delta t\, \mathbf{v}_1 \cdot \nabla_{\mathbf{r}_1} f(\mathbf{r}_1, \mathbf{v}_1, t) + \cdots \tag{6}$$

where we neglect higher-order terms.

To compute Γ_+ and Γ_- we need to consider binary collision processes in some detail.* Let us begin by discussing the collisions that take place between molecules with velocity $\mathbf{v}_1$ and molecules with some other definite velocity $\mathbf{v}_2$, say, which we will refer to as $(\mathbf{v}_1, \mathbf{v}_2)$ collisions. We now pick one molecule with velocity $\mathbf{v}_1$ and imagine that a coordinate system has been placed at its center with z axis in the direction of $\mathbf{v}_2 - \mathbf{v}_1$, as illustrated in Fig. 3. In this coordinate system, a $(\mathbf{v}_1, \mathbf{v}_2)$ collision occurs whenever the center of the molecule with velocity $\mathbf{v}_2$ is within a distance a from the origin.

For central forces the relative motion of the two particles takes place in a fixed plane that contains both the z axis, in the direction of $\mathbf{v}_2 - \mathbf{v}_1$, and the trajectory of particle 2 before its center intersects the sphere of radius a (the

*A critical review of derivations of Γ_+ and Γ_- can be found in Dahlberg.[43]

action sphere) about particle 1. We measure the rotation of the plane about some fixed plane that contains the z axis, by an azimuthal angle ε, as illustrated in Fig. 4. The final quantity that specifies the collision, given $\mathbf{v}_1$, $\mathbf{v}_2$, and ε, is the impact parameter b, which is the perpendicular distance between the trajectory of particle 2 before the collision and the z axis. For $b > a$ no collision can take place, since the center of particle 2 never penetrates the action sphere.

Denoting by $\mathbf{v}_1'$ and $\mathbf{v}_2'$ the velocities of particles 1 and 2 after collision, we can show that the magnitude of the relative velocity after collision $|\mathbf{v}_2' - \mathbf{v}_1'|$, is equal to $|\mathbf{v}_2 - \mathbf{v}_1|$, the magnitude of the relative velocity before collision. For, using the laws of conservation of kinetic energy and of momentum, we have

$$\mathbf{v}_1^2 + \mathbf{v}_2^2 = \mathbf{v}_1'^2 + \mathbf{v}_2'^2 \tag{7}$$

$$\mathbf{v}_1 + \mathbf{v}_2 = \mathbf{v}_1' + \mathbf{v}_2' \tag{8}$$

Then by squaring (8) and subtracting (7), we find

$$\mathbf{v}_1 \cdot \mathbf{v}_2 = \mathbf{v}_1' \cdot \mathbf{v}_2' \tag{9}$$

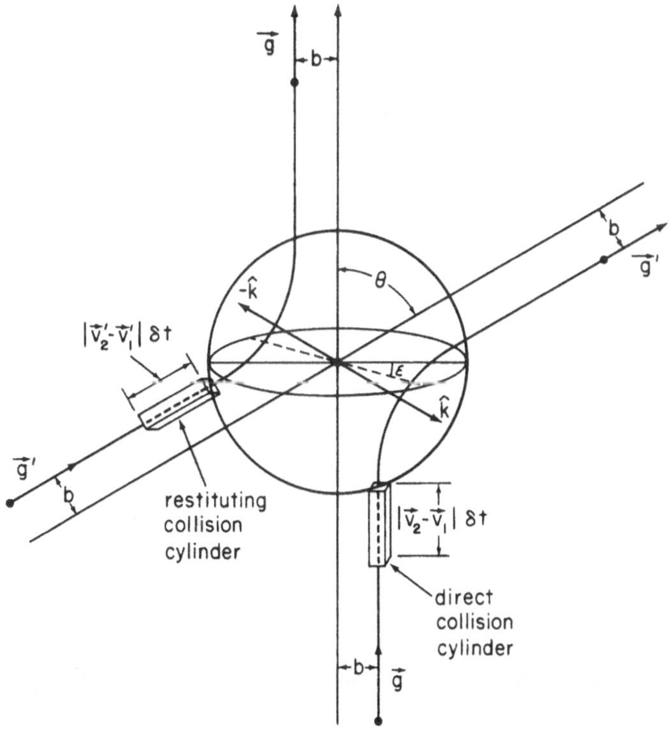

Fig. 4. Direct and restituting collisions in the relative coordinate system centered on particle 1. The scattering angle is θ, and both collisions take place in a plane that makes an angle ϵ with respect to a fixed plane in space.

which leads to

$$\mathbf{g}^2 = \mathbf{g}'^2 \qquad \text{or} \qquad |\mathbf{g}| = |\mathbf{g}'| \tag{10}$$

where $\mathbf{g} = \mathbf{v}_2 - \mathbf{v}_1$ and $\mathbf{g}' = \mathbf{v}_2' - \mathbf{v}_1'$.

Therefore, as a result of the collision the relative velocity rotates by an angle θ in the relative coordinate system, where θ depends on the impact parameter and the magnitude of $\mathbf{g} = \mathbf{v}_2 - \mathbf{v}_1$, but the magnitude of the relative velocity does not change. In addition the perpendicular distance of the trajectory of particle 2 after the collision from the line through the origin in the direction of $\mathbf{g}'$ in the relative frame is equal to the impact parameter b, as a consequence of the conservation of angular momentum. Finally, if we are given $\mathbf{v}_1$, $\mathbf{v}_2$, b, and ε, the velocities $\mathbf{v}_1'$ and $\mathbf{v}_2'$ are determined in terms of the potential $\phi(r)$.

It will be useful below to consider another way to determine $\mathbf{v}_1'$ and $\mathbf{v}_2'$ in terms of $\mathbf{v}_1$ and $\mathbf{v}_2$. We note that since the forces are central, the orbit of particle 2 in the relative coordinate frame is symmetrical about the apse line. This is the line from the origin to the center of particle 2, at the point of closest approach.* Let $\hat{\mathbf{k}}$ be a unit vector in the direction of the apse line. If $\mathbf{g}$ and $\hat{\mathbf{k}}$ are known, then $\mathbf{g}'$ can be found by the relation

$$\mathbf{g}' = \mathbf{g} - 2(\mathbf{g}\cdot\hat{\mathbf{k}})\hat{\mathbf{k}} \tag{11}$$

since $|\mathbf{g}'| = |\mathbf{g}|$, and $\mathbf{g}' - \mathbf{g}$ must be in the direction of $\hat{\mathbf{k}}$. Using Eq. (11) and the fact that the velocity of the center of mass $\mathbf{V} = (\mathbf{v}_1 + \mathbf{v}_2)/2$ does not change in the course of the collision, we can write the final velocities $\mathbf{v}_1'$ and $\mathbf{v}_2'$ in terms of $\mathbf{v}_1$, $\mathbf{v}_2$, and $\hat{\mathbf{k}}$ by

$$\begin{aligned}\mathbf{v}_1' &= (\mathbf{v}_1' + \mathbf{v}_2')/2 - (\mathbf{v}_2' - \mathbf{v}_1')/2 \\ &= (\mathbf{v}_1 + \mathbf{v}_2)/2 - \mathbf{g}/2 + (\mathbf{g}\cdot\hat{\mathbf{k}})\hat{\mathbf{k}} \\ &= \mathbf{v}_1 + (\mathbf{g}\cdot\hat{\mathbf{k}})\hat{\mathbf{k}}\end{aligned} \tag{12a}$$

$$\mathbf{v}_2' = \mathbf{v}_2 - (\mathbf{g}\cdot\hat{\mathbf{k}})\hat{\mathbf{k}} \tag{12b}$$

Consequently the final velocities $\mathbf{v}_1'$ and $\mathbf{v}_2'$ are determined by $\mathbf{v}_1$, $\mathbf{v}_2$, and $\hat{\mathbf{k}}$. Of course, $\hat{\mathbf{k}}$ is a function of b, ε, $\mathbf{v}_1$, $\mathbf{v}_2$, and $\phi(r)$.

We now turn to a computation of Γ_-. We first consider $(\mathbf{v}_1, \mathbf{v}_2)$ collisions, which in the relative coordinate frame take place with impact parameter in the range b to $b + db$, and azimuthal angle in the range ε to $\varepsilon + d\varepsilon$. We see that such a $(\mathbf{v}_1, \mathbf{v}_2)$ collision will be initiated in the time interval δt if there is a molecule with velocity $\mathbf{v}_2$ situated within the little collision cylinder of volume $|\mathbf{v}_2 - \mathbf{v}_1| b\, db\, d\varepsilon\, \delta t$ attached to the action sphere as shown in Fig. 4. We imagine

*To see that the orbit must be symmetrical about the apse line, consider the time-reversed motion in the relative frame. By rotating the coordinate system about the origin, one should be able to make the time-reversed trajectory coincide with that of the forward trajectory.

that each molecule with velocity $\mathbf{v}_1$ in $\delta\mathbf{r}_1$ has attached to it such a collision cylinder, which we call a $(\mathbf{v}_1, \mathbf{v}_2, b, \varepsilon)$ collision cylinder, and we will suppose that the number of $(\mathbf{v}_1, \mathbf{v}_2)$ collisions taking place in time interval δt is equal to the number of particles with velocity $\mathbf{v}_2$ present inside these cylinders at time t.* That is, that each particle with velocity $\mathbf{v}_2$ in a $(\mathbf{v}_1, \mathbf{v}_2, b, \varepsilon)$ collision cylinder actually leads to a $(\mathbf{v}_1, \mathbf{v}_2)$ collision. According to Boltzmann, this number is found by combining

(a) the total volume occupied by the $(\mathbf{v}_1, \mathbf{v}_2, \mathrm{b}, \varepsilon)$ collision cylinders† in $\delta\mathbf{r}_1$:

$$f(\mathbf{r}_1, \mathbf{v}_1, t)\delta\mathbf{r}_1\delta\mathbf{v}_1 b\, db\, d\varepsilon\, |\mathbf{v}_2 - \mathbf{v}_1|\, \delta t$$

[since each of the $f(\mathbf{r}_1, \mathbf{v}_1, t)\, \delta\mathbf{r}_1\, \delta\mathbf{v}_1$ particles has such a $(\mathbf{v}_1, \mathbf{v}_2, b, \varepsilon)$ collision cylinder attached to it], with

(b) the number of molecules with velocity $\mathbf{v}_2$ (in range $\delta\mathbf{v}_2$) per unit volume about the point‡ $\mathbf{r}_1$, $f(\mathbf{r}_1, \mathbf{v}_2, t)\, \delta\mathbf{v}_2$.

Thus, according to this prescription, the number of $(\mathbf{v}_1, \mathbf{v}_2, b, \varepsilon)$ collisions taking place in $\delta\mathbf{r}_1$ in the time interval δt is

$$f(\mathbf{r}_1, \mathbf{v}_1, t) f(\mathbf{r}_1, \mathbf{v}_2, t)|\mathbf{v}_2 - \mathbf{v}_1| b\, db\, d\varepsilon\, \delta\mathbf{v}_1\, \delta\mathbf{v}_2\, \delta\mathbf{r}_1\, \delta t$$

Integrating over all $\mathbf{v}_2$, b, ε, we are led to an expression for Γ_- as

$$\Gamma_- = \delta\mathbf{v}_1\, \delta\mathbf{r}_1\, \delta t \int d\mathbf{v}_2 \int_0^a b\, db \int_0^{2\pi} d\varepsilon\, |\mathbf{v}_2 - \mathbf{v}_1| f(\mathbf{r}_1, \mathbf{v}_1, t) f(\mathbf{r}_1, \mathbf{v}_2, t) \quad (13)$$

The assumption that the number of collisions is found by combining the total volume of collision cylinders with the number of particles of a specified type per unit volume is called the *Stosszahlansatz*.[14] It must be emphasized that this way of computing Γ_- does *not* follow from the laws of mechanics, and we will see in the next section that the probabilistic argument used to arrive at the *Stosszahlansatz* leads to results that are not consistent with mechanics.§ The expression for Γ_- just obtained can also be written in terms of $\hat{\mathbf{k}}$, the unit vector in the direction of the apse line. To do this we note that $\hat{\mathbf{k}}$ makes an angle of $(\pi + \theta)/2$ with the z axis, as illustrated in Fig. 5, so

$$\begin{aligned} d\hat{\mathbf{k}} &= \sin[(\pi + \theta)/2]\, d[(\theta + \pi)/2]\, d\varepsilon \\ &= \tfrac{1}{2}\cos(\theta/2)\, d\theta\, d\varepsilon \\ &= \tfrac{1}{2}\cos(\theta/2)|d\theta/db|\, db\, d\varepsilon \end{aligned} \quad (14)$$

*We ignore contributions from collisions taking place at time t.

†Here we assume that none of these collision cylinders overlap, since the gas is dilute.

‡Here we ignore variations in f over a region of the order $\delta\mathbf{r}_1$ (and also regions of the order of a collision cylinder).

§Although we treat the dynamics of binary collisions correctly by using the laws of mechanics, we abandon mechanics when we attempt to compute the number of such collisions.

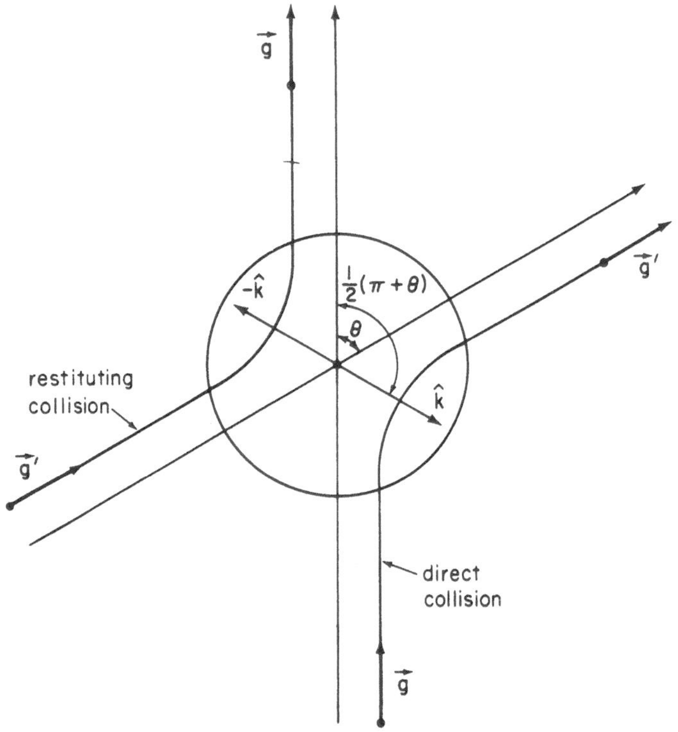

Fig. 5. The direct and restituting collisions in the collision plane.

Therefore we may write

$$|\mathbf{v}_2 - \mathbf{v}_1| b\, db\, d\varepsilon = B(\mathbf{g}, \hat{\mathbf{k}})\, d\hat{\mathbf{k}} \tag{15a}$$

where

$$B(\mathbf{g}, \hat{\mathbf{k}}) = 2gb|db/d\theta|[\cos(\theta/2)]^{-1} \tag{15b}$$

$B(\mathbf{g}, \hat{\mathbf{k}})$ is determined by the intermolecular potential, since one needs to know the potential in order to compute the scattering angle θ for a given value of b and g, but it does not depend on the azimuthal angle ε. Using Eq. (15a) we may write Γ_- as

$$\Gamma_- = \delta\mathbf{r}_1\, \delta\mathbf{v}_1\, \delta t \int d\mathbf{v}_2 \int d\hat{\mathbf{k}}\, B(\mathbf{g}, \hat{\mathbf{k}}) f(\mathbf{r}_1, \mathbf{v}_1, t) f(\mathbf{r}_1, \mathbf{v}_2, t) \tag{16}$$

The region of the $\hat{\mathbf{k}}$ integration is determined by the intermolecular potential. If the potential is purely repulsive, we see that $\hat{\mathbf{k}} \cdot \mathbf{g} < 0$ for all collisions. However for more complicated potentials such as that illustrated in Fig. 1, the impact paramter b may have more than one value for a given θ and g, and we must then redefine $B(\mathbf{g}, \hat{\mathbf{k}})\, d\hat{\mathbf{k}}$ to be the sum of all possible $gb\, db\, d\varepsilon$ for which the impact parameters b lead to the same $\hat{\mathbf{k}}$.

To compute Γ_+ we must consider what are called the *restituting* collisions. To construct them we first consider a direct collision between particles 1 and 2 specified by quantities $\mathbf{v}_1$, $\mathbf{v}_2$, and $\hat{\mathbf{k}}$. In the case that one value of $\hat{\mathbf{k}}$ may correspond to several impact parameters we also specify one particular impact parameter b. We then construct a new collision where particle 1 with velocity $\mathbf{v}_1'$ collides with particle 2 with velocity $\mathbf{v}_2'$, with apse line in the direction $-\hat{\mathbf{k}}$, and with the same impact parameter b as illustrated in Figs. 4–6. This is a restituting collision, that is, particle 1 has velocity $\mathbf{v}_1$ after the collision. To prove it, we use the fact that if the initial velocities in a collision are $\mathbf{v}_1'$, $\mathbf{v}_2'$ and if the collision is specified by unit vector $-\hat{\mathbf{k}}$, the final velocities of particles 1 and 2, $\mathbf{v}_1''$ and $\mathbf{v}_2''$, are given by Eqs. (12a) and (12b)

$$\mathbf{v}_1'' = \mathbf{v}_1' + [(\mathbf{v}_2' - \mathbf{v}_1') \cdot \hat{\mathbf{k}}] = \mathbf{v}_1' + (\mathbf{g}' \cdot \hat{\mathbf{k}})\hat{\mathbf{k}} \tag{17a}$$

$$\mathbf{v}_2'' = \mathbf{v}_2' - [(\mathbf{v}_2' - \mathbf{v}_1') \cdot \hat{\mathbf{k}}]\hat{\mathbf{k}} = \mathbf{v}_2' - (\mathbf{g}' \cdot \hat{\mathbf{k}})\hat{\mathbf{k}} \tag{17b}$$

Now use $\mathbf{g}' \cdot \hat{\mathbf{k}} = -\mathbf{g} \cdot \hat{\mathbf{k}}$ as follows from Eq. (11), and Eqs. (12a) and (12b) for $\mathbf{v}_1'$ and $\mathbf{v}_2'$. It then follows immediately that

$$\mathbf{v}_1'' = \mathbf{v}_1, \qquad \mathbf{v}_2'' = \mathbf{v}_2$$

Therefore the collision specified by $\mathbf{v}_1'$, $\mathbf{v}_2'$, $-\hat{\mathbf{k}}$ is the restituting collision that corresponds to the direct collision specified by $\mathbf{v}_1$, $\mathbf{v}_2$, $\hat{\mathbf{k}}$.

To obtain Γ_+ we first consider the number of particles with velocity $\mathbf{v}_1$ that are produced in collisions specified by $\mathbf{v}_1'$, $\mathbf{v}_2'$, $-\hat{\mathbf{k}}$ taking place in time δt. Consider therefore small collision cylinders of volume

$$B(\mathbf{g}', -\hat{\mathbf{k}})\, d\hat{\mathbf{k}}\, \delta t = gb\, db\, d\varepsilon'\, \delta t \tag{18}$$

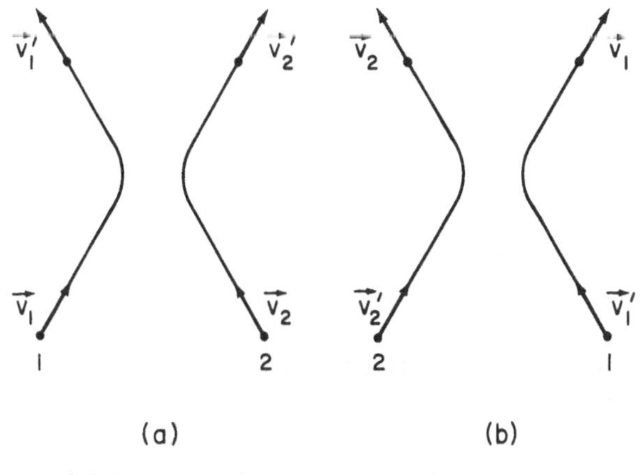

Fig. 6. Schematic illustration of (a) the direct collision and (b) the restituting collision.

attached to particles moving with velocity $\mathbf{v}_1'$ in $\delta\mathbf{r}_1$, as illustrated in Fig. 4.* Here ε' is an azimuthal angle measured in a plane perpendicular to $\mathbf{g}'$.† The total volume of all such collision cylinders in $\delta\mathbf{r}_1$ is

$$f(\mathbf{r}_1, \mathbf{v}_1', t)\,\delta\mathbf{r}_1\,\delta\mathbf{v}_1' B(\mathbf{g}', -\hat{\mathbf{k}})\,d\hat{\mathbf{k}}\,\delta t$$

Using the *Stosszahlansatz*, we find that the total number of $(\mathbf{v}_1', \mathbf{v}_2')$ collisions with $-\hat{\mathbf{k}}$ taking place in $\delta\mathbf{r}_1$ in δt is

$$f(\mathbf{r}_1, \mathbf{v}_1', t) f(\mathbf{r}_1, \mathbf{v}_2', t) B(\mathbf{g}', -\hat{\mathbf{k}})\,d\hat{\mathbf{k}}\,\delta\mathbf{v}_1'\,\delta\mathbf{v}_2'\,\delta\mathbf{r}_1\,\delta t \tag{19}$$

To obtain Γ_+ we must integrate this expression over a region R' of $\mathbf{v}_1', \mathbf{v}_2'$, and $\hat{\mathbf{k}}$, which is determined by the requirement that particle 1 has velocity in the region $\delta\mathbf{v}_1$ about $\mathbf{v}_1$ after the $(\mathbf{v}_1', \mathbf{v}_2')$ collision. Thus

$$\Gamma_+ = \delta\mathbf{r}_1\,\delta t \int d\mathbf{v}_1' \int_{R'} d\mathbf{v}_2' \int d\hat{\mathbf{k}}\, B(\mathbf{g}', -\hat{\mathbf{k}}) f(\mathbf{r}_1, \mathbf{v}_1', t) f(\mathbf{r}_1, \mathbf{v}_2', t) \tag{20}$$

This integration is most easily performed if we show that

$$d\mathbf{v}_1'\,d\mathbf{v}_2' = d\mathbf{v}_1\,d\mathbf{v}_2 \qquad \text{for fixed } \hat{\mathbf{k}} \tag{21}$$

This can be proved by noting that from Eqs. (17a) and (17b) we can compute the Jacobian $J(\mathbf{v}_1, \mathbf{v}_2/\mathbf{v}_1', \mathbf{v}_2')$. It is easy to see that this Jacobian depends only on the components of $\hat{\mathbf{k}}$ and that its value is unity. Moreover, from the facts that $|\mathbf{g}'| = |\mathbf{g}|$ and that $-\hat{\mathbf{k}}$ makes an angle of $\frac{1}{2}(\pi + \theta)$ with $\mathbf{g}'$, it follows that‡

$$B(\mathbf{g}', -\hat{\mathbf{k}}) = B(\mathbf{g}, \hat{\mathbf{k}}) \tag{22}$$

Therefore we can write expression (20) as

$$\Gamma_+ = \delta\mathbf{r}_1\,\delta t \int d\hat{\mathbf{k}} \int_R d\mathbf{v}_1 \int d\mathbf{v}_2\, B(\mathbf{g}, \hat{\mathbf{k}}) f(\mathbf{r}_1, \mathbf{v}_1', t) f(\mathbf{r}_1, \mathbf{v}_2', t) \tag{23a}$$

$$= \delta\mathbf{r}_1\,dt \int_{\delta\mathbf{v}_1} d\mathbf{v}_1 \int d\mathbf{v}_2 \int d\hat{\mathbf{k}}\, B(\mathbf{g}, \hat{\mathbf{k}}) f(\mathbf{r}_1, \mathbf{v}_1', t) f(\mathbf{r}_1, \mathbf{v}_2', t) \tag{23b}$$

In the right-hand side of Eq. (23a), R denotes the region of $\mathbf{v}_1, \mathbf{v}_2, \hat{\mathbf{k}}$ where $\mathbf{v}_1$ and $\mathbf{v}_2$ are the finial velocities after the $(\mathbf{v}_1', \mathbf{v}_2')$ collision, and where $-\hat{\mathbf{k}}$ is in the direction of the apse line of the restituting collision. To obtain Eq. (23a) we had to order the integration in expression (20) for Γ_+ so that the final integration is

*In the case that more than one impact parameter corresponds to a given $-\hat{\mathbf{k}}$, we consider the sum of all small collision cylinders corresponding to $-\hat{\mathbf{k}}$, and replace the right-hand side of Eq. (15) by the sum over all such impact parameters, $\Sigma g\, b_n\, db_n\, d\varepsilon'\, \delta t$.

†Notice that we have used $d\hat{\mathbf{k}} = d(-\hat{\mathbf{k}})$, and we have used the fact that the same vector $\hat{\mathbf{k}}$ can be described in a coordinate system with z axis in the direction of $\mathbf{g}$, as in the direct collisions, or in a coordinate system with z axis in the direction of $\mathbf{g}'$, as in the restituting collisions discussed here.

‡Here again this proof is simple if b is a one-valued function of θ. Otherwise we must consider all the values of b for a $\hat{\mathbf{k}}$ separately and then obtain Eq. (23) after summation.

the integration of $\hat{\mathbf{k}}$, since Eq. (21) holds only for fixed $\hat{\mathbf{k}}$, but once Eq. (23a) is obtained we can rearrange the order of integration to obtain Eq. (23b). Then for small enough $\delta\mathbf{v}_1$, we obtain

$$\Gamma_+ = \delta\mathbf{r}_1\,\delta\mathbf{v}_1\,\delta t\int dv_2\int d\hat{\mathbf{k}}\,f(\mathbf{r}_1,\mathbf{v}_1',t)f(\mathbf{r}_1,\mathbf{v}_2',t)B(\mathbf{g},\hat{\mathbf{k}}) \tag{23c}$$

Here we integrate over all $\mathbf{v}_2$ and over the allowed range of $\hat{\mathbf{k}}$, which is determined by the intermolecular potential.

We have now computed Γ_f, Γ_+, and Γ_- in Eq. (4), so that we may combine Eqs. (4), (6), (13), and (23) to obtain the *Boltzmann equation*

$$\begin{aligned}\partial f(\mathbf{r}_1,\mathbf{v}_1,t)/\partial t = -\mathbf{v}_1\cdot\nabla_{\mathbf{r}_1} f(\mathbf{r}_1,\mathbf{v}_1,t) + \int d\mathbf{v}_2\int d\mathbf{k}\,B(\mathbf{g},\hat{\mathbf{k}})(f_1'f_2'-f_1f_2)\\ +\Gamma_w/\delta\mathbf{r}_1\,\delta\mathbf{v}_1\,\delta t\end{aligned} \tag{24}$$

where we have dropped all higher derivatives, and adopted the notation

$$f_1 = f(\mathbf{r}_1,\mathbf{v}_1,t),\qquad f_1' = f(\mathbf{r}_1,\mathbf{v}_1',t) \tag{25a}$$

$$f_2 = f(\mathbf{r}_1,\mathbf{v}_2,t),\qquad f_2' = f(\mathbf{r}_1,\mathbf{v}_2',t) \tag{25b}$$

Before completing the derivation of the Boltzmann equation by deriving an expression for Γ_w, we summarize the essential assumptions that have been used to derive Γ_f, Γ_-, and Γ_+:

(a) The gas consists of monatomic particles that interact with short-range, central forces.

(b) The gas is sufficiently dilute that only binary collisions need to be taken into account.

(c) The number of direct and of restituting collisions taking place in a short time interval δt is computed by means of the *Stosszahlansatz*.

(d) The distribution function varies slowly enough with position that the difference in position of the centers of a colliding pair of particles immediately before collision can be neglected when formulating the *Stosszahlansatz*. In this connection, it should be mentioned that the Boltzmann equation is often applied to systems that interact with infinite-ranged potentials, such as the Lennard-Jones 6–12 potential.[8] To do this one needs to assume that (a) the potential function $\phi(r)$ decreases sufficiently rapidly that the Boltzmann collision integral appearing on the right-hand side of Eq. (24) exists, and (b) $f(\mathbf{r}_1,\mathbf{v}_1,t)$ does not vary much over a distance that characterizes the range of the potential.*

To complete the derivation of the Boltzmann equation, we must compute Γ_w, the change in the distribution function due to collision of the molecules with the walls of the container or with any surface in contact with the gas.[44–47] As usual, we consider particles with velocity $\mathbf{v}_1$, and write

$$\Gamma_w = \Gamma_{w+} - \Gamma_{w-} \tag{26}$$

*For example, the distance to the minimum of the well, for a Lennard-Jones potential.

where Γ_{w-} is the number of particles with velocity $\mathbf{v}_1$ located in the region $\delta\mathbf{r}_1$ about $\mathbf{r}_1$ that collide with a boundary in time interval δt and then change their velocity, and Γ_{w+} is the number of molecules in $\delta\mathbf{r}_1$ that acquire velocity $\mathbf{v}_1$ in time δt through collisions with the boundaries. To compute Γ_{w-} we fix our attention on a small section of the surface located with respect to some fixed coordinate system by a vector $\boldsymbol{\rho}_s$, as illustrated in Fig. 7. Let dS be the area of the small section of the surface and $\hat{\mathbf{n}}$ be a unit vector normal to the surface and pointing into the region occupied by the gas. A molecule with velocity $\mathbf{v}_1$ can only collide with the surface if $\mathbf{v}_1 \cdot \hat{\mathbf{n}} < 0$. Now the probability that this small element of surface is contained in the region $\delta\mathbf{r}_1$ about $\mathbf{r}_1$ is given by $\delta(\mathbf{r}_1 - \boldsymbol{\rho}_s)\,\delta\mathbf{r}_1$ where $\delta(\mathbf{r}_1 - \boldsymbol{\rho}_s)$ is the Dirac delta function. By considering a small collision cylinder of base area dS and height $|\mathbf{v}_1 \cdot \hat{\mathbf{n}}|\,\delta t$, we see that the number of collisions of particles with velocity $\mathbf{v}_1$ in range $\delta\mathbf{v}_1$, in region $\delta\mathbf{r}_1$ about $\mathbf{r}_1$ with the surface element in time δt is

$$\delta\mathbf{r}_1\,\delta\mathbf{v}_1\,\delta t\,\delta(\mathbf{r}_1 - \boldsymbol{\rho}_s)\Theta(-\mathbf{v}_1 \cdot \hat{\mathbf{n}})|\mathbf{v}_1 \cdot \hat{\mathbf{n}}| f(\mathbf{r}_1, \mathbf{v}_1, t)\,dS \tag{27}$$

where the function $\Theta(x)$ is defined by

$$\Theta(x) = \begin{cases} 1, & \text{if } x \geq 0 \\ 0, & \text{if } x < 0 \end{cases} \tag{28}$$

To obtain Γ_{w-}, we integrate over the entire surface bounding the gas so that

$$\Gamma_{w-} = \delta\mathbf{r}_1\,\delta\mathbf{v}_1\,\delta t \int dS\,\Theta(-\mathbf{v}_1 \cdot \hat{\mathbf{n}})\,\delta(\mathbf{r}_1 - \boldsymbol{\rho}_s)|\mathbf{v}_1 \cdot \hat{\mathbf{n}}| f(\mathbf{r}_1, \mathbf{v}_1, t) \tag{29}$$

To compute Γ_{w+} we make the following assumptions:

(a) If a molecule collides with a boundary it leaves the surface instantly at the point of the collision. That is, the molecule does not spend any time at the surface.

(b) When a particle with velocity $\mathbf{v}_1'$ hits the surface, the probability that it leaves the surface with velocity $\mathbf{v}_1$ in the range $\delta\mathbf{v}_1$ is given by a function

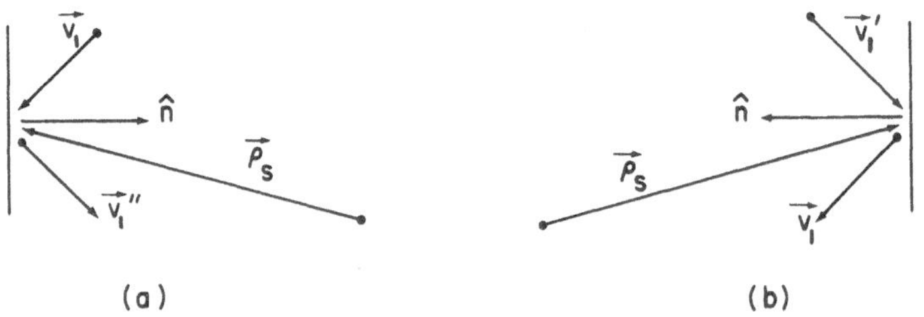

Fig. 7. Schematic illustration of molecule–wall collisions. In (a) the number of molecules with velocity $\mathbf{v}_1$ is diminished when one of them collides with the wall. In (b) this number is increased when the velocity of a molecule changes from $\mathbf{v}_1'$ to $\mathbf{v}_1$ by means of a collision with the wall.

$P(\mathbf{v}_1, \mathbf{v}_1')\,\delta\mathbf{v}_1$. P is taken to be independent of time, and to satisfy the following conditions:

(c) $P(\mathbf{v}_1, \mathbf{v}_1')$ vanishes unless $\mathbf{v}_1' \cdot \hat{\mathbf{n}} < 0$ and $\mathbf{v}_1 \cdot \hat{\mathbf{n}} > 0$. That is, $\mathbf{v}_1'$ must be incoming at the wall and $\mathbf{v}_1$ must be an outgoing velocity. We then write

$$P(\mathbf{v}_1, \mathbf{v}_1') = K(\mathbf{v}_1, \mathbf{v}_1')\Theta(\mathbf{v}_1 \cdot \hat{\mathbf{n}})\Theta(-\mathbf{v}_1' \cdot \hat{\mathbf{n}}) \tag{30}$$

(d) $P(\mathbf{v}_1, \mathbf{v}_1')$ is properly normalized. That is, if the wall is *not* a source or sink of gas particles, then the probability that a particle is reemitted must be unity, i.e.,

$$\int_{\mathbf{v}_1 \cdot \mathbf{n} > 0} d\mathbf{v}_1\, P(\mathbf{v}_1, \mathbf{v}_1') = 1 \qquad \text{for all } \mathbf{v}_1' \text{ for which } \mathbf{v}_1' \cdot \hat{\mathbf{n}} < 0 \tag{31}$$

Although very little is known about the actual form of $P(\mathbf{v}_1, \mathbf{v}_1')$, it must depend on the material composing the boundaries and on the particular gas in the vessel.[6–8,46–50] In the discussions here we will avoid as much as possible having to specify this function. We will assume, though, that Eq. (31) is always satisfied.* An additional condition that $P(\mathbf{v}_1, \mathbf{v}_1')$ must satisfy in order that the system have an equilibrium state will be discussed in Section 2.2.

Using assumptions (a)–(c) and the by now familiar collision cylinder constructions, we may easily obtain Γ_{w+} as

$$\Gamma_{w+} = \delta\mathbf{r}_1\, \delta\mathbf{v}_1\, \delta t \int dS\, \delta(\mathbf{r}_1 - \boldsymbol{\rho}_s)$$
$$\times \int d\mathbf{v}_1'\, |\mathbf{v}_1' \cdot \hat{\mathbf{n}}|\Theta(-\mathbf{v}_1' \cdot \hat{\mathbf{n}})\Theta(\mathbf{v}_1 \cdot \hat{\mathbf{n}})K(\mathbf{v}_1, \mathbf{v}_1')f(\mathbf{r}_1, \mathbf{v}_1', t) \tag{32}$$

This argument used to compute Γ_{w+} can also be used to express the distribution function at the wall for outgoing particles, i.e., $f(\boldsymbol{\rho}_s, \mathbf{v}_1, t)$ for $\mathbf{v}_1 \cdot \hat{\mathbf{n}} > 0$, in terms of $K(\mathbf{v}_1, \mathbf{v}_1')$ and the distribution function of the incoming particles. To do this, we require that the rate at which particles with velocity $\mathbf{v}_1$ leave the surface at the point $\boldsymbol{\rho}_s$ be equal to the rate at which they are produced there, or

$$|\mathbf{v}_1 \cdot \hat{\mathbf{n}}|f(\boldsymbol{\rho}_s, \mathbf{v}_1, t)\Theta(\mathbf{v}_1 \cdot \hat{\mathbf{n}})$$
$$= \int d\mathbf{v}_1'\, \Theta(-\mathbf{v}_1' \cdot \hat{\mathbf{n}})\Theta(\mathbf{v}_1 \cdot \hat{\mathbf{n}})K(\mathbf{v}_1, \mathbf{v}_1') \cdot |\mathbf{v}_1' \cdot \hat{\mathbf{n}}|f(\boldsymbol{\rho}_s, \mathbf{v}_1', t) \tag{33}$$

We will use this expression for $f(\boldsymbol{\rho}_s, \mathbf{v}_1, t)$ in Section 2.2.†

We may now combine Eqs. (24), (26), (29), and (32) to write the Boltzmann equation, including interactions with the wall, as

$$\frac{\partial f}{\partial t} + \mathbf{v}_1 \cdot \boldsymbol{\nabla}_{\mathbf{r}_1} f = J(f, f) + \bar{\mathbf{T}}f \tag{34a}$$

*One could relax Eq. (31) and thereby construct a theoretical description for processes where molecules are gained or lost at the boundaries. An example is a theory for droplet growth in a vapor.[51]

†It is worth mentioning that in the usual discussions of the Boltzmann equation Γ_W is not explicitly taken into account. Instead, Eq. (33) is imposed as a condition that the distribution function must satisfy at the boundary of the container.

where $J(f, f)$ is the molecule–molecule collision term

$$J(f, f) = \int d\mathbf{v}_2 \int d\hat{\mathbf{k}}\, B(\mathbf{g}, \hat{\mathbf{k}})(f_1' f_2' - f_1 f_2)$$

and $\bar{\mathbf{T}}f$ is the molecule–boundary collision term

$$\bar{\mathbf{T}}f = \int dS\, \delta(\mathbf{r}_1 - \boldsymbol{\rho}_s)\Big\{\Theta(\mathbf{v}_1 \cdot \hat{\mathbf{n}}) \int d\mathbf{v}_1'\, \Theta(-\mathbf{v}_1' \cdot \hat{\mathbf{n}})|\mathbf{v}_1' \cdot \hat{\mathbf{n}}| K(\mathbf{v}_1, \mathbf{v}_1') f(\mathbf{r}_1, \mathbf{v}_1', t) - \Theta(-\mathbf{v}_1 \cdot \hat{\mathbf{n}})|\mathbf{v}_1 \cdot \hat{\mathbf{n}}| f(\mathbf{r}_1, \mathbf{v}_1, t)\Big\} \tag{34b}$$

Equation (34a) may be regarded as being valid everywhere in space, even outside the container. For if the particles are initially confined to the interior of the container, none of them will ever leave the container, since we have assumed that the boundaries are not sources or sinks of particles. Consequently we expect that Eq. (34a) should have solutions that vanish identically outside the container.

Finally, if there is a velocity-independent external force $\mathbf{F}$ acting on the particles in the gas, one can see that Γ_f given by Eq. (5) should be changed to

$$\Gamma_f = [-\mathbf{v}_1 \cdot \nabla_{\mathbf{r}_1} f - (\mathbf{F}/m) \cdot \nabla_{\mathbf{v}_1} f]\, \delta\mathbf{r}_1\, \delta\mathbf{v}_1\, \delta t \tag{35}$$

where $\nabla_{\mathbf{v}}$ denotes the gradient with respect to velocity, and Eq. (34) becomes

$$\frac{\partial f}{\partial t} + \mathbf{v} \cdot \nabla_{\mathbf{r}_1} f + (\mathbf{F}/m) \cdot \nabla_{\mathbf{v}_1} f = J(f, f) + \bar{\mathbf{T}}f \tag{36}$$

2.1.2. Gas Mixtures

It is easy to extend Boltzmann's derivation to cover a mixture of monatomic gases whose particles all interact with pairwise additive central forces. To do this we denote the distribution function for gas molecules of species α by $f_\alpha(\mathbf{r}_1, \mathbf{v}_1, t)$. Then let $J_{\alpha\beta}(f_\alpha, f_\beta)$ be the collision integral for collisions of particles of species α with particles of species β, let $\mathbf{F}_\alpha$ be an external force acting on particles of species α, and let $\bar{\mathbf{T}}_\alpha f_\alpha$ denote the collision term for collisions of particles of species α with the boundaries. Then f_α satisfies

$$\frac{\partial f_\alpha(\mathbf{r}_1, \mathbf{v}_1, t)}{\partial t} + \mathbf{v}_1 \cdot \nabla_{\mathbf{r}_1} f_\alpha(\mathbf{r}_1, \mathbf{v}_1, t) + (\mathbf{F}_\alpha/m_\alpha) \cdot \nabla_{\mathbf{v}_1} f_\alpha(\mathbf{r}_1, \mathbf{v}_1, t) = \sum_{\beta=1}^{\kappa} J_{\alpha\beta}(f_\alpha, f_\beta) + \bar{\mathbf{T}}_\alpha f_\alpha \tag{37}$$

where m_α is the mass of a particle of species α and κ the total number of different components in the gas mixture. We will not consider gas mixtures in any detail in this chapter; instead we refer the reader to the literature.[5-13]

2.2. The *H*-Theorem and the Approach to Equilibrium

According to experiment, and as follows from the study of equilibrium statistical mechanics, when a dilute gas confined to a container is in an equilibrium state in an external potential field and its center of mass is at rest, the distribution function $f_{eq}(\mathbf{r}, \mathbf{v})$ is

$$f_{eq}(\mathbf{r}, \mathbf{v}) = W_w(\mathbf{r}) n_0 C(\beta m/2\pi)^{3/2} \exp(-\tfrac{1}{2}\beta m \mathbf{v}^2 - \beta V(\mathbf{r})) \tag{38}$$

with

$$C = V\left[\int d\mathbf{r}'\, W_w(\mathbf{r}') \exp(-\beta V(\mathbf{r}'))\right]^{-1}$$

where $\beta = (k_B T)^{-1}$, k_B is Boltzmann's constant, T the thermodynamic temperature of the equilibrium state, n_0 the (overall) number density of the gas, $V(\mathbf{r})$ the potential energy of the external force field $\mathbf{F}(\mathbf{r})$, $W_w(\mathbf{r})$ is a function that vanishes for $\mathbf{r}$ outside the container and is unity for $\mathbf{r}$ inside, and V is the volume.

It is therefore natural to pose the following two questions of the Boltzmann equation:

1. Can one show that a gas that is not in equilibrium at some time will, as time progresses, approach an equilibrium state where the distribution function is given by Eq. (38) and that once in an equilibrium state the system remains in that state?
2. Can one describe in detail how the approach to equilibrium proceeds?

To answer the second question, we must be able to construct a solution to the Boltzmann equation corresponding to some initial nonequilibrium state and then follow its development in the course of time. This is, in general, an extremely difficult procedure, and we shall describe its main features in the next section. In this section we consider the first question and outline the solution along the lines given by Boltzmann.[3-13,52]

Boltzmann showed that one can give an affirmative answer to the first question without actually having to construct a solution to the equation, if one assumes that a solution exists. To do this, he proved a theorem, the Boltzmann H-theorem, which states that the function $H(t)$ defined by

$$H(t) = \int_V d\mathbf{r} \int d\mathbf{v}\, f(\mathbf{r}, \mathbf{v}, t)[\log f(\mathbf{r}, \mathbf{v}, t) - 1] \tag{39}$$

is a monotonically decreasing function of time, provided that $f(\mathbf{r}, \mathbf{v}, t)$ is a solution to the Boltzmann equation, and that the molecules make elastic, specular collisions with the walls. The H-theorem was suggested by Clausius's form of the second law of thermodynamics, which states that the entropy increases in a spontaneous irreversible process between two equilibrium states of an isolated system. Boltzmann's choice of H was motivated by the fact that if f in Eq. (39) is replaced by f_{eq}, given by Eq. (38), then $H = -\tilde{S}/k_B + \text{const}$, where $\tilde{S}$ is the equilibrium entropy of an ideal gas of N particles in volume V at temperature T.

Here we will prove a stronger version of Boltzmann's H-theorem, which holds for more general gas–wall interactions and which gives a nonequilibrium analog of Clausius's formula

$$d\tilde{S} \geq dQ/T$$

for the change in the entropy $\tilde{S}$ for an irreversible process between two equilibrium states of a system in contact with a heat reservoir maintained at temperature T, and in which an amount of heat dQ is transferred from the reservoir to the system.

To obtain the analog of Clausius's formula from the Boltzmann equation we will impose an additional condition on the interactions of the gas molecules with the walls. We will assume that there exists a local temperature of the wall $T_w(\boldsymbol{\rho}_s)$ at every point $\boldsymbol{\rho}_s$ such that if the distribution of molecules incident on the surface at $\boldsymbol{\rho}_s$ is a Maxwell–Boltzmann distribution with temperature $T_w(\boldsymbol{\rho}_s)$, then the distribution of the reflected molecules is also a Maxwell–Boltzmann distribution with the same temperature.* Using Eq. (33) we may express this *thermostat condition* as a condition on $K(\mathbf{v}, \mathbf{v}')$ as

$$\begin{aligned}\Theta(\mathbf{v}\cdot\hat{\mathbf{n}})&|\mathbf{v}\cdot\hat{\mathbf{n}}|\phi_w(\mathbf{v}, \boldsymbol{\rho}_s)\\ &= \Theta(\mathbf{v}\cdot\hat{\mathbf{n}})\int d\mathbf{v}'\,\Theta(-\mathbf{v}'\cdot\hat{\mathbf{n}})|\mathbf{v}'\cdot\hat{\mathbf{n}}|K(\mathbf{v}, \mathbf{v}')\phi_w(\mathbf{v}', \boldsymbol{\rho}_s) \qquad \text{(40a)}\end{aligned}$$

where $\phi_w(\mathbf{v}, \boldsymbol{\rho}_s)$ is a Maxwell–Boltzmann distribution with temperature $T_w(\boldsymbol{\rho}_s)$,

$$\phi_w(\mathbf{v}, \boldsymbol{\rho}_s) = \left(\frac{m}{2\pi k_B T_w(\boldsymbol{\rho}_s)}\right)^{3/2} \exp\left(-\frac{m\mathbf{v}^2}{2k_B T_w(\boldsymbol{\rho}_s)}\right) \qquad \text{(40b)}$$

We will now prove that if the function $K(\mathbf{v}, \mathbf{v}')$ satisfies the thermostat condition, and if $f(\mathbf{r}, \mathbf{v}, t)$ is a solution of the Boltzmann equation, Eq. (36), then $H(t)$ defined by Eq. (39) satisfies

$$\frac{dH}{dt} \leq -\frac{1}{k_B}\int dS \frac{1}{T_w(\boldsymbol{\rho}_s)}\frac{dQ(\boldsymbol{\rho}_s)}{dt} \qquad \text{(41)}$$

where $dQ(\boldsymbol{\rho}_s)/dt$ is the energy flux† into the gas at the boundary point $\boldsymbol{\rho}_s$ of the container, and $T_w(\boldsymbol{\rho}_s)$ is defined by the thermostat condition; The integral is carried out over the boundary surface S.[6,7,53] In the special case that the molecules are specularly reflected from the walls, there is no energy exchange with the walls and the right-hand side of Eq. (41) vanishes.

*In the case that the molecules are specularly reflected from the wall, this condition is satisfied for a Maxwell–Boltzmann distribution with any temperature, so there is no unique wall temperature. This case will be discussed separately below. The condition for specular reflection is $K(\mathbf{v},\mathbf{v}') = \delta(\mathbf{v}-\mathbf{v}'+2(\mathbf{v}'\cdot\hat{\mathbf{n}})\hat{\mathbf{n}})$.

†Note that the walls of the container are held fixed. Then there is no work done on the gas and the energy flow is identical to the heat flow.

The proof of Eq. (41) is as follows: Let

$$h(\mathbf{r}, t) = \int d\mathbf{v}\, f(\mathbf{r}, \mathbf{v}, t)[\log f(\mathbf{r}, \mathbf{v}, t) - 1] \tag{42}$$

Then

$$\frac{\partial h(\mathbf{r}, t)}{\partial t} = \int d\mathbf{v} \frac{\partial f(\mathbf{r}, \mathbf{v}, t)}{\partial t} \log f(\mathbf{r}, \mathbf{v}, t) \tag{43}$$

which, using Eq. (36), is equal to

$$\frac{\partial h(\mathbf{r}, t)}{\partial t} = -\nabla \cdot \mathbf{j}_h + \int d\mathbf{v} \log f\, J(f, f) + \int d\mathbf{v} \log f\, \bar{\mathbf{T}} f \tag{44a}$$

where*

$$\mathbf{j}_h = \int d\mathbf{v}\, \mathbf{v} f \log f \tag{44b}$$

and we assume that $f(\mathbf{r}, \mathbf{v}, t)$ vanishes sufficiently rapidly as $|\mathbf{v}| \to \infty$ that†

$$\int d\mathbf{v} \nabla_{\mathbf{v}} [f(\log f - 1)] = 0 \tag{45}$$

The quantity $\mathbf{j}_h$ represents the flux of h due to the free motion of the particles, the integral $\int d\mathbf{v} \log f\, J(f, f)$ represents the change of h due to molecule–molecule collisions, and $\int d\mathbf{v} \log f\, \bar{\mathbf{T}} f$ the change due to molecule–wall collisions.

To compute $\int d\mathbf{v} \log f\, J(f, f)$ we first establish a general result:

$$\int d\mathbf{v}\, \psi(\mathbf{v}) J(f, f) = \tfrac{1}{4} \int d\mathbf{v} \int d\mathbf{v}_1 \int d\hat{\mathbf{k}}\, B(\mathbf{g}, \hat{\mathbf{k}}) \times [\psi(\mathbf{v}) + \psi(\mathbf{v}_1) - \psi(\mathbf{v}') - \psi(\mathbf{v}_1')](f' f_1' - f f_1) \tag{46}$$

We start with

$$\int d\mathbf{v}\, \psi(\mathbf{v}) J(f, f) = \int d\mathbf{v}_1 \int d\hat{\mathbf{k}} B(\mathbf{g}, \hat{\mathbf{k}}) \psi(\mathbf{v})(f' f_1' - f f_1) \tag{47}$$

which, upon interchanging $\mathbf{v}$ and $\mathbf{v}_1$, summing, and dividing by two, leads to

$$\int d\mathbf{v}\, \psi(\mathbf{v}) J(f, f) = \tfrac{1}{2} \int d\mathbf{v} \int d\mathbf{v}_1 \int d\hat{\mathbf{k}}\, B(\mathbf{g}, \hat{\mathbf{k}})\, [\psi(\mathbf{v}) + \psi(\mathbf{v}_1)](f' f_1' - f f_1) \tag{48}$$

*We have simplified the notation so that $(\mathbf{r}_1, \mathbf{v}_1) \to (\mathbf{r}, \mathbf{v})$; $(\mathbf{r}_2, \mathbf{v}_2) \to (\mathbf{r}_1, \mathbf{v}_1)$.

†In fact, we shall impose the constraints on f that

$$\int d\mathbf{r} \int d\mathbf{v}\, f(\mathbf{r}, \mathbf{v}, t) = N < \infty$$

and

$$\int d\mathbf{r} \int d\mathbf{v} \frac{m\mathbf{v}^2}{2} f(\mathbf{r}, \mathbf{v}, t) = E_k(t) < \infty$$

representing the conditions that the total number of particles and the total kinetic energy of the gas be finite for all times. From these contraints one can easily prove that H is bounded.

Now, using Eqs. (21) and (22) to replace $d\mathbf{v}_1\, d\mathbf{v}_2$ by $d\mathbf{v}_1'\, d\mathbf{v}_2'$, $B(\mathbf{g}, \hat{\mathbf{k}})$ by $B(\mathbf{g}', -\hat{\mathbf{k}})$ and letting $\hat{\mathbf{k}} \to -\hat{\mathbf{k}}$, we have

$$\int d\mathbf{v}\, \psi(\mathbf{v}) J(f, f) = \tfrac{1}{2} \int d\mathbf{v}' \int d\mathbf{v}_1' \int d\hat{\mathbf{k}}\, B(\mathbf{g}', \hat{\mathbf{k}})\, [\psi(\mathbf{v}) + \psi(\mathbf{v}_1)](f' f_1' - f f_1) \tag{49}$$

We now remove the primes in the integration variables in the right-hand side of Eq. (49) and use the fact that $(\mathbf{v}', \mathbf{v}_1')$ are the initial velocities in the restituting collision, which leads to $(\mathbf{v}, \mathbf{v}_1)$ to obtain*

$$\int d\mathbf{v}\, \psi(\mathbf{v}) J(f, f) = \tfrac{1}{2} \int d\mathbf{v} \int d\mathbf{v}_1 \int d\hat{\mathbf{k}}\, B(\mathbf{g}, \hat{\mathbf{k}}) [\psi(\mathbf{v}') + \psi(\mathbf{v}_1')](f f_1 - f' f_1') \tag{50}$$

which, together with Eq. (48), leads to Eq. (46).

As a consequence of Eq. (46) we can write Eq. (44) as

$$\frac{\partial h}{\partial t} + \nabla \cdot \mathbf{j}_h = \frac{1}{4} \int d\mathbf{v} \int d\mathbf{v}_1 \int d\hat{\mathbf{k}}\, B(\mathbf{g}, \hat{\mathbf{k}}) \log \frac{f f_1}{f' f_1'} (f' f_1' - f f_1)$$
$$+ \int d\mathbf{v} \log f(\mathbf{r}, \mathbf{v}, t) \bar{\mathbf{T}} f(\mathbf{r}, \mathbf{v}, t) \tag{51}$$

The first term on the right-hand side of Eq. (51) is negative, since the form $(b-a) \log a/b \leq 0$ for any real positive a, b, the equality holding only if $a = b$. Therefore we may write

$$\frac{\partial h}{\partial t} + \nabla \cdot \mathbf{j}_h \leq \int d\mathbf{v} \log f(\mathbf{r}, \mathbf{v}, t) \bar{\mathbf{T}} f(\mathbf{r}, \mathbf{v}, t) \tag{52}$$

Now we integrate Eq. (52) over all space, including over the region exterior to the container, and use the fact that f vanishes outside the container, to obtain

$$\frac{d}{dt} H(t) \leq \int d\mathbf{r} \int d\mathbf{v} \log f(\mathbf{r}, \mathbf{v}, t) \bar{\mathbf{T}} f(\mathbf{r}, \mathbf{v}, t) \tag{53a}$$

$$\frac{d}{dt} H(t) \leq \int dS \int d\mathbf{v} \log f(\boldsymbol{\rho}_s, \mathbf{v}, t) \Big\{ \Theta(\mathbf{v} \cdot \hat{\mathbf{n}})$$
$$\times \int d\mathbf{v}' \Theta(-\mathbf{v}' \cdot \hat{\mathbf{n}}) |\mathbf{v}' \cdot \hat{\mathbf{n}}| K(\mathbf{v}, \mathbf{v}') f(\boldsymbol{\rho}_s, \mathbf{v}', t)$$
$$- \Theta(-\mathbf{v} \cdot \hat{\mathbf{n}}) |\mathbf{v} \cdot \hat{\mathbf{n}}| f(\boldsymbol{\rho}_s, \mathbf{v}, t) \Big\} \tag{53b}$$

where we have used Eq. (34b) for $\bar{\mathbf{T}} f$.

*Notice that the integration regions for the variables $\mathbf{v}'$, $\mathbf{v}_1'$, and $-\hat{\mathbf{k}}$ are the same as those for $\mathbf{v}_1$, $\mathbf{v}_2$, and $\hat{\mathbf{k}}$.

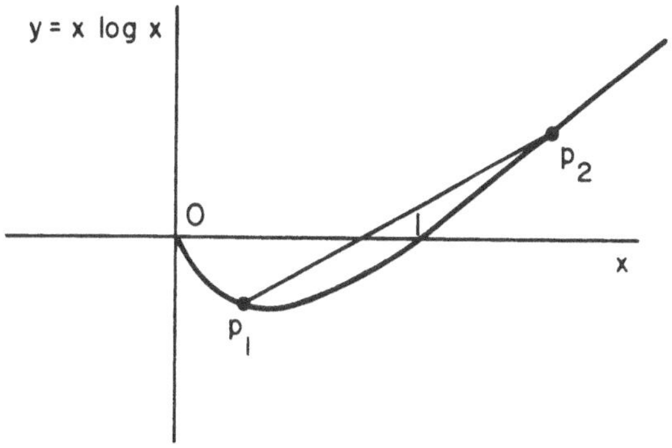

Fig. 8. A plot of $y = x \log x$ vs x for $x \geq 0$. Note that for any p_1 and p_2 on the curve, the values of the function $x \log x$ between p_1 and p_2 lie below the straight line joining p_1 and p_2.

Using the normalization condition, Eq. (31), and expression (30) for $K(\mathbf{v}, \mathbf{v}')$, we write Eq. (53b) as

$$\frac{d}{dt}H(t) \leq \int dS \int d\mathbf{v} \int d\mathbf{v}'\, \Theta(\mathbf{v}\cdot\hat{\mathbf{n}})\Theta(-\mathbf{v}'\cdot\hat{\mathbf{n}})K(\mathbf{v}, \mathbf{v}')|\mathbf{v}'\cdot\hat{\mathbf{n}}|$$
$$\times [f(\boldsymbol{\rho}_s, \mathbf{v}', t)\log f(\boldsymbol{\rho}_s, \mathbf{v}, t) - f(\boldsymbol{\rho}_s, \mathbf{v}', t)\log f(\boldsymbol{\rho}_s, \mathbf{v}', t)] \tag{54}$$

Now we insert into the right-hand side of Eq. (54) the Maxwell–Boltzmann distribution $\phi_w(\mathbf{v})$ defined by Eq. (40b), and we write

$$\frac{d}{dt}H(t) \leq \int dS \int d\mathbf{v} \int d\mathbf{v}'\, \Theta(\mathbf{v}\cdot\hat{\mathbf{n}})\Theta(-\mathbf{v}'\cdot\hat{\mathbf{n}})K(\mathbf{v}, \mathbf{v}')|\mathbf{v}'\cdot\hat{\mathbf{n}}|\phi_w(\mathbf{v}')$$
$$\times [\hat{f}(\boldsymbol{\rho}_s, \mathbf{v}', t)\log \hat{f}(\boldsymbol{\rho}_s, \mathbf{v}, t) - \hat{f}(\boldsymbol{\rho}_s, \mathbf{v}', t)\log \hat{f}(\boldsymbol{\rho}_s, \mathbf{v}', t)$$
$$+ \hat{f}(\boldsymbol{\rho}_s, \mathbf{v}', t)\log \phi_w(\mathbf{v}) - \hat{f}(\boldsymbol{\rho}_s, \mathbf{v}', t)\log \phi_w(\mathbf{v}')] \tag{55a}$$

where

$$\hat{f}(\boldsymbol{\rho}_s, \mathbf{v}, t) = f(\boldsymbol{\rho}_s, \mathbf{v}, t)/\phi_w(\mathbf{v}) \tag{55b}$$

The crucial step in the argument is to use the fact that the function $x \log x$ is convex for $x \geq 0$. That is, for every two points p_1 and p_2 on the curve $y = x \log x$, the points on the arc of y between p_1 and p_2 lie below the chord joining p_1 and p_2, as illustrated in Fig. 8. For such a function Jensen's inequality applies.[6] This inequality states that if f and g are positive functions, then

$$\frac{\int dx\, f(x)g(x)}{\int dx\, g(x)}\log\left[\frac{\int dx\, f(x)g(x)}{\int dx\, g(x)}\right] \leq \frac{\int dx\, g(x)f(x)\log f(x)}{\int dx\, g(x)} \tag{56}$$

provided all the integrals involved exist. To apply Jensen's inequality to the right-hand side of Eq. (55a) we set the integration region to be over all $\mathbf{v}'$ with $\mathbf{v}'\cdot\hat{\mathbf{n}} < 0$, and we set

$$g(\mathbf{v}') = |\mathbf{v}'\cdot\hat{\mathbf{n}}|K(\mathbf{v}, \mathbf{v}')\phi_w(\mathbf{v}')\Theta(\mathbf{v}\cdot\hat{\mathbf{n}})$$
$$f(\mathbf{v}') = \hat{f}(\boldsymbol{\rho}_s, \mathbf{v}', t)$$

Then from Eq. (56), we obtain the inequality

$$\int d\mathbf{v}'\,\Theta(\mathbf{v}\cdot\hat{\mathbf{n}})\Theta(-\mathbf{v}'\cdot\hat{\mathbf{n}})|\mathbf{v}'\cdot\hat{\mathbf{n}}|K(\mathbf{v},\mathbf{v}')\phi_{\mathrm{w}}(\mathbf{v}')\hat{f}(\boldsymbol{\rho}_s,\mathbf{v},t)\log\hat{f}(\boldsymbol{\rho}_s,\mathbf{v},t)$$
$$\leq \int d\mathbf{v}'\,\Theta(-\mathbf{v}'\cdot\hat{\mathbf{n}})\Theta(\mathbf{v}\cdot\hat{\mathbf{n}})|\mathbf{v}'\cdot\hat{\mathbf{n}}|K(\mathbf{v},\mathbf{v}')\phi_{\mathrm{w}}(\mathbf{v}')\hat{f}(\boldsymbol{\rho}_s,\mathbf{v}',t)\log\hat{f}(\boldsymbol{\rho}_s,\mathbf{v}',t) \tag{57}$$

where we have used (a) Eq. (33) to relate the distribution function of the reflected molecules at the surface to the incident distribution function; (b) the thermostat condition, Eq. (40a); and (c) the fact that

$$\Theta(\mathbf{v}\cdot\hat{\mathbf{n}})\log\frac{|\mathbf{v}\cdot\hat{\mathbf{n}}|\Theta(\mathbf{v}\cdot\hat{\mathbf{n}})f(\boldsymbol{\rho}_s,\mathbf{v},t)}{|\mathbf{v}\cdot\hat{\mathbf{n}}|\Theta(\mathbf{v}\cdot\hat{\mathbf{n}})\phi_{\mathrm{w}}(\mathbf{v})}=\Theta(\mathbf{v}\cdot\hat{\mathbf{n}})\log\hat{f}(\boldsymbol{\rho}_s,\mathbf{v},t)$$

If we now compare the integrand on the right-hand side of Eq. (55a) with the terms appearing in Eq. (57) we see immediately that

$$\frac{d}{dt}H(t)\leq\int dS\int d\mathbf{v}\int d\mathbf{v}'\,\Theta(\mathbf{v}\cdot\hat{\mathbf{n}})\Theta(-\mathbf{v}'\cdot\hat{\mathbf{n}})K(\mathbf{v},\mathbf{v}')|\mathbf{v}'\cdot\hat{\mathbf{n}}|\phi_{\mathrm{w}}(\mathbf{v}')$$
$$\times[\hat{f}(\boldsymbol{\rho}_s,\mathbf{v}',t)\log\phi_{\mathrm{w}}(\mathbf{v})-\hat{f}(\boldsymbol{\rho}_s,\mathbf{v}',t)\log\phi_{\mathrm{w}}(\mathbf{v}')] \tag{58}$$

Now if we insert expression (40b) for $\phi_{\mathrm{w}}(\mathbf{v})$ into the right-hand side of Eq. (58), and use Eqs. (31) and (33), we obtain

$$\frac{d}{dt}H(t)\leq-\frac{1}{k_{\mathrm{B}}}\int dS\,\frac{1}{T_{\mathrm{w}}(\boldsymbol{\rho}_s)}\int d\mathbf{v}\,\mathbf{v}\cdot\hat{\mathbf{n}}\frac{m\mathbf{v}^2}{2}f(\boldsymbol{\rho}_s,\mathbf{v},t) \tag{59}$$

where the velocity integration in Eq. (59) is over all $\mathbf{v}$. Thus we have derived the H-theorem stated in Eq. (41) with the heat flux into the system at the point $\boldsymbol{\rho}_s$ on the walls given by

$$\frac{dQ(\boldsymbol{\rho}_s)}{dt}=\int d\mathbf{v}\,(\mathbf{v}\cdot\hat{\mathbf{n}})\frac{m\mathbf{v}^2}{2}f(\boldsymbol{\rho}_s,\mathbf{v},t) \tag{60}$$

Boltzmann's form of the H-theorem, $dH/dt\leq 0$, is recovered whenever the boundary conditions are such that $dQ/dt=0$. If the molecules are specularly reflected at the walls, then this condition is certainly satisfied, as it is if the distribution function at the walls is a Maxwell–Boltzmann distribution with the same temperature for both incident and reflected particles. In addition, we can also verify that if the walls are at a uniform temperature T_{w} and if the thermostat condition is satisfied, then the equilibrium distribution function, Eq. (38), is a solution of the Boltzmann equation,* Eq. (36), and for this distribution $\partial f/\partial t=0$ and $dH/dt=0$. Therefore, according to the Boltzmann equation, if in the course of time the system reaches an equilibrium state with f given by Eq. (38), it will remain in equilibrium for all later times.

*To show this explicitly one needs to use the fact that $\nabla_{\mathbf{r}}W_{\mathrm{w}}(\mathbf{r})=\hat{\mathbf{n}}\,\delta(\mathbf{r}-\boldsymbol{\rho}_s)$.

Returning to our original problem, we would like to prove that for any initial distribution $f(\mathbf{r}, \mathbf{v}, t)$ and for boundary interactions that satisfy the thermostat condition, Eq. (40a), with a constant wall temperature at all points, the gas will eventually reach a state of total equilibrium in which the distribution function has the Maxwell–Boltzmann form given by Eq. (38) with the temperature equal to the wall temperature.

To prove this we consider a time-dependent Helmholtz free energy[46,47]

$$F(t) \equiv E(t) - T_w\tilde{S}(t) \equiv E_k(t) + \int d\mathbf{r} \int d\mathbf{v}\; V(\mathbf{r}) f(\mathbf{r}, \mathbf{v}, t) + k_B T_w H(t)$$

where T_w is the constant wall temperature, and E_k is the total kinetic energy of the gas.* From Eqs. (31), (33), and (59), one can easily show that

$$dF(t)/dt \leq 0$$

Then if $E_k(t)$ is bounded for all times t, the system approaches an equilibrium state, as is shown below.

Let $f(\mathbf{r}, \mathbf{v}, t)$ be a solution to the Boltzmann equation for a system satisfying the conditions

(a) The total number of particles is a finite number, i.e.,

$$\int d\mathbf{r} \int d\mathbf{v}\, f(\mathbf{r}, \mathbf{v}, t) = N$$

(b) The total kinetic energy of the gas is bounded, i.e.,

$$\int d\mathbf{r} \int d\mathbf{v} \frac{m\mathbf{v}^2}{2} f(\mathbf{r}, \mathbf{v}, t) = E_k(t) < M_E < \infty$$

where M_E is some finite bound.

Then as we have mentioned above, it follows from Eq. (59) that $F(t)$ decreases in time. However under conditions (a) and (b), $F(t)$ is bounded from below. The proof is left to the reader. Now since $F(t)$ is bounded from below and yet is a monotonically decreasing function, $F(t)$ must eventually approach a limit defined by $dF(t)/dt = 0$. An inspection of the various inequalities used in deriving the H-theorem, Eqs. (52) and (57), shows that $dF(t)/dt = 0$ if and only if

$$f(\mathbf{r}, \mathbf{v}, t) f(\mathbf{r}, \mathbf{v}_1, t) = f(\mathbf{r}, \mathbf{v}', t) f(\mathbf{r}, \mathbf{v}_1', t) \tag{61}$$

in Eq. (51), and if the inequality in Eq. (57) is satisfied as an equality. We first study Eq. (61). Taking logarithms we see that

$$\log f + \log f_1 = \log f' + \log f_1' \tag{62}$$

*If the walls can move, then one can generalize the argument by defining the appropriate Gibbs free energy and showing that it is a monotonically decreasing function. We also insist that $V(\mathbf{r})$ not depend on time.

Therefore log f must be a quantity that is conserved in a binary collision, usually called a "collision invariant." The only collision invariants in a binary collision are (a) the number of particles, or the total mass of the particles, (b) the total linear momentum, (c) the total kinetic energy, and (d) the total angular momentum. Therefore, to satisfy Eq. (62), $\log f$ must be of the form*

$$\log f(\mathbf{r}, \mathbf{v}, t) = a(\mathbf{r}, t) + \mathbf{b}(\mathbf{r}, t) \cdot \mathbf{v} + \tfrac{1}{2}c(\mathbf{r}, t)m\mathbf{v}^2 \tag{63}$$

where a, $\mathbf{b}$, and c might depend on $\mathbf{r}$ and t. Hence as $t \to \infty$,

$$f \to f^* = W_w(\mathbf{r}) \exp\{a(\mathbf{r}, t) + \mathbf{b}(\mathbf{r}, t) \cdot \mathbf{v} + \tfrac{1}{2}c(\mathbf{r}, t)m\mathbf{v}^2\} \tag{64}$$

and for this distribution function

$$J(f^*, f^*) = 0 \tag{65}$$

However the condition that f^* also be a solution of the Boltzmann equation in the interior of the container implies that

$$\frac{\partial f^*}{\partial t} + \mathbf{v} \cdot \nabla_r f^* + \frac{\mathbf{F}}{m} \cdot \nabla_v f^* = 0 \tag{66}$$

and from this condition it follows that if the container is at rest, and if the external potential is independent of velocity f^* must have the form[52]

$$f^*(\mathbf{r}, \mathbf{v}, t) = A\, W_w(\mathbf{r}) \exp\{-\tfrac{1}{2}\beta m\mathbf{v}^2 - \beta V(\mathbf{r})\} \tag{67}$$

where A and β are constants. Furthermore we may identify A as

$$\begin{aligned} A &= C(\beta)\frac{N}{V}\left(\frac{\beta m}{2\pi}\right)^{3/2} = C(\beta)n_0\left(\frac{\beta m}{2\pi}\right)^{3/2} \\ C(\beta) &= V\left[\int d\mathbf{r}\, W_w(\mathbf{r}) \exp(-\beta V(\mathbf{r}))\right]^{-1} \end{aligned} \tag{68}$$

since the number of particles in the container is N. Since the thermostat condition Eq. (40a) must also be satisfied when $dF/dt = 0$, we must also have that $\beta = \beta_w = (k_B T_w)^{-1}$. Consequently

$$f^* = W_w(\mathbf{r})C(\beta_w)n_0(\beta_w m/2\pi)^{3/2} \exp\{-\tfrac{1}{2}\beta_w m\mathbf{v}^2 - \beta_w V(\mathbf{r})\} \tag{69}$$

In the case when the molecules make specular collisions with the wall, the distribution function approaches a Maxwell–Boltzmann distribution function but with a temperature that is not determined by the walls but by the total energy of the gas.

Thus we have shown by means of the H-theorem that if the walls are maintained at a constant wall temperature T_w, and the molecule–wall interaction satisfies the thermostat condition, or if there is no energy flow into or out of

*Notice that since the spatial arguments of f in (61) are all the same, the terms in $\log f$ coming from the conservation of linear momentum and angular momentum are equivalent, i.e., $\mathbf{r}\times\mathbf{v}_1 + \mathbf{r}\times\mathbf{v} = \mathbf{r}\times\mathbf{v}_1' + \mathbf{r}\times\mathbf{v}'$ follows from $\mathbf{v} + \mathbf{v}_1 = \mathbf{v}' + \mathbf{v}_1'$.

the container at the walls, the distribution function for the gas approaches the Maxwell–Boltzmann equilibrium distribution as $t \to \infty$.

It is very satisfactory from a macroscopic point of view that the Boltzmann equation, through the H-theorem, predicts the approach to equilibrium of an initial nonequilibrium state of the gas. However, one can raise serious objections to the H-theorem, and to the Boltzmann equation, from a microscopic point of view. The fundamental difficulty is that the Boltzmann equation is inconsistent with the laws of mechanics. The laws of mechanics require that any equation of motion describing the gas be invariant under time reversal if the particles make specular collisions with the walls. Otherwise any dynamical processes that do not involve collisions with the walls must be time reversal invariant. That is, the form of the equations of motion must be invariant if $\mathbf{v} \to -\mathbf{v}$ and $t \to -t$. It is clear from an inspection of the Boltzmann equation for points far from the walls,

$$\partial f/\partial t + \mathbf{v} \cdot \nabla_{\mathbf{r}} f + (\mathbf{F}/m) \cdot \nabla_{\mathbf{v}} f = J(f, f) \tag{70}$$

that under this transformation, the left-hand side of Eq. (70) changes sign, but the right-hand side does not. The H-theorem is also in conflict with the time reversal invariance of the equations of motion. In addition, when the particles make specular collisions with the walls, the laws of mechanics require that starting from almost any initial state of the gas, eventually the gas will come arbitrarily close to the given initial state.[14,52,54]* The result that the gas reaches an equilibrium state and stays in equilibrium thereafter is thus not in accord with mechanics.

The source of these conflicts between the laws of mechanics and the Boltzmann equation is the *Stosszahlansatz*, for we have used mechanics correctly in describing what happens in *one* binary collision, but we have probabilistic and not mechanical arguments to compute the number of binary collisions taking place in the gas. As a result, the *Stosszahlansatz* can only represent the result of a statistical average for the number of collisions, taken over a large number of systems. Therefore, the Boltzmann equation cannot be thought to give a *complete* description for one particular bottle of gas, but instead it describes the average behavior in time of a large collection of identically prepared bottles of gas. To prove that this is the case, we will need to consider the time evolution of an ensemble of systems and then to show that if one chooses a reasonable ensemble distribution at some initial time, then the single-particle distribution function $f(\mathbf{r}, \mathbf{v}, t)$, which determines the *average* number of particles in $d\mathbf{r}\, d\mathbf{v}$ at time t, satisfies the Boltzmann equation. We will consider this approach to the Boltzmann equation in more detail in Section 3. However, we can see the main features of these considerations by means of a

*The time it takes to return, however, is usually enormously long. For a discussion of this point see Chandrasekhar.[54]

simple example. Consider a special initial state of the gas where the molecules are all confined to a small region of the container, a corner, say. Then prepare an ensemble of systems such that in each member of the ensemble all the molecules are confined to the same corner, but the members may differ in the placement of specific molecules, their velocities, etc. It is reasonably clear that as time progresses most of the ensemble of the system will behave in such a way that molecules move into the other, initially unoccupied regions of the container.* This flow of the gas into the container, which characterizes the behavior of most of the members of the ensemble, should be correctly described by the Boltzmann equation, if the gas is dilute. As a consequence, the Boltzmann equation should describe to a good approximation the time development of most of the members of the ensemble, in particular, the member of the ensemble that we happen to be observing in the laboratory, if we were to perform the experiment.

2.3. Solutions of the Boltzmann Equation

2.3.1. General Remarks

The Boltzmann equation is a nonlinear, integrodifferential equation. As such it is extremely difficult to solve and, in fact, almost no exact solutions are known, apart from the Maxwell–Boltzmann equilibrium solution.† Furthermore, only a few existence theorems are known; notable are the theorems of Carleman,[55] later extended by Wild[56] and by Morgenstern,[57,58] proving the existence of a solution of the nonlinear Boltzmann equation for special intermolecular potentials‡ in the case that the system is spatially uniform, i.e., that the distribution function does not depend on $\mathbf{r}$.§ However, there are a number of circumstances where the system is close enough to equilibrium that the distribution function may be written

$$f(\mathbf{r}, \mathbf{v}, t) = f_{\text{eq}}(\mathbf{r}, \mathbf{v})[1 + \Phi(\mathbf{r}, \mathbf{v}, t)]$$

and where terms proportional to Φ^2 may be neglected, so that the nonlinear Boltzmann equation reduces to a linear equation for Φ. This equation takes the form

$$[\partial/\partial t + \mathbf{v} \cdot \nabla_{\mathbf{r}} + (\mathbf{F}/m) \cdot \nabla_{\mathbf{v}}] f_{\text{eq}} \Phi = f_{\text{eq}} L \Phi + \bar{\mathrm{T}} f_{\text{eq}} \Phi \tag{71}$$

*There may be some members of the ensemble for which their subsequent motion involves a further compression of the gas into the same corners.

†The known existence theorems and solutions are discussed in some detail by Brush.[1]

‡Such as the hard-sphere potential (Carleman)[55] or an r^{-4} potential with a cutoff for scattering at small angles (Wild and Morgenstern).[56–58]

§Except for its $\mathbf{r}$ dependence through the function $W_{\text{w}}(\mathbf{r})$.

where

$$L\Phi = n_0 \int d\mathbf{v}_1 \int d\hat{\mathbf{k}}\, B(\mathbf{g}, \hat{\mathbf{k}})\varphi_0(\mathbf{r}, \mathbf{v}_1)[\Phi_1' + \Phi' - \Phi_1 - \Phi] \tag{72}$$

and

$$\varphi_0(\mathbf{r}, \mathbf{v}_1) = C(\beta)(\beta m/2\pi)^{3/2} \exp[-\beta m\mathbf{v}^2/2 - \beta V(\mathbf{r})]$$

and we take $\beta_w = \beta$ in the equilibrium distribution function. Equation (71) is linear in Φ and, especially in the case that the external force vanishes, there are a number of existence and uniqueness theorems for initial value and boundary value problems for a number of different intermolecular potentials.(6,7) In addition, exact solutions have been constructed for some model linearized Boltzmann equations where the collision operator L appearing in Eq. (71) is replaced by an integral operator with a simple set of eigenfunctions and eigenvalues. The most well known of these models is the BGK model, which we will consider in Section 2.3.4(6,7) In this section we will discuss one case of special physical interest where an approximate solution to the full nonlinear Boltzmann equation has been constructed. This is the so-called "normal solution" of the Boltzmann equation(1,5–13,59) which describes the final stages of the approach of a dilute gas to equilibrium and makes clear the relation between the Boltzmann equation and the macroscopic hydrodynamic equations of fluid mechanics. In the next section we will compare some of the results of the normal solution method with experiment, and in the following section we will give a brief discussion of some other approximate solutions of the Boltzmann equation that are of physical interest.

Before constructing the normal solutions, we find it convenient to rewrite the Boltzmann equation, Eq. (36), as two equations, one of which determines the distribution function for points in the interior of the container and the other can be interpreted as a matching condition for the distribution function at the walls. To obtain these two equations we look for solutions $f(\mathbf{r}, \mathbf{v}, t)$ that vanish outside the container. That is, we write

$$f(\mathbf{r}, \mathbf{v}, t) = W_w(\mathbf{r})\tilde{f}(\mathbf{r}, \mathbf{v}, t) \tag{73}$$

where $W_w(\mathbf{r})$ is defined below Eq. (38). It is equal to unity for $\mathbf{r}$ inside the container and vanishes whenever $\mathbf{r}$ is outside. Next we suppose that the physical properties of the gas do not change discontinuously as we approach the walls from within the container. That is, we assume that $\tilde{f}(\mathbf{r}, \mathbf{v}, t)$ is a continuous function of $\mathbf{r}$ as $\mathbf{r}$ approaches the boundaries from within. Under these circumstances the Boltzmann equation can be put in the form

$$W_w(\mathbf{r})[\partial\tilde{f}/\partial t + \mathbf{v}\cdot\nabla_r\tilde{f} + (\mathbf{F}/m)\cdot\nabla_v\tilde{f} - J(\tilde{f}, \tilde{f})] = \mathbf{T}\tilde{f} \tag{74}$$

where*

$$\mathbf{T}\tilde{f} = \bar{\mathbf{T}}\tilde{f} - \tilde{f}\mathbf{v}\cdot\nabla_r W_w(\mathbf{r})$$

*Here we assume that $\bar{\mathbf{T}}W_w(\mathbf{r})\tilde{f} = \bar{\mathbf{T}}\tilde{f}$, that is, the delta functions appearing in the $\bar{\mathbf{T}}$ operators are evaluated as one approaches the walls from within the container.

or

$$\mathbf{T}\tilde{f} = \int dS\, \Theta(\mathbf{v}\cdot\hat{\mathbf{n}})\, \delta(\mathbf{r}-\boldsymbol{\rho}_s)\left\{\int d\mathbf{v}'\, \Theta(-\mathbf{v}'\cdot\hat{\mathbf{n}})|\mathbf{v}'\cdot\hat{\mathbf{n}}|K(\mathbf{v},\mathbf{v}') - |\mathbf{v}\cdot\hat{\mathbf{n}}|\tilde{f}(\mathbf{r},\mathbf{v},t)\right\} \tag{75a}$$

where we have used

$$\nabla_{\mathbf{r}} W_w(\mathbf{r}) = \int dS\, \hat{\mathbf{n}}\, \delta(\mathbf{r}-\boldsymbol{\rho}_s) \tag{75b}$$

If we examine Eq. (74), we see that the left-hand side cannot contain delta functions evaluated at $\mathbf{r}=\boldsymbol{\rho}_s$ since we have assumed that $\tilde{f}$ is continuous at the walls. However, the right-hand side does contain $\delta(\mathbf{r}-\boldsymbol{\rho}_s)$. Therefore, Eq. (74) can only be satisfied if both the right- and left-hand sides separately vanish. We are thus led to the two equations

$$\partial\tilde{f}/\partial t + \mathbf{v}\cdot\nabla_{\mathbf{r}}\tilde{f} + (\mathbf{F}/m)\cdot\nabla_{\mathbf{v}}\tilde{f} = J(\tilde{f},\tilde{f}) \qquad \text{for } \mathbf{r}\in V \tag{76a}$$

where V denotes points in the interior of the vessel, and

$$\mathbf{T}\tilde{f} = 0$$

or

$$|\mathbf{v}\cdot\hat{\mathbf{n}}|\tilde{f}(\boldsymbol{\rho}_s,\mathbf{v},t) = \int d\mathbf{v}'\, |\mathbf{v}'\cdot\hat{\mathbf{n}}|\Theta(-\mathbf{v}'\cdot\hat{\mathbf{n}})K(\mathbf{v},\mathbf{v}')\tilde{f}(\boldsymbol{\rho}_s,\mathbf{v}',t) \qquad \text{for } \mathbf{v}\cdot\hat{\mathbf{n}}>0 \tag{76b}$$

Here we see that f must satisfy (a) the Boltzmann equation without the gas–wall collision term for points in V, and (b) a condition that relates the distribution function for outgoing molecules at the wall to the distribution function for incoming molecules.*

Our procedure now is to construct the normal solution to Eq. (76a) for points in the interior and then to see if the solution also satisfies the boundary condition Eq. (76b).

2.3.2. *The Normal Solution of the Boltzmann Equation*

The normal solution of the Boltzmann equation, Eq. (77a), was constructed by Enskog and Chapman[1,5] in order to provide a microscopic foundation for the Navier–Stokes equations of fluid mechanics. Although these equations were derived from phenomenological arguments for *continuum* fluids,[60] they accurately describe heat and viscous flows in dilute gases under a variety of circumstances. We will show in this section that the method used in constructing the normal solution leads to a derivation of the Navier–Stokes

*Notice that, Eq. (76b) is identical to Eq. (33). In other discussions of the Boltzmann equation, Eq. (36) is replaced by the two equations (76a) and (76b). Since they are identical if f satisfies the above-mentioned conditions, one can use either Eq. (36) or (76a) and (76b).

equations; it provides expressions for constants appearing in these equations, such as the coefficients of shear viscosity and thermal conductivity, in terms of the intermolecular potential of the molecules; and it provides a means to determine the range of validity of these equations. In addition, we will see that the method also leads to generalizations of the Navier–Stokes equations that are useful in describing flow phenomena with rapid space and time variations, such as sound propagation at high frequencies.

The hydrodynamic equations describe the space and time dependence of the local density, local velocity, and local temperature of the fluid. For a dilute gas we consider these quantities to be determined by appropriate moments of the distribution function $f(\mathbf{r}, \mathbf{v}, t)$. For points in V, the local density $n(\mathbf{r}, t)$ is

$$n(\mathbf{r}, t) = \int d\mathbf{v}\, \tilde{f}(\mathbf{r}, \mathbf{v}, t) \tag{77a}$$

and the local velocity $\mathbf{u}(\mathbf{r}, t)$ is

$$\mathbf{u}(\mathbf{r}, t) = \frac{1}{n(\mathbf{r}, t)} \int d\mathbf{v}\, \mathbf{v}\tilde{f}(\mathbf{r}, \mathbf{v}, t) \tag{77b}$$

The local temperature $T(\mathbf{r}, t)$ is obtained by first defining the local kinetic energy density $\varepsilon_k(\mathbf{r}, t)$ by

$$\varepsilon_k(\mathbf{r}, t) = \frac{1}{n(\mathbf{r}, t)} \int d\mathbf{v}\, \frac{m\mathbf{v}^2}{2} \tilde{f}(\mathbf{r}, \mathbf{v}, t) \tag{78a}$$

Then ε_k can be written as the sum of two terms,

$$n(\mathbf{r}, t)\varepsilon_k(\mathbf{r}, t) = n(\mathbf{r}, t)\frac{m}{2}\mathbf{u}^2(\mathbf{r}, t) + \frac{m}{2}\int d\mathbf{v}\, |\mathbf{v}-\mathbf{u}|^2 \tilde{f}(\mathbf{r}, \mathbf{v}, t) \tag{78b}$$

where the first term on the right represents the mean kinetic energy of the fluid flow at $\mathbf{r}$, t, and the second term represents the mean kinetic energy of the molecules as measured in a coordinate system moving with the mean velocity of the fluid $\mathbf{u}$. This latter energy is used to define the nonequilibrium gas temperature, since by analogy with equilibrium the temperature should be a measure of the average kinetic energy of the gas molecules in the center-of-mass coordinate system. Therefore we define the local temperature $T(\mathbf{r}, t)$ by

$$\frac{3}{2} n(\mathbf{r}, t) k_B T(\mathbf{r}, t) = \frac{m}{2} \int d\mathbf{v}\, \mathbf{V}^2 \tilde{f}(\mathbf{r}, \mathbf{v}, t) \tag{79}$$

with $\mathbf{V} = \mathbf{v} - \mathbf{u}$.

To derive equations of motion for n, $\mathbf{u}$, and T, we multiply the Boltzmann equation, Eq. (76a), by 1, $\mathbf{v}$, and $\frac{1}{2}m\mathbf{v}^2$, respectively, use Eq. (46), and the fact that 1, $\mathbf{v}$, and $\frac{1}{2}m\mathbf{v}^2$ are collision invariants, i.e., 1, $\sum \mathbf{v}_i$, and $\sum \frac{1}{2}m\mathbf{v}_i^2$ are conserved in a binary collision, and we obtain

$$dn/dt + n\nabla_{\mathbf{r}} \cdot \mathbf{u} = 0 \tag{80a}$$

$$nm\,d\mathbf{u}/dt + \nabla_{\mathbf{r}} \cdot \mathsf{P} = n\mathbf{F} \tag{80b}$$

and

$$\frac{3}{2} nk_B \frac{dT}{dt} + \nabla_{\mathbf{r}} \cdot \mathbf{J}_T + \mathsf{P} : \mathsf{D} = 0 \tag{80c}$$

where the "substantial" time derivative given by

$$\frac{d}{dt} = \frac{\partial}{\partial t} + \mathbf{u} \cdot \nabla_{\mathbf{r}} \tag{80d}$$

represents the change in time of a quantity measured in a coordinate frame moving with the mean velocity $\mathbf{u}$; P is the pressure tensor, given by

$$\mathsf{P} = m \int d\mathbf{v}\, \mathbf{V}\mathbf{V} \tilde{f}(\mathbf{r}, \mathbf{v}, t) \tag{81a}$$

$\mathbf{J}_T$ is the heat flow vector, given by

$$\mathbf{J}_T = \frac{m}{2} \int d\mathbf{v}\, V^2 \mathbf{V} \tilde{f}(\mathbf{r}, \mathbf{v}, t) \tag{81b}$$

and D is the rate of strain tensor with components D_{ij} given by

$$D_{ij} = \frac{1}{2}\left(\frac{\partial u_i}{\partial x_j} + \frac{\partial u_j}{\partial x_i}\right) \tag{81c}$$

Equations (80a)–(80c) are called *conservation equations*, since their form is a direct consequence of the conservation of number of particles, momentum, and energy in the binary collisions taking place in the gas. In the phenomenological theories of fluid dynamics, equations in the form of Eqs. (80a)–(80c) are derived from the fact that mass, momentum and energy are conserved in the fluid, but in these theories one does not express the heat flow vector $\mathbf{J}_T$ and the pressure tensor P in terms of a microscopic quantity, like the distribution function $f(\mathbf{r}, \mathbf{v}, t)$. Instead, one relates $\mathbf{J}_T$ and P to n, $\mathbf{u}$, and T by means of the so-called linear laws.[60]* For a one-component fluid with no internal structure, these linear laws are Fourier's law of heat conduction

$$\mathbf{J}_T = -\lambda \nabla_{\mathbf{r}} T \tag{82a}$$

and Newton's law of friction

$$\mathsf{P} = p(\mathbf{r}, t)\mathsf{I} - 2\eta[\mathsf{D} - \tfrac{1}{3}\mathsf{I}(\nabla_{\mathbf{r}} \cdot \mathbf{u})] - \zeta \mathsf{I}(\nabla_{\mathbf{r}} \cdot \mathbf{u}) \tag{82b}$$

Here λ is called the coefficient of thermal conductivity, η the coefficient of shear viscosity, and ζ the coefficient of bulk viscosity; $p(\mathbf{r}, t)$ is the local

*The tensorial forms of the linear laws are determined, to a certain extent, by the symmetry properties of the fluid.[60]

hydrostatic pressure; and $\mathbf{I}$ is the unit tensor. In the phenomenological theory, the system of equations is completed by assuming that the relation between $p(\mathbf{r}, t)$ and the local density and temperature is the same as between p, n, and T in total equilibrium.* For the case of a dilute gas this assumption leads to the relation

$$p(\mathbf{r}, t) = n(\mathbf{r}, t)k_B T(\mathbf{r}, t) \tag{82c}$$

Equations (82a)–(82c) together with the conservation laws (80a)–(80c) then lead to the Navier–Stokes equations.[60] In the phenomenological approach the constants λ, η, and ζ must be determined from experiment.

From these remarks we see that there are three main ingredients in a hydrodynamic description of fluid flow, which we would like to examine with the aid of the Boltzmann equation for a dilute gas. These are (a) *conservation laws* for number (or mass), momentum, and energy, (b) *linear laws* relating fluxes in particles, momentum, and energy to gradients of n, $\mathbf{u}$, and T, and (c) an *equilibrium-like relation*, connecting the hydrostatic pressure $p(\mathbf{r}, t)$ to $n(\mathbf{r}, t)$ and $T(\mathbf{r}, t)$.†

The fact that a local equilibrium assumption is an essential part of the hydrodynamic description of fluid flow suggests that we look for solutions of the Boltzmann equation where the distribution function is close to a local Maxwellian distribution $\tilde{f}_{\text{le}}(\mathbf{r}, \mathbf{v}, t)$ given by

$$\tilde{f}_{\text{le}}(\mathbf{r}, \mathbf{v}, t) = n(\mathbf{r}, t)\left[\frac{\beta(\mathbf{r}, t)m}{2\pi}\right]^{3/2} \exp\left[-\frac{\beta(\mathbf{r}, t)m}{2}\mathbf{V}^2\right] \tag{83}$$

Here $n(\mathbf{r}, t)$, $\mathbf{u}(\mathbf{r}, t)$ and $T(\mathbf{r}, t) = (k_B\beta(\mathbf{r}, t))^{-1}$ are the local density, velocity, and temperature of the gas at the point $\mathbf{r}$ at time t. If $\tilde{f}$ were equal to $\tilde{f}_{\text{le}}$ then the equation for the local hydrostatic pressure, Eq. (82c), would be exact. However, $\tilde{f}_{\text{le}}$ is not a solution of the Boltzmann equation, if n, β and $\mathbf{u}$ depend on $\mathbf{r}$ and t, and it does not describe a state of a gas where either heat flow or viscous flow is taking place.‡ Nevertheless, under the following circumstances $\tilde{f}_{\text{le}}$ would be a good first approximation to the actual distribution function. We imagine that the state of the gas is characterized by two length scales, l_c, and l_h, where l_c is a length on the order of a mean free path between collisions for a typical molecule, and l_h is a length on the order of the characteristic size of the container.§ Associated with l_c and l_h are two time scales $t_c = l_c/\langle v\rangle$ and $t_h = l_h/\langle v\rangle$, where $\langle v\rangle$ is the average velocity of molecules in the gas. Let us now suppose that‖ $l_h \gg l_c$, or equivalently that $t_h \gg t_c$. Then the following picture of the time evolution of the gas seems plausible¶: Consider a time t after some

*This is often called the "hypothesis of local equilibrium state."

†In fact, we have used an "equilibrium-like relation" to define the local temperature.

‡For $\tilde{f} = \tilde{f}_{\text{le}}$, $\mathbf{J}_T = 0$ and $\mathbf{P} = nk_B T\mathbf{1}$.

§Here the subscript c stands for "collision" and h for "hydrodynamic."

‖In Section 2.3.4 we consider the other case, where $l_c \gg l_h$. This is the limit of rarefied gas dynamics.

¶The argument is due to Cohen.[59]

initial state of the gas. If t is in the range $t_h \gg t \gg t_c$, then in any region $\delta\mathbf{r}$ about a point $\mathbf{r}$ in the gas, the molecules will have collided a number of times. By means of these collisions the gas approaches a local equilibrium state, where the distribution function is Maxwellian, but with parameters n, $\mathbf{u}$, and T that may vary from one region of the gas to another and may be time dependent. Then after a time $t \gg t_h$, every particle has traveled distances on the order of the size of the container several times. This allows any local variation in n, $\mathbf{u}$, and T to be smoothed out to constant values over the entire container, and the gas is then in a state of *total* equilibrium.

Thus we distinguish between two processes: (a) a fast process on the time scale of t_c, where after a time the system is, to a good approximation, in a local equilibrium state; and (b) a slow process on the time scale t_h, where hydrodynamics gives a good description of the time evolution of the gas and, after $t \gg t_h$, the system has reached total equilibrium.

To incorporate these ideas into a method of solving the Boltzmann equation, we assume that we are interested in a time $t \gg t_c$ after some initial state, and that $\tilde{f}(\mathbf{r}, \mathbf{v}, t)$ has the form

$$\tilde{f}(\mathbf{r}, \mathbf{v}, t) = \tilde{f}_{le}(\mathbf{r}, \mathbf{v}, t)[1 + \Phi(\mathbf{r}, \mathbf{v}, t)] \tag{84}$$

where $\tilde{f}_{le}$ is the local equilibrium distribution function given by Eq. (83) and Φ represents the deviation from local equilibrium. Substituting Eq. (84) into the Boltzmann equation (76a), we obtain

$$\begin{aligned}[\partial/\partial t + \mathbf{v}\cdot\nabla_{\mathbf{r}} + (\mathbf{F}/m)\cdot\nabla_{\mathbf{v}}]\tilde{f}_{le}(1+\Phi) &= J(\tilde{f}_{le}(1+\Phi), \tilde{f}_{le}(1+\Phi)) \\ &= \tilde{f}_{le}L(\mathbf{V})\Phi(\mathbf{r}, \mathbf{v}, t) + J(\tilde{f}_{le}\Phi, \tilde{f}_{le}\Phi)\end{aligned} \tag{85}$$

where*

$$L(\mathbf{V})\Phi = \int d\mathbf{v}_1 \int d\hat{\mathbf{k}}\, B(\mathbf{g}, \hat{\mathbf{k}})\tilde{f}_{le}(\mathbf{r}, \mathbf{v}_1, t)(\Phi_1' + \Phi' - \Phi_1 - \Phi) \tag{86a}$$

and we have set

$$J(\tilde{f}_{le}, \tilde{f}_{le}) = 0 \tag{86b}$$

We observe that the left-hand side of this equation is, roughly speaking, characterized by the time scale t_h and length scale l_h, over which the quantities n, $\mathbf{u}$, and T change with time and position and the external potential changes with $\mathbf{r}$. The collision operator appearing on the right-hand side of Eq. (85) is characterized, on the other hand, by the collision scales t_c and l_c. Thus, the left-hand side is of order t_h^{-1}, while the right-hand side is of order $t_c^{-1}\Phi$. A solution of this equation might be found if we make the further assumption that

*Notice that when $\mathbf{u} = 0$, β is constant and $n(\mathbf{r}) = n_0 C(\beta)\exp[-\beta V(\mathbf{r})]$, $L(\mathbf{V})$ reduces to the linearized collision operator L defined by Eq. (72).

Φ is of order $\mu = t_c/t_h$.* In fact, we will assume that Φ can be expanded in a series in μ as

$$\Phi(\mathbf{r}, \mathbf{v}, t) = \mu\Phi_1(\mathbf{r}, \mathbf{v}, t) + \mu^2\Phi_2(\mathbf{r}, \mathbf{v}, t) + \cdots \tag{87}$$

and this in turn leads to a μ-expansion of the right-hand side of Eq. (85). A corresponding μ-expansion of the left-hand side of Eq. (85) is obtained by writing

$$\begin{aligned}
&[\partial/\partial t + \mathbf{v}\cdot\nabla_{\mathbf{r}} + (\mathbf{F}/m)\cdot\nabla_{\mathbf{v}}]\tilde{f}_{\text{le}}(1+\Phi) \\
&= \left(\frac{\partial \tilde{f}_{\text{le}}}{\partial n}\frac{\partial n}{\partial t} + \frac{\partial \tilde{f}_{\text{le}}}{\partial \mathbf{u}}\cdot\frac{\partial \mathbf{u}}{\partial t} + \frac{\partial \tilde{f}_{\text{le}}}{\partial T}\frac{\partial T}{\partial t}\right) + [\mathbf{v}\cdot\nabla_{\mathbf{r}} + (\mathbf{F}/m)\cdot\nabla_{\mathbf{v}}]\tilde{f}_{\text{le}} \\
&+ \sum_{j=1}^{\infty} \mu^j\left[\frac{\partial}{\partial t} + \mathbf{v}\cdot\nabla_{\mathbf{r}} + (\mathbf{F}/m)\cdot\nabla_{\mathbf{v}}\right]\hat{f}_{\text{le}}\Phi_j(\mathbf{r}, \mathbf{v}, t)
\end{aligned} \tag{88}$$

The quantities $\partial n/\partial t$, $\partial \mathbf{u}/\partial t$, and $\partial T/\partial t$ each have a μ-expansion that is obtained by substituting Eqs. (84) and (87) into the expressions for $\mathbf{P}$ and $\mathbf{J}_T$, Eqs. (81a) and (81b), and then substituting the resulting μ-expansions for $\mathbf{P}$ and $\mathbf{J}_T$ into the conservation equations (80b) and (80c). Next we assume that since the external force is equal to $-\nabla V(\mathbf{r})$, where $V(\mathbf{r})$ is the external potential, then $\mathbf{F}$, when expressed in terms of dimensionless units, is of order μ. We make the further assumptions that (a) the spatial gradient of any quantity of order μ^n is, in turn, of order μ^{n+1}, and that (b) the time derivative of any quantity of order μ^n is at least of order μ^{n+1}. These two assumptions are motivated by the fact that we are looking for solutions of the Boltzmann equation where the distribution function changes only over distances of order l_h and over times on the order of t_h, so that spatial gradients or time derivatives of the distribution function should be, respectively, in dimensionless units, of order l_c/l_h or t_c/t_h smaller than the distribution function itself.†

As a consequence of these assumptions, we obtain a μ-expansion for both the right- and left-hand sides of Eq. (85).‡ The next step in the construction of the normal solution of Eq. (76a) consists of equating the coefficients of equal powers of μ on the right- and left-hand side of Eq. (85).§ We will write the

*From here on we use the symbol μ in two ways. In discussions, the symbol μ always represents a small quantity of order l_c/l_h. In equations, μ is used to label the relative orders of magnitude of the various terms. Once the relative orders of magnitude of the terms have been established, we set $\mu = 1$ in the equation since the relative orders of magnitude will then be clear.

†Here, by assumption (a) we have chosen the actual expansion parameter to be the (dimensionless) spatial gradient. We will then see that time derivatives of a quantity of order μ^n lead to terms of order μ^{n+2}, μ^{n+3}, ... as well as terms of order μ^{n+1}. This accounts for the "at least" in assumption (b).

‡However, we will see shortly that the coefficients of each power of μ depend on n, $\mathbf{u}$, and T, and these quantities in turn depend on μ.

§In view of the fact that the coefficients of each power of μ also depend on μ through n, $\mathbf{u}$, and T, the equating of equal powers of μ is not a mathematical necessity. This point will be discussed later in this section.

order μ^0 and μ^1 equations:

$$O(\mu^0):\quad 0 = J(\tilde{f}_{\text{le}}, \tilde{f}_{\text{le}}) \tag{89}$$

$$O(\mu^1):\quad \mu\frac{\partial \tilde{f}_{\text{le}}}{\partial n}\left(\frac{\partial n}{\partial t}\right)_1 + \mu\frac{\partial \tilde{f}_{\text{le}}}{\partial \mathbf{u}}\cdot\left(\frac{\partial \mathbf{u}}{\partial t}\right)_1 + \mu\frac{\partial \tilde{f}_{\text{le}}}{\partial T}\left(\frac{\partial T}{\partial t}\right)_1 + \mu[\mathbf{v}\cdot\nabla_{\mathbf{r}} + (\mathbf{F}/m)\cdot\nabla_{\mathbf{v}}]\tilde{f}_{\text{le}} = \mu\tilde{f}_{\text{le}}L(\mathbf{V})\Phi_1 \tag{90}$$

Here $(\partial n/\partial t)_1$, $(\partial\mathbf{u}/\partial t)_1$, and $(\partial T/\partial t)_1$ are the coefficients of the first power of μ appearing in the μ-expansion of the conservation laws. These μ-expansions are obtained by using Eqs. (80), (81), (84), and (87) and are given by

$$\partial n/\partial t = \mu(\partial n/\partial t)_1 + \mu^2(\partial n/\partial t)_2 + \cdots \tag{91a}$$

with

$$\mu(\partial n/\partial t)_1 = -\nabla_{\mathbf{r}}\cdot(n\mathbf{u}) \tag{91b}$$

and

$$\mu^j(\partial n/\partial t)_j = 0, \qquad j = 2, 3, \ldots \tag{91c}$$

by

$$\partial\mathbf{u}/\partial t = \mu(\partial\mathbf{u}/\partial t)_1 + \mu^2(\partial\mathbf{u}/\partial t)_2 + \cdots \tag{92a}$$

with

$$\mu\left(\frac{\partial\mathbf{u}}{\partial t}\right)_1 = \frac{1}{nm}[-(\mathbf{u}\cdot\nabla_{\mathbf{r}})\mathbf{u} + n\mathbf{F} + \nabla_{\mathbf{r}}(nk_{\text{B}}T)] \tag{92b}$$

and

$$\mu^j\left(\frac{\partial\mathbf{u}}{\partial t}\right)_j = \mu^{j-1}\frac{1}{nm}\nabla\cdot\mathsf{P}_{j-1} \tag{92c}$$

and by

$$\partial T/\partial t = \mu(\partial T/\partial t)_1 + \mu^2(\partial T/\partial t)_2 + \cdots \tag{93a}$$

with

$$\mu(\partial T/\partial t)_1 = -\frac{2}{3}T\nabla_{\mathbf{r}}\cdot\mathbf{u} - \mathbf{u}\cdot\nabla_{\mathbf{r}}T \tag{93b}$$

and

$$\mu^j\left(\frac{\partial T}{\partial t}\right)_j = -\frac{2}{3nk_{\text{B}}}\mu^{j-1}[\nabla_{\mathbf{r}}\cdot\mathbf{J}_{T,j-1} + \mathsf{P}_{j-1} : \mathsf{D}] \tag{93c}$$

respectively. Here

$$\mu^j\mathsf{P}_j = m\mu^j\int d\mathbf{v}\,\mathbf{V}\mathbf{V}\tilde{f}_{\text{le}}\Phi_j \tag{94a}$$

and

$$\mu^j \mathbf{J}_{T,j} = \mu^j \int d\mathbf{v} \frac{m\mathbf{V}^2}{2} \mathbf{V} \tilde{f}_{le} \Phi_j \tag{94b}$$

Using Eqs. (91b), (92b), and (93b), and the explicit form of $\tilde{f}_{le}$, we write Eq. (90) as*

$$\left[\left(\frac{m\mathbf{V}^2}{2k_B T} - \frac{5}{2}\right)\mathbf{V} \cdot \nabla \log T + \beta m\left(\mathbf{V}\mathbf{V} - \frac{\mathbf{V}^2}{3}\mathbf{I}\right) : \mathbf{D}\right] = L(\mathbf{V})\Phi_1 \tag{95}$$

and this equation must be solved in order to obtain Φ_1. Once we have obtained an expression for Φ_1, we can proceed to obtain explicit forms for $(\partial n/\partial t)_2$, $(\partial \mathbf{u}/\partial t)_2$, and $(\partial T/\partial t)_2$, as well as the order μ^2 equation that follows from equating the coefficients of order μ^2 on the right and left sides of Eq. (85). To solve Eq. (95) we need to use the following properties of the linearized Boltzmann collision operator $L(\mathbf{V})$[7,52,61–64]:

1. Consider the eigenvalue equation

$$L(\mathbf{V})\psi_\lambda(\mathbf{V}) = \lambda \psi_\lambda(\mathbf{V}) \tag{96}$$

Then (i) there are five eigenfunctions corresponding to $\lambda = 0$; these are 1, $\mathbf{V}$, $\frac{1}{2}m\mathbf{V}^2$, as follows from the definition of L given by (72) and the conservation of number of particles, momentum, and energy in a binary collision; and (ii) all other eigenvalues of L are strictly negative. This follows from the fact that one can show, using Eq. (46), that

$$\int d\mathbf{v}\, \psi^*(\mathbf{V}) \tilde{f}_{le} L(\mathbf{V}) \psi(\mathbf{V}) < 0 \tag{97}$$

unless $\psi(\mathbf{V}) = 1, \mathbf{V}, \frac{1}{2}m\mathbf{V}^2$, or a linear combination of these. Hence if $\psi_\lambda(\mathbf{V})$ is an eigenfunction orthogonal to 1, $\mathbf{V}$, and $\mathbf{V}^2$, then†

$$\int d\mathbf{v}\, \psi_\lambda^*(\mathbf{V}) \tilde{f}_{le} L \psi_\lambda(\mathbf{V}) = \lambda \int d\mathbf{v}\, |\psi_\lambda|^2 \tilde{f}_{le} < 0 \tag{98a}$$

Hence

$$\lambda < 0 \qquad \text{if } \lambda \neq 0 \tag{98b}$$

The spectrum of $L(\mathbf{V})$ depends on the intermolecular potential of the gas molecules. For hard-sphere molecules the spectrum of L has a discrete and a continuous part, with the continuum ranging from $-m$ to $-\infty$ with m a positive constant.[65–68] If the potential is repulsive and varies with r as $\phi(r) = Kr^{-s}$ with $s > 2$, then the spectrum is discrete.[69] For the special case $s = 4$, the so-called Maxwell molecules, the eigenvalues and eigenfunctions of L are completely known. However, nothing is known about the spectrum of L

*From here on we set $\mu = 1$ in the equations. See footnote on p. 99.

†The inner product between two functions ψ and ϕ is $(\psi, \phi) = \int d\mathbf{v}\, \psi^*(\mathbf{v}) \tilde{f}_{le}(\mathbf{v}) \phi(\mathbf{v})$.

for other force laws. For a fictitious intermolecular potential that varies with r as Kr^{-s} and where the scattering cross section is set equal to zero for near-grazing collisions, i.e., those collisions where $|\mathbf{g}\cdot\hat{\mathbf{k}}|$ is near zero, Grad[65] has proved that the spectrum of L has a continuum and discrete part with the continuum separated from zero for $s>4$, and extending to zero for $s<4$.* It is essential for most applications of the Boltzmann equation that there be a gap between zero and the first nonzero eigenvalue of L. For hard spheres, and for potentials that vary as Kr^{-s}, with and without angular cutoff for $s\geqslant 4$, and without cutoff for $2<s<4$, such a gap always exists.[65–69] The inverse of the width of the gap then provides a convenient measure for the mean free time between collisions t_c. In our further discussions we will assume such a gap exists.

From part (i) above it follows that the homogeneous equation

$$L(\mathbf{V})\Phi=0 \tag{99a}$$

has the solution

$$\Phi=a+\mathbf{b}\cdot\mathbf{V}+c\mathbf{V}^2 \tag{99b}$$

where a, $\mathbf{b}$, and c do not depend on $\mathbf{V}$.

2. L is an isotropic operator.[61–64] That is, if $Y_l^m(\hat{\mathbf{V}})$ is a spherical harmonic depending on the unit vector in the direction of $\mathbf{V}$, $\hat{\mathbf{V}}=(\mathbf{v}-\mathbf{u})/|\mathbf{v}-\mathbf{u}|$, and if $g(\mathbf{V}^2)$ is some function depending on $\mathbf{V}^2$ only, then the action of $L(\mathbf{V})$ on the product of these functions is of the form

$$L(\mathbf{V})g(\mathbf{V}^2)Y_l^m(\hat{\mathbf{V}})=h(\mathbf{V}^2)Y_l^m(\hat{\mathbf{V}}) \tag{100}$$

where $h(\mathbf{V}^2)$ is some other function depending only on $\mathbf{V}^2$.

Returning to Eq. (95), we see that a necessary condition for it to have a solution is that the inhomogeneous term on the left-hand side be orthogonal to 1, $\mathbf{V}$, and $\mathbf{V}^2$. It is easy to verify that this is so.† We then see that the term on the left-hand side of (95) involving $\nabla\log T$ is proportional to $Y_1^m(\hat{\mathbf{V}})$, and the term involving D is proportional to $Y_2^m(\hat{\mathbf{V}})$. Using the isotropy of $L(\mathbf{V})$, we can immediately write the solution Φ_1 in the form

$$\begin{aligned}\Phi_1=&A(\mathbf{V}^2)\mathbf{V}\cdot\nabla\log T+B(\mathbf{V}^2)\beta m(\mathbf{V}\mathbf{V}-\tfrac{1}{3}\mathbf{V}^2\mathsf{I}):\mathsf{D}\\&+c_1+\mathbf{c}_2\cdot\mathbf{V}+c_3\mathbf{V}^2\end{aligned} \tag{101}$$

where $A(\mathbf{V}^2)$ and $B(\mathbf{V}^2)$ are determined by constructing the particular solutions to the equation

$$L(\mathbf{V})A(\mathbf{V}^2)\mathbf{V}=\left(\frac{m}{2kT}\mathbf{V}^2-\frac{5}{2}\right)\mathbf{V} \tag{102a}$$

*The case where $s>4$ and there is an angular cutoff is called a "hard potential in the sense of Grad." When $s=4$, the spectrum is discrete, even with an angular cutoff.

†In fact, we see from Eq. (90) that this solubility condition leads to the conservation equations to order μ^1, Eqs. (91b), (92b), and (93b).

and

$$L(\mathbf{V})B(\mathbf{V}^2)(\mathbf{VV}-\tfrac{1}{3}\mathbf{V}^2\mathbf{I}) = (\mathbf{VV}-\tfrac{1}{3}\mathbf{V}^2\mathbf{I}) \tag{102b}$$

The constants c_1, $\mathbf{c}_2$, and c_3 in Eq. (101) are not yet determined. We fix their values and consequently arrive at a unique solution for Φ_1 by requiring that the functions n, $\mathbf{u}$, and T appearing in the local equilibrium distribution function be the actual values of the local density, mean velocity, and temperature of the gas.* That is, we require that

$$n(\mathbf{r}, t) = \int d\mathbf{v}\, \tilde{f}(\mathbf{r}, \mathbf{v}, t) = \int d\mathbf{v}\, \tilde{f}_{\text{le}}(\mathbf{r}, \mathbf{v}, t) \tag{103a}$$

$$n(\mathbf{r}, t)\mathbf{u}(\mathbf{r}, t) = \int d\mathbf{v}\, \mathbf{v}\tilde{f}(\mathbf{r}, \mathbf{v}, t) = \int d\mathbf{v}\mathbf{v}\tilde{f}_{\text{le}}(\mathbf{r}, \mathbf{v}, t) \tag{103b}$$

$$\frac{3}{2}nk_{\text{B}}T(\mathbf{r}, t) = \int d\mathbf{v}\frac{m}{2}\mathbf{V}^2\tilde{f}(\mathbf{r}, \mathbf{v}, t) = \int d\mathbf{v}\frac{m}{2}\mathbf{V}^2\tilde{f}_{\text{le}}(\mathbf{r}, \mathbf{v}, t) \tag{103c}$$

From this, it follows from Eqs. (84) and (87), that

$$\int d\mathbf{v}\, \tilde{f}_{\text{le}}\Phi_i \begin{Bmatrix} 1 \\ \mathbf{V} \\ \mathbf{V}^2 \end{Bmatrix} = 0 \qquad \text{for } i = 1, 2, \ldots \tag{104}$$

Therefore we can set c_1, c_2, and $c_3 = 0$ in (101) and require that

$$\int d\mathbf{v}\, \mathbf{V}^2 A(\mathbf{V}^2)\tilde{f}_{\text{le}} = 0 \tag{105}$$

Explicit expressions for the quantities $A(\mathbf{V}^2)$ and $B(\mathbf{V}^2)$ appearing in Eq. (101) for Φ_1 can be found by expanding $A(\mathbf{V}^2)$ and $B(\mathbf{V}^2)$ in terms of a complete set of orthonormal polynomials. A set of functions that is particularly convenient for this purpose is the Sonine polynomials, $S_m^{(n)}(x)$, defined for a given argument x by[5,6]†

$$S_m^{(n)}(x) = \sum_{p=0}^{n} (-x)^p \frac{(m+n)!}{(m+p)!(n-p)!} \tag{106}$$

*This is not as innocent as it first might appear. If one can expand $f(\mathbf{r}, \mathbf{v}, t)$ in a convergent power series in μ, one can show that this requirement can be dispensed with and the *same* solution for f results. This is a consequence of the so-called Hilbert uniqueness theorem.[6,9–11] However, in the case that f cannot be so expanded, this requirement is sufficient for obtaining one solution, assuming the series of approximations converges.

†Here $m! \equiv \Gamma(m+1)$ if m is not an integer. These polynomials satisfy the orthonormality relation

$$\int_0^\infty dx e^{-x} x^m S_m^{(p)}(x) S_m^{(r)}(x) = \frac{\Gamma(m+p+1)}{p!}\delta_{p,r}$$

These polynomials have been used because the quantities

$$V^l S^{(r)}_{l+1/2}(\tfrac{1}{2}\beta m \mathbf{V}^2)\, Y_l^m(\hat{\mathbf{V}})$$

are eigenfunctions of the linearized collision operator[5-13,52] $L(\mathbf{V})$ for Maxwell molecules where the intermolecular potential $\phi(r) = Kr^{-4}$. For a general potential one expands $A(\mathbf{V}^2)\mathbf{V}$ as

$$A(\mathbf{V}^2)\mathbf{V} = \sum_{r=0}^{\infty} a_r S^{(r)}_{3/2}(\tfrac{1}{2}\beta m \mathbf{V}^2)\mathbf{V} \tag{107a}$$

and $B(\mathbf{V}^2)(\mathbf{VV} - \frac{1}{3}\mathbf{V}^2\mathbf{I})$ as

$$B(\mathbf{V}^2)(\mathbf{VV} - \tfrac{1}{3}\mathbf{V}^2\mathbf{I}) = \sum_{r=0}^{\infty} b_r S^{(r)}_{5/2}(\tfrac{1}{2}\beta m \mathbf{V}^2)(\mathbf{VV} - \tfrac{1}{3}\mathbf{V}^2\mathbf{I}) \tag{107b}$$

Then, after inserting these expressions in the integral equations (102a) and (102b) we obtain an infinite set of equations for determining the coefficients a_r and b_r. In the case of Maxwell molecules only a_1 and b_0 do not vanish, since the right-hand sides of (102a) and (102b) are proportional to $S^{(1)}_{3/2}(\frac{1}{2}\beta m \mathbf{V}^2)\mathbf{V}$ and $S^{(0)}_{5/2}(\frac{1}{2}\beta m \mathbf{V}^2)(\mathbf{VV} - \frac{1}{3}\mathbf{V}^2\mathbf{I})$, respectively, which in turn are eigenfunctions of $L(\mathbf{V})$. For general potentials only the first few terms in Eqs. (107a) and (107b) need to be taken into account since the series converge rapidly.[5,8]

Having determined $A(\mathbf{V}^2)$ and $B(\mathbf{V}^2)$, we thus know Φ_1, and can compute $\mathbf{J}_{T,1}$ and $\mathbf{P}_1$, the $O(\mu)$ terms in the expansion of the heat flow vector and pressure tensor, given by (94b) and (94a), respectively. By this procedure, we obtain Fourier's law given in Eq. (82a),

$$\mathbf{J}_{T,1} = -\lambda \nabla T \tag{108}$$

with an explicit expression for the thermal conductivity λ,

$$\begin{aligned} \lambda &= -\frac{k_B}{3}\int d\mathbf{v}\left(\frac{\beta m}{2}\mathbf{V}^2 - \frac{5}{2}\right)\mathbf{V}^2 A(\mathbf{V}^2)\tilde{f}_{le} \\ &= \frac{5}{2}\frac{k_B^2}{m} T n a_1 \end{aligned} \tag{109}$$

where we have used Eqs. (105) and (107a). Similarly, we obtain Newton's law of friction given in Eq. (82b),

$$\mathbf{P}_1 = -2\eta[\mathbf{D} - \tfrac{1}{3}\mathbf{I}(\nabla \cdot \mathbf{u})] - \zeta \mathbf{I}(\nabla \cdot \mathbf{u}) \tag{110}$$

with an explicit expression for the coefficient of shear viscosity

$$\eta = -\frac{m^2}{15 k_B T}\int d\mathbf{v}\, \mathbf{V}^4 B(\mathbf{V}^2)\tilde{f}_{le} = -n k_B T b_0 \tag{111}$$

where we have used Eq. (107b). For the coefficient of bulk viscosity, we find

$$\zeta = 0 \tag{112}$$

Notice that all of the a_r and b_r appearing in Eqs. (107a) and (107b) are inversely proportional to the density n. Hence λ and η are independent of the density.

Expressions for a_1 and b_0 in Eqs. (109) and (111) that are good to within a few percent can be obtained by regarding the first nonvanishing terms in Eqs. (107a) and (107b) to be approximate eigenfunctions of the operator $L(\mathbf{V})$. Then one obtains expressions for η and λ in terms of the intermolecular potential as[5,8]

$$\eta = \tfrac{5}{8} k_B T / \Omega^{(2,2)}(T) \tag{113a}$$

$$\lambda = \frac{5}{2}\eta c_v = \frac{75}{32m}\frac{k_B^2 T}{\Omega^{(2,2)}(T)} \tag{113b}$$

where $c_v = 3k_B/2m$ is the specific heat per unit mass of a monatomic gas, and $\Omega^{(2,2)}(T)$ is one of a set of integrals that depend on the intermolecular potential through the scattering function $B(\mathbf{g}, \hat{\mathbf{k}})$ and are given in detail by Chapman and Cowling.[5] For hard-sphere molecules of diameter a we can easily obtain explicit expressions[5]

$$\eta = \frac{5}{16}\frac{(\pi m k_B T)^{1/2}}{\pi a^2} \tag{114}$$

$$\lambda = \frac{75}{64}\frac{(\pi m k_B T)^{1/2} k_B}{m \pi a^2} \tag{115}$$

If we now keep only terms of order μ and μ^2 on the right-hand sides of Eqs. (91a), (92a), and (93a), we obtain the Navier–Stokes equations,*

$$\frac{dn}{dt} + n\nabla\cdot\mathbf{u} = 0 \tag{116}$$

$$mn\frac{d\mathbf{u}}{dt} = n\mathbf{F} - \nabla(nk_B T) + \nabla\cdot\left[2\eta\left(\mathbf{D} - \frac{1}{3}\mathbf{I}(\nabla\cdot\mathbf{u})\right)\right] \tag{117}$$

$$\frac{3}{2}nk_B\frac{dT}{dt} + nk_B T(\nabla\cdot\mathbf{u}) = \nabla\cdot\lambda\nabla T + 2\eta\left[\mathbf{D}:\mathbf{D} - \frac{1}{3}(\nabla\cdot\mathbf{u})^2\right] \tag{118}$$

It is clear that the general procedure used to derive the Navier–Stokes equations can be used to obtain the corrections to the hydrodynamic equations to higher order in the uniformity parameter. The order μ^3 equations—the Burnett equations—have been worked out in detail,[5,61–64,70,71] but as we shall discuss below the applications of these higher-order equations are limited.

*The equations obtained by keeping terms up to order μ are called the Euler ideal-fluid equations. They can be obtained from Eqs. (116)–(118) by setting η and λ equal to zero.

Before considering if the normal solution for $\tilde{f}$ just obtained satisfies the boundary condition Eq. (76b), we will briefly summarize here the principal assumptions and the main results obtained in this section.

1. The system evolves on two time scales, a collision time t_c and a hydrodynamic time t_h with corresponding length scales l_c and l_h, respectively.
2. The time and length scales are such that $t_h \gg t_c$ and $l_h \gg l_c$.
3. For $t_h \gg t \gg t_c$, the system is close to a local equilibrium state, and the distribution $\tilde{f}(\mathbf{r}, \mathbf{v}, t)$ may be expanded in a power series in the uniformity parameter $\mu = l_c/l_h$ as

$$\tilde{f}(\mathbf{r}, \mathbf{v}, t) = \tilde{f}_{le}(1 + \mu\Phi_1 + \mu^2\Phi_2 + \cdots)$$

where $\tilde{f}_{le}(\mathbf{r}, \mathbf{v}, t)$ is the local equilibrium distribution function.
4. The spatial gradient of $\mu^j\Phi_j$ is of order μ^{j+1} and the time derivative of $\mu^j\Phi_j$ is of order μ^{j+1} and higher.

With the aid of these assumptions, we were able to decompose both sides of Eq. (85) into formal power series in μ, and we then equated coefficients of equal powers of μ. Since these coefficients depend on n, $\mathbf{u}$, and T, which may also depend on μ, the equations obtained this way contain in general more than one order of μ. Therefore, the equating of equal powers of μ in Eq. (85) is not done in order to collect the coefficients of a single power of μ in one equation consistently, but instead to generate at every stage an equation for each Φ_j that is soluble, without having to assume that all quantities are analytic functions of μ. This is accomplished by arranging terms in such a way that the solubility conditions for the equation determining Φ_j, say, are identical with the conservation laws, written to order μ^j. Consequently we have no *a priori* guarantee that the resulting hydrodynamic equations give a good description of the fluid unless it can be proved that the μ-expansion of the hydrodynamic equations is convergent. Such a proof was given by McLennan(72) for the linearized hydrodynamic equations that result when the normal solution method is applied to the linearized Boltzmann collision operator, in the special case that the system consists of hard-sphere molecules in an unbounded region. In the general case such a proof is not available. Moreover, Grad(10,73–76) has argued that the μ-expansion is asymptotic in μ* and it seems likely that, in general, the μ-expansion does not converge. Thus in such a case the higher-order hydrodynamic equations would only improve the Navier–Stokes equation when the latter already provide a good description of the flow.

Having obtained and solved each equation for Φ_j, we found that we could always add an arbitrary solution of the homogeneous linearized Boltzmann equation to the particular solution of the inhomogeneous equation. By requir-

*For points sufficiently far from the boundaries, and for sufficiently long times.

ing that n, $\mathbf{u}$, and T appearing in the local equilibrium distribution function be the actual local values of the density, mean velocity, and temperature, respectively, we were able to discard the solution of homogeneous equations and keep only the particular solution. This requirement provides us with a definite solution to the Boltzmann equation,* which corresponds to the physical circumstances where a hydrodynamic description of the fluid should apply.

We are therefore led to an expression for $\tilde{f}(\mathbf{r}, \mathbf{v}, t)$ of the form

$$\tilde{f}(\mathbf{r}, \mathbf{v}, t) = \tilde{f}_{\mathrm{le}}(\mathbf{r}, \mathbf{v}, t)[1 + A(\mathbf{V}^2)\mathbf{V} \cdot \nabla \log T + \beta m B(\mathbf{V}^2) \times (\mathbf{V}\mathbf{V} - \tfrac{1}{3}\mathbf{V}^2\mathbf{I}) : \mathbf{D} + O(\mu^2)] \tag{119}$$

This is the Chapman–Enskog normal solution of the Boltzmann equation. When the solution is inserted into the expressions for $\mathbf{P}$ and $\mathbf{J}_T$ in the conservation laws, it leads to the Navier–Stokes hydrodynamic equations, which involve the first and second spatial derivatives of the functions n, $\mathbf{u}$, and T. If we use the order $\mu^2, \mu^3, \ldots$ terms in Eq. (119), we are led to the Burnett, super-Burnett, . . . hydrodynamic equations, all with transport coefficients that are determined by the intermolecular potential. These higher-order hydrodynamic equations involve the third, fourth, . . . spatial derivatives of the variables n, $\mathbf{u}$, and T.

Notice that $\tilde{f}(\mathbf{r}, \mathbf{v}, t)$ in Eq. (119) depends on $\mathbf{r}$ and t only through the $\mathbf{r}$- and t-dependence of n, $\mathbf{u}$, and T. That is, as far as the space and time dependence is concerned, $\tilde{f}(\mathbf{r}, \mathbf{v}, t)$ is a *functional* of n, $\mathbf{u}$, and T. This result is a direct consequence of the above set of assumptions, especially assumption 4.† This functional dependence of $\tilde{f}$ on n, $\mathbf{u}$, and T has a very clear physical origin, which is connected to the existence of the time scales t_c and t_h mentioned earlier in this section.[59] Consider the evolution of the gas from some initial state; then for $t > t_c$ the particles will have collided several times. Each collision changes the velocities of the participating particles and therefore almost all functions of the velocities of the particles in the vicinity of a point $\mathbf{r}$, say, will change rapidly with time as the collisions take place in the gas. However, the functions $n(\mathbf{r}, t)$, $\mathbf{u}(\mathbf{r}, t)$, and $T(\mathbf{r}, t)$ will not change in time as a result of the collisions taking place near $\mathbf{r}$, since n, $\mathbf{u}$, and T represent the average values of the number of particles, velocity, and energy, and these quantities do not change through collisions. Consequently, we might expect that all rapidly changing quantities will adjust to the prevailing values of n, $\mathbf{u}$, and T at the point $\mathbf{r}$ at time t, and will become dependent on n, $\mathbf{u}$, and T after a time $t > t_c$. This process is therefore the cause of the functional dependence of $\tilde{f}$ on n, $\mathbf{u}$, and T exhibited in Eq. (119).

*Instead of a whole class of solutions.

†In other presentations,[59] assumption 4 is replaced by the assumption that $\tilde{f}$ is a functional of n, $\mathbf{u}$, and T.

To complete our discussion of the normal solution of the Boltzmann equation, we must check to see if $\tilde{f}(\mathbf{r}, \mathbf{v}, t)$ given by Eq. (119) satisfies the boundary condition, Eq. (76b). To do this we must give specific forms for the quantity $K(\mathbf{v}, \mathbf{v}')$ appearing on the right-hand side of Eq. (76b). Although there are a number of possible forms for the function that describe the molecule–wall interaction, we consider here two typical examples.[6,7]

(a) Specular reflections, where the molecules are reflected elastically and specularly from the walls. In this case $K = K_s$, where

$$K_s(\mathbf{v}, \mathbf{v}') = \delta(\mathbf{v} - \mathbf{v}' + 2(\mathbf{v}' \cdot \hat{\mathbf{n}})\hat{\mathbf{n}}) \tag{120}$$

(b) Diffuse reflection, where a molecule is absorbed at the surface and is instantly reemitted with a velocity determined by a Maxwell–Boltzmann distribution with a wall temperature $T_w(\boldsymbol{\rho}_s) = (k_B\beta_w)^{-1}$. Here $K(\mathbf{v}, \mathbf{v}') = K_D$ is given by

$$K_D(\mathbf{v}, \mathbf{v}') = |\mathbf{v} \cdot \hat{\mathbf{n}}||\mathbf{v}' \cdot \hat{\mathbf{n}}|(2\pi m\beta_w)^{1/2}\phi_w(\mathbf{v}, \boldsymbol{\rho}_s) \tag{121}$$

with $\phi_w(\mathbf{v}, \boldsymbol{\rho}_s)$ given by Eq. (40b). In addition, one can consider linear combinations of K_D and K_s of the form

$$K = (1 - \alpha)K_s + \alpha K_D \tag{122}$$

The coefficient α is called the "accommodation coefficient." This condition is called the Maxwell boundary condition.[6,7]

If we now insert $\tilde{f}$ given by Eq. (119) into Eq. (76b), use one of the above three possible forms for $K(\mathbf{v}, \mathbf{v}')$, and compare like powers of μ on the right- and left-hand sides of Eq. (76b), we find that[7,9,44,45,73–81]

(a) For specular reflections the order μ^0 equation leads to

$$\mathbf{u}(\boldsymbol{\rho}_s, t) \cdot \hat{\mathbf{n}} = 0 \tag{123a}$$

and the order μ^1 equation gives

$$(\hat{\mathbf{n}}\hat{\mathbf{n}}_\perp + \hat{\mathbf{n}}_\perp\hat{\mathbf{n}}) : \nabla\mathbf{u}(\boldsymbol{\rho}_s, t) = 0 \tag{123b}$$

$$\hat{\mathbf{n}} \cdot \nabla \log T(\boldsymbol{\rho}_s, t) = 0 \tag{123c}$$

where $\hat{\mathbf{n}}_\perp$ is any unit vector perpendicular to $\hat{\mathbf{n}}$. Equation (123a) expresses the fact that there is no net flow of the fluid out of or into the container, Eq. (123b) says there is no tangential stress exerted on the wall by the fluid at $\boldsymbol{\rho}_s$, and Eq. (123c) says there is no temperature gradient normal to the walls. Equation (123a) should hold for any boundary condition where particles cannot leave the container, while Eqs. (123b) and (123c) are valid only for specular reflections, where the molecules make elastic collisions with the walls, and where there is no transfer of momentum to the walls in a plane tangent to the walls.

The boundary conditions (123a)–(123c) are sufficient for finding solutions to the Navier–Stokes equations for a given geometrical arrangement of the boundaries.* To find the boundary conditions that are to be used with the Burnett and higher-order hydrodynamic equations,† one can investigate Eq. (76b) to order μ^2 and higher. However, to order μ^2 and beyond, one can show that the normal solution no longer satisfies the boundary conditions and that the boundary conditions that are to be used with the Burnett, . . . equations cannot be determined this way.[77,78]‡ This implies that there is a region close to the walls where the normal solution is no longer a solution to the complete Boltzmann equation, including interactions at the walls. In this region, called the "kinetic boundary layer," the quantities n, $\mathbf{u}$, and T vary over distances of the order of l_c instead of over distances on the order of l_h as was assumed in constructing the normal solution.

(b) For diffuse reflection the order μ^0 Eq. (76b) leads to

$$\mathbf{u}(\boldsymbol{\rho}_s, t) = 0 \tag{124a}$$

$$T(\boldsymbol{\rho}_s, t) = T_w(\boldsymbol{\rho}_s) \tag{124b}$$

These are the so-called "stick" boundary conditions that are usually imposed when the Navier–Stokes equations are solved. For this reflection mechanism, the normal solution breaks down near the walls already at order μ^1. As a result the kinetic boundary layer in the stick case is of order μ, whereas it is of order μ^2 in the slip case.[77,78] The existence of such a boundary layer leads to corrections of order μ in the boundary conditions (124). The boundary conditions for the tangential velocity and the temperature at the wall actually become of the form[6,7,77,78]§

$$\mathbf{u} \cdot \hat{\mathbf{n}}_\perp = \zeta_s(\hat{\mathbf{n}}\hat{\mathbf{n}}_\perp + \hat{\mathbf{n}}_\perp\hat{\mathbf{n}}) : \nabla\mathbf{u} + \omega\left(\frac{2}{m\beta}\right)^{1/2}(\hat{\mathbf{n}}_\perp \cdot \nabla) \log T \tag{125a}$$

$$\frac{T - T_w}{T_w} = \tau\hat{\mathbf{n}} \cdot \nabla \log T + \chi\left(\frac{m\beta}{2}\right)^{1/2}(3\hat{\mathbf{n}}\hat{\mathbf{n}} - \mathbf{I}) : \nabla\mathbf{u} \tag{125b}$$

Here ζ_s, ω, τ, and χ are microscopic constants: ζ_s, ω, and τ are called the slip coefficient, the thermal creep coefficient, and the temperature jump distance, respectively.[7] All four constants have the dimension of a length and their

*These are called "slip" boundary conditions, since the gas slips along the walls.

†Additional boundary conditions are needed for these higher-order equations, since they involve third- and higher-order spatial gradients of the hydrodynamic variables.

‡In fact, the higher-order equations put conditions on the hydrodynamic variables n, $\mathbf{u}$, and T that are not consistent with the lower-order conditions and there are more conditions than there are parameters to be determined.[44,45,77,78]

§We have not been able to find Eq. (125b) in this form in the literature. (Cercignani[7] gives expressions for the hydrodynamic boundary condition that are not entirely equivalent to ours.) It is derived in Ref. (78). For the case of stationary flow, where $\nabla \cdot \mathbf{u} = 0$, a boundary condition equivalent to (125b) has been derived.[73–75]

magnitude is of the order of the mean free path. Since the gradient is of order l_h^{-1}, the corrections to (124) are indeed of order μ, and vanish in the so-called continuum limit, where l_c/l_h approaches zero.

The determination of the constants ζ_s, ω, τ, and χ is a complicated problem requiring a complete solution of the Boltzmann equation, including the kinetic boundary layer. Exact solutions have been found only for certain modeled Boltzmann equations, like the BGK equation, in flows with a simple geometry[6,7] (e.g., stationary shear flow along a flat plate in a semi-infinite space, the so-called Kramers problem). Approximate results have been obtained by using variational methods and moment expansions.[6,7,9,81]

The Maxwell boundary condition [Eq. (122)] also gives rise to hydrodynamic boundary conditions of the form (125) with ζ_s, ω, τ, and χ roughly proportional to α^{-1}. So, as long as α is of order unity (more specifically $\alpha \gg l_c/l_h$), the corrections to stick boundary conditions remain small, and only if α becomes of order l_c/l_h boundary conditions differing appreciably from stick are obtained.[77–80,82] This is an indication why stick boundary conditions are for most purposes a very good approximation in hydrodynamic theory; a reflection mechanism that is almost specular is not very likely to occur in nature, due to irregularities in surface structures and thermal motion of the surface molecules.

The boundary conditions to be used with the Burnett equations have also been determined for a BGK model by Sone,[83] and for more general models by de Wit[84] using variational methods, but in fact this set of boundary conditions is not complete. The Burnett and higher-order hydrodynamic equations have nonphysical solutions showing spatial variations on the length scale of the mean free path.[85] One would like to have boundary conditions that could be used to reject these unphysical solutions. However, the available set of boundary conditions is not sufficient for that. Instead one must postulate that the rapidly varying solutions are absent and then use the available boundary conditions to determine the remaining hydrodynamic solution.

In most cases of physical interest the higher-order hydrodynamic equations give only a small improvement, if any, over the Navier–Stokes equations. However in Section 2.3.3 we will discuss one case, sound propagation, where the Burnett and higher-order equations do successfully improve the description of experimental results.[61–64]

We should also mention that the normal solution of the Boltzmann equation discussed here, together with the H-theorem discussed in the previous section, can be used to provide a derivation of the principles of nonequilibrium thermodynamics.[59,86,87] For mixtures, one can show that the various diffusion coefficients that occur in the Navier–Stokes equations can be expressed in a form where Onsager reciprocal relations are satisfied. However, both for mixtures and for pure gases the relation between the normal solution and irreversible thermodynamics only holds if one does not go beyond $\mu\Phi_1$ in the μ-expansion of the distribution function.[59,86,87]

We conclude this discussion of the normal solution of the Boltzmann equation by reiterating the point that the normal solution method has shown us how to connect the macroscopic theory of continuum fluid flow to a microscopic theory of molecular motions. In the next section we will discuss how well the predictions of the normal solution method are supported by experiment.

2.3.3. *Comparison of the Results of the Normal Solution Method with Experiment*

The normal solution method just outlined leads to two principal results, both of which can be tested experimentally.* These are (i) explicit expressions for the coefficients of viscosity η and thermal conductivity λ for dilute monatomic gases, in terms of the intermolecular potential $\phi(r)$, and (ii) an explicit form for the Burnett and higher-order corrections to the Navier–Stokes equation, together with expressions for the associated (higher-order) transport coefficients in terms of the intermolecular potential.

(i) *The Transport Coefficients η and λ.* In the previous section, we noted that one could make two predictions about η and λ that are independent of the exact form of the intermolecular potential. These are that η and λ are independent of density and that, to a good approximation,

$$\lambda/\eta c_v = \tfrac{5}{2} \tag{126}$$

Both of these predictions are completely in agreement with experiment. In fact, the prediction that η and λ should be independent of density was one of the first important results of the kinetic theory of gases.[1,2] The correctness of this prediction, too, has been known for some time. In Table 1 we give some typical values for $\lambda/\eta c_v$ for several noble gases and at various temperatures obtained prior to 1969. The agreement with the predicted value of $\frac{5}{2}$ is within a few percent. Further on we will discuss more recent and more accurate determinations of $\lambda/\eta c_v$ than those in Table 1, which even improve the agreement between theory and experiment.

If we want to compare the exact values of η and λ given by the theory with experimental results, we must know $\phi(r)$. Ideally one would want to determine $\phi(r)$ by quantum mechanical calculations alone or, failing that, by molecular scattering experiments. However, there are serious difficulties and uncertainties in determining $\phi(r)$ by either of these procedures. Nevertheless, one can still test the theoretical results for η and λ by either or both of the following methods:

(*a*) *The model potential method.* We assume that $\phi(r)$ has a certain analytic form, with undetermined numerical coefficients. We compute an

*Here, as earlier, we are restricting our attention to systems of pure monatomic gases. The theory as well as the comparison with experiment can be generalized to include dilute gas mixtures and dilute gases of polyatomic molecules (cf. Sections 2.4.4.1 and 2.4.4.2).

Table 1. $\lambda/\eta c_v$ *for Several Monatomic Gases*[a]

Gas	Temperature T, K	$\dfrac{\lambda}{\eta c_v}$	Gas	Temperature T, K	$\dfrac{\lambda}{\eta c_v}$
A	90.1	2.493	Ne	90.18	2.507
	194.6	2.527		194.6	2.508
	273.1	2.508		273.15	2.536
	298.1	2.503		298.15	2.492
	373.1	2.515		373.15	2.516
	491.0	2.489		491.15	2.485
	579.0	2.482		579.05	2.460
Kr	273.15	2.508			
	373.15	2.497			
	491.15	2.488			
	579.05	2.488			

[a]Data from H. J. M. Hanley (ed.), *Transport Phenomena in Fluids*, Marcel Dekker, New York (1969).

equilibrium property such as the second virial coefficient $B(T)$, defined by[(8)]

$$B(T) = -\tfrac{1}{2}\int d\mathbf{r}\,(e^{-\beta\phi(r)}-1) \tag{127}$$

for this potential function in terms of the undetermined coefficients. Then by fitting the experimental values obtained for $B(T)$ over a range of temperatures to the theoretical results in terms of $\phi(r)$, we determine the best values of the numerical coefficients in the potential, appropriate for that gas. With the potential now completely determined by the fit with equilibrium data, we compute η and λ and compare with nonequilibrium data.

(*b*) *The corresponding states method.* We assume that $\phi(r)$ has the same analytic form for a number of similar gases, but that the parameters that characterize it may vary from one gas to another. By dimensional analysis we see that any potential must be of the form†

$$\phi(r) = \varepsilon f(r/\sigma) \tag{128}$$

where ε and σ are constants with the dimensions of energy and length, respectively. If the form of the function $f(r/\sigma)$ is the same for a number of gases we can easily show that η and λ obey certain scaling relations. That is, if we define a dimensionless temperature by

$$T^* = k_B T/\varepsilon \tag{129a}$$

then the quantities‡

$$\eta^*(T^*) = \sigma^2\eta/(mk_BT)^{1/2}, \qquad \lambda^*(T^*) = \sigma^2\lambda/(k_B^3 m^{-1} T)^{1/2} \tag{129b}$$

†Strictly speaking, $\phi(r)$ must be of the form $\phi(r) = \sum_i \varepsilon_i f_i(r/\sigma_i)$ with a number of different energy and length constants ε_i and σ_i with different functions f_i. Here we make the simplifying assumption that there is only one energy parameter and one length parameter in $\phi(r)$.

‡This can easily be checked for hard spheres by considering the first Sonine polynomial approximation to η and λ, as in Eqs. (114) and (115).

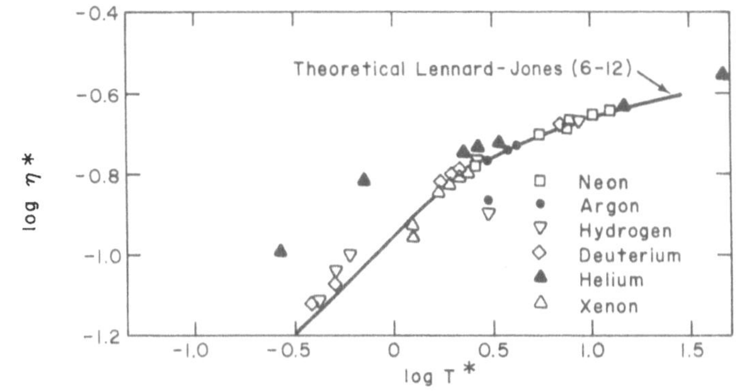

Fig. 9. A plot of the reduced viscosity η^* vs the reduced temperature T^*. The solid curve represents the theoretical value for η^* obtained using the Lennard-Jones potential and the first Sonine polynomial approximation, Eq. (113a). (From J. O. Hirschfelder, C. F. Curtiss, and R. B. Bird, *Molecular Theory of Gases and Liquids.*[8])

should be functions only of the reduced temperature T^* and should have the same values for all gases whose potential has the same analytic form. Consequently, we try to find two parameters ε and σ for each gas such that the experimental data for η and λ can be reduced to universal functions for both η^* and λ^*.†

Before discussing the results of these methods for comparing the theory with experiment, it should be mentioned that any such comparison should take into account the inaccuracies inherent in the experiments themselves. In particular, measurements of λ are generally not as accurate as measurements of η because convection in the thermal conduction cells causes errors in the determination of λ.[88–90] Consequently the most convincing tests of the Boltzmann equation results for transport coefficients have, until recently, come from a comparison of experimental measurements of η with theoretical values, over a large temperature range.

In the model potential method, one of the simplest and most useful models is the Lennard-Jones 6–12 potential[8]

$$\phi_{LJ}(r) = 4\varepsilon[(\sigma/r)^{12} - (\sigma/r)^6] \tag{130}$$

Here ε and σ have the dimensions of energy and length, respectively, and their values for a particular gas are obtained by fitting the temperature dependence of the second virial coefficient. We can then compute T^*, η^*, and λ^* from the experimental values of T, η, and λ, and the values obtained for ε and σ. Then we should expect that a plot of η^*, say, vs T^* should yield the same curve for all gases for which the Lennard-Jones potential leads to a good fit of $B(T)$ as a function of T, and in addition should agree with the theoretical values of η^* as a function of T^* obtained using Eq. (113a). In Fig. 9 we show the theoretical

†In addition, $B^*(T^*) = B(T)/\sigma^3$ should be a function only of T^*. In practice, one determines ε and σ by finding universal relations for $B^*(T^*)$, $\eta^*(T^*)$, and $\lambda^*(T^*)$. Notice that this method does not test predictions of the magnitude of η and λ, but only the form of the temperature dependence.

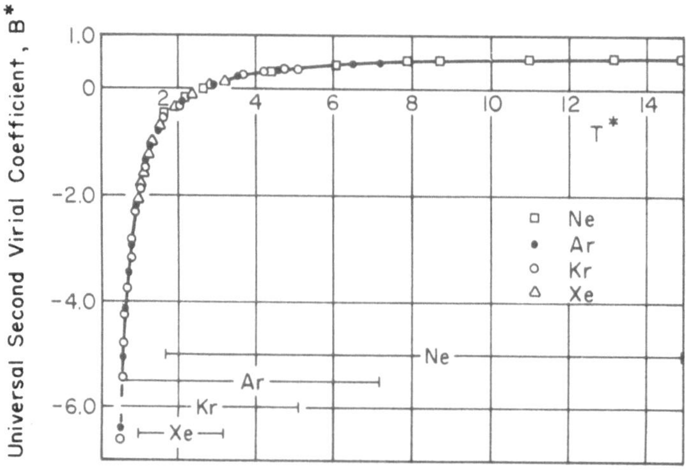

Fig. 10. A plot of the reduced second virial coefficient $B^*(T^*) = B(T)/[\frac{2}{3}\pi\sigma^3]$ vs T^*, where $T^* = k_B T/\epsilon$. The quantities σ and ϵ are the length and energy scaling factors determined using the hypothesis that the law of corresponding states is valid. [From J. Kestin and E. A. Mason, *in*: *Transport Phenomena* (J. Kestin, ed.).[90]]

curve of η^* vs T^* for a Lennard-Jones potential, and compare it with the experimental values obtained by the above procedure for the noble gases, hydrogen, and deuterium. The agreement is quite satisfactory except at low temperatures, where quantum effects, which we have neglected here, have to be taken into account.[8]

Figures 10 and 11 show the results of Kestin, Ro, and Wakeham,[90–92] who used the corresponding states method to compare theory and experiment

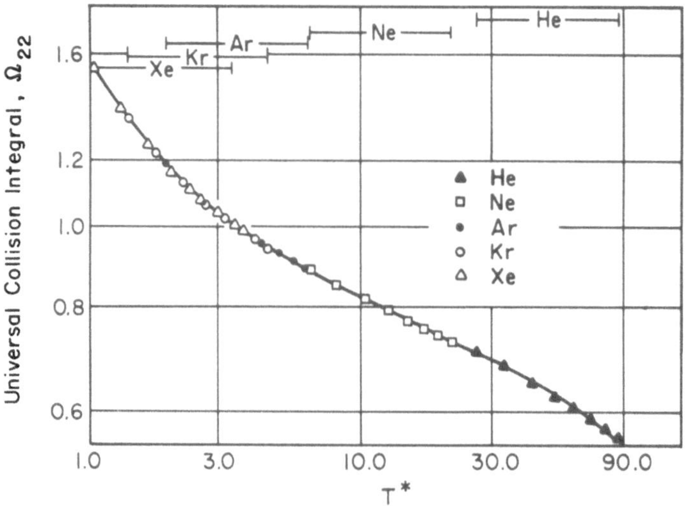

Fig. 11. A plot of $\Omega_{22}(T^*) = \frac{5}{16}(k_B m T/\pi)^{1/2}(1/\sigma^2\eta)$ vs T^*, $T^* = k_B T/\epsilon$. Here σ and ϵ are determined as in Fig. 10. [From J. Kestin and E. A. Mason, *in*: *Transport Phenomena* (J. Kestin, ed.).[90]]

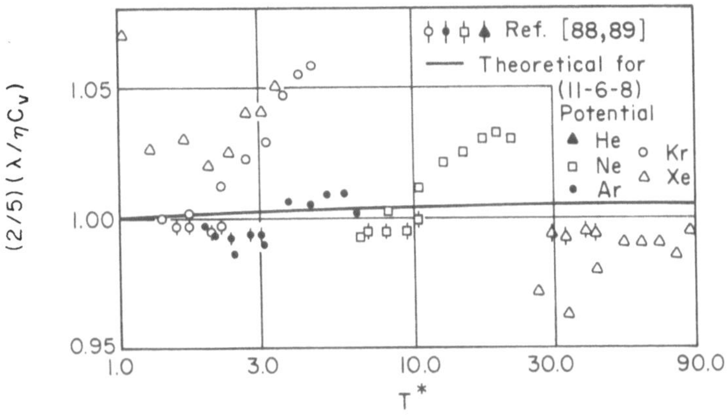

Fig. 12. A plot of $\frac{2}{5}(\lambda/\eta c_V)$ vs T^*, where $T^* = k_B T/\epsilon$, and is determined as in Fig. 10. The solid line represents the value for this quantity obtained by using the 11–6–8 potential. Note that Haarman's results for the thermal conductivity[88,89] combined with Kestin's results for the viscosity[90–92] lead to a value that is in very good agreement with the theory. [From J. Kestin and E. A. Mason, *in*: *Transport Phenomena* (J. Kestin, ed.).[90]]

for the noble gases, independent of a particular form of the function $f(r/\sigma)$ in Eq. (128). Here $B^*(T^*)$ vs T^* and $\eta^{*-1}(T^*)$ vs T^* are plotted; the values of ε and σ used in constructing B^*, η^*, and T^* were obtained by requiring that all noble gas data for B^* and η^* as a function of T^* can be placed as nearly as possible on a universal curve. Here the agreement of the data with the predictions that B^* and η^* should be universal functions of T^* is quite striking. In Fig. 12 we show a plot of $2\lambda/5\eta c_v$ vs T^*. As mentioned earlier, if the Boltzmann equation results are correct, $2\lambda/5\eta c_v$ should be almost unity for all T^*.* Notice that the data fluctuate about this value by as much as 7%, except for one set of data measured by Haarman.[88,89] These large fluctuations are attributed to errors in the measurement of λ due to thermal convection, which seem to be avoided in Haarman's measurements.

In conclusion, the agreement of experiment with the predictions of the Boltzmann equation is so good that these predictions are regarded as verified.[93] The emphasis of studies along these lines now is to use the comparison of theoretical values for η and λ with experiment as a test of various model potential energy functions. A potential energy function proposed recently by Klein and Hanley,[94] called the 11–6–8 potential, having the form

$$\phi(r) = \frac{\varepsilon}{5}(6+2\gamma)\left(\frac{r_m}{r}\right)^{11} - \frac{\varepsilon}{5}(11-3\gamma)\left(\frac{r_m}{r}\right)^{6} - \gamma\varepsilon\left(\frac{r_m}{r}\right)^{8} \tag{131}$$

*The value of $2\lambda/5\eta c_v$ is only unity in the first Sonine polynomial approximation. A more complete calculation shows that this quantity differs from unity by a few percent and the deviations depend on temperature as well as on the interaction potential.[5,8] In Fig. 12, the higher Sonine polynomial corrections have been taken into account for the 11–6–8 potential.

with *three* adjustable parameters γ, ε, and r_m, seems to provide the best overall fit of theory with experiment to date.[94–96] In Fig. 12 we also show the values for $2\lambda/5\eta c_v$ vs $k_B T/\varepsilon$ obtained by assuming the 11–6–8 potential.

(ii) *The Burnett and Super-Burnett Equations.* An experimental check of the higher-order corrections to the Navier–Stokes hydrodynamic equations is faced with a number of serious difficulties. Foremost among these is that since the higher-order hydrodynamic equations such as the Burnett and super-Burnett equations involve third, fourth, etc., spatial derivatives of the hydrodynamic variables n, $\mathbf{u}$, and T, more boundary conditions than needed for the Navier–Stokes equation must be specified in order to determine the proper solution for a given flow. However, it is not clear how to determine the additional boundary conditions. To avoid this difficulty, one looks for possible applications of the higher-order equations to hydrodynamic flows where the boundary effects do not play an important role. An example of such application is the determination of the dispersion relation for propagation of sound in a gas.[61–64] To apply the equations of hydrodynamics to sound propagation, we consider the propagation of a sound wave of sufficiently small amplitude, such that the disturbance of the gas from equilibrium is very small. Under these circumstances we can consider the sound wave to be a superposition of plane waves, each of which has a pressure variation $\delta p(\mathbf{r}, t)$ of the form

$$\delta p(\mathbf{r}, t) = a \exp i(\mathbf{k} \cdot \mathbf{r} - \omega t) \tag{132}$$

where $\mathbf{k}$ is a vector in the direction of propagation; $k = 2\pi\lambda_s^{-1}$, where λ_s is the wavelength and ω the angular frequency of the wave. We then look for the solutions of the linearized* hydrodynamic equations, where $p(\mathbf{r}, t) = p_{eq} + \delta p(\mathbf{r}, t)$ and $\delta p(\mathbf{r}, t)$ has the form given in Eq. (132). The general hydrodynamic equations only have solutions with a pressure deviation of this form if there is an analytic relation between ω and k, called a dispersion relation. If we consider the linearized hydrodynamic equations to arbitrary order in μ, we find that for small ω and k, the dispersion relation has the form

$$\omega = \pm v_0 k + Ak^2 \pm Bk^3 + Ck^4 + \cdots \tag{133}$$

Here v_0 is the adiabatic sound velocity for an ideal gas. The coefficient A is completely determined by the transport coefficients η and λ appearing in the Navier–Stokes equations; B depends on these as well as on the additional transport coefficients that appear in the Burnett hydrodynamic equations; C depends on all of the transport coefficients in A and B, as well as the super-Burnett transport coefficients, and so on.†

*The linearized hydrodynamic equations of a certain order are obtained from the hydrodynamic equations of that order by writing $n = n_{eq} + \delta n$, $\mathbf{u} = \mathbf{u}_{eq} + \delta \mathbf{u}$, $T = T_{eq} + \delta T$, and dropping quadratic and higher-order terms in the deviations from the equilibrium values.

†That is, the order μ^n corrections to the hydrodynamic equations (μ^1 = Euler, μ^2 = Navier–Stokes, etc.) contribute to the coefficie ts of order k^n and higher in Eq. (133). Actually one can obtain Eq. (133) directly from the Boltzmann equation without having first to derive the linearized hydrodynamic equation. For further details see Ford and Foch.[61]

The various coefficients appearing in the dispersion relation are known for any particular molecular model, and we can compare the predictions obtained from Eq. (133) with experimental results. In an experiment one fixes ω (experimentally) and then determines k as a function of ω; theoretically, this can be done by inverting the expansion given in (133), with the result that $k(\omega)$ has a real and a complex part,

$$k(\omega) = k_R(\omega) + ik_I(\omega) \tag{134}$$

Here $k_R(\omega)$ describes the propagation of the sound wave; that is, the phase velocity of the sound wave is

$$v_{ph} = \omega / k_R(\omega)$$

while k_I describes the damping of the wave. For example, if we consider a wave propagating in the z direction, $\delta p(\mathbf{r}, t)$ has the form

$$\delta p(\mathbf{r}, t) = a \exp[i(k_R(\omega)z - \omega t) - k_I(\omega)z] \tag{135}$$

In Figs. 13 and 14 we compare the results of experimental measurement of v_{ph} and k_I as a function of frequency for sound in neon with the theoretical value of these quantities obtained using the Lennard-Jones molecular model.[61–64] In these figures ξ is a dimensionless frequency

$$\xi = \omega\eta / v_0^2 n_0 m \tag{136}$$

which is such that $\xi \sim O(\omega t_c)$, where t_c is the mean free time between collisions. Hence we expect the theory will be good for $\xi < 1$, which seems to be confirmed reasonably well. Notice that the theoretical results obtained by including the Burnett and super-Burnett corrections to the Navier–Stokes equation lead to better agreement with the data than the results obtained from the Navier–Stokes equation alone. This improvement over the Navier–Stokes equation

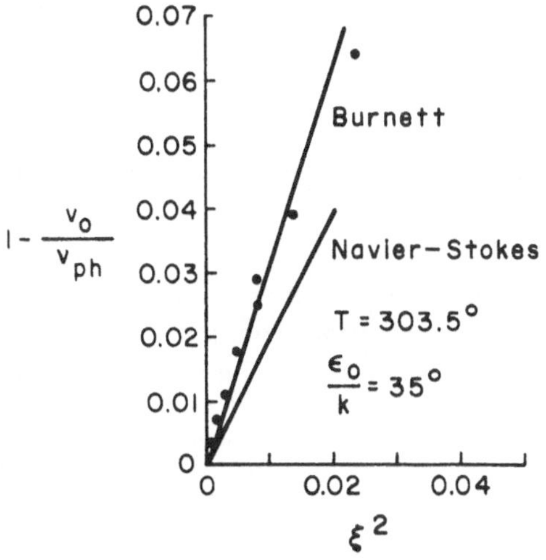

Fig. 13. The dispersion of sound in neon. Here V_0 is the ideal gas sound speed and ξ the dimensionless frequency $\xi = (3\eta/5n_0k_BT)\omega$. The theoretical curves are obtained by assuming that the gas molecules interact according to a Lennard-Jones potential with energy parameter ε_0. [From J. D. Foch and G. W. Ford, *in*: *Studies in Statistical Mechanics*, Vol. 5 (J. deBoer and G. E. Uhlenbeck, eds.).[61]]

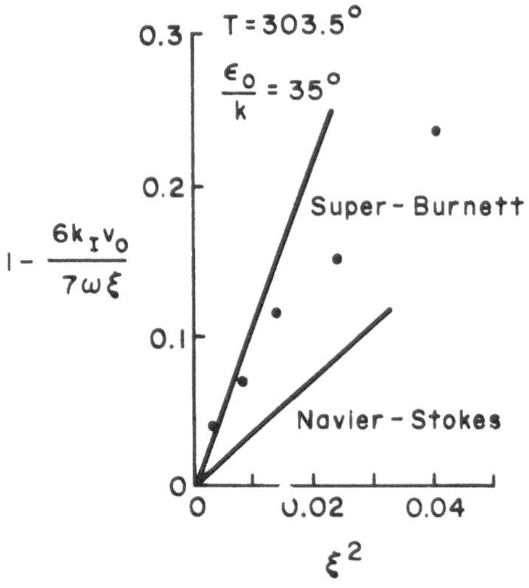

Fig. 14. The absorption of sound in neon. k_I is the inverse of the absorption length, i.e., the distance over which the amplitude of the sound wave decreases by e^{-1}. The solid curves are determined using the same Lennard-Jones potential as in Fig. 13. [From J. D. Foch and G. W. Ford, *in*: *Studies in Statistical Mechanics*, Vol. 5 (J. deBoer and G. E. Uhlenbeck, eds.).[61]]

results cannot be used as evidence to argue that the higher-order hydrodynamic equations will *always* improve the Navier–Stokes results. For the special case of the propagation of small-amplitude sound waves in hard-sphere gases one can show[72] that the decomposition into various orders in μ discussed in the previous section is a consistent ordering in powers of μ, but in more general circumstances this may not always be true.

2.3.4. Other Solutions and Other Applications of the Boltzmann Equation

Besides the normal solution of the Boltzmann equation, a number of other solutions have been constructed that extend the application of this equation to a wider variety of situations. Among these are:

(*a*) *Moment Expansions and Variational Methods.* Here one expands the distribution function in terms of a complete set of polynomial functions of velocity, with coefficients that depend on space and time. Using the Boltzmann equation, one can derive an infinite set of coupled space- and time-dependent equations for these coefficients,[5–13,97,98] which can only be solved to some degree of approximation. Usually one truncates the set of equations at some order n by setting all but the first n coefficients equal to zero. In the Grad 13 moment method,[99] one expands f in generalized Hermite polynomials in $\mathbf{V}$, and keeps the first 13 coefficients.* Moment expansions have been used to discuss the decay of an initial nonequilibrium disturbance of the gas to a state where the normal solution is valid, to discuss flow phenomena where the collision time t_c and the macroscopic time t_h are on the same order, and to

*The number 13 is chosen so as to include the moments corresponding to n, $\mathbf{u}$, T, $\mathbf{J}_T$, and $\mathbf{P}$. Using the fact that the hydrodynamic pressure is $\frac{1}{3}\mathrm{Tr}\mathbf{P}$, one sees that there are 13 independent moments.

discuss various boundary value problems, such as Couette and Poiseuille flows and the flow of a gas around a fixed sphere.[7,9–11,100]

Another method that has been applied to a number of problems of gas flows is based on a variational principle that can be derived from the linearized Boltzmann equation.[7,81,85] One particularly useful feature of the variational method is that quantities of physical interest such as the drag on an object or the flow rate can be directly related to the stationary point in the variational procedure. One particularly striking application of this method is the computation of the force on a sphere in a gas stream at low Mach numbers for all values of the ratio of the mean free path to the diameter of the sphere. The results are in very good agreement with the experimental results from the Millikan oil drop experiment.[101–103]

(*b*) *The Rarefied Gas Solution.* When the distance traveled by a typical molecule between collisions, l_c, is large compared to some macroscopic length l_m which characterizes the boundaries for the flow being considered, the flow is called rarefied gas flow. The characteristic feature of rarefied gas flows is that the dominant physical effects are due to the collisions of the molecules with the boundaries, while the molecule–molecule collisions provide small corrections to these effects. Therefore it is customary in treating rarefied gas flows, to expand the quantities of physical interest, particularly the distribution function, in powers of the inverse Knudsen number, K_n^{-1}, where $K_n = l_c/l_m$. In such expansions the lowest-order term takes into account only the molecule–boundary collisions, while the higher powers of K_n^{-1} take into account successively greater numbers of molecule–molecule collisions. Such expansions can be obtained from the Boltzmann equation (36), by regarding the collision term $J(f, f)$ as a small perturbation to the other terms in the equation.[51,77–80,104–116]

For example, if one considers the flow of a rarefied gas around a fixed sphere of radius R, one can use the expansion of K_n^{-1} of the distribution function to expand the force $\mathbf{F}$ exerted by the gas on the sphere, in powers of $K_n^{-1} = R/l_c$ as

$$\mathbf{F} = \mathbf{F}_0[1 + a_1 K_n^{-1} + a_2 K_n^{-2} + \cdots] \tag{137a}$$

where $\mathbf{F}_0$ is the force exerted by the gas on the sphere when collisions between the molecules are neglected and is called the free molecular flow force. A typical dynamical event that contributes to $\mathbf{F}_0$ is illustrated in Fig. 15a. The coefficient a_1 in Eq. (137a) is determined by collision processes involving two molecules and the sphere, such as those illustrated in Fig. 15b,c,d, where a molecule, having collided with the sphere, either is (a) reflected back to the sphere by another molecule as in Fig. 15b, (b) collides with a second molecule in such a way that the second molecule hits the sphere as in Fig. 15c, or (c) prevents a second molecule, which is on the way to a collision with the sphere, from actually hitting it, as in Fig. 15d.[112] The coefficient a_2 is determined by collision processes involving three molecules and the sphere, and so on.

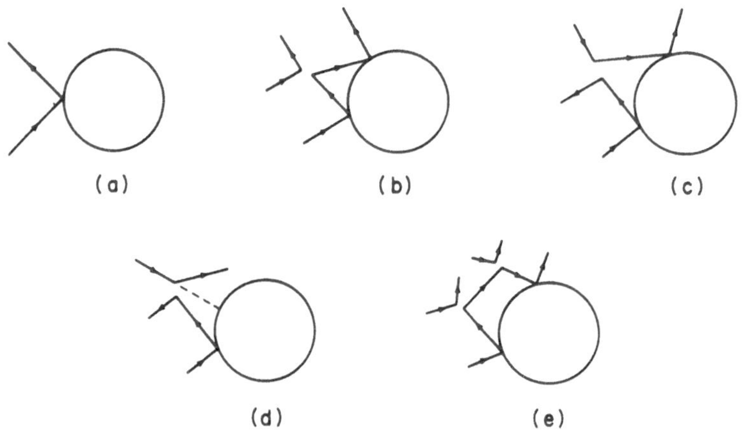

Fig. 15. The collisions of gas particles with a sphere, which are taken into account in the expansion of the force on the sphere in powers of the inverse Knudsen number. (a) Collisions that are responsible for the free molecular flow force; (b, c, d) dynamical events that contribute to the K_n^{-1} correction to this value; (d) represents a process where the second gas particle does not hit the sphere, but would have, had the second collision not taken place; (e) represents one of the type of events that contribute to order $K_n^{-2} \log K_n$.

Actually $a_2, a_3, \ldots$ are all infinite for reasonable intermolecular potentials $\phi(r)$, and the divergence of these coefficients can be traced to the assumption that the distribution function and $\mathbf{F}$ have an expansion purely in powers of K_n^{-1}.[44,45] A more refined analysis shows that a correct expansion of $\mathbf{F}$ is of the form

$$\mathbf{F} = \mathbf{F}_0[1 + a_1 K_n^{-1} + a_2' K_n^{-2} \log K_n^{-1} + a_2'' K_n^{-2} + \cdots] \tag{137b}$$

The coefficient a_2' is determined by collision processes in which three gas particles participate, such as that illustrated in Fig. 15e, while the higher terms in the series are determined by collisions involving three or more particles. The divergence in the power series expansion in K_n^{-1} as well as the resulting $\log K_n^{-1}$ terms in $\mathbf{F}$ are closely related to divergences and $\log n$ terms that appear in the density expansion of the transport coefficients for a dense gas.[23–26] We will discuss this problem in some detail in the next section.

Well-behaved, approximate solutions to the Boltzmann equation, valid for small K_n^{-1} have been obtained for a large variety of rarefied gas problems,* and we refer the reader to the literature for more details.[7,9,44,45,51,77–81,101–116]

(*c*) ***Solutions of the Linearized Boltzmann Equation for Special Forms of the Linearized Collision Operator.*** As we discussed in Section 2.3.1, if the deviation of the system from total equilibrium is small enough that terms

*One should consult the series *Rarefied Gas Dynamics*, Academic Press, New York, for discussions of a large number of problems.

quadratic in the deviations can be neglected, the nonlinear Boltzmann equation can be replaced by the linear equation (71), which for convenience we write again here:

$$[\partial/\partial t + \mathbf{v}\cdot\nabla_{\mathbf{r}} + (\mathbf{F}/m)\cdot\nabla_{\mathbf{v}}]f_{\text{eq}}\Phi = f_{\text{eq}}L\Phi + \bar{\mathbf{T}}f_{\text{eq}}\Phi \tag{138a}$$

Or writing $f_{\text{eq}} = W_{\text{w}}(\mathbf{r})n_0\varphi_0(\mathbf{r},\mathbf{v})$, we can decompose Eq. (138a) into the two equations

$$[\partial/\partial t + \mathbf{v}\cdot\nabla_{\mathbf{r}} + (\mathbf{F}/m)\cdot\nabla_{\mathbf{v}}]\Phi = L\Phi \qquad \text{for } \mathbf{r}\in V \tag{138b}$$

$$\mathbf{T}\Phi f_{\text{eq}} = 0 \tag{138c}$$

Since Eq. (138a) is a linear equation for Φ, it is much easier to solve than the nonlinear equation. In spite of this simplification, it is still difficult to produce explicit solutions to this equation unless the operator L has a simple form and the geometry of the boundaries is simple enough. The operator L can be characterized by its spectrum, and only for Maxwell molecules is this spectrum known explicitly. For this case the spectrum is discrete and the corresponding eigenfunctions can be expressed in terms of Sonine polynomials and spherical harmonics in $\mathbf{v}$.[(6–8,52,61–64)]* However, even for this special potential, it is still difficult to solve the linearized Boltzmann equation.

To avoid the difficulties in solving Eq. (138b) due to the complicated spectrum of L, it has been suggested that L in the equation be replaced by a model integral operator that has most of the properties of the actual collision operator L but that has a simple spectrum. The most commonly used model for L is the BGK model[(6,7,117)] given [when $V(\mathbf{r}) = 0$] by

$$L_{\text{BGK}}\Phi = -\nu\Phi + \nu\int d\mathbf{v}_1[1 + \beta m\mathbf{v}\cdot\mathbf{v}_1 + \tfrac{2}{3}(\tfrac{1}{2}\beta m\mathbf{v}^2 - \tfrac{3}{2})(\tfrac{1}{2}\beta m\mathbf{v}_1^2 - \tfrac{3}{2})]\varphi_0(\mathbf{v}_1)\Phi(\mathbf{v}_1) \tag{139}$$

where ν is a positive constant.† It can easily be seen that L_{BGK} has five zero eigenvalues, corresponding to the five eigenfunctions $1, (\beta m)^{1/2}\mathbf{v}$, and $(\tfrac{2}{3})^{1/2}(\tfrac{1}{2}\beta m\mathbf{v}^2/2 - \tfrac{3}{2})$, and it has an eigenvalue $-\nu$ with infinite degeneracy, such that any function of velocity that is orthogonal to 1, $\mathbf{v}$, and $\beta m\mathbf{v}^2 - 3$ is an eigenfunction of L_{BGK} with eigenvalue $-\nu$.

If one replaces L in Eq. (138b) by L_{BGK}, the linearized Boltzmann equation can be solved for a number of interesting cases. The simplest case where Eqs. (138b) and (138c) have been solved completely is the so-called Kramers problem. Here one considers the flow of a gas in a semi-infinite space bounded by a plane wall with which the molecules make diffusive collisions. For this problem one can show that there is a kinetic boundary layer near the wall and that the Chapman–Enskog normal solution is correct for points that

*The eigenfunctions are of the form $v^l S_{l+\frac{1}{2}}^{(r)}(m\mathbf{v}^2/2k_BT)Y_l^m(\hat{\mathbf{v}})$, where $S_{l+\frac{1}{2}}^{(s)}$ is a Sonine polynomial and $Y_l^m(\hat{\mathbf{v}})$ a spherical harmonic in $\hat{\mathbf{v}}$.

†$\varphi_0(\mathbf{v})$ is obtained from $\varphi_0(\mathbf{r},\mathbf{v})$, Eq. (72), by setting $V(\mathbf{r}) = 0$.

are greater than several mean free paths distant from the wall. Other problems that have been studied using the linear BGK operator are Couette flow, Poiseuille flow, the flow around spheres, cylinders, and other subjects, and sound propagation and light scattering in gases.[6,7,81,85,101–103,115,116]

(*d*) *Other Applications.* This brief summary of some solutions that have been obtained for the Boltzmann equation or simple models of it can hardly do justice to the wide range of problems that have been treated using the Boltzmann equation. Among the applications that we have only barely touched upon, or not yet mentioned at all are:

(i) The theory of dilute gas mixtures. The Boltzmann equation for mixtures of monatomic gases was briefly described in Section 2.1.2. This equation has been used to discuss transport phenomena, particularly diffusion, in such mixtures. Among the important predictions of the Boltzmann equation is the phenomenon of thermal diffusion, where a temperature gradient in the gas mixture can produce a diffusive flow.[5–13,118–121]

(ii) The theory of polyatomic molecules. The Boltzmann equation has been generalized to treat transport phenomena in polyatomic gases by taking into account the fact that polyatomic molecules have internal degrees of freedom and are not spherically symmetric.[5–13,122–127] One can no longer ignore quantum effects, since they must be taken into account when describing the internal state of the polyatomic molecules and in determining the internal states of the molecules after collisions if one knows their internal states before collision.* The theory for polyatomic molecules has been applied to obtain expressions for transport coefficients for such systems, and to describe the transport properties of paramagnetic polyatomic molecules in magnetic fields.[128–130]

(iii) The theory of Brownian motion. When a heavy particle is placed in a gas of lighter particles, the collisions of the heavy particle with the light ones cause irregular changes in the velocity of the heavy particle. The resulting motion of the heavy particle is usually called Brownian motion. If the gas is dilute enough, one can derive, using the now familiar arguments, a Boltzmann equation for the probability distribution function $f_B(\mathbf{r}, \mathbf{v}, t)$ of the Brownian particle, in terms of the distribution function of the light particles and the collision cross section for a collision of the Brownian particle and a light one. This Boltzmann equation then forms a convenient starting point for describing a number of interesting properties of Brownian motion, such as the friction exerted by the gas on the Brownian particle, and the approach of the velocity distribution of the Brownian particle to equilibrium.[11,54,131,132]

(iv) The theory of neutron transport. The diffusion of neutrons through a material system is usually described by means of a Boltzmann-like transport

*A semiclassical generalization of the Boltzmann equation to include polyatomic molecules was given by Wang Chang *et al.*[122] This equation was later extended to a more correct quantum version by Waldmann[123–125] and Snider.[126,127]

equation. Under most circumstances the neutrons interact only with the atoms of the material and not with each other, so the Boltzmann equation is a linear equation for the neutron distribution function. The equation only becomes nonlinear if the effects of the neutrons on the properties of the medium are taken into account. The theory of the neutron transport equation is very similar to that for the linearized Boltzmann equation. However, in many applications of the theory, especially to reactor problems, one must take into account that neutrons can be absorbed or produced by the atoms of the medium.[133]

(v) The theory of ionized gases. The kinetic theory for partially and fully ionized gases is an area where the Boltzmann equation and the basic ideas that underlie it have had a large variety of applications—from a theory for the mobility of heavy ions in neutral gases to a theory for transport processes in fully ionized gases where the particles interact with a Coulomb potential. There is an enormous literature on this subject, and we refer the reader to some of the surveys of the physics of ionized gases for more details.[6,121,134–141]

(vi) The theory of the rates of chemical reactions. In the theory of transport processes in dilute gases developed here, we have not taken into account the possibility that collisions of molecules could lead to chemical reactions. This is a restriction that can be relaxed without serious difficulty.[121,142,143] Thus the Boltzmann equation can be generalized to include the possibility that when two particles collide there may be a chemical reaction of some sort—one molecule might break up into two, or the two molecules might form a temporarily bound pair, there might be an exchange of atoms, and so on. The essential difficulty in treating chemical reactions is, of course, computing the relevant cross sections for the various processes that can take place.

We hope that this brief survey will give the reader some indication of the many fruitful applications of the Boltzmann equation to a wide variety of problems in the physics of fluids.

3. The Kinetic Theory of Dense Gases

3.1. Introduction

Our discussion of transport phenomena in gases has been based on the Boltzmann equation and is therefore restricted in application to dilute gases where only binary collisions need to be taken into account. To describe transport processes in dense gases we must generalize the Boltzmann equation to higher order in the density so as to take into account dynamical processes that involve three, four, or more particles.

The first important attempt to extend the Boltzmann equation to higher densities was made by Enskog in 1922.[1,5,8] He modified the Boltzmann equation by incorporating into the collision integral the difference in position between the centers of a colliding pair of molecules, and he corrected the *Stosszahlansatz* by supposing that an equilibrium-like pair correlation function could be used to compute the probability that two molecules would be close enough to collide. Enskog's modification of the Boltzmann equation leads to the Navier–Stokes equations, where contributions to the heat flow and to the pressure tensor come from the transport of momentum and energy by means of the intermolecular potential as well as by means of the motion of the molecules. The transport coefficients that result from Enskog's theory are in good agreement with experimental values over a wide range of densities, including densities high enough that the mean free path l_c is small compared to the molecular diameter a. In spite of this good agreement, the Enskog theory cannot be regarded as a complete theory for dense gases, for a number of reasons. First, the Enskog theory applies only to hard sphere-molecules.* Second, it ignores the possibility that the velocities of two particles may be correlated before they collide and thereby fails to take into account a number of dynamical effects that should be included in a correct theory of irreversible processes in dense gases.

To develop a more systematic kinetic theory for gases in which the *Stosszahlansatz* can be understood in terms of the properties of an ensemble of similar systems, and in which all the effects of microscopic processes taking place in the gas on its transport properties are accounted for, one must employ the general methods of statistical mechanics, using the Liouville equation as the starting point. The first important attempt to derive the Boltzmann equation from the Liouville equation and to find the correct generalization of the Boltzmann equation to higher order in the density was made in 1947 by Bogoliubov.[15] Although Bogoliubov's original theory has some serious difficulties, his ideas provided the foundation for the modern kinetic theory of dense gases, which has been considerably clarified and further developed by Cohen[16–19,23] and Green[20–22] and their co-workers.

In this section we will give a survey of the main features of the kinetic theory of dense gases as it has been developed so far. In Section 3.2 we will give a brief outline of the Enskog theory for dense gases of hard spheres, since this theory is intuitively simple and leads to results that are in good agreement with experiment. In addition, we shall discuss some recent modifications of this theory. In Section 3.3 we will discuss how one derives the Boltzmann equation from the Liouville equation and obtains an expansion for the corrections to the Boltzmann equation in powers of the density. Then in Section 3.4 we will discuss divergences that appear in this power series, and the rearrangements of the series that have been made in an attempt to eliminate these divergences. Finally, we will discuss the comparisons that have been made of the theoretical

*To compare the Enskog theory with experiment, one must determine an "effective hard-sphere" diameter for the molecules, which is, in general, a function of temperature.

prediction of the density dependence of the transport coefficients with experimental results.

3.2. The Enskog Theory and Its Modifications

3.2.1. The Enskog–Boltzmann Equation

Consider a dense gas of hard spheres, all with mass m and diameter a. Since the collisions of hard-sphere molecules are instantaneous, the probability is zero that any particle will collide with more than one particle at a time. Hence we still suppose that the dynamical events taking place in the gas are made up of binary collisions, and that to derive an equation for the single-particle distribution function $f(\mathbf{r}, \mathbf{v}, t)$ we need only take binary collisions into account. However, the *Stosszahlansatz* used in deriving the Boltzmann equation for a dilute gas should be modified to take into account any spatial and velocity correlations that may exist between the colliding spheres. The Enskog theory continues to ignore the possibility of correlations in the velocities before collision, but attempts to take into account the spatial correlations. In addition, the Enskog theory takes into account the variation of the distribution function over distances of the order of the molecular diameter, which also leads to corrections to the Boltzmann equation.

To construct Enskog's extension of the Boltzmann equation to higher densities, we consider only the change in $f(\mathbf{r}, \mathbf{v}, t)$ with time due to collisions, since only this term is affected by the density of the gas. Let us first consider the change in $f(\mathbf{r}, \mathbf{v}, t)$ due to the *direct* collisions. For a dilute gas we have argued that the number of direct collisions taking place in $\delta\mathbf{r}$ in time δt between molecules with velocity $\mathbf{v}$ and molecules with velocity $\mathbf{v}_1$, with apse line in direction $d\hat{\mathbf{k}}$ about $\hat{\mathbf{k}}$ is

$$f(\mathbf{r}, \mathbf{v}, t) f(\mathbf{r}, \mathbf{v}_1, t) B(\mathbf{g}, \hat{\mathbf{k}})\, d\hat{\mathbf{k}}\, \delta\mathbf{v}_1\, \delta\mathbf{v}\, \delta\mathbf{r}\, \delta t \tag{140}$$

For hard spheres, $B(\mathbf{g}, \hat{\mathbf{k}})$ is given by

$$B(\mathbf{g}, \hat{\mathbf{k}}) = a^2 |\mathbf{g} \cdot \hat{\mathbf{k}}| \tag{141}$$

and $\mathbf{g} \cdot \hat{\mathbf{k}} < 0$. However, at the instant of the collision the centers of the two particles are separated by a distance a along the apse line in the direction of $\hat{\mathbf{k}}$ as illustrated in Fig. 16a. We also take into account the variation in f over a molecular size, and replace expression (140) by

$$f(\mathbf{r}, \mathbf{v}, t) f(\mathbf{r} + a\hat{\mathbf{k}}, \mathbf{v}_1, t) a^2 |\mathbf{g} \cdot \hat{\mathbf{k}}|\, d\hat{\mathbf{k}}\, \delta\mathbf{v}_1\, d\mathbf{v}\, \delta\mathbf{r}\, \delta t \tag{142}$$

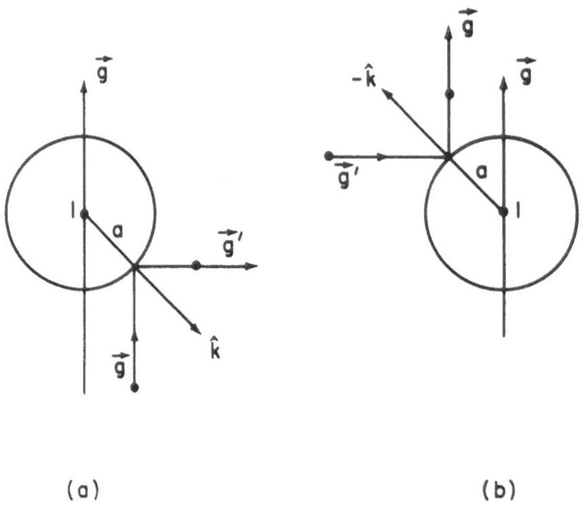

Fig. 16. Direct (a) and restituting (b) collisions for a collision of two hard spheres. We consider a relative coordinate system centered on particle 1 with z axis in the direction of $\mathbf{g}=\mathbf{v}_1-\mathbf{v}$. In (a) the apse line is in the direction $\hat{\mathbf{k}}$, and in (b), the apse line is in the direction $-\hat{\mathbf{k}}$. The action sphere of radius a about particle 1 is indicated by the circle.

assuming, of course, that δt may be made as small as one wants.* † However, Eq. (142) does not take into account that the probability of finding a molecule at $\mathbf{r}+a\hat{\mathbf{k}}$, given that there is a molecule at $\mathbf{r}$, is influenced by the presence of the other molecules in the gas. At high densities the probability for a collision will be much higher than predicted by Eq. (142) because the free volume available to any particle is reduced by the presence of the other particles. In equilibrium, the actual collision rate is obtained by multiplying Eq. (142) by the equilibrium pair correlation function $g(\mathbf{r}, \mathbf{r}+a\hat{\mathbf{k}})\equiv g(a)$, for two particles, evaluated when the two particles are in contact.[144]

Enskog proposed that since the gas is not in equilibrium we should multiply Eq. (142) by a factor $\chi(\mathbf{r}, \mathbf{r}+a\hat{\mathbf{k}})$, which is the nonequilibrium pair correlation function for two particles in contact. Since the form of this function is not known, he made the following assumption: The analytic dependence of $\chi(\mathbf{r}, \mathbf{r}+a\hat{\mathbf{k}})$ on density is exactly the same as that of $g(a)$, but the density to be used in computing $\chi(\mathbf{r}, \mathbf{r}+a\hat{\mathbf{k}})$ is the local density at a point midway between the two spheres, i.e., $n(\mathbf{r}+\frac{1}{2}a\hat{\mathbf{k}})$. Since $g(a)$ has the density expansion[145]

$$g(a)=1+0.4167\pi a^3 n_0+0.1275(\pi a^3 n_0)^2+\cdots \tag{143}$$

Enskog's assumption is that $\chi(\mathbf{r}, \mathbf{r}+a\hat{\mathbf{k}})$ is given by

$$\chi(\mathbf{r}, \mathbf{r}+a\hat{\mathbf{k}})=1+0.4167\pi a^3 n(\mathbf{r}+\tfrac{1}{2}a\hat{\mathbf{k}})+0.1275[\pi a^3 n(\mathbf{r}+\tfrac{1}{2}a\hat{\mathbf{k}})]^2+\cdots \tag{144}$$

*Otherwise the length of the collision cylinders must be taken into account.

†The variation of f over a molecular size provides corrections to the Boltzmann equation roughly of order a/l_c. At low densities these corrections are of order na^3, and at high densities $a/l_c \gg 1$, so these corrections are not insignificant.

This expression for the nonequilibrium pair correlation function is based on the supposition that even in a nonequilibrium state the pair correlation function should be determined solely by the physical requirement that hard spheres cannot overlap. Thus $\chi(\mathbf{r}, \mathbf{r}+a\hat{\mathbf{k}})$ should take into account the same kind of excluded volume effects as are incorporated in the equilibrium quantity $g(a)$. To do this, Enskog assumes that $\chi(\mathbf{r}, \mathbf{r}+a\hat{\mathbf{k}})$ has the same dependence on the local density at the point of contact, $\mathbf{r}+\frac{1}{2}a\hat{\mathbf{k}}$, as $g(a)$ has on the equilibrium density n_0. The point of contact $\mathbf{r}+\frac{1}{2}a\hat{\mathbf{k}}$ is used, since the pair correlation function should be invariant upon interchange of the two particles.

A similar analysis can be given for the restituting collision and we can immediately write the modification of the Boltzmann equation proposed by Enskog:

$$\begin{aligned}
&\frac{\partial f(\mathbf{r}, \mathbf{v}, t)}{\partial t}+\mathbf{v}\cdot\nabla_{\mathbf{r}} f(\mathbf{r}, \mathbf{v}, t)+(\mathbf{F}/m)\cdot\nabla_{\mathbf{v}} f(\mathbf{r}, \mathbf{v}, t) \\
&\quad=\int d\mathbf{v}_1\int d\hat{\mathbf{k}}\, a^2|\mathbf{g}\cdot\hat{\mathbf{k}}|\Theta(-\mathbf{g}\cdot\hat{\mathbf{k}}) \\
&\qquad\times\{\chi(\mathbf{r}, \mathbf{r}-a\hat{\mathbf{k}})f(\mathbf{r}, \mathbf{v}', t)f(\mathbf{r}-a\hat{\mathbf{k}}, \mathbf{v}_1', t)-\chi(\mathbf{r}, \mathbf{r}+a\hat{\mathbf{k}})f(\mathbf{r}, \mathbf{v}, t)f(\mathbf{r}+a\hat{\mathbf{k}}, \mathbf{v}_1, t)\} \\
&\qquad+\bar{\mathbf{T}}f(\mathbf{r}, \mathbf{v}, t)
\end{aligned}\tag{145}$$

where $\chi(\mathbf{r}, \mathbf{r}-a\hat{\mathbf{k}})$ is given by Eq. (144) with $\hat{\mathbf{k}}$ replaced by $-\hat{\mathbf{k}}$.

3.2.2. The Transport Coefficients in the Enskog Theory

Having obtained the Enskog–Boltzmann equation, we can now proceed to derive the conservation laws for particles, momentum, and energy. Then by constructing the normal solutions to the Enskog–Boltzmann equation, we can derive the hydrodynamic equations and obtain explicit expressions for the transport coefficients for a dense gas of hard spheres according to this theory. The Enskog theory takes into account an additional mechanism for the transport of energy and momentum that is not included in the low-density theory. This is the *collisional transfer* of momentum and energy whereby some of the energy and momentum of a particle whose center is at a point $\mathbf{r}$, say, is transferred at a collision to another sphere whose center is at $\mathbf{r}+a\hat{\mathbf{k}}$. Hence there is an instantaneous flow of energy and momentum across a distance a, which is due entirely to the finite size of the molecule. To see the effect in a simple example, consider the head-on collision of a moving sphere with a sphere at rest. Then when the collision takes place, all the momentum and energy is transferred from one sphere to the other over a distance a, as illustrated in Fig. 17.

By carrying out the indicated calculations, one finds that in the Enskog theory, the conservation laws have the same form as those given by the

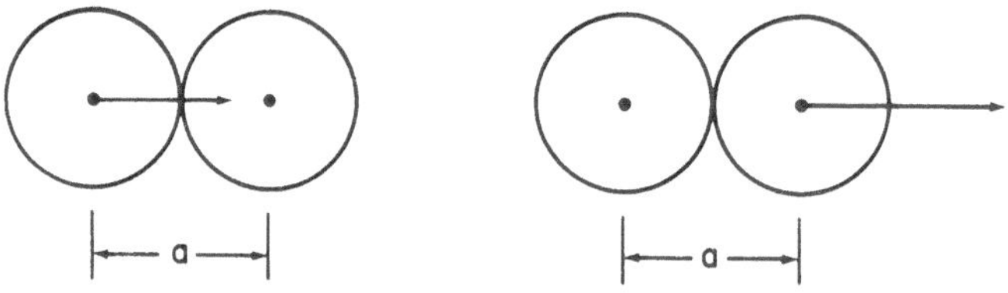

Fig. 17. The instantaneous collisional transfer of momentum and energy across a distance a when two hard spheres, each of diameter a, collide head on. The figure on the left represents the velocity and positions of the two spheres immediately before the collision, and the figure on the right represents them immediately after the collision.

Boltzmann equation, except that $\mathbf{J}_T$ and $\mathbf{P}$ are replaced by $\mathbf{J}_T^E$ and $\mathbf{P}^E$ given by

$$\mathbf{J}_T^E = \mathbf{J}_{T,k} + \mathbf{J}_{T,c} \tag{146a}$$

$$\mathbf{P}^E = \mathbf{P}_k + \mathbf{P}_c \tag{146b}$$

where $\mathbf{J}_{T,k}$ and $\mathbf{P}_k$ are the kinetic parts of the heat flow vector and the pressure tensor, due to the motion of molecules through the gas. These are equal to the Boltzmann equation values of $\mathbf{P}$ and $\mathbf{J}_T$ given by Eqs. (81a) and (81b). $\mathbf{J}_{T,c}$ and $\mathbf{P}_c$ are the contributions to $\mathbf{J}_T$ and $\mathbf{P}$ coming from the collisional transfer of energy and momentum, respectively.[5,6]

Here we will not go through the detailed calculations that lead to the Enskog theory values for the transport coefficients of shear viscosity, bulk viscosity, and thermal conductivity appearing in the Navier–Stokes hydrodynamic equations. Instead we shall merely cite the results obtained and refer the reader to the literature for more details.[1,5,6,18,19] One finds that the coefficient of shear viscosity η^E is given by*

$$\eta^E(n, a, T) = \frac{1}{\chi}\left(1 + \frac{4}{15}\pi n a^3 \chi\right)^2 \eta_0(T) + \frac{3}{5}\zeta^E \tag{147a}$$

the bulk viscosity ζ^E is given by

$$\zeta^E(n, a, T) = \frac{4}{9} n^2 a^4 \chi \left(\frac{\pi m}{\beta}\right)^{1/2} \tag{147b}$$

and the thermal conductivity is given by

$$\lambda^E(n, a, T) = \frac{1}{\chi}\left(1 + \frac{2}{3}\pi n a^3 \chi\right)^2 \lambda_0(T) + \frac{k_B}{2m}\zeta^E \tag{147c}$$

Here η_0 and λ_0 are the coefficients of shear viscosity and thermal conductivity as given by the Boltzmann equation, and $\chi = \chi(n(\mathbf{r}, t))$, where the local density

*The superscript E denotes Enskog theory values.

is evaluated at $\mathbf{r}$ at time t. The local equilibrium hydrostatic pressure $p(\mathbf{r}, t)$ is given by

$$p(\mathbf{r}, t) = nk_B T(1 + \tfrac{2}{3}\pi n a^3 \chi) \tag{148}$$

The quantities n and T appearing in Eqs. (147a)–(147c) and (148) are the local number density and local temperature evaluated at $\mathbf{r}$ at time t. If the gas is sufficiently close to equilibrium so that only the linearized hydrodynamic equations need be considered, then the quantities n, T, and χ appearing in (147a)–(147c) and (118) can be replaced by their equilibrium values.

3.2.3. Comparison of the Enskog Theory with Experiment

Despite the fact that real molecules are not hard spheres, the Enskog theory has been used to describe transport properties of real fluids over a wide range of densities and temperatures with a considerable degree of success. To apply the Enskog theory to real systems one must assume that (a) the mechanisms for the transport of energy and momentum in a real system do not differ in any essential way from the mechanisms of transport in a hard-sphere fluid, and (b) the expressions for the transport coefficients of a real fluid at a given temperature and density are identical to those of a hard-sphere fluid at the same density, provided one replaces a and $\chi(a)$ in the hard-sphere expressions by quantities $\tilde{a}$ and $\tilde{\chi}(T)$ where $\tilde{a}$ is an effective hard-sphere diameter of the molecules at temperature T, and $\tilde{\chi}(T)$ is an effective radial distribution function that takes into account the temperature dependence of the collision frequency in the real fluid.[6,146–148]

The simplest way to determine the effective hard-sphere diameter $\tilde{a}$ of the fluid molecules at temperature T and the corresponding value of $\tilde{\chi}$ is to fit $\eta_0^{\exp}(T)$, the experimental value for the coefficient of shear viscosity at low density, to the theoretical formula for hard-sphere molecules. That is, $\tilde{a}$ is determined by using the first Sonine polynomial approximation* to η_0 for a gas of hard spheres [Eq. (115a)],

$$\tilde{a}^2 = 5(\pi m k_B T)^{1/2}/16\pi\eta_0^{\exp}(T) \tag{149}$$

Once $\tilde{a}$ is determined, then $\tilde{\chi}(T)$ is found by using the density expansion for χ for hard spheres,

$$\tilde{\chi}(T) = 1 + 0.625\tilde{b}\rho + 0.2869(\tilde{b}\rho)^2 + 0.115(\tilde{b}\rho)^3 + \cdots$$

where $\rho = nm$, and

$$\tilde{b} = 2\pi\tilde{a}^3/3m \tag{150}$$

*In more accurate comparisons, higher-order Sonine polynomials are also taken into account.

In Fig. 18, we show a plot of the shear viscosity η given by the simple Enskog theory as a function of the mass density ρ, and compare with the experimental values for argon at temperature 348 K. One can see that the Enskog theory represents the data to within 10% for densities up to about $0.6\rho_c$, where ρ_c is the critical density.

The Enskog theory can be used to describe the transport properties of the fluid over a much wider range of temperatures and densities if one determines $\tilde{a}$ and $\tilde{\chi}$ in consistent way from equilibrium properties of the fluid.[1,6,8,146–148] In the simple method the attractive forces are only accounted for in $\tilde{a}$, but $\tilde{\chi}$ is determined as if the fluid were a hard-sphere fluid. In the method we describe now, the "modified" Enskog theory (MET), we take into account the effect of the attractive forces on $\tilde{\chi}$ also.

To determine $\tilde{a}$ and $\tilde{\chi}$ from p, V, T data, we follow a suggestion originally made by Enskog,[1] which is to replace p in the hard-sphere equation of state

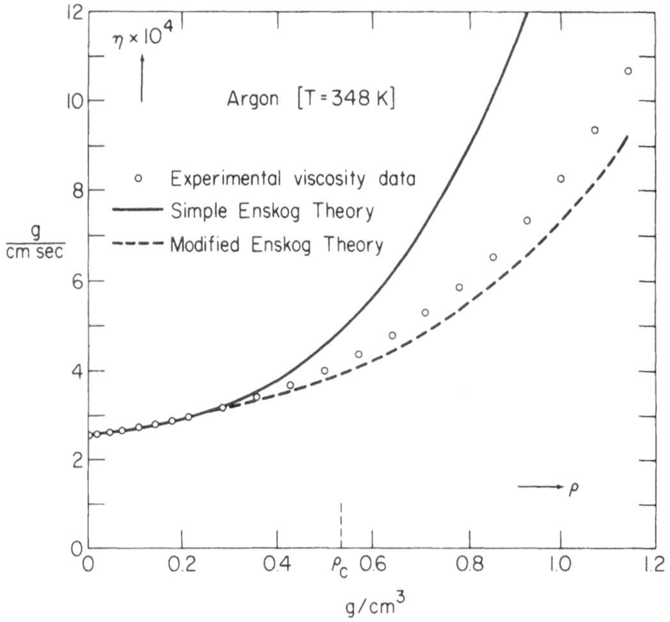

Fig. 18. A comparison of the experimental viscosity data for argon as a function of density with the predictions of the simple Enskog theory and of the modified Enskog theory. In the simple Enskog theory the effective hard-sphere diameter is determined by fitting the low-density limit of the viscosity to the Boltzmann equation result for hard spheres. Then χ is calculated for all densities by using this diameter in Eq. (144). In the modified Enskog theory, both χ and the hard-sphere diameter are determined from the equation of state data. (Figure courtesy of J. V. Sengers.)

Eq. (148), by the thermal pressure $T(\partial p/\partial T)_V$, so that $\tilde{\chi}$ is determined by

$$(\partial p/\partial T)_V = nk_B(1+\tilde{b}\rho\tilde{\chi}) \tag{151}$$

where $\tilde{b}$ is given by Eq. (150).*

To determine a consistent value of $\tilde{a}$, or equivalently of $\tilde{b}$, we require that $\chi \to 1$ as $\rho \to 0$. Then $\tilde{b}$ is determined from

$$\tilde{b} = \lim_{\rho\to 0} \frac{1}{\rho}\left[\frac{1}{nk_B}\left(\frac{\partial p}{\partial T}\right)_V - 1\right] \tag{152}$$

This limit can be evaluated using the virial expansion of the pressure

$$p/nk_BT = 1 + B(T)\rho + C(T)\rho^2 + \cdots \tag{153}$$

as

$$\tilde{b} = \frac{d}{dT}[TB(T)] \tag{154}$$

Thus $\tilde{b}$ can be found using experimental data of the temperature dependence of the second virial coefficient $B(T)$ and $\tilde{b}$ is independent of density.†

In Fig. 18 we also show the density dependence of the viscosity η for argon as computed by this modified Enskog theory (MET) at 348 K, where χ and b in Eq. (147a) are replaced by $\tilde{\chi}$ and $\tilde{b}$ as determined from Eqs. (151) and (154), respectively. We see that the MET agrees with experiment to within 10% over a much wider range of densities than does the simple Enskog theory discussed earlier. This improvement appears to be due to the fact that the MET takes into account the actual density dependence of the collision frequency, while the simple Enskog theory approximates the collision frequency by the hard-sphere theory, extrapolated from low densities.

Finally, we mention that although hard-sphere fluids do not exist in nature, they can be simulated and studied on a computer. Such studies have been carried out by Alder *et al.*[38–42] and we give in Fig. 19 the plot of η/η^E as a

*Enskog's argument[1] for this is that if the potential energy consists of a hard core plus a weak, long-range attractive part, then an approximate form of the equation of state is the van der Waals form $p+\alpha\rho^2 = nk_BT(1+b\rho\chi)$, where α is a constant determined by the attractive part of the potential energy, and χ is the same as in the hard-sphere equation of state. From this, Eq. (151) follows. Thus the thermal pressure takes into account mainly the short-range forces.

†Note that if the pair potential of the fluid consists of a hard-sphere core plus a weak, slowly varying (with r) attractive part $\varphi_a(r)$, then $\tilde{b}$ is equal to

$$\frac{d}{dT}\left[\frac{2}{3}\pi a^3 T - \frac{2\pi}{k_B}\int_a^\infty \varphi_a(r) r^2\, dr\right] = \frac{2}{3}\pi a^3$$

Thus we might expect that the MET can be used as long as the pair potential is such that $\beta\phi(r) \ll 1$ for large r, and $\int_a^\infty d\mathbf{r}\, \varphi(r) < \infty$ (for some finite a).

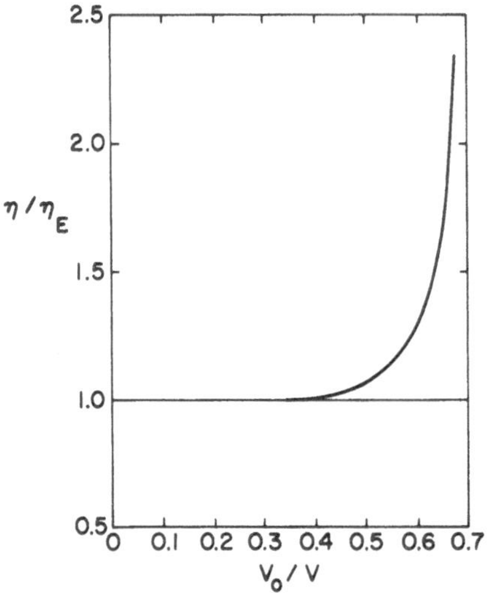

Fig. 19. A plot of η/η^E as a function of density for hard-sphere molecules. Here η is the actual viscosity as determined by computer-simulated molecular dynamics, η^E the Enskog theory value of the viscosity at the same density, V_0 the volume of the system of spheres at close packing, and V the actual volume of the system. The computer data on which this curve is based are good to about $\pm 5\%$. [From B. J. Alder, D. M. Gass, and T. E. Wainwright, *J. Chem. Phys.* **53**, 3813 (1970).]

function of density for a gas of hard spheres studied in a computer.* Here V_0 is the volume of the hard-sphere fluid at close packing and V is the actual volume of the fluid. Note that up to densities of the order $V_0/V \approx 0.3$ the computer values for η do not differ significantly from η^E. However, for higher densities η is larger than η^E by as much as a factor of 2 at the highest densities. The deviation from the Enskog theory results can only be accounted for by taking into account the velocity correlations that were ignored in the Enskog theory, and that will be discussed in Section 3.3.

3.2.4. *The Modified Enskog Equation*

As was already noted by Enskog, there is a difficulty with his equation if there is a strong density gradient in the gas, or a strong external field, which results in the presence of a preferred direction in the gas. If there is such a preferred direction, then one should expect that $\chi(\mathbf{r}, \mathbf{r} \pm a\hat{\mathbf{k}})$ should depend on the gradients of the local density, as well as on the local density itself. Such a possible dependence of χ on the gradient in $n(\mathbf{r}, t)$ is not taken into account in Eq. (144). However, as χ must be a scalar quantity, any dependence of χ on ∇n must necessarily† involve quantities of order $(\nabla n)^2$ or $\nabla^2 n$, and if we only intend to derive the Navier–Stokes equations from the Enskog–Boltzmann equation, these additional gradient terms in χ can be dispensed with. However, they should be included if one wants to derive Burnett and higher-order hydrodynamic equations. Moreover, the Enskog theory as it now stands cannot be extended to mixtures in an entirely satisfactory way.[149] For if one tries to

*The computer results for η are obtained by using the time correlation functions described in Section 4.

†That is, if we assume that χ is independent of the collision parameter of the colliding pair.

extend it to mixtures, and to derive hydrodynamic equations, the resulting set of (Navier–Stokes) diffusion coefficients and thermal-diffusion coefficients do not satisfy the Onsager reciprocal relations. As these relations are derived for fluids on the basis of very general arguments, the failure of the Enskog theory for mixtures to predict them is a serious difficulty. Here the difficulty is not connected with the neglect of gradient terms,* but rather with the assumed relation between χ and the local density.

Recently a modification of Enskog's relation between χ and $n(\mathbf{r}, t)$ has been derived from the more general theory to be discussed in the next section, which does take the density gradients into account in computing χ for pure gases, and which for mixtures leads to transport coefficients that satisfy the Onsager reciprocal relations.(150) Since the modified form of the Enskog–Boltzmann equation can be obtained from a simple generalization of Eq. (144), it seems worthwhile to present it here.

We begin by using the fact that if the gas were in equilibrium then χ is given by the pair distribution function $g(\mathbf{r}_1, \mathbf{r}_2)$ evaluated at $|\mathbf{r}_1 - \mathbf{r}_2| = a$. Consider now the density expansion of $g(\mathbf{r}_1, \mathbf{r}_2)$ in terms of Mayer f-functions. This expansion for general $\mathbf{r}_1, \mathbf{r}_2$ is given for hard spheres by†

$$g(\mathbf{r}_1, \mathbf{r}_2) = W(\mathbf{r}_1, \mathbf{r}_2)\Big[1 + n_0 \int d\mathbf{r}_3\, V(\mathbf{r}_1, \mathbf{r}_2|\mathbf{r}_2) + \frac{n_0^2}{2!}\int d\mathbf{r}_3 \int d\mathbf{r}_4\, V(\mathbf{r}_1, \mathbf{r}_2|\mathbf{r}_3, \mathbf{r}_4) + \cdots\Big] \tag{155}$$

where

$$W(\mathbf{r}_1, \mathbf{r}_2) = e^{-\beta\phi(|\mathbf{r}_1 - \mathbf{r}_2|)} = \begin{cases} 1, & \text{for } |\mathbf{r}_1 - \mathbf{r}_2| \geqslant a \\ 0, & \text{for } |\mathbf{r}_1 - \mathbf{r}_2| < a \end{cases} \tag{156}$$

$$V(\mathbf{r}_1, \mathbf{r}_2|\mathbf{r}_3) = f(\mathbf{r}_1, \mathbf{r}_3) f(\mathbf{r}_2, \mathbf{r}_3) \tag{157}$$

with the Mayer f-function $f(\mathbf{r}_2, \mathbf{r}_j)$ given by

$$f(\mathbf{r}_i, \mathbf{r}_j) = e^{-\beta\phi(|\mathbf{r}_i - \mathbf{r}_j|)} - 1 \tag{158}$$

and we refer to the literature for the expressions for $V(\mathbf{r}_1, \mathbf{r}_2|\mathbf{r}_3, \ldots, \mathbf{r}_n)$ in terms of Mayer f-functions.(16–19,144) The $\mathbf{r}_3, \mathbf{r}_4, \ldots$ integrations in Eq. (155) refer to the presence of a particle at $\mathbf{r}_3$, another at $\mathbf{r}_4$, and so on. In equilibrium, the density of particles at any point is the uniform value n_0. However, if we want to base a construction of the factor $\chi(\mathbf{r}, \mathbf{r} \pm a\hat{\mathbf{k}})$ on the density expansion of the equilibrium pair distribution function as was done by Enskog, then instead of replacing each factor of n_0 in Eq. (155) by $n(\mathbf{r} + \frac{1}{2}a\hat{\mathbf{k}}, t)$, it would seem more

*Since even Navier–Stokes coefficients are affected.

†A derivation of this result, for a general potential, is given in Section 3.4. See Eq. (191).

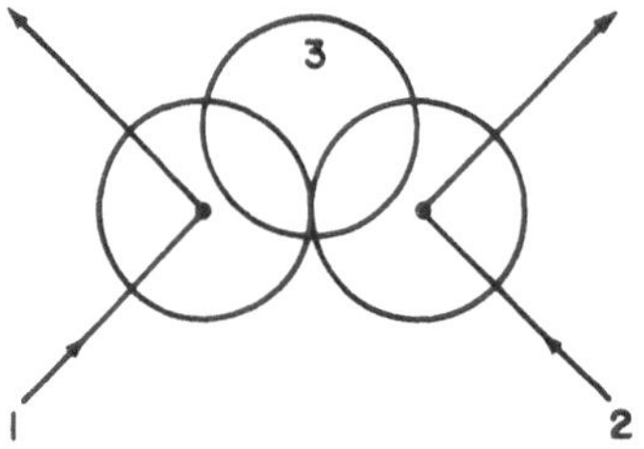

Fig. 20. An excluded-volume correction to the Boltzmann equation incorporated in the Enskog theory. Here particle 3 is overlapping both particles 1 and 2 at the instant of the (1,2) collision. This particular excluded volume correction is accounted for in the term $\int d\mathbf{r}_3 f_{13} f_{23}$ appearing in the density expansion of χ.

appropriate to use the local density of particles at $\mathbf{r}_3, \mathbf{r}_4, \ldots$, etc. Hence the modified expression for χ, χ_M, should be taken to be*

$$\chi_M(\mathbf{r}_1, \mathbf{r}_2) = 1 + \int d\mathbf{r}_3\, n(\mathbf{r}_3, t)\, V(\mathbf{r}_1, \mathbf{r}_2|\mathbf{r}_3) + \frac{1}{2}\int d\mathbf{r}_3 \int d\mathbf{r}_4\, n(\mathbf{r}_3, t) n(\mathbf{r}_4, t)\, V(\mathbf{r}_1, \mathbf{r}_2|\mathbf{r}_3, \mathbf{r}_4) + \cdots \tag{159}$$

and $\chi_M(\mathbf{r}, \mathbf{r}\pm a\hat{\mathbf{k}})$ is obtained by setting $\mathbf{r}_1 = \mathbf{r}$, $\mathbf{r}_2 = \mathbf{r}\pm a\hat{\mathbf{k}}$. Consequently the modified Enskog equation is identical to Eq. (145) except χ in Eq. (145) is replaced by χ_M given by Eq. (159). By using the expression for χ_M in terms of the Mayer f-functions, we see that the first density correction to the Boltzmann equation comes from configurations of three particles where the center of particle 3 is inside the action spheres of both particles 1 and 2 at the instant when these two particles collide, as illustrated in Fig. 20. This is clearly an excluded volume effect.

Notice that if $n(\mathbf{r}_i, t)$ in Eq. (159) are expanded about $n(\mathbf{r}\pm\frac{1}{2}a\hat{\mathbf{k}}, t)$, χ_M will be equal to Enskog's χ, Eq. (144), plus additional terms involving the gradients of the density.† It has been shown that if one entirely ignores velocity correlations between the colliding pair of particles, then the modified Enskog equation

$$\frac{\partial f(\mathbf{r}, \mathbf{v}, t)}{\partial t} + \mathbf{v}\cdot\nabla_{\mathbf{r}} f(\mathbf{r}, \mathbf{v}, t) + \frac{\mathbf{F}}{m}\cdot\nabla_{\mathbf{v}} f(\mathbf{r}, \mathbf{v}, t)$$
$$= \int d\mathbf{v}_1 \int d\hat{\mathbf{k}}\, a^2 |\mathbf{g}\cdot\hat{\mathbf{k}}| \Theta(-\mathbf{g}\cdot\hat{\mathbf{k}}) \{\chi_M(\mathbf{r}, \mathbf{r}-a\hat{\mathbf{k}}) f(\mathbf{r}, \mathbf{v}', t) f(\mathbf{r}-a\hat{\mathbf{k}}, \mathbf{v}_1', t)$$
$$- \chi_M(\mathbf{r}, \mathbf{r}+a\hat{\mathbf{k}}) f(\mathbf{r}, \mathbf{v}, t) f(\mathbf{r}+a\hat{\mathbf{k}}, \mathbf{v}_1, t)\} + \bar{\mathbf{T}} f(\mathbf{r}, \mathbf{v}, t) \tag{160}$$

is a consequence of the more systematic theory to be developed in the next section.

*We set $W(\mathbf{r}_1, \mathbf{r}_2) = 1$, since the colliding particles can never overlap with each other.
†The term proportional to $\nabla_{\mathbf{r}} n$ vanishes in the one-component system.

The modified Enskog equation can easily be generalized to apply to dense mixtures of hard-sphere gases.[150] The transport coefficients that result satisfy the Onsager reciprocal relations. Thus the principal difficulty in generalizing the Enskog theory to mixtures has now been removed.

We conclude this section by mentioning that if the system is close to equilibrium so that $f(\mathbf{r}, \mathbf{v}, t)$ can be replaced by $f_{\text{eq}}(1+\delta h)$, then Eq. (160) can be shown to reduce to*

$$\begin{aligned}\frac{\partial}{\partial t}\delta h(\mathbf{r}, \mathbf{v}, t)+\mathbf{v}\cdot\nabla_{\mathbf{r}}\,\delta h(\mathbf{r}, \mathbf{v}, t)&\\ = n_0 g(a)\int d\mathbf{v}_1\int d\hat{\mathbf{k}}\, a^2|\mathbf{g}\cdot\hat{\mathbf{k}}|\Theta(-\mathbf{g}\cdot\hat{\mathbf{k}})\phi_0(\mathbf{v}_1)&\\ \times[\delta h(\mathbf{r}, \mathbf{v}', t)+\delta h(\mathbf{r}-a\hat{\mathbf{k}}, \mathbf{v}_1', t)-\delta h(\mathbf{r}, \mathbf{v}, t)-\delta h(\mathbf{r}+a\hat{\mathbf{k}}, \mathbf{v}_1, t)]&\\ +\mathbf{v}\cdot\nabla_{\mathbf{r}}\int d\mathbf{r}_1\,[C(\mathbf{r}, \mathbf{r}_1)-g(a)f(\mathbf{r}, \mathbf{r}_1)]\,\delta n(\mathbf{r}_1, t)&\end{aligned} \tag{161}$$

where

$$\delta n(r, t)=\int d\mathbf{v}\, n_0\varphi_0(\mathbf{v})\,\delta h(\mathbf{r}, \mathbf{v}, t) \tag{162}$$

$f(\mathbf{r}, \mathbf{r}_1)$ is the Mayer f-function defined by Eq. (158) and $C(\mathbf{r}, \mathbf{r}_1)$ is the direct correlation function for a gas of hard spheres in equilibrium.[151] Equation (161) for $\delta h(\mathbf{r}, \mathbf{v}, t)$ was derived by several authors by using a number of different methods.[152–155] It finds application in the theory of time correlation functions, and is discussed by Mazenko and Yip in this volume.

3.3. The Liouville Equation and the BBGKY Hierarchy Equations

Our discussion of the kinetic theory of gases has raised two fundamental questions so far: (1) Can one show that the Boltzmann equation describes the "most probable" time development of a dilute gas, starting from some reasonable initial state, in spite of the fact that the Boltzmann equation is not consistent with the laws of mechanics? (2) Can one derive in a systematic way the generalization of the Boltzmann equation for dense gases?

In order to answer these questions, we must turn to the basic equation of nonequilibrium statistical mechanics, the Liouville equation.[156] Therefore, we no longer consider one particular container of a gas, but we consider an ensemble of similarly prepared† containers. We construct the $2Nd$-dimensional phase space of positions and momenta of the N-particles in

*For simplicity we drop the external force term and the molecule–wall collision term.

†The ensemble will be more clearly specified in the discussion that follows.

d-dimensions or Γ-space. A point in Γ-space will be denoted by $(x_1, \ldots, x_N)$, where $x_i = (\mathbf{r}_i, \mathbf{p}_i)$ denotes the position and momentum of particle i. Let $D_N(x_1, \ldots, x_N, t)$ be the distribution function in Γ-space, normalized as

$$\int d^N x\, D_N(x_1, \ldots, x_N, t) = V^N \tag{163}$$

where

$$d^N x = dx_1\, dx_2 \cdots dx_N = d\mathbf{r}_1\, d\mathbf{p}_1\, d\mathbf{r}_2\, d\mathbf{p}_2 \cdots d\mathbf{r}_N\, d\mathbf{p}_N$$

denotes a differential volume element in Γ-space and V is the volume of the container. This normalization is chosen in such a way that $D_N(d\mathbf{r}_1/V)\, d\mathbf{p}_1 \cdots (d\mathbf{r}_N/V)\, d\mathbf{p}_N$ represents the probability that a particular system will be located in the volume $d\mathbf{r}_1\, d\mathbf{p}_1 \cdots d\mathbf{r}_N\, d\mathbf{p}_N$ about the point $(x_1, \ldots, x_N)$ in Γ-space.* $D_N(x_1, \ldots, x_N, t)$ should be a symmetric function of all the particle indices, if all the particles are identical. The equation that determines the behavior of $D_N(x_1, \ldots, x_N, t)$ with time is Liouville's equation, which reads

$$\frac{\partial D_N(x_1, \ldots, x_N, t)}{\partial t} + \mathscr{H}_N D_N(x_1, \ldots, x_N, t) - \sum_{i=1}^{N} \bar{\mathbf{T}}_{\mathrm{w}}(i) D_N(x_1, \ldots, x_N, t) = 0 \tag{164}$$

where the Liouville operator $\mathscr{H}_N$ is a linear differential operator,†

$$\mathscr{H}_N = \sum_{i=1}^{N} \frac{\mathbf{p}_i}{m} \cdot \nabla_{\mathbf{r}_i} - \sum_{i<j}^{N} \theta_{ij} \tag{165a}$$

with

$$\theta_{ij} = \frac{\partial \phi(r_{ij})}{\partial \mathbf{r}_i} \cdot \left(\frac{\partial}{\partial \mathbf{p}_i} - \frac{\partial}{\partial \mathbf{p}_j} \right) \tag{165b}$$

and $\phi(r_{ij})$ is the intermolecular potential. Here $\bar{\mathbf{T}}_{\mathrm{w}}(i)$ gives the change in D_N due to a collision of particle i with the walls. Using arguments similar to those used in Section 2.1, we can give a convenient representation of the operator $\bar{\mathbf{T}}_{\mathrm{w}}(i)$ as

$$\hat{\mathbf{T}}_{\mathrm{w}}(i) f(x_1, \ldots, x_N) = \int dS\, \delta(\mathbf{r}_i - \boldsymbol{\rho}_s)$$
$$\times \left\{ \int d\mathbf{p}_i'\, \hat{P}(\mathbf{p}_i, \mathbf{p}_i') \left| \frac{\mathbf{p}_i}{m} \cdot \hat{\mathbf{n}} \right| f(x_1, \ldots, x_i', \ldots, x_N) - \Theta(-\mathbf{p}_i \cdot \hat{\mathbf{n}}) \left| \frac{\mathbf{p}_i}{m} \cdot \hat{\mathbf{n}} \right| f(x_1, \ldots, x_N) \right\} \tag{166a}$$

where

$$\hat{P}(\mathbf{p}_i, \mathbf{p}_i') = m^{-d} P(\mathbf{v}_i, \mathbf{v}_i') \tag{166b}$$

*We will usually call $D_N(x_1, \ldots, x_N, t)$ the N-particle distribution function, in accordance with a notation to be developed as we proceed.

†We have dropped the external force term for convenience. It can easily be included. We also now use the symbol n to represent the overall density N/V.

with $P(\mathbf{v}_i, \mathbf{v}_i')$ given by Eq. (30); $f(x_1, \ldots, x_N)$ is a function of $x_1, \ldots, x_N$; and $x_i' = \mathbf{r}_i, \mathbf{p}_i'$. Since we are including the wall interactions explicitly, we note that $D_N(x_1, \ldots, x_N, t)$ is defined for all values of $\mathbf{r}_i$ and vanishes if the coordinates of any particles lie outside the container. It will also be convenient for us to use the (formally) integrated form of Liouville's equation, which reads

$$D_N(x_1, \ldots, x_N, t) = \exp(-t\bar{\mathscr{H}}_N) D_N(x_1, \ldots, x_N, 0) \tag{167}$$

where $D_N(x_1, \ldots, x_N, 0)$ is the N-particle distribution function at some initial time $t = 0$, and the operator $\bar{\mathscr{H}}_N$ is

$$\bar{\mathscr{H}}_N = \mathscr{H}_N - \sum_{i=1}^{N} \bar{\mathbf{T}}_{\mathrm{w}}(i) \tag{168}$$

From our discussions of the Boltzmann equation and the Enskog equation, we know that a quantity of special interest to us is $f(\mathbf{r}, \mathbf{v}, t)$, which we now interpret as the ensemble average value of the number of particles in $d\mathbf{r}\, d\mathbf{v}$ about $\mathbf{r}, \mathbf{v}$ at time t. Since the Γ-space is properly formulated in $\mathbf{r}$ and $\mathbf{p}$ instead of $\mathbf{r}$ and $\mathbf{v}$, we define a new function $\hat{f}(\mathbf{r}, \mathbf{p}, t)$ such that $\hat{f}(\mathbf{r}, \mathbf{p}, t)\, d\mathbf{r}\, d\mathbf{p}$ is the number of particles in $d\mathbf{r}\, d\mathbf{p}$ about $\mathbf{r}, \mathbf{p}$ at time t. The relation between $f(\mathbf{r}, \mathbf{v}, t)$ and $\hat{f}(\mathbf{r}, \mathbf{p}, t)$ is clearly

$$f(\mathbf{r}, \mathbf{v}, t)\, d\mathbf{r}\, d\mathbf{v} = \hat{f}(\mathbf{r}, \mathbf{p}, t)\, d\mathbf{r}\, d\mathbf{p} \tag{169}$$

Now $\hat{f}(\mathbf{r}, \mathbf{p}, t)$ can be expressed as the average value

$$\hat{f}(\mathbf{r}, \mathbf{p}, t) = \frac{1}{V^N} \int d^N x \sum_{i=1}^{N} \delta(\mathbf{r}_i - \mathbf{r})\, \delta(\mathbf{p}_i - \mathbf{p}) D_N(x_1, \ldots, x_N, t) \tag{170}$$

If we use the fact that $D_N(x_1, \ldots, x_N, t)$ is a symmetric function of the particle indices, then $\hat{f}(\mathbf{r}, \mathbf{p}, t)$ is given by

$$\hat{f}(\mathbf{r}, \mathbf{p}, t) = \frac{N}{V} F_1(x_1, t) \tag{171a}$$

where $F_1(x_1, t)$ is

$$F_1(x_1, t) = \frac{1}{V^{N-1}} \int dx_2 \cdots \int dx_N\, D_N(x_1, x_2, \ldots, x_N, t) \tag{171b}$$

and $x_1 = (\mathbf{r}, \mathbf{p})$. The function $F_1(x_1, t)$ is called the reduced one-particle distribution function. From Eq. (171b) we see that $(1/V)F_1(x_1, t)\, dx_1$ represents the probability that one arbitrarily chosen molecule will be found in dx_1 about the point x_1, at time t. As we are interested in the time dependence of $F_1(x_1, t)$, we derive an equation for $\partial F_1/\partial t$ by integrating the Liouville equation over the phases of all but one of the particles. Then using the fact that the distribution function $D_N(x_1, \ldots, x_N, t)$ vanishes as $|\mathbf{r}_i| \to \infty$ and as $|\mathbf{p}_i| \to \infty$,* and that the walls of the container are neither sources nor sinks of particles, we obtain

$$\frac{\partial F_1(x_1, t)}{\partial t} + \mathcal{H}_0(1) F_1(x_1, t) - \bar{\mathbf{T}}_w(1) F_1(x_1, t) = \frac{N-1}{V} \int dx_2\, \theta_{12} F_2(x_1, x_2, t) \quad (172)$$

$\mathcal{H}_0(1)$ is the one-particle Liouville operator $\mathcal{H}_0(1) = (\mathbf{p}/m) \cdot \nabla_{\mathbf{r}}$. Here we have introduced the two-particle reduced distribution function $F_2(x_1, x_2, t)$ by

$$F_2(x_1, x_2, t) = \frac{1}{V^{N-2}} \int dx_3 \cdots \int dx_N\, D_N(x_1, x_2, \ldots, x_N, t) \quad (173)$$

If we compare Eq. (172) with the Boltzmann equation (36) and use the relation between $F_1(x_1, t)$ and $f(\mathbf{r}, \mathbf{v}, t)$, we see that Eq. (172) differs from the Boltzmann equation in that in Eq. (172) the Boltzmann collision integral is replaced by

$$\frac{N-1}{V} \int dx_2\, \theta_{12} F_2(x_1, x_2, t)$$

Therefore, to proceed further we must determine $F_2(x_1, x_2, t)$. If we now integrate the Liouville equation over the phases of all but two of the particles, in order to obtain an equation for $F_2(x_1, x_2, t)$, we arrive at

$$\frac{\partial F_2(x_1, x_2, t)}{\partial t} + \bar{\mathcal{H}}_2(x_1, x_2) F_2(x_1, x_2, t) = \frac{N-2}{V} \int dx_3 (\theta_{13} + \theta_{23}) F_3(x_1, x_2, x_3, t) \quad (174)$$

where the two-particle Liouville operator

$$\bar{\mathcal{H}}_2(x_1, x_2) = \mathcal{H}_2(x_1, x_2) - \bar{\mathbf{T}}_w(1) - \bar{\mathbf{T}}_w(2) \quad (175a)$$

with

$$\mathcal{H}_2(x_1, x_2) = \mathcal{H}_0(1) + \mathcal{H}_0(2) - \theta_{12} \quad (175b)$$

and $F_3(x_1, x_2, x_3)$, is the reduced three-particle distribution function

$$F_3(x_1, x_2, x_3, t) = \frac{1}{V^{N-3}} \int dx_4 \cdots \int dx_N\, D_N(x_1, x_2, \ldots, x_N, t) \quad (176)$$

Thus, to determine $F_1(x_1, t)$, we must know $F_2(x_1, x_2, t)$, and to determine this function, we must know $F_3(x_1, x_2, x_3, t)$, and so on. So we have an open ended chain of equations, given by Eqs. (172) and (174) and their generalizations to equations for F_3, F_4, and so on, all obtained by integrating the Liouville equation. This set of equations is called the BBGKY hierarchy

*D_N vanishes as $|\mathbf{r}_i| \to \infty$ since the wall interactions prevent any particles from being outside the container. D_N vanishes as $|\mathbf{p}_i| \to \infty$ since the normalization of D_N requires that all velocity integrals be finite.

equations.[15,157–159]*† If we ignore the collisions with the walls, the hierarchy equations are time reversible, unlike the Boltzmann equation. It is clear that not very much new can be learned simply by integrating the Liouville equation. If we are to derive the Boltzmann equation from the first hierarchy equation we must try to find the circumstances under which we can express the two-particle distribution function $F_2(x_1, x_2, t)$ appearing in Eq. (172) in terms of the single-particle function $F_1(x_1, t)$. We will see in the next section that if the initial N-particle distribution function has certain properties, and if the density of the system is sufficiently low that $n_0 a^3$ can be regarded as a small expansion parameter, then we can express $F_2(x_1, x_2, t)$ in terms of $F_1(x, t)$ as a power series in $n_0 a^3$ and proceed to derive the Boltzmann equation and its generalization to higher order in the density.

3.4. The Cluster Expansion Method and the Generalized Boltzmann Equation

In equilibrium statistical mechanics, there is a well-established procedure for obtaining an expression for the two-particle distribution function for a dilute or moderately dense gas in terms of the single-particle distribution function as a power series in the density. This procedure involves the introduction of so-called cluster expansions, whereby properties of the gas are expanded in a series of terms each of which is determined by the properties of a system composed of a small number of particles.[144]‡

Here we will show that the cluster expansion method can be extended to apply also to dilute or moderately dense gases not in equilibrium.[16–24] In order to see how the method will be developed, let us first consider the case of equilibrium. For equilibrium, the N-particle distribution function is the canonical distribution, which, using the normalization in Eq. (163), is

$$D_N^{eq}(x_1, \ldots, x_N) = V^N e^{-\beta H_N}/Z_N \tag{177}$$

where

$$Z_N = \int d^N x e^{-\beta H_N} \tag{178a}$$

and, as in Eq. (1),

$$H_N = \sum_{i=1}^{N} \frac{\mathbf{p}_i^2}{2m} + \sum_{i<j}^{N} \phi(r_{ij}) + \sum \Phi_w(\mathbf{r}_i) \tag{178b}$$

*After Bogoliubov, Born, Green, Kirkwood, and Yvon, all of whom derived them.[15,157–159]
†Equation (172) is called the first hierarchy equation, Eq. (174) the second, etc.
‡The virial expansion of the pressure is well-known example of such a series.[52]

where $\Phi_w(\mathbf{r}_i)$ is a wall potential that keeps the particles in the container, so that

$$e^{-\beta\Phi_w(\mathbf{r}_i)} = \begin{cases} 1, & \text{for } \mathbf{r} \in V \\ 0, & \text{for } \mathbf{r} \notin V \end{cases} \tag{179}$$

The presence of the walls of the container is a complication that we wish to ignore for the moment. Therefore, to eliminate the boundary effects we will consider the thermodynamic limit where $N \to \infty$ and $V \to \infty$ such that the density N/V remains constant.

The one-particle distribution function, in the thermodynamic limit, in equilibrium is

$$F_1^{\text{eq}}(x_1) = \lim_{\substack{N\to\infty \\ V\to\infty \\ N/V=n}} \frac{1}{V^{N-1}} \int dx_2 \cdots \int dx_N \frac{V^N e^{-\beta H_N}}{Z_N} = \phi_0(\mathbf{p}) \tag{180}$$

where $\varphi_0(\mathbf{p})$ is the Maxwell–Boltzmann momentum distribution

$$\phi_0(\mathbf{p}) = \left(\frac{\beta}{2\pi m}\right)^{d/2} \exp\left(-\frac{\beta \mathbf{p}^2}{2m}\right) \tag{181}$$

Now we begin the cluster expansion method by introducing a set of unnormalized S-particle distribution functions

$$\begin{aligned} W_s(x_1, \ldots, x_S) &= \exp\left[-\beta \sum_{i<j}^{s} \phi(r_{ij})\right] \\ W_1(x_1) &\equiv 1 \end{aligned} \tag{182}$$

where $\phi(r_{ij})$ is the pair potential, and a set of cluster functions, $U^{\text{eq}}(x_1)$, $U^{\text{eq}}(x_1|x_2)$, $U^{\text{eq}}(x_1|x_2, x_3), \ldots$, defined by

$$\begin{aligned} W_s(x_1, \ldots, x_s) &= U^{\text{eq}}(x_1) W_{s-1}(x_2, \ldots, x_s) \\ &+ \sum_{j=2}^{s} U^{\text{eq}}(x_1|x_j) W_{s-2}(x_2, \ldots, x_{j-1}, x_{j+1}, \ldots, x_s) \\ &+ \sum_{2\leq j<k\leq s} U^{\text{eq}}(x_1|x_j, x_k) W_{s-3}(x_2, \ldots, x_{j-1}, x_{j+1}, \ldots, x_{k-1}, x_{k+1}, \ldots, x_s) \\ &+ \cdots + U^{\text{eq}}(x_1|x_2, \ldots, x_s) \end{aligned} \tag{183}$$

The cluster functions can be obtained by writing this equation in detail for $s = 1, 2, 3, \ldots$, so as to obtain

$$U^{\text{eq}}(x_1) = W_1(x_1) \equiv 1 \tag{184a}$$

$$U^{eq}(x_1|x_2) = W_2(x_1, x_2) - W_1(x_1)W_1(x_2) \tag{184b}$$

$$U^{eq}(x_1|x_2, x_3) = W_3(x_1, x_2, x_3) - W_2(x_1, x_2)W_1(x_3) - W_2(x_1, x_3)W_1(x_2) - W_2(x_2, x_3)W_1(x_1) + 2W_1(x_1)W_1(x_2)W_1(x_3) \tag{184c}$$

etc. Now inserting the cluster expansion Eq. (183) for W_N in expression (180) for $F_1^{eq}(x_1)$, we obtain

$$nF_1^{eq}(x_1) = \phi_0(\mathbf{p})\left[z + z^2 \int dx_2\, \varphi_0(\mathbf{p}_2)U^{eq}(x_1|x_2) + \frac{z^3}{2!}\int dx_2 \int dx_3\, \phi_0(\mathbf{p}_2)\varphi_0(\mathbf{p}_3)U^{eq}(x_1|x_2, x_3) + \cdots\right] \tag{185}$$

where z is the fugacity, defined by

$$\left(\frac{\beta}{2\pi m}\right)^{ds/2} z^s = \lim_{\substack{N,V\to\infty \\ N/V=n}} \frac{N(N-1)\cdots(N-s+1)}{Z_N} Z_{N-s} \tag{186}$$

Equation (185) is simply the relation between the density n and the fugacity z for a gas in equilibrium.[16–19,144] Now consider the two-particle distribution function $F_2^{eq}(x_1, x_2)$ defined by

$$F_2^{eq}(x_1, x_2) = \lim_{\substack{N,V\to\infty \\ N/V=n}} \frac{1}{V^{N-2}} \int dx_3 \cdots \int dx_N \frac{V^N e^{-\beta H_N}}{Z_N} \tag{187}$$

and define a new set of cluster functions $U^{eq}(x_1, x_2|x_3, \ldots, x_5)$, $s = 2, \ldots$ by

$$W_s(x_1, \ldots, x_s) = U^{eq}(x_1, x_2)W_{s-2}(x_3, \ldots, x_s) + \sum_{i=3}^{s} U^{eq}(x_1, x_2|x_i)W_{s-3}(x_3, \ldots, x_{i-1}, \ldots, x_s) + \cdots + U^{eq}(x_1, x_2|x_3, \ldots, x_s) \tag{188}$$

from which we obtain

$$U^{eq}(x_1, x_2) = W_2(x_1, x_2) \tag{189a}$$

$$U^{eq}(x_1, x_2|x_3) = W_3(x_1, x_2, x_3) - W_2(x_1, x_2)W_1(x_3) \tag{189b}$$

$$U^{eq}(x_1, x_2|x_3, x_4) = W_4(x_1, x_2, x_3, x_4) - W_3(x_1, x_2, x_3)W_1(x_4) - W_3(x_1, x_2, x_4)W_1(x_3) - W_2(x_1, x_2)W_2(x_3, x_4) + 2W_2(x_1, x_2)W_1(x_3)W_1(x_4) \tag{189c}$$

etc. Then using the cluster expansion Eq. (188) for W_N in expression (187) for $F_2^{eq}(x_1, x_2)$, we obtain

$$n^2F_2^{eq}(x_1, x_2) = z^2\phi_0(\mathbf{p}_1)\phi_0(\mathbf{p}_2)U^{eq}(x_1, x_2) + z^3 \int dx_3\, U^{eq}(x_1, x_2|x_3) \prod_{i=1}^{3} \phi_0(\mathbf{p}_i) + \frac{z^4}{2!}\int dx_3 \int dx_4\, U^{eq}(x_1, x_2|x_3, x_4) \prod_{i=1}^{4} \phi_0(\mathbf{p}_i) + \cdots \tag{190}$$

We have now obtained z expansions for both $nF_1^{\text{eq}}(x_1)$ and $n^2F_2^{\text{eq}}(x_1, x_2)$ and we want to obtain an expression for $F_2^{\text{eq}}(x_1, x_2)$ in terms of $F_1^{\text{eq}}(x)$. To do this we solve Eq. (185) for $z\phi_0(\mathbf{p}_i)$ in terms of $nF_1(x_i)$ and we substitute the expression for $z\phi_0(\mathbf{p}_i)$ in Eq. (190), to obtain

$$\begin{aligned}
F_2^{\text{eq}}(x_1, x_2) = {} & F_1^{\text{eq}}(x_1)F_1^{\text{eq}}(x_2)U^{\text{eq}}(x_1, x_2) \\
& + nF_1^{\text{eq}}(x_1)F_1^{\text{eq}}(x_2) \int dx_3\, F_1^{\text{eq}}(x_3)[U^{\text{eq}}(x_1, x_2|x_3) \\
& - U^{\text{eq}}(x_1, x_2)U^{\text{eq}}(x_1|x_3) - U^{\text{eq}}(x_1, x_2)U^{\text{eq}}(x_2|x_3)] + \cdots \\
= {} & W_2(x_1, x_2)F_1^{\text{eq}}(x_1)F_1^{\text{eq}}(x_2) + n \int dx_3\, [W_3(x_1, x_2, x_3) \\
& - W_2(x_1, x_2)W_2(x_1, x_3) - W_2(x_1, x_2)W_2(x_2, x_3) \\
& + W(x_1, x_2)] \prod_{i=1}^{3} F_l^{\text{eq}}(x_i) + \cdots
\end{aligned} \tag{191}$$

which is the density expansion of the two-particle distribution function in equilibrium. The utility of this expansion depends on the fact that the integrands appearing on the right-hand side of (191) vanish whenever the field particles 3, 4, ... are more than a few molecular diameters from particles 1 and 2. This property of the integrands is assured by the fact that the $W_s(x_1, \ldots, x_s)$ have a *factorization* property. That is, if a group of s_1 particles is sufficiently far from the remaining $s - s_1$ particles, then $W_s(x_1, \ldots, x_s) \to W_{s_1} W_{s-s_1}$.

We have just shown that for equilibrium systems, it is possible to express the two-particle distribution function as a power series in the density in terms of the single-particle distribution function, by using Mayer's cluster expansion method. We are now going to show that the same method can also be extended to nonequilibrium systems, so that one can express $F_2(x_1, x_2, t)$ as a power series in the density in terms of $F_1(x_i, t)$, similar to Eq. (191). This nonequilibrium expansion of F_2 in terms of F_1, when inserted into the first hierarchy equation, will then enable us to derive the Boltzmann equation and to extend it to higher densities. We begin by constructing a set of functions $D_s(x_1, x_2, \ldots, x_s, t)$ that satisfy the s-particle Liouville equation* for $s = 1, 2, 3, \ldots$, which we will use to derive cluster expansions for $F_s(x_1, \ldots, x_s, t)$ similar to those for $F_s^{\text{eq}}(x_1, \ldots, x_3)$ in terms of the $W_s(x_1, \ldots, x_s)$. The functions D_s satisfy

$$\frac{\partial D_s(x_1, \ldots, x_s, t)}{\partial t} + \bar{\mathcal{H}}_S(x_1, \ldots, x_s)D_s(x_1, \ldots, x_s, t) = 0 \tag{192}$$

*We show below that D_s functions are determined for all times once the initial values of the F_s are given.

and are normalized according to

$$\int d^s x\, D_s(x_1, \ldots, x_s, t) = V^s \tag{193}$$

Apart from the difference in normalization, the D_s will be the nonequilibrium analogs of the W_s. Now we define a set of nonequilibrium cluster functions $U_t(x_1|x_2, \ldots, x_s)$ and $U_t(x_1, x_2|x_3, \ldots, x_s)$ by

$$\begin{aligned} D_s(x_1, x_2, \ldots, x_s, t) = {} & U_t(x_1)D_{s-1}(x_2, \ldots, x_s, t) \\ & + \sum_{i=2}^{s} U_t(x_1|x_i)D_{s-2}(x_2, \ldots, x_{i-1}, x_{i+1}, \ldots, x_s, t) \\ & + \sum_{2\leq i<j\leq s} U_t(x_1|x_i, x_j)D_{s-3} + \cdots + U_t(x_1|x_2, \cdots, x_s) \end{aligned} \tag{194a}$$

and

$$\begin{aligned} D_s(x_1, x_2, \ldots, x_s, t) = {} & U_t(x_1, x_2)D_{s-2}(x_3, \ldots, x_s, t) + \sum_{i=3}^{s} U_t(x_1, x_2|x_i)D_{s-3} \\ & + \cdots + U_t(x_1, x_2|x_3, \ldots, x_s) \end{aligned} \tag{194b}$$

As in equilibrium, we obtain explicit forms for the U_t functions as*

$$U_t(x_1) = D_1(x_1, t) \tag{195a}$$

$$U_t(x_1|x_2) = D_2(x_1, x_2, t) - D_1(x_1, t)D_1(x_2, t) \tag{195b}$$

$$\begin{aligned} U_t(x_1|x_2, x_3) = {} & D_3(x_1, x_2, x_3, t) - D_2(x_1, x_2, t)D_1(x_3, t) \\ & - D_2(x_1, x_3, t)D_1(x_2, t) - D_2(x_2, x_3, t)D_1(x_1, t) \\ & + 2D_1(x_1, t)D_1(x_2, t)D_1(x_3, t) \end{aligned} \tag{195c}$$

etc., and

$$U_t(x_1, x_2) = D_2(x_1, x_2, t) \tag{196a}$$

$$U_t(x_1, x_2|x_3) = D_3(x_1, x_2, x_3, t) - D_2(x_1, x_2, t)D_1(x_3, t) \tag{196b}$$

$$\vdots$$

Then using Eqs. (194a) and (194b) for $s = N$, the normalization condition (193) for the D_s and the definitions (171b) and (173) for $F_1(x_1, t)$ and $F_2(x_1, x_2, t)$, we obtain

$$F_1(x_1, t) = D_1(x_1, t) + \frac{N-1}{V}\int dx_2\, [D_2(x_1, x_2, t) - D_1(x_1, t)D_1(x_2, t)] + \cdots \tag{197}$$

*Notice that the U_t are related to D_s in the same way that the U^{eq} are related to the W_s.

$$F_2(x_1, x_2, t) = D_2(x_1, x_2, t) + \frac{N-2}{V} \int dx_3 \, [D_3(x_1, x_2, x_3, t) - D_2(x_1, x_2, t) D_1(x_1, t)] + \cdots \tag{198}$$

Consequently we have expressed $F_1(x_1, t)$ and $F_2(x_1, x_2, t)$ in terms of density expansions involving the functions $D_s(x_1, \ldots, x_S, t)$. These functions can, in turn, be obtained if we know the initial ensemble distribution $D_N(x_1, x_2, \ldots, x_N, 0)$. From this function we can then obtain $F_1(x_1, 0)$ and $F_2(x_1, x_2, 0), \ldots, F_s(x_1, x_2, \ldots, x_s, 0)$ and if we expand these functions in powers of the density,* the lowest-order terms are $D_1(0)$, $D_2(0), \ldots, D_s(0), \ldots$, respectively. Finally, we can relate $D_s(x_1, \ldots, x_s, t)$ to $D_s(x_1, \ldots, x_s, 0)$ by using the fact that $D_s(x_1, \ldots, x_s, t)$ satisfies the Liouville equation, which can be formally integrated, to yield the relation

$$D_s(x_1, \ldots, x_s, t) = \bar{S}_{-t}(x_1, x_2, \ldots, x_s) D_s(x_1, x_2, \ldots, x_s, 0) \tag{199}$$

with the "streaming operator" $\bar{S}_{-t}(x_1, \ldots, x_s)$ defined by

$$\bar{S}_{-t}(x_1, \ldots, x_s) = \exp(-t\mathcal{H}_s) \tag{200}$$

So far, everything we have done has been correct, no matter what the initial distribution function $D_N(x_1, \ldots, x_N, 0)$ happens to be. However, we are now going to introduce a new assumption into the theory, which will take the place of the *Stosszahlansatz*. We will assume that the initial state is such that each of the $D_s(x_1, \ldots, x_s, 0)$ may be expressed as

$$D_s(x_1, \ldots, x_s, 0) = A_s(x_1, \ldots, x_s, 0) \prod_{i=1}^{s} D_i(x_i, 0) \tag{201}$$

where each of the $A_s(x_1, \ldots, x_s, 0)$ has the same factorization property as the equilibrium $W_s(x_1, \ldots, x_s)$ functions. That is, if the initial phase of the particles are such that a group of s_1 particles are several molecular diameters from all the other s–s_1 particles, then $A_s(x_1, \ldots, x_s, 0) \to A_{s_1}(0) A_{s-s_1}(0)$.† This assumption implies that any correlation between the particles in the initial state of the gas extends only over distances of a few molecular diameters. Consequently, our theory will only describe the time development of a gas starting from some initial state where the particles have only short-range correlations. As we proceed, we will see that the collisions taking place in the gas produce

*Here we assume that it is possible to expand these reduced distribution functions as a power series in the density at time $t = 0$. If this is not possible, one can still derive equations similar to (197) and (198) by making a cluster expansion of the N-particle streaming operator $\exp(-t\mathcal{H}_N)$. In the modified cluster expansions only the functions $F_1(x_i, 0)$, $F_2(x_i, x_j, 0), \ldots$, appear.

†As a result of this property, it follows that if at $t = 0$ all the s particles are more than a few molecular diameters from each other then $A_s = 1$.

long-range correlations after some time has elapsed, but our assumption is that in the initial state, at least, such correlations are absent.

If we now use the fact that we are considering a large system so that $(N-1)/V \approx n$, etc., we can write Eqs. (197) and (198) for $F_1(x_1, t)$ and $F_2(x_1, x_2, t)$, using Eqs. (199) and (201), as

$$F_1(x_1, t) = \bar{S}_{-t}(x_1)D_1(x_1, 0) + n \int dx_2 \, [\bar{S}_{-t}(x_1, x_2)A_2(x_1, x_2, 0) - \bar{S}_{-t}(x_1)\bar{S}_{-t}(x_2)]D_1(x_1, 0)D_1(x_2, 0) + O(n^2) \tag{202a}$$

and

$$F_2(x_1, x_2, t) = \bar{S}_{-t}(x_1, x_2)A_2(x_1, x_2, 0)D_1(x_1, 0)D_1(x_2, 0) + n \int dx_3 \, [\bar{S}_{-t}(x_1, x_2, x_3)A_3(x_1, x_2, x_3, 0) - \bar{S}_t(x_1, x_2)\bar{S}_{-t}(x_3)A_2(x_1, x_2, 0)] \prod_{i=1}^{3} D_1(x_i, 0) + O(n^2) \tag{202b}$$

Then we solve Eq. (202a) for $D_1(x_i, 0)$ in terms of $F_1(x_i, t)$ and use this relation to eliminate the $D_1(x_i, 0)$ in $F_2(x_1, x_2, t)$ in favor of the $F_1(x_i, t)$ in the same way as the fugacity z is eliminated in the corresponding equilibrium expansions.* We find

$$F_2(x_1, x_2, t) = \bar{\bar{\mathscr{S}}}_t(x_1, x_2)F_1(x_1, t)F_1(x_1, t) + n \int dx_3 \, \bar{\bar{\mathscr{T}}}_t(x_1, x_2|x_3) \prod_{i=1}^{3} F_1(x_i, t) + O(n^2) \tag{203}$$

where

$$\bar{\bar{\mathscr{S}}}_t(x_1, \ldots, x_s) = \bar{S}_{-t}(x_1, \ldots, x_s)A_s(x_1, \ldots, x_s, 0) \prod_{i=1}^{s} \bar{S}_{+t}(x_i) \tag{204}$$

One might ask why any rearrangement of Eqs. (202a) and (202b) is necessary, i.e., why can't these equations be used to determine F_1 and F_2 for all times? The answer was given by Bogoliubov, who showed that the lth term in each expansion grows with time as $(t/t_{\mathrm{mfp}})^{l-1}$. For example, the second term on the right-hand side of Eq. (202a) grows with time as (t/t_{mfp}) due to the fact that contributions to this term come from regions of phase space for particle 2, where particle 2 collides with particle 1 at some time during a time interval of length t. The volume of this phase space is proportional to the volume of the collision cylinder for such a (1,2) collision, which in turn is proportional to time t. In general, the $(t/t_{\mathrm{mfp}})^{l-1}$ dependence of the lth term in (202a) comes from sequences of $l-1$ binary collisions among l-particles. A similar discussion can be made for (202b). For this reason Eqs. (202a) and (202b) cannot be used for $t \gtrsim t_{\mathrm{mfp}}$. Therefore, some rearrangement is necessary, and this provides the fundamental motivation for the rearrangement carried out here, which then leads to a derivation of the Boltzmann equation. Roughly speaking, using the Boltzmann equation is equivalent to summing the $(t/t_{\mathrm{mfp}})^{l-1}$ series contained in (202a) to an exponential of form $\exp-(t/t_{\mathrm{mfp}})$. A more complete discussion is given by Cohen.[18,19,160,161] and where $\bar{S}_{+t}(x_i)$ is defined by the relation

$$\bar{S}_{+t}(x_i)\bar{S}_{-t}(x_i) = 1 \tag{205}$$

and

$$\begin{aligned}\bar{\bar{\mathcal{T}}}_t(x_1, x_2|x_3) = \bar{\bar{\mathcal{S}}}_t(x_1, x_2, x_3) - \bar{\bar{\mathcal{S}}}_t(x_1, x_2)\bar{\bar{\mathcal{S}}}_t(x_1, x_3)\\ - \bar{\bar{\mathcal{S}}}_t(x_1, x_2)\bar{\bar{\mathcal{S}}}_t(x_2, x_3) + \bar{\bar{\mathcal{S}}}_t(x_1, x_2)\end{aligned} \tag{206}$$

We have now succeeded in expressing $F_2(x_1, x_2, t)$ in terms of $F_1(x_i, t)$.† Therefore, the first hierarchy equation now takes the form

$$\begin{aligned}\frac{\partial F_1(x_1, t)}{\partial t} + \frac{\mathbf{p}}{m} \cdot \nabla_{\mathbf{r}} F_1(x_1, t) - \bar{\mathbf{T}}_w(1)F_1(x_1, t)\\ = n\int dx_2\, \theta_{12}\bar{\bar{\mathcal{S}}}_t(x_1, x_2)F_1(x_2, t)F_1(x_2, t)\\ + n^2 \int dx_2 \int dx_3\, \theta_{13}\, \bar{\bar{\mathcal{T}}}_t(x_1, x_2|x_3) \prod_{i=1}^{3} F_1(x_i, t) + O(n^3)\end{aligned} \tag{207}$$

This equation is the first important result of our use of the nonequilibrium generalization of the cluster expansion method. It expresses the time rate of change of the single-particle distribution function as a density expansion whose terms depend successively on the dynamics of a system of two, three, etc., particles in the container.

Although interactions with the walls are also taken into account in this equation, it is again convenient to simplify our discussion by ignoring them. To do this, we again make use of the thermodynamic limit $N \to \infty$, $V \to \infty$, $N/V = n$, fixed. In this limit, all boundary effects disappear and the first hierarchy equation becomes

$$\begin{aligned}\frac{\partial F_1(x_1, t)}{\partial t} + \frac{\mathbf{p}}{m} \cdot \nabla_{\mathbf{r}} F_1(x_1, t)\\ = n\int dx_2\, \theta_{12}\tilde{\mathcal{S}}_t(x_1, x_2)F_1(x_1, t)F_1(x_2, t)\\ + n^2 \int dx_2 \int dx_3\, \theta_{12}\tilde{\mathcal{T}}_t(x_1, x_2|x_3) \prod_{i=1}^{3} F_1(x_2, t) + \cdots\end{aligned} \tag{208}$$

where the $\tilde{\mathcal{S}}_t$ are obtained from the $\bar{\bar{\mathcal{S}}}_t$ by dropping the $\bar{\mathbf{T}}_w$ collision operators in the streaming operators $\bar{S}_{-t}(x_1, \ldots, x_s)$, and in $\bar{S}_{+t}(x_i)$. The first term on the right-hand side of Eq. (208) represents the contribution to the change in $F_1(x_1, t)$ due to binary collisions, and it is of interest to us to see if there are circumstances where this term is equal‡ to the Boltzmann collision integral

*In the case that we neglect the wall interactions then $\bar{S}_t(x_i) = \exp(t\mathcal{H}_0(x_i))$. If we include the wall interactions, some care is needed to define $\bar{S}_t(x_i)$ properly. We will not pursue this here.

†Note the close structural similarity between the equilibrium expansion for $F_2^{\text{eq}}(x_1, x_2)$, Eq. (191), and the nonequilibrium expansion, Eq.(203). The two equations can be made to correspond by replacing $\bar{\bar{\mathcal{S}}}_t(x_1, \ldots, x_s)$ in (203) by $W_s(x_1, \ldots, x_s)$.

‡Apart from the slight difference between $f(\mathbf{r}, \mathbf{v}, t)$ and $F(x, t)$.

$J(f, f)$, discussed in Section 2. This term was first obtained by Bogoliubov[15] and he was able to show that if one neglects differences in position that are of the order of a molecular size and that if $t \gg t_d$, where t_d is the average *duration* of a binary collision, then

$$\int dx_2\, \theta_{12}\tilde{\mathcal{S}}_t(x_1, x_2)F_1(x_1, t)F_1(x_2, t)$$

$$= \int d\mathbf{p}_2 \int b\, db\, d\varepsilon \left|\frac{\mathbf{p}_2}{m} - \frac{\mathbf{p}_1}{m}\right| [F_1(\mathbf{r}_1, \mathbf{p}_1', t)F_1(\mathbf{r}_1, \mathbf{p}_2', t) - F_1(\mathbf{r}_1, \mathbf{p}_1, t)F_1(\mathbf{r}_1, \mathbf{p}_2, t)]$$

$$\equiv J(F_1, F_1) \tag{209}$$

where $(\mathbf{p}_1', \mathbf{p}_2')$ are the restituting momenta. The proof is simple and instructive. We begin by noting that the θ_{12} operator appearing in the integrand on the left-hand side of Eq. (209) only has nonzero values inside the action sphere of radius a about the center of particle 1. Therefore, we may write the left-hand side of (209) as

$$\int d\mathbf{p}_2 \int d\mathbf{r}_2\, \theta_{12}\tilde{\mathcal{S}}_t(x_1, x_2)F(x_1, t)F(x_2, t), \qquad |\mathbf{r}_{12}| < a \tag{210}$$

Consider next the action of the operator $\tilde{\mathcal{S}}_t(x_1, x_2)$ acting on some function $g(x_1, x_2)$ of the phases x_1, x_2 of particles 1 and 2. First of all, the operator* $S_{-t}(x_1, x_2)$ acting on some function $g(x_1, x_2)$ of the phases of particles 1 and 2 replaces this function by $g(x_1(-t), x_2(-t))$, i.e.,

$$S_{-t}(x_1, x_2)g(x_1, x_2) = g(x_1(-t), x_2(-t)) \tag{211}$$

where $x_1(-t)$ and $x_2(-t)$ are defined as follows. If particles 1 and 2 start from the phases $x_1(-t)$ and $x_2(-t)$ and move only under the influence of their mutual potential energy, then after a time t they will reach the phases x_1 and x_2, respectively. This situation is illustrated in Fig. 21a. The operators $S_{+t}(x_i)$ trace the free-particle motion forward for a time t,† starting at the phase x_i. Consequently

$$\begin{aligned}
\tilde{\mathcal{S}}_t(x_1, x_2)g(x_1, x_2) &= S_{-t}(x_1, x_2)A_2(x_1, x_2, 0)S_{+t}(x_1)S_{+t}(x_2)g(x_1, x_2) \\
&= A_2(x_1(-t), x_2(-t), 0)S_{-t}(x_1, x_2)S_{+t}(x_1)S_{+t}(x_2)g(x_1, x_2) \\
&= A_2(x_1(-t), x_2(-t), 0)S_{+t}(x_1(-t))S_{+t}(x_2(-t))g(x_1(-t), x_2(-t)) \\
&= A_2(x_1(-t), x_2(-t))g(x_1^*, x_2^*)
\end{aligned} \tag{212}$$

where x_1^* and x_2^* are the phases of particles 1 and 2 that are reached from $x_1(-t)$ and $x_2(-t)$ by following the free-particle motion of each particle

*We denote by $S_{-t}(x_1, \ldots, x_s)$ the result of $\bar{S}_{-t}(x_1, \ldots, x_s)$, where the $\bar{\mathbf{T}}_w(x_i)$ are dropped, i.e., $S_{-t}(x_1, \ldots, x_s) = \exp[-t\mathcal{H}_s(x_1, \ldots, x_s)]$.

†Notice that the action of the streaming operators is such that we first consider the leftmost operator acting on all functions and operators to its right. Then we consider the action of the next leftmost operator acting on every thing to its right, etc.

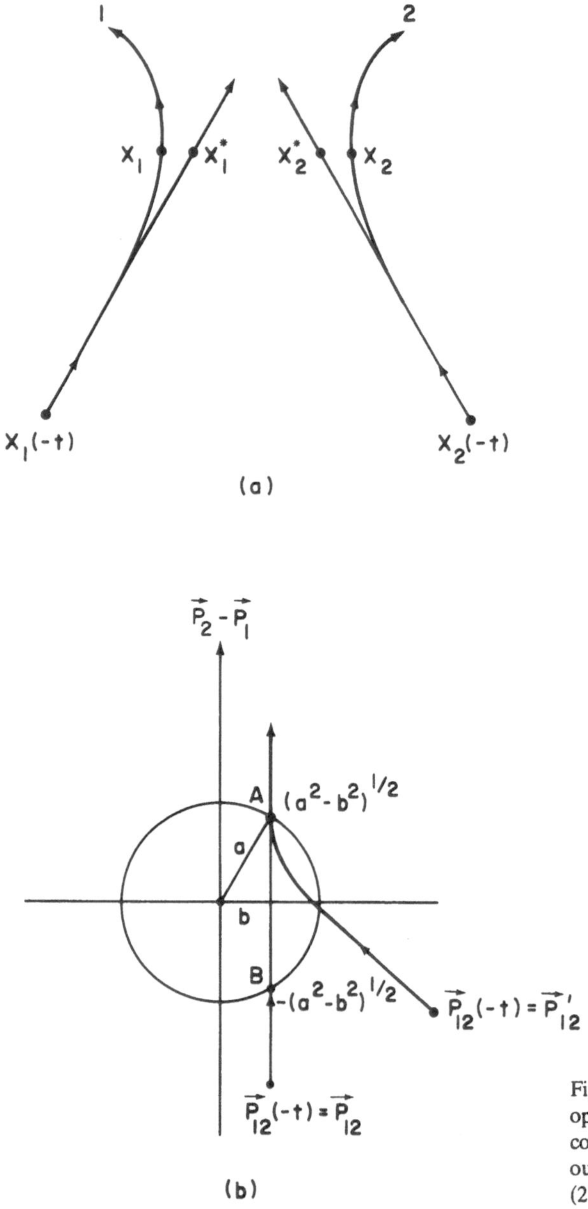

Fig. 21. (a) The action of the operator $\mathscr{S}_t(x_1, x_2)$ on r_{12}. (b) The coordinate system used in carrying out the $\mathbf{r}_{21}$ integration in Eqs. (220)–(222).

forward in time over an interval t. We have then†

$$\tilde{\mathscr{S}}_t(x_1, x_2)F_1(x_1, t)F_1(x_2, t) = A_2(x_1(-t), x_2(-t))F_1(\mathbf{r}_1^*, \mathbf{p}_1(-t), t) \times F_1(\mathbf{r}_2^*, \mathbf{p}_2(-t), t) \tag{213}$$

†From here on we write $A_s(x_1, \ldots, x_s, 0)$ as $A_s(x_1, \ldots, x_s)$.

where we have used

$$\mathbf{p}_1^* = \mathbf{p}_1(-t), \qquad \mathbf{p}_2^* = \mathbf{p}_2(-t) \tag{214}$$

since the single-particle forward streaming operators $S_{+t}(x_i)$ do not change the momentum of the particle. That is,

$$S_{+t}(x_i)f(x_i) = S_{+t}(x_i)f(\mathbf{r}_i, \mathbf{p}_i) = f\left(\mathbf{r}_i + \frac{\mathbf{p}_i}{m}t, \mathbf{p}_i\right) \tag{215}$$

Now we use the facts that if the phases x_1 and x_2 are such that $|r_1 - r_2| < a$, and if $t \gg t_d$, where t_d is the duration of the binary collision of particles 1 and 2, then the phases $x_1(-t), x_2(-t)$ will be such that $|r_1(-t) - r_2(-t)| \gg a$, as illustrated as Fig. 21a. Under these circumstances,

$$A_2(x_1(-t), x_2(-t)) = 1 \tag{216}$$

since we have assumed that the $A_s(x_1, \ldots, x_s)$ have the same factorization property that the $W_S(x_1, \ldots, x_s)$ have in equilibrium. Using this property, we may write

$$\int_{|\mathbf{r}_{12}|<a} dx_2\, \theta_{12}\mathcal{\tilde{S}}_t(x_1, x_2)F_1(x_1, t)F_1(x_2, t)$$
$$= \int_{|\mathbf{r}_{12}|<a} dx_2\, \theta_{12}F_1(\mathbf{r}_1^*, \mathbf{p}_1(-t), t)F_1(\mathbf{r}_2^*, \mathbf{p}_2(-t), t) \tag{217}$$

Now we neglect the difference in position between $\mathbf{r}_1^*$ and $\mathbf{r}_1$ and between $\mathbf{r}_2^*$ and $\mathbf{r}_1$, since both $|\mathbf{r}_1^* - \mathbf{r}_1| \lesssim a$ and $|\mathbf{r}_2^* - \mathbf{r}_1| \lesssim a$. Therefore we replace both $\mathbf{r}_1^*$ and $\mathbf{r}_2^*$ in the right-hand side of (217) by $\mathbf{r}_1$ and write†

$$\int_{|\mathbf{r}_{12}|<a} dx_2\, \theta_{12}\mathcal{\tilde{S}}_t(x_1, x_2)F_1(x_1, t)F_1(x_2, t)$$
$$= \int_{|\mathbf{r}_{12}|<a} dx_2\, \theta_{12}F_1(\mathbf{r}_1, \mathbf{p}_1(-t), t)F_1(\mathbf{r}_1, \mathbf{p}_2(-t), t)$$
$$= \int_{|\mathbf{r}_{12}|<a} dx_2\, \theta_{12}S_{-t}(x_1, x_2)F_1(\mathbf{r}_1, \mathbf{p}_1, t)F_1(\mathbf{r}_1, \mathbf{p}_2, t) \tag{218}$$

where the operator $S_{-t}(x_1, x_2)$ acts *only* on $\mathbf{p}_1$ and $\mathbf{p}_2$, and we regard $\mathbf{r}_1$ as a *fixed* parameter in the single-particle distribution function, since we have already

†We see here that the two-particle collision operator in Eq. (208) does take into account the differences in position of the colliding particles. As in Enskog theory, these differences in position must be taken into account in computing density corrections to the Boltzmann equation transport coefficients. They give rise to the collisional transfer contributions.

taken into account the action of the streaming operator on $\mathbf{r}_1$ and $\mathbf{r}_2$. To obtain the Boltzmann operator $J(f, f)$, we use the fact that, for $|\mathbf{r}_{12}| \le a$,

$$\theta_{12}S_{-t}(x_1, x_2) = S_{-t}(x_1, x_2)(\theta_{12} - \mathscr{H}_0(x_1) - \mathscr{H}_0(x_2)) + (\mathscr{H}_0(x_1) + \mathscr{H}_0(x_2))S_{-t}(x_1, x_2) \tag{219}$$

where we have used the fact that the operator $\mathscr{H}_2(x_1, x_2) = \mathscr{H}_0(x_1) + \mathscr{H}_0(x_2) - \theta_{12}$ commutes with the streaming operator $S_{-t}(x_1, x_2)$. Thus

$$\int_{|\mathbf{r}_{12}|<a} dx_2\, \theta_{12}S_{-t}(x_1, x_2)F(\mathbf{r}_1, \mathbf{p}_1, t)F(\mathbf{r}_1, \mathbf{p}_2, t)$$
$$= \int_{|\mathbf{r}_{12}|<a} d\mathbf{r}_2 \int d\mathbf{p}_2 \frac{1}{m}\mathbf{p}_{21} \cdot \frac{\partial}{\partial \mathbf{r}_{21}} F_1(\mathbf{r}_1, \mathbf{p}_1(-t), t)F_1(\mathbf{r}_1, \mathbf{p}_2(-t), t) \tag{220}$$

where we have used

$$\mathscr{H}_0(x_i)F_1(\mathbf{r}_1, \mathbf{p}_i, t) = (\mathbf{p}_i/m) \cdot \nabla_{\mathbf{r}_i} F_1(\mathbf{r}_1, \mathbf{p}_i, t) = 0 \tag{221}$$

since the operators do not act on the position coordinates in the distribution function, for $t \gg t_d$, and we have used the fact that if $|\mathbf{r}_{12}| < a$ then $\theta_{12}(x_1(-t), x_2(-t)) = 0$; $|\mathbf{r}_1(-t) - \mathbf{r}_2(-t)| \gg a$ and the force between the particles vanishes. To perform the integration in (220), set up a cylindrical coordinate system with *z axis* in the direction of $\mathbf{p}_{21}$ and write

$$d\mathbf{r}_2 = d\mathbf{r}_{21} = b\, db\, d\epsilon\, dz$$

where b, ϵ measure the polar coordinates in a plane perpendicular to the z axis as illustrated in Fig. 21b. Consequently,

$$\int_{|\mathbf{r}_{12}|<a} dx_2 \theta_{12} S_{-t}(x_1, x_2)F_1(\mathbf{r}_1, \mathbf{p}_1, t)F_1(\mathbf{r}_1, \mathbf{p}_2, t)$$
$$= \int d\mathbf{p}_2 \int_0^a b\, db \int_0^{2\pi} d\epsilon \frac{1}{m}|\mathbf{p}_2 - \mathbf{p}_1|\, F_1(\mathbf{r}_1, \mathbf{p}_1(-t), t)F_1(\mathbf{r}_1, \mathbf{p}_2(-t), t)\Big|_{z=-(a^2-b^2)^{1/2}}^{z=(a^2-b^2)^{1/2}} \tag{222}$$

When $z = (a_2 - b_2)^{1/2}$, then particles 1 and 2 have just finished colliding, so that if $t \gg t_d$, $\mathbf{p}_1(-t) = \mathbf{p}_1'$, $\mathbf{p}_2(-t) = \mathbf{p}_2'$, where $\mathbf{p}_1'$ and $\mathbf{p}_2'$ are the restituting momenta of particles 1 and 2, respectively. When $z = -(a^2 - b^2)^{1/2}$, particles 1 and 2 are just about to collide, so $\mathbf{p}_1(-t) = \mathbf{p}_1$ and $\mathbf{p}_2(-t) = \mathbf{p}_2$. Thus

$$\int_{|\mathbf{r}_{12}|<a} dx_2\, \theta_{12}\tilde{\mathscr{S}}_t(x_1, x_2)F_1(x_1, t)F_1(x_2, t) = \int d\mathbf{p}_2 \int_0^a b\, db \int_0^{2\pi} d\epsilon \frac{1}{m}|\mathbf{p}_1 - \mathbf{p}_2|$$
$$\times [F_1(\mathbf{r}_1, \mathbf{p}_1', t)F_1(\mathbf{r}_1, \mathbf{p}_2', t) - F_1(\mathbf{r}_1, \mathbf{p}_1, t)F_2(\mathbf{r}_1, \mathbf{p}_2, t)]$$
$$= J(F_1, F_1) \tag{223}$$

Thus we have shown that if the correlations between the particles in the initial state are short ranged, if $t \gg t_d$, and if we neglect the change in positions of two particles during a binary collision, then the binary collision term in Eq. (208) is equal to the Boltzmann collision integral.

In the analysis of the higher-order collision terms, in the generalized Boltzmann equation, it is important to determine:

(a) the dynamical events that contribute to each term;

(b) whether or not the initial correlation functions $A_l(x_1, \ldots, x_l)$ can all be set equal to unity in the collision integrals for times longer than some microscopic time;

(c) whether or not the collision operators

$$\int dx_2 \cdots \int dx_l\, \theta_{12}\tilde{\mathcal{T}}_t(x_1, x_2|x_3, \ldots, x_l)$$

reach a time-dependent form,

$$\int dx_2 \cdots \int dx_l\, \theta_{12}\mathcal{T}_\infty(x_1, x_2|x_3, \ldots, x_l),$$

for times t greater than some microscopic time; and

(d) whether the generalized Boltzmann equation predicts that as $t \to \infty$, the single-particle distribution function will approach its equilibrium value.

We would like to prove that the initial state is forgotten, and that the collision operators reach an asymptotic form for times long compared to some microscopic time. If this were true, then the normal solution method, when applied to the generalized Boltzmann equation, would lead to expressions for the transport coefficients for a dense gas that would (a) be independent of the precise initial state of the gas, (b) be independent of the time elapsed since the initial state of the gas, and (c) have a density expansion of the form

$$\eta/\eta_0 = 1 + \eta_1(T)na^3 + \eta_2(T)(na^3)^2 + \cdots \tag{224}$$

for the viscosity, say, where $\eta_1(T)$ is determined by the dynamics of three particles, $\eta_2(T)$ by the dynamics of four particles, and so on, and η_0 is the low-density value of the viscosity.*

We consider first $\theta_{12}\tilde{\mathcal{T}}_t(x_1, x_2|x_3)$, the three-particle collision operator.(162–167)† Because of the presence of the θ_{12} operator, we are again only interested, for the generalized Boltzmann equation, in phases x_1 and x_2 of particles 1 and 2, where $|\mathbf{r}_{12}| < a$. For these phases of particles 1 and 2, a dynamical analysis of the operator $\tilde{\mathcal{T}}_t(x_1, x_2|x_3)$ similar to that given for $\mathcal{S}_t(x_1, x_2)$ shows that the following dynamical events contribute to this collision operator.(164–167)

(a) A triple collision where all three particles are within each other's action spheres at the same time, as illustrated in Fig. 22a. For a gas of hard spheres, the triple collision reduces to the three-body excluded volume term

*For the binary collision contribution to the generalized Boltzmann equation, we have just proved that the initial state is forgotten and that the binary collision contribution to η is the Boltzmann equation result.

†The three-body collision integral was first discussed in some detail by Choh and Uhlenbeck,(162) and it is often referred to as the Choh–Uhlenbeck term.

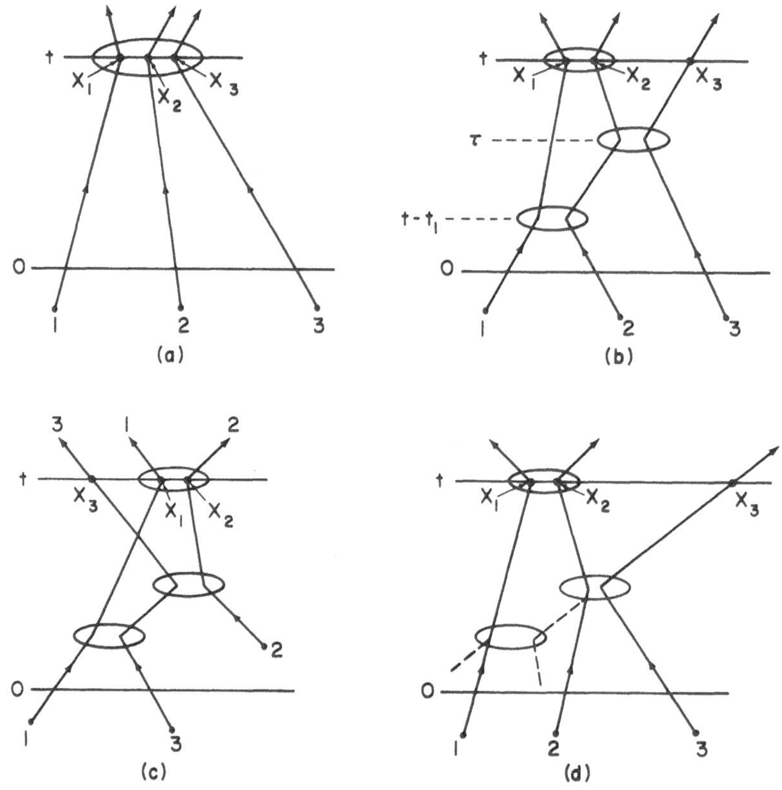

Fig. 22. Some of the collision events taking place among three particles that contribute to $\tilde{\mathcal{T}}_t(x_1, x_2|x_3)$. (a) represents a genuine triple collision taking place among three particles. For the case of hard spheres this would also include "overlapping" events like that illustrated in Fig. 20. (b), (c), and (d) represent three successive binary collisions among the three particles. The quantities x_1, x_2, x_3 represent the phases of the three particles at time t. In (b) (the recollision event) we indicate the times at which the various binary collisions take place, for use in the discussion of this event in the text and in Fig. 23.

incorporated in the Enskog theory, illustrated in Fig. 20, where the center of particle 3 is within the action sphere of both particles 1 and 2 when these two particles collide.

(b) Successions of three or more binary collisions taking place between the three particles. These binary collisions must be ordered in time in such a way that they take place in the time interval from the initial state $t=0$, to the time t when we are evaluating the distribution function, and the final binary collision is the one taking place at time t when the phases of particles 1 and 2 are x_1 and x_2 with $|\mathbf{r}_{12}|<a$, as required by the θ_{12} operator. Examples of three binary collisions among the three particles that contribute to $\tilde{\mathcal{T}}_t(x_1, x_2|x_3)$ are illustrated in Fig. 22b-d. In Figs. 22b and 22c we give some examples where

three binary collisions actually take place between time $t = 0$ and t, and in Fig. 22d, we illustrate a collision sequence where two collisions actually take place, and the third would have occurred, when the motion of the particles is traced backward in time, except for the interference of particle 2.* Although the (1,3) collision never actually occurs, the Boltzmann equation has taken it into account. Therefore, the three-body collision integral contains a term that corrects the mistake of the Boltzmann equation. Similar corrections are taken into account in the higher-order collision operators. For hard-sphere molecules at least, there are, in addition, sequences of four binary collisions among three particles that we have not illustrated here.

We see that if particles 1 and 2 have suffered earlier collisions with each other, as illustrated in Fig. 22b, or with the same third particle, as in Figs. 22c and 22d, then there will be both spatial and velocity correlations between particles 1 and 2 before the onset of their collision at time t. These dynamical correlations are not taken into account in the Boltzmann equation, since the *Stosszahlansatz* assumes they do not exist, nor are they taken into account by the Enskog theory for hard spheres, which ignores all velocity correlations. Here we see that dynamical correlations do exist in the gas and that they are accounted for in the generalized Boltzmann equation since it takes the "concerted" action of three or more particles to produce such correlations.

We can now apply the information about the dynamical events that contribute to $\theta_{12}\tilde{\mathcal{T}}_t(x_1, x_2|x_3)$ to answer the questions posed earlier about whether or not the initial state is eventually forgotten, and if the three-body collision integral ever reaches an asymptotic form. In formulating the answer to these questions, we must bear in mind that we would like to construct the normal solution to the generalized Boltzmann equations. Like the normal solution to the Boltzmann equation, the generalized normal solution would depend on time only through the time variations of the hydrodynamic variables and would decay to the equilibrium distribution function only very slowly, i.e., on the hydrodynamic time scale. As we mentioned earlier, to be useful the generalized Boltzmann equation should reach an asymptotic form after a microscopic time. Therefore, we are forced to rely on dynamical arguments to prove that the collision operators reach an asymptotic form in the proper time interval, and we cannot use any arguments based on the eventual decay of the distribution function to its equilibrium form.[168]

For the two-particle collision integral, we were able to use dynamical arguments to show that in this term the initial state is forgotten and the asymptotic form is reached after a time long compared to the duration of a collision. In the case of the three-particle collision term, it is possible to show

*In the Liouville equation derivation of the Boltzmann equation, one is forced to follow the streaming of the particles backward in time. Here one would say that since particles 1 and 3 appear to have collided in the past, the Boltzmann equation treats this collision as if it had actually taken place.

that if the distribution function $F_1(x_1, t)$ is not the Maxwell–Boltzmann equilibrium distribution function* then[(22,23,25,26,115,116)]:

(a) The initial correlations between the three particles decay with time as $(t_d/t)^2$.

(b) The difference between the correct three-body collision integral $\int dx_2 \int dx_3\, \theta_{12}\bar{\mathcal{T}}_t(x_1, x_2|x_3)\cdot\Pi_{i=1}^3 F_1(x_i, t)$, and the form obtained by replacing the collision operator by its asymptotic form, $\theta_{12}\mathcal{T}_\infty(x_1, x_2|x_3)$, decays in time as t_d/t. Here, t_d is the average time it takes a molecule to travel a distance a, i.e., $t_d = a/\langle v\rangle$.

These slow time decays are due to the correlated binary collision sequences that take place among the three particles. A rough argument will reveal their origin. For reasons similar to those given earlier for the binary collision term, it is easy to see that the contributions to the three-particle collision term from genuine triple collisions and from sequences of binary collisions where all the binary collisions take place within a few molecular diameters from each other reach their asymptotic form and become independent of the initial state after a time t greater than a few t_d. Let us now consider the contribution to the three-particle term from sequences of three or more binary collisions where the time interval between the first binary collision and the final one at time t can vary between T and t, where T is a time on the order of several t_d. To estimate the order of magnitude of these contributions, we consider the dynamical events illustrated in Fig. 22b, the so-called recollisions, and first estimate the volume of that region of the phase space of particle 3 for which the time between the first and last (1,2) collision lies in the interval t_1 to $t_1 + dt_1$. To be precise, we must consider the motion of the three particles 1, 2, 3 backward in time from their phases x_1, x_2, x_3 at time t, where $|\mathbf{r}_{12}| < a$, to a configuration at time $t - t_1$, where the particles 1 and 2 are colliding for the first time, i.e., $|\mathbf{r}_{12}(t - t_1)| < a$. We estimate the phase volume $\int dx_3$ for particle 3 at time t, for which the first (1,2) collision takes place in the time interval $(t - t_1 - dt_1)$ to $(t - t_1)$ in the following way. Consider a coordinate system in which particle 2 is at rest immediately before the (1,2) collision at time t. If we follow the motion of the three particles in the system backward in time, it appears that particle 1 moves away from particle 2 with some relative velocity $\mathbf{v}_{12}$, say, and that particle 3 collides with particle 2 at some time τ, say, and knocks particle 2 into particle 1, so that a (1,2) collision takes place at time $t - t_1$.† One can see that particle 2 must be scattered into a solid angle of order

*For which special cancellations occur that invalidate the arguments used to prove the estimates given here.

†That is, we consider the time-reversed motion of the particles starting from their phases x_1, x_2, x_3 at time t. In the time-reversed motion it appears that particle 3 knocks 2 into 1, so that a (1,2) collision takes place at time $t - t_1$.

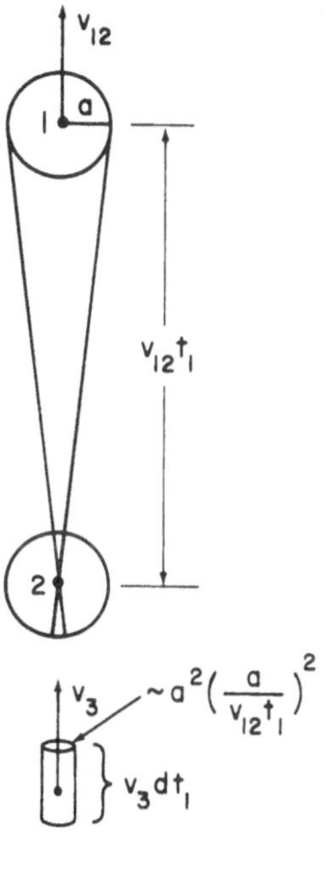

Fig. 23. The recollision event as seen in a coordinate system centered on particle 2. For the purpose of illustration, we take the velocity of particle 3 to be in the direction of $\mathbf{v}_{12}$. The small cylinder of volume $\sim a^2(a/v_{12}t_1)^2 v_3\, dt_1$ represents the region of configuration space for particle 3, so that the second (1,2) collision takes place in the time interval t_1 to t_1+dt_1, after the first (1,2) collision.

of magnitude $(a/v_{12}t_1)^2$, provided $v_{12}t_1 \gg a$. The phase space volume of particle 3 is proportional to this solid angle for the scattering of particle 2, and if the time interval between the two (1,2) collisions varies between t_1 and t_1+dt_1, the phase space volume for particle 3, $d\Gamma_3(t_1)$ is proportional to*

$$d\Gamma_3(t_1) \sim (a/v_{12}t_1)^2\, dt_1 \sim (t_d/t_1)^2\, dt_1 \tag{225}$$

Therefore the contribution to the three-particle collision integral from recollision events taking place over a time interval from T to t is proportional to

$$\Gamma_3(t) \sim \int_T^t (t_d/t_1)^2\, dt_1 \sim [(t_d/T)-(t_d/t)] \tag{226}$$

A similar result is obtained for each of the sequences of three collisions that contribute to the three-particle term.(22–26) From the result for the phase space of particle 3, we see that the three-particle term reaches an asymptotic form

*To see this, consider the case illustrated in Fig. 23, where $\mathbf{v}_3$ is in the direction of $\mathbf{v}_{12}$. Then for fixed v_3 the range of the z coordinate of particle 3 is proportional to $v_3\, dt_1$, and the ranges of the x and y coordinates are each proportional to $a(a/v_{12}t_1)$. Thus the phase volume in coordinate space for particle 3 is proportional to $v_3 a^2(a/v_{12}t_1)^2\, dt_1$. If $\mathbf{v}_3$ is not in the direction of $\mathbf{v}_{12}$, the geometry of the region of configuration space for particle 3 is more complicated, but the phase volume is of the same order, as long as the velocity of particle 2 after the (2,3) collision is sufficient for particle 2 to catch up to and collide with particle 1. We assume also that all velocity integrals are well behaved.

very slowly, the correction being of order t_d/t, as mentioned earlier. Moreover, to estimate how rapidly the initial correlations decay, we note that, for the recollision event, the initial correlations between particles 1 and 2 affect the (1,2) collision at time t only if particles 1 and 2 were also within a molecular distance at $t=0$.* The previous arguments show that the phase space volume of particle 3 for a recollision of particles 1 and 2 with a time interval of t between the first and last collision is of order $(t_d/t)^2$, which gives the rate at which the initial correlations decay in time. Although these results have been obtained here by phase space estimates, they have been confirmed by more careful calculations carried out by Alle and Pomeau[169] and by Dorfman and Kan[115,116] for hard-sphere molecules.

These phase space estimates of the three-body term suggest that if the same calculations were carried out for a hypothetical two-dimensional gas, the three-body collision integral would be logarithmically diverging for long time t, since the solid angle $(t_d/t_1)^2$ would be replaced by a plane angle t_d/t_1 and the corresponding phase space volume of particle 3 would be proportional to

$$\int_T^t dt_1\, t_d/t_1 \sim t_d \log t/T \sim t_d \log(t/t_d) \tag{227}$$

where we have used $T \approx t_d$. This logarithmic divergence of the three-body term in two dimensions has been confirmed by Sengers for a gas of hard-disk molecules.[170,171]

Similar phase space estimates for the time dependence of the higher-order collision terms for both two- and three-dimensional systems lead to the following results[23–26,115,116]:

(a) For two-dimensional systems, only the two-body collision integral reaches a finite asymptotic value, but the l-body collision terms grow with time as $(t/t_d)^{l-3}$ for $l>3$, and $\log(t/t_d)$ for $l=3$.

(b) For three-dimensional systems, the two- and three-body collision integrals reach finite asymptotic values for long times, but the l-body collision integrals grow with time as $(t/t_d)^{l-4}$ for $l>4$, and as $\log(t/t_d)$ for $l=4$.

(c) These secularly growing terms are associated with dynamical events where there are l successive binary collisions taking place between the particles. Some examples of four binary collisions taking place among four particles that contribute to $\theta_{12}\bar{\mathscr{T}}_t(x_1, x_2|x_3, x_4)$ are illustrated in Fig. 24.

The dynamical events that lead to the secular growth of the collision integrals are usually referred to in the literature as the ring events, in accordance with a description that can be given for them in terms of ring graphs.[172,173] Statements (a)–(c) are based on phase space estimates of the type discussed here. Only in the case that the molecules are hard spheres or

*Otherwise the $A_2(x_1, x_2)$ appearing in the $\bar{\mathscr{T}}_t(x_1, x_2|x_3)$ operator can be set equal to unity, and there is no effect of the initial correlations between particles 1 and 2 on the recollision events.

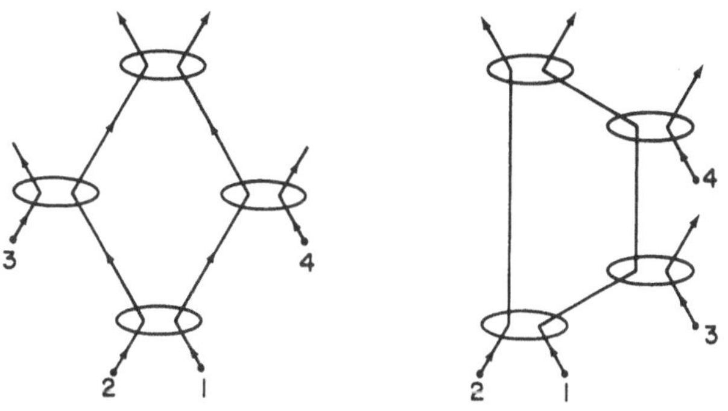

Fig. 24. Some of the four successive binary collision events taking place among four particles (four-body ring events) that are responsible for the divergence of the four-body collision operator.

hard disks[24–26,115,116,169–171] or for simple gas models such as hard-sphere or hard-disk Lorentz models[174–176] has it been possible to verify them in detail, but for more general potentials no rigorous proof that they are true exists.

As a result of the secular growth of the l-body collision integrals with time, we are compelled to conclude that, although the cluster expansion method can be used successfully to derive the Boltzmann equation from the Liouville equation and to obtain corrections to the Boltzmann equation, there are serious difficulties in trying to represent these corrections as a power series in the density. An example of the difficulties that appear if one attempts to apply the generalized Boltzmann equation as it stands now to a problem of some interest is provided by the calculation of the density expansion of the coefficient of shear viscosity. By constructing normal solutions to the generalized Boltzmann equation, one finds that the viscosity η has the expansion of the form mentioned in Eq. (224),

$$\eta/\eta_0(T) = 1 + \eta_1(T)na^3 + \eta_2(T, t)(na^3)^2 + \cdots \tag{228}$$

where $\eta_0(T)$ is the Boltzmann equation result. The first correction to this result, $\eta_1(T)$[161–167] comes from the three-body collision term as well as from collisional transfer effects in binary collisions,* and is independent of the time t. However, the next correction, $\eta_2(T, t)$, coming from the four-body collision term and from collisional transfer effects among three particles, grows with time t as

$$\eta_2(T, t) = \tilde{\eta}_2(T) + \eta_2' \log(t/t_d) \tag{229}$$

where $\tilde{\eta}_2(T)$ is a finite, time-independent quantity. This logarithmic growth of $\eta_2(T, t)$ with time is not confirmed by experimental measurements, and it indicates the presence of a fundamental difficulty in the theory.

*For a gas of hard spheres, Sengers and co-workers have computed $\eta_1(T)$. It differs from the result of the Enskog theory by only a few percent.[164–167]

The divergences that are appearing in the theory are due to the fact that in the collision integrals the particles are allowed to travel arbitrarily long distances between the collisions, in each collision sequence. In a real gas system no particle can travel along a straight line for more than a few mean free paths between successive collisions, since the other particles will interfere with it. However, if we consider small groups of particles in isolation, as we do in the various l-body collision terms, then there is no mechanism for cutting off the lengths of the paths between collisions. Therefore, the expansion of the generalized Boltzmann equation in powers of the density, where each term involves only the dynamics of a small group of molecules, leads to divergences because a collective effect, the mean free path, is not properly taken into account in any individual term.

Therefore, one must make a further rearrangement of the density expansion for the generalized collision integral so that the mean free path damping of the particle trajectories is taken into account. To do this, one sums the most divergent terms in each order of the density in the density expansion of the generalized Boltzmann equation, to obtain a resummed collision operator that incorporates the mean free path damping. For a linearized kinetic theory this resummation was carried out by Kawasaki and Oppenheim(172,173) and by Weinstock(177), and for the full nonlinear theory by Frieman and Goldman,(178,179) and by Dorfman and Cohen.(180,181)

Since the leading divergences in the generalized Boltzmann equation are associated with sequences of binary collisions, the resummations are usually carried out by first expressing the S-particle streaming operators $S_{-t}(x_1, \ldots, x_s)$ in Eq. (208) in terms of sequences of binary collisions that take place between the s particles. This is accomplished by means of the binary collision expansion, which has proven to be one of the most useful tools in the kinetic theory of gases.(182–184)

When we construct normal solutions of the generalized Boltzmann equation using the resummed collision operator and computes the transport coefficients for a moderately dense gas (in three dimensions),* we find that the viscosity, say, has the expansion(25,26,37,115,116,172,173)

$$\eta/\eta_0 = 1 + \eta_1(T)na^3 - \eta_2'(T)(na^3)^2 \log(na^3) + \eta_2''(T)(na^3)^2 + \cdots \quad (230)$$

where the $\log(t/t_d)$ divergence in the original expression of η, Eq. (228), is replaced in the resummed theory by $\log(t_c/t_d)$, which is equal to $\log(na^3)^{-1}$. Thus, the resummed theory predicts that the transport coefficients cannot be

*As we will discuss in the next section, this resummation still does not lead to finite values for transport coefficients in two dimensions.

expanded solely in powers of n, but that terms proportional to $n^2 \log n$ should appear. The presence of these terms can be understood if we note that inclusion of the mean free path damping in the phase space estimates replaces the logarithmically divergent integral (227) by

$$\int_T^t \frac{dt_1}{t_1} e^{-t_1/t_c} \to O(\log(t_c/t_d)) \tag{231}$$

For large times t, the exponential factor e^{-t_1/t_c} in Eq. (231), which takes the damping of the particle trajectories into account, restricts the time that each particle spends between collisions to be on the order of t_c, the mean free time. Upon comparing (231) with (137), the expansion of the force on a sphere in a rarefied gas in powers of the inverse Knudsen number, one can see that these two expansions have a remarkably similar structure. This similarity has its source in the fact that the coefficients in the density expansion of η/η_0 and in the K_N^{-1} expansion of F/F_0 are determined by dynamical events of the same basic types, as may be seen by comparing Fig. 15 with Figs. 22 and 24. The dynamical events that contribute to F/F_0 differ from those which contribute to η/η_0 only in that in F/F_0 one of the gas particles is replaced by a macroscopic object, in this case a sphere.

So far there has been no convincing experimental confirmation of the presence of logarithmic terms in the density expansion of the transport coefficients.(185–188) However, a theoretical value for the coefficient η_2' has been computed for hard-sphere molecules,(25,26,115,116) and if its size is any indication of the order of magnitude of this coefficient for real non-hard-sphere molecules, it is not too surprising that these terms have not been unambiguously detected in the analysis of the experimental data. This is because in the density range where the $n^2 \log n$ terms are important, the terms of order n^3 and higher are also important and it is difficult to separate them.* There are, however, computer studies on simple models, e.g., on the diffusion of a light particle through a system of fixed, hard-disk scatterers† placed with average density n, which clearly show the presence of a $\log n$ term in the density expansion of the diffusion coefficient D.(189)‡

In the next section we will discuss recent computer studies of time correlation functions for dense gases that also provide a striking confirmation of the resummation techniques discussed here.

*There is some experimental evidence that the $n^2 \log n$ terms are important for the viscosity of a dense gas at low temperatures.(187)

†This is the hard-disk Lorentz gas model, and the density n is the density of the scatterers.

‡For another model, the Ehrenfest wind–tree model,(175,176) there are also computer studies that confirm the correctness of the resummation methods discussed here.(33)

4. Time Correlation Functions

4.1. Basic Theory

As we discussed earlier, the generalized Boltzmann equation leads to a density expansion of the transport coefficients of a dense gas. However, general expressions for transport coefficients of a fluid that are not in the form of an expansion can be derived by another technique, the time correlation function method.(27,28) This approach has provided a general framework by means of which one can make detailed comparisons between theoretical results,(37,190) the results of computer-simulated molecular dynamics,(33) and experimental results.(31,32)

There are a number of different formulations of the time correlation function method, all of which lead to the same results for the linearized hydrodynamic equations.(27,28) One way is to generalize the Chapman–Enskog normal solution method so as to apply it to the Liouville equations, and obtain the N-particle distribution function for a system near a local equilibrium state.(191–195) Expressions for the heat current and pressure tensor for a general fluid system can be obtained, which have the form of the macroscopic linear laws, with explicit expressions for the various transport coefficients. These expressions for the transport coefficients have the form of time integrals of equilibrium correlation functions of microscopic currents, *viz.*, a transport coefficient τ is given by

$$\tau = \int_0^\infty dt \langle \mathcal{J}^{(\tau)}(0)\mathcal{J}^{(\tau)}(t)\rangle \tag{232}$$

where $\mathcal{J}^{(\tau)}$ represents a microscopic current associated with the transport coefficient, $\mathcal{J}^{(\tau)}(0)$ is its value at some time, $\mathcal{J}^{(\tau)}(t)$ is its value at a time t later, and the angular brackets denote an average over an equilibrium ensemble. Expressions have been derived for the coefficients of shear and bulk viscosity and thermal conductivity for a pure system, and for mixtures, and for the additional transport coefficients that appear in mixtures.* If one tries to evaluate these general expressions for a dense gas, one can use the cluster expansion method described earlier, and obtain expressions for the transport coefficients in terms of density expansions. In this way one can show that the time correlation function method leads to expressions for transport coefficients appearing in the linearized hydrodynamic equations(29,30)† that are identical to

*The time correlation function method has mainly been used to derive the linearized hydrodynamic equations and their associated transport coefficients. The extension of the theory to nonlinear hydrodynamic equations is still in an early state of development.

†It should be pointed out, however, that the proofs of the equivalence of the time correlation function expressions for transport coefficients with those from the generalized Boltzmann equation were given before the true extent of the divergence difficulties and "long-time tail" difficulties, discussed here, were known. In view of this, it would be worthwhile to reexamine these proofs.

those given by the generalized Boltzmann equation. The principal advantage of the time correlation function method is that it introduces a new quantity, the time correlation function $\langle \mathcal{J}^{(\tau)}(0)\mathcal{J}^{(\tau)}(t)\rangle$, which is more sensitive to the microscopic properties of the fluid than the transport coefficients and can be studied using a number of different methods. The comparison between the results obtained using these methods has proven to be a very fruitful stimulus to further studies in each area.

Here we will study a somewhat simpler nonequilibrium process than we have considered so far, the diffusion of a tagged particle through a gas of particles that are mechanically identical to the tagged one. This diffusion process is called *self-diffusion.*[190,191] Although this process cannot be studied in the laboratory,* it can be studied on a computer, and we refer the reader to the article of Wood and Erpenbeck in this volume for further details on how these studies are performed.[33]

To obtain a macroscopic theory for self-diffusion, we define a quantity $P(\mathbf{r}, t)$ that is the probability density for finding the tagged particle at the point $\mathbf{r}$ at time t. Since the number of tagged particles is conserved, $P(\mathbf{r}, t)$ satisfies a conservation law of the form

$$\partial P(\mathbf{r}, t)/\partial t + \nabla_{\mathbf{r}} \cdot \mathbf{J}(\mathbf{r}, t) = 0 \tag{233}$$

where $\mathbf{J}$ is the probability current. In the hydrodynamic description $\mathbf{J}(\mathbf{r}, t)$ is related to $P(\mathbf{r}, t)$ by Fick's law of diffusion,

$$\mathbf{J}(\mathbf{r}, t) = -D\nabla_{\mathbf{r}} P(\mathbf{r}, t) \tag{234}$$

where D is called the coefficient of self-diffusion, and $P(\mathbf{r}, t)$ then satisfies

$$\partial P(\mathbf{r}, t)/\partial t = D\nabla_{\mathbf{r}}^2 P(\mathbf{r}, t) \tag{235}$$

if we assume that D does not depend on $\mathbf{r}$. One of the consequences of this diffusion equation is that if the particle is released at time $t = 0$ from the point $\mathbf{r}_0$, then the average mean square displacement of the particle $\langle|\Delta\mathbf{r}(t)|^2\rangle = \langle|\mathbf{r}(t) - \mathbf{r}(0)|^2\rangle$ satisfies[190,191]

$$\langle|\Delta\mathbf{r}(t)|^2\rangle = 2dDt \tag{236}$$

where d is the number of dimensions, and the angular brackets denote an average taken with respect to $P(\mathbf{r}, t)$.

The microscopic theory based on the Liouville equation leads to diffusion equation (235) for $t \gg t_{\text{mic}}$, with the coefficient of self-diffusion D given by an equation identical to (236) except that in the microscopic theory the angular bracket is taken to be the average over an equilibrium ensemble.[190,191] To be

*Since tagging a particle changes its mechanical properties.

precise, if the indicated limits exist, D is given by*

$$\begin{aligned} D &= \lim_{t\to\infty} \frac{1}{2dt}\langle|\mathbf{r}_1(t)-\mathbf{r}_1(0)|^2\rangle \\ &= \lim_{t\to\infty} \frac{1}{d}\langle \mathbf{v}_1(t)\cdot(\mathbf{r}_1(t)-\mathbf{r}_1(0))\rangle \\ &= \lim_{t\to\infty} \frac{1}{d}\int_0^t d\tau\langle \mathbf{v}_1(\mathrm{t})\cdot \mathbf{v}_1(\tau)\rangle \\ &= \int_0^\infty d\tau\langle v_{1x}(0)v_{1x}(\tau)\rangle \end{aligned} \tag{237}$$

where the subscript 1 denotes the tagged particle. We have used Liouville's theorem to replace $\langle \mathbf{v}_1(t)\cdot\mathbf{v}_1(\tau)\rangle$ by $\langle \mathbf{v}_1(t-\tau)\cdot\mathbf{v}_1(0)\rangle$, and we have used $\langle v_{1x}(0)v_{1x}(t)\rangle = \langle v_{1y}(0)v_{1y}(t)\rangle$, etc., due to spatial isotropy. Here we see that the coefficient of self-diffusion is given by the time integral of the velocity autocorrelation function. Similar correlation function expressions can also be given for the transport coefficients that appear in the higher-order hydrodynamic equations for this self-diffusion process. For example, for this process the super-Burnett equation takes the form[190,191]

$$\partial P(\mathbf{r}, t)/\partial t = D\nabla_\mathbf{r}^2 P(\mathbf{r}, t) + D_2\nabla_\mathbf{r}^2\nabla_\mathbf{r}^2 P(\mathbf{r}, t)$$

and the super-Burnett self-diffusion coefficient D_2 is given by

$$D_2 = \lim_{t\to\infty} (1/3!)[\langle v_{1x}(t)(\Delta r_{1x}(t))^3\rangle - 3\langle v_{1x}\Delta r_{1x}(t)\rangle\langle(\Delta r_{1x}(t))^2\rangle] \tag{238}$$

Reverting to the velocity autocorrelation function, we see that if the self-diffusion coefficient is to be finite for a particular system, then $\langle v_{1x}(t)v_{1x}(0)\rangle$ must approach zero more rapidly than t^{-1} for large t. Since this correlation function measures how fast a particle "forgets" its initial velocity, we might say that the coefficient of self-diffusion is finite for those systems where a particle forgets its initial velocity sufficiently rapidly with time. One might expect that in a dense gas the collisions that a particle experiences will tend to randomize its velocity, so that the initial velocity will be forgotten after a time on the order of a few mean free times. One of the great surprises resulting from the computer studies of the velocity correlation function is that this expectation is not realized.[38–42]

4.2. Computer Results for the Velocity Autocorrelation Function

Of all the time correlation functions, the velocity autocorrelation function has been studied most extensively. Of particular interest to us here is the work

*One must also take the thermodynamic limit of the ensemble averages, before letting $t\to\infty$, to avoid difficulties with the recurrence behavior of finite mechanical systems.[52]

of Alder, Wainwright, and co-workers,[38–42] and of Wood and Erpenbeck[33] on the velocity autocorrelation function for systems of hard spheres or hard disks. These results have the most direct bearing on the long-time behavior of this function, which is of crucial interest for the theory.

In Fig. 25 we give a rough sketch of the computer results for the dimensionless velocity autocorrelation function $\rho_{\mathrm{D}}^{(d)}(t)$,

$$\rho_{\mathrm{D}}^{(d)}(t)=\frac{\langle v_{1x}(0)v_{1x}(t)\rangle}{\langle v_{1x}^2(0)\rangle}=\beta m\langle v_{1x}(0)v_{1x}(t)\rangle \tag{239}$$

for hard disks or hard spheres.

For times t less than a few t_0, where t_0 is the mean free time between collisions, $\rho_{\mathrm{D}}^{(d)}(t)$ decays exponentially. After several t_0, however, the decay of $\rho_{\mathrm{D}}^{(d)}(t)$ is no longer exponential. If the density is not too high ($V/V_0 \geq 1.8$, where V_0 is the volume at close packing) $\rho_{\mathrm{D}}^{(d)}(t)$ remains positive over the time intervals studied on the computer, $0 \leq t \leq 80t_0$ to $\sim 100t_0$, and appears to decay to zero as an inverse power of t. In two dimensions, $\rho_{\mathrm{D}}^{(2)}(t)$ decays as t^{-1} for $t > 10t_0$, and in three dimensions $\rho_{\mathrm{D}}^{(3)}(t)$ decays as $t^{-3/2}$ for $t > 25t_0$. For very high densities, $\rho_{\mathrm{D}}^{(d)}(t)$ becomes negative for $t > 5t_0$, and there is some evidence that for longer times, $t > 30t_0$, $\rho_0^{(d)}(t)$ may become positive again and decay to zero through positive values. In any event, the $t^{-d/2}$ decay observed in the computer calculations, first by Alder and Wainwright and later by Wood and Erpenbeck, has serious implications for the derivation of hydrodynamic equations, which we will discuss later. It is worth noting also that the same general features of $\rho_{\mathrm{D}}^{(d)}(t)$ have been observed in computer studies of particles that interact according to other potentials. In particular, the $t^{-d/2}$ decay has been observed in systems where the particles interact with the repulsive part of the Lennard-Jones potential.[196] In addition, computer studies made on particles that interact according to a full Lennard-Jones potential or an 11–6–8 potential indicate that the velocity autocorrelation function may decay very slowly in time, but the precise form of the slow decay has not yet been determined.[197,198]

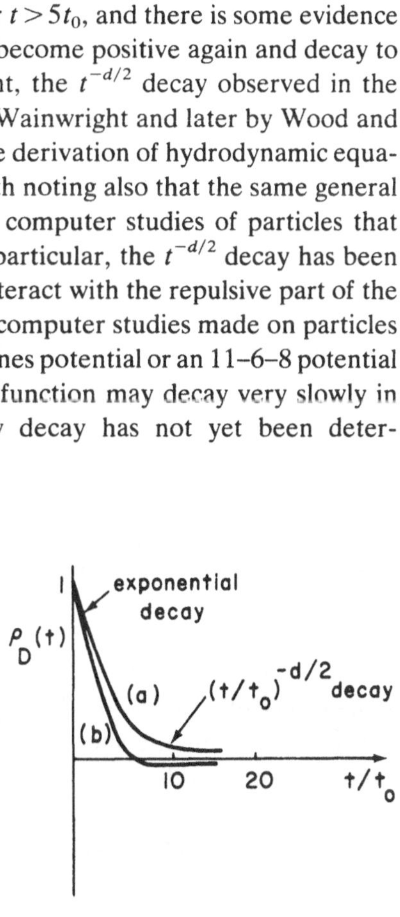

Fig. 25. A rough sketch of the main features of the computer results for the normalized velocity autocorrelation function, $\rho_{\mathrm{d}}^{(d)}(t)$, as a function of time for hard-sphere or hard-disk systems. Curve (a) shows the main features for systems at low densities, (b) at high densities. Here t_0 is the mean free time between collisions.

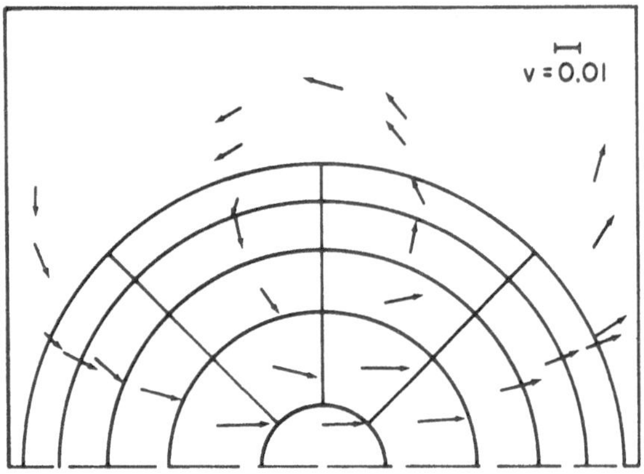

Fig. 26. The velocity correlation between a central particle and its neighborhood at a density corresponding to $\frac{1}{2}$ of close packing for 224 hard disks. Because of symmetry, only the upper half of the neighborhood is shown. The size of the central particle is shown by the smallest half-circle, and the sizes of the other concentric circles are chosen so as to include about six neighboring particles. These circles have been partitioned into four parts. The arrows give the direction and magnitude of the average velocity in each section after about 10 collision times. The scale of the velocity is indicated as 0.01 of the initial velocity. A number of additional arrows are included in the figure to show the extension of the vortex pattern outside the areas covered by the molecular dynamics calculations. These arrows are obtained by a hydrodynamic simulation of the molecular flow. [After B. J. Alder and T. E. Wainwright, *Phys. Rev.* **AI**, 18 (1970).]

Finally, we mention the fact that the molecular dynamics calculation of $\rho_{\mathrm{D}}^{(2)}(t)$ for a gas of hard disks exhibited a vortex type of velocity correlation between the tagged molecule and the surrounding molecules that is very similar to the hydrodynamic flow field surrounding a moving volume element in a fluid initially at rest.[38–42] This vortex pattern, illustrated in Fig. 26, suggests that a fraction of the momentum transferred by the tagged particle to the particles in front of it is eventually returned to it from behind. This process causes the velocity autocorrelation function to be larger than it would be if these vortices did not occur, and it is connected with the slow decay of the velocity autocorrelation function.

4.3. Applications of the Kinetic Theory of Gases to the Time Correlation Functions

The kinetic theory of gases is, in principle, capable of determining the time dependence of $\rho_{\mathrm{D}}^{(d)}(t)$ over all time intervals and for all densities of the gas. So far, however, only the short- and long-time behaviors have been studied in detail. Here we will concentrate on the long-time behavior of the velocity autocorrelation function, since it is of interest for the existence of the transport coefficients.[34–37] However, there is a large literature on the short-time behavior of this function to which we refer the reader for more details.(152–155,199–201) The intermediate region in time, say, for $5t_0 \le t \le 25$–$30t_0$ has not been studied very thoroughly, and there is as yet no convincing calculation of the negative part of $\rho_D^{(d)}(t)$ at high densities using kinetic theory.*

To apply the kinetic theory methods developed in the previous section to the calculations of the velocity autocorrelation function, we write $\rho_D^{(d)}(t)$ in terms of a single-particle distribution function $\Phi_D^{(d)}(\mathbf{v}_1, t)$ by†,‡,§

$$\rho_D^{(d)}(t) = \int d\mathbf{v}_1\, v_{1x}\Phi_D^{(d)}(\mathbf{v}_1, t) \tag{240}$$

with

$$\Phi_D^{(d)}(\mathbf{v}_1, t) = \lim_{\substack{N, V\to\infty \\ N/V=n}} \frac{m^d V}{Z_N}\beta m \int dx_2 \cdots \int dx_N \exp(-t\mathcal{H}_N)\exp(-\beta H_N)\, v_{1x} \tag{241a}$$

with

$$Z_N = \int d^N x e^{-\beta H_N} \tag{241b}$$

where we take the equilibrium ensemble to be a canonical ensemble. We have also written

$$\rho_D^{(d)}(t) = \beta m \langle v_{1x}(0) v_{1x}(-t)\rangle \tag{242}$$

with t replaced by $-t$ so as to simplify the correspondence between the single-particle distribution function discussed here and the single-particle distribution function discussed in Section 3.4. Using Eq. (241a), one can also show that $\Phi_D^{(d)}(\mathbf{v}_1, t)$ satisfies the first hierarchy equation

$$\partial\Phi_D^{(d)}(\mathbf{v}_1, t)/\partial t = n\int dx_2\, \theta_{12}\Phi_D^{(d)}(x_1, x_2, t) \tag{243}$$

where the two-particle distribution function $\Phi_D^{(d)}(x_1, x_2, t)$ is given by

$$\Phi_D^{(d)}(x_1, x_2, t) = \lim_{\substack{N, V\to\infty \\ N/V=n}} m^d\beta m\frac{V^2}{Z_N}\int dx_3 \cdots \int dx_N \exp(-t\mathcal{H}_N)\exp(-\beta H_N) v_{1x} \tag{244}$$

*See however the interesting quasi-hydrodynamical approach of Zwanzig and Bixon.(202)

†Strictly speaking $\Phi_D^{(d)}(\mathbf{v}_1, t)$ is not a distribution function, since it can take negative values, and is not properly normalized. However $\Phi_D^{(d)}(\mathbf{v}_1, t)$ can be related to the deviations of a nonequilibrium distribution function from the total equilibrium distribution function. For this reason we will continue to refer to $\Phi_D^{(d)}(\mathbf{v}_1, t)$ and related functions as distribution functions.

‡We ignore the walls, and therefore all quantities will be defined in the thermodynamic limit, $N\to\infty$, $V\to\infty$, $N/V = n$, fixed.

§The $\mathbf{r}_1$ integral simply gives a factor of V in the right-hand side of (241a), in the thermodynamic limit.

The method discussed in Section 3 for the derivation of the generalized Boltzmann equation can be carried over, with some simple modifications, to derive an analogous kinetic equation for $\Phi_D^{(d)}(\mathbf{v}_1, t)$. However, due to the special role of the tagged particle and the fact that the corresponding initial N-particle distribution is proportional to v_{1x}, one obtains a linear equation for $\Phi_D^{(d)}(\mathbf{v}_1, t)$ that for low densities reduces to the Lorentz–Boltzmann equation[5,34-36]:

$$\frac{\partial\Phi_D^{(d)}(\mathbf{v}_1, t)}{\partial t} = n\int d\mathbf{v}_2 \int d\hat{\mathbf{k}}\, B(\mathbf{g}, \hat{\mathbf{k}})[\phi_0(\mathbf{v}_2')\Phi_D^{(d)}(\mathbf{v}_1', t) - \Phi_D^{(d)}(\mathbf{v}_1, t)\phi_0(\mathbf{v}_2)] + O(n^2)$$

$$= L_D\Phi_D^{(d)}(\mathbf{v}_1, t) + O(n^2) \tag{245}$$

where $\phi_0(\mathbf{v}) = (\beta m/2\pi)^{d/2} \exp[-(\beta m/2)\mathbf{v}^2]$. The higher terms in the density expansion on the right-hand side of Eq. (245), like the generalized Boltzmann equation, are determined successively by the dynamics of three, four, five, etc., particles in infinite space. The first term in the expansion, the two-body term, given by Eq. (245), is of the form

$$\frac{\partial\Phi_{D,0}^{(d)}(\mathbf{v}_1, t)}{\partial t} = L_D\Phi_{D,0}^{(d)}(\mathbf{v}_1, t) \tag{246}$$

where the operator L_D is called the Lorentz–Boltzmann collision operator and the subscript zero denotes the low-density value.

The solution of Eq. (246) is

$$\Phi_{D,0}^{(d)}(\mathbf{v}_1, t) = e^{tL_D}\phi_0(\mathbf{v}_1)\beta m v_{1x} \tag{247}$$

where we have used the fact that $\Phi_D^{(d)}(\mathbf{v}_1, 0) = \phi_0(\mathbf{v}_1)\beta m v_{1x}$.

The corresponding value for the velocity autocorrelation function is

$$\rho_{D,0}^{(d)}(t) = \beta m \int d\mathbf{v}_1\, v_{1x} e^{tL_D} v_{1x}\phi_0(\mathbf{v}_1) \tag{248}$$

Since v_{1x} is orthogonal to the eigenfunction of L_D corresponding to eigenvalue zero,* $\rho_{D,0}^{(d)}(t)$ can be expressed as a sum of damped exponentials corresponding to the nonzero eigenvalues of L_D, all of which are negative.[6,7] Hence the two-body term in Eq. (245) for $\Phi_D^{(d)}(\mathbf{v}_1, t)$ leads to an exponential form for $\rho_D^{(d)}(t)$ that decays after a few mean free times.

For a gas of hard disks or hard spheres, if one neglects all the effects of correlated successive binary collisions and includes only the excluded volume

*This eigenfunction is simply a constant, since the Lorentz–Boltzmann collision operator conserves the number of tagged particles, but not the momentum or energy of the tagged particles.

corrections to the binary collision taking place at time t in Eq. (245), then one obtains a Lorentz–Enskog equation for $\Phi_{\mathrm{D}}^{(d)}(\mathbf{v}_1, t)$ of the form

$$\partial\Phi_{\mathrm{D,E}}^{(d)}(\mathbf{v}_1, t)/\partial t = g(a)L_{\mathrm{D}}\Phi_{\mathrm{D,E}}^{(d)}(\mathbf{v}_1, t) \tag{249}$$

where $g(a)$ is given by Eq. (143), and where the subscript E denotes Enskog. This equation leads to the expression

$$\rho_{\mathrm{D,E}}^{(d)}(t) = \beta m \int d\mathbf{v}_1\, v_{1x} e^{g(a)tL_{\mathrm{D}}} v_{1x}\phi_0(\mathbf{v}_1) \tag{250}$$

which is also exponentially damped over a few mean free times, but the mean free time is now the mean free time at arbitrary density defined by

$$t_0(n) = t_{c,0}/g(a)$$

where $t_{c,0}$ is the mean free time between collisions as it would follow from the Boltzmann equation. For short times, Eq. (250) is in very good agreement with the computer results for $\rho_{\mathrm{D}}^{(d)}(t)$ for all densities studied so far.[33,38–42] One can show that the initial slope of $\rho_{\mathrm{D}}^{(d)}(t)$ predicted by Eq. (250) is *exact* for a gas of hard spheres or disks.[34–36,201,203] That is,

$$\left.\frac{d\rho_{\mathrm{D}}^{(d)}(t)}{dt}\right|_{t=0} = \left.\frac{d\rho_{\mathrm{D,E}}^{(d)}(t)}{dt}\right|_{t=0} = \beta m g(a) \int d\mathbf{v}_1\, v_{1x} L_{\mathrm{D}} v_{1x}\phi_0(\mathbf{v}_1) \tag{251}$$

Moreover, one can also show that the time integral of (248) for $\rho_{\mathrm{D},0}^{(d)}(t)$ leads to exactly the same values for the coefficient of self-diffusion D as given by the Boltzmann equation.* Similarly the time integral of $\rho_{\mathrm{D,E}}^{(d)}(t)$, Eq. (250), leads to an expression for D that is identical to the Enskog theory result.[5]

If one now considers the contributions to $\partial\Phi_{\mathrm{D}}^{(d)}(\mathbf{v}_1, t)/\partial t$ coming from groups of more than two particles, one encounters divergence difficulties similar to the difficulties in the expansion of the generalized Boltzmann collision operator in powers of the density. The origin of these difficulties is the same, i.e., the failure of the terms in the power series in the density expansion to incorporate the mean free path damping into the theory. The solution is also the same. The density expansion of the collision operator must be rearranged by summing the most divergent terms,† which are the contributions from sequences of l-binary collisions among l-particles, for $l = 3, 4, \ldots$, for $d = 2$, or $l = 4, 5, \ldots$, for $d = 3$.

*The Boltzmann equation result for D is obtained by determining the mutual diffusion coefficient in a dilute binary mixture and then letting the two species be mechanically identical.[5]
†That is, the "ring" terms.

Using a method originated by Pomeau,[204,205] (and by Goldman[178,179]) one can analyze the contribution to $\rho_{\mathrm{D}}^{(d)}(t)$ coming from the resummed collision operator. The resummed theory[34-37,190] predicts that for times longer than several mean free times,

$$\rho_{\mathrm{D}}^{(d)}(t) \approx \alpha_{\mathrm{D}}^{(d)}(\rho)(t_0/t)^{d/2} \tag{252}$$

where $\rho = na^d$, which is precisely the time dependence found in the computer experiments. To lowest order in the density, one finds that the coefficient $\alpha_{\mathrm{D}}^{(d)}(\rho)$ is given by

$$\alpha_{\mathrm{D},0}^{(2)} = \left[8\pi n\left(D_0 + \frac{\eta_0}{nm}\right)t_{c,0}\right]^{-1} \tag{253}$$

in two dimensions, and by

$$\alpha_{\mathrm{D},0}^{(3)} = \frac{1}{12n}\left[\pi\left(D_0 + \frac{\eta_0}{nm}\right)t_{c,0}\right]^{-3/2} \tag{254}$$

in three dimensions. Here D_0 and η_0 are the coefficients of self-diffusion and shear viscosity, respectively, as determined from the Boltzmann equations appropriate for a gas of hard disks or hard spheres,[5,170] and m is the mass of a particle. These expressions for $\alpha_{\mathrm{D},0}^{(d)}$ are in good agreement with the computer results when they are extrapolated to low density. An attempt has been made to extend the results of kinetic theory to higher densities by incorporating into the theory excluded volume corrections to each of the collisions in the "ring" collision sequences that were taken into account in the resummation.[34-36] This leads to an expression for $\alpha_{\mathrm{D}}^{(d)}(\rho)$ denoted by $\alpha_{\mathrm{D,E}}^{(d)}(\rho)$, which is applicable to higher densities than is $\alpha_{\mathrm{D},0}^{(d)}$. In fact, $\alpha_{\mathrm{D,E}}^{(d)}(\rho)$ is identical to $\alpha_{\mathrm{D},0}^{(d)}$ given by Eqs. (253) and 254) except that D_0 and η_0 are replaced by D_{E} and η_{E}, where these are the coefficients of self-diffusion and of shear viscosity as determined by the Enskog theory,[5,206] and $t_{c,0}$ replaced by $t_0(n)$, the mean free time at the given density of the gas. The comparison of $\alpha_{\mathrm{D,E}}^{(d)}(\rho)$ with the computer results is shown in Fig. 27. We see that the theoretical values $\alpha_{\mathrm{D,E}}^{(d)}(\rho)$ and the computer values for $\alpha_{\mathrm{D}}^{(d)}(\rho)$ are in very good quantitative agreement over the entire range of densities for which the computer results are available. Thus the quantitative agreement of the resummed theory with the computer results can be taken as a vindication of the resummation and as a confirmation of the general approach to the kinetic theory of gases that we have outlined here.

Similar results have been obtained for the time correlation functions that determine the coefficients of shear viscosity, thermal conductivity, and bulk viscosity for a pure system.[34-36,190] That is, they all decay exponentially for short times and as $t^{-d/2}$ for long times, i.e.,

$$\rho_{\tau}^{(d)}(t) \approx \alpha_{\tau}^{(d)}(\rho)(t_c/t)^{d/2}$$

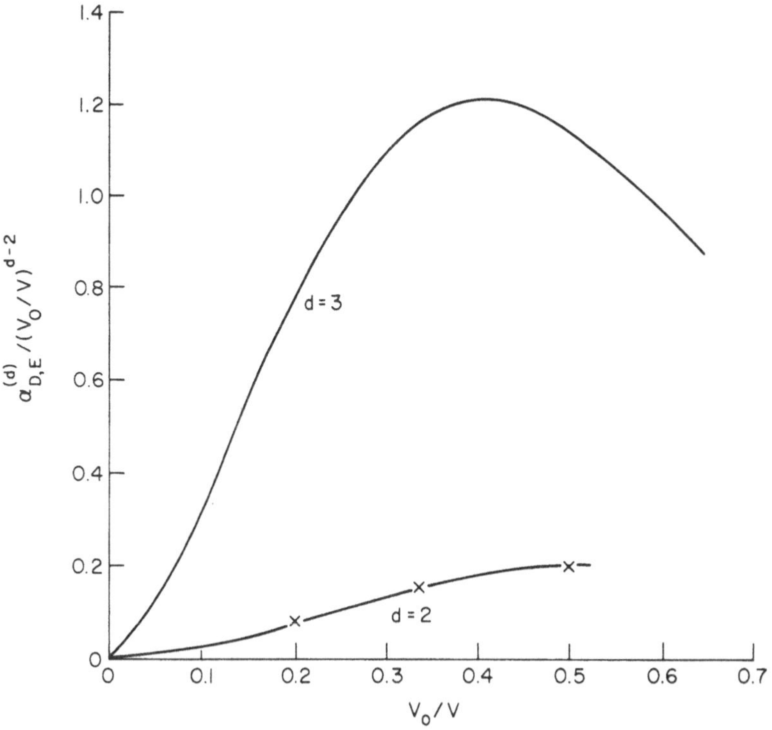

Fig. 27. $\alpha_{D,E}^{(d)}/(V_0/V)^{d-2}$ plotted as a function of the reduced density V_0/V, where V_0 is the volume at close packing, for $d = 2$ and $d = 3$. The crosses indicate the results of Alder and Wainwright[38] for $d = 2$.

where $\rho_\tau^{(d)}(t)$ is the normalized time correlation function for a transport coefficient τ. For hard spheres or hard disks, theoretical expressions for $\alpha_\tau^{(d)}(\rho)$ have been obtained that should hold over a wide range of densities.[37–38] Although the theory has not been worked out sufficiently for an estimation of the full range of time for which the $t^{-d/2}$ decay is dominant, it does appear that this decay should hold at least over the time intervals of interest for comparison with the computer results mentioned earlier.* It would also seem that the $t^{-d/2}$ behavior of the time correlation functions should hold for a wider class of potentials than those for hard spheres or hard disks, but so far only the low-density values for the coefficient have been worked out on the basis of kinetic theory for a general potential.[34–36]

In view of the fact that $\rho_D^{(2)}(t) \sim t^{-1}$, there is now some question as to whether the transport coefficients associated with the linearized Navier–Stokes hydrodynamic equations are finite in two dimensions. For if the t^{-1} decay persists for asymptotically long times, time integrals like that in Eq. (232) will

*For three-dimensional systems of hard spheres, it would appear from recent work that the $t^{-3/2}$ decay should persist for asymptotically long times.[37,207,208]

not exist. As a result, the resummation of the most divergent terms in the virial expansion of the generalized Boltzmann equation does *not* seem to provide a mean free path cutoff for the divergent three-body term in two dimensions. In three dimensions, although the asymptotic $t^{-3/2}$ behavior will ensure the existence of the Navier–Stokes transport coefficients, it appears that the transport coefficients associated with the linear Burnett and higher-order (linear) hydrodynamic equations do not exist.(209–213) For example, if one computes D_2, given by Eq. (238), for a gas of hard spheres, one finds that the summation of the ring terms leads to the result that

$$D_2 = \lim_{t\to\infty} \frac{-D_0^2 t^{1/2}}{10\pi^{3/2}\beta m(D_0 + \eta_0/nm)^{5/2}} \tag{255}$$

This result is consistent with recent computer studies on this function by Alder, Wainwright, Wood, and Erpenbeck.(214) However, we should point out that in order to determine the asymptotic behavior of the correlation functions as $t\to\infty$, and then to establish whether or not the above-mentioned transport coefficients diverge, all other dynamical events that also give divergent contributions to the collision integrals* should be taken into account, since their sum could be as large as or even larger than the sum of the ring terms, for very long times.(37,190,207,208)

In view of the very slow time decay of the time correlation functions, it is not clear to what extent the Navier–Stokes transport coefficients can be used even in three dimensions to describe phenomena that vary on a time scale of $\sim 50t_c$, for on this time scale there is not yet a clear separation of microscopic and macroscopic effects. However, usually the Navier–Stokes equations are applied to phenomena that vary on a much longer time scale, and then the slow decay of the correlation functions does not interfere with the hydrodynamic processes. Nevertheless, the divergences of the Burnett and higher-order transport coefficients do appear to have experimental consequences even for three-dimensional systems. In particular, it appears that the dispersion relation for the sound wave frequency ω in terms of the wave number k can no longer be expressed as a power series in k as was done in Eq. (133) but instead that fractional powers of the form $k^{3-1/2n}$ for $n = 1, 2, \ldots$ appear.(36,209,210,215,216) Unfortunately the coefficients of these powers are so small that these terms have not yet been detected in experiments on sound propagation in dense gases.

Finally, we should mention that there are other approaches to determining the long-time behavior of the time correlation functions, which are based on solutions of the linearized hydrodynamic equations.(37,190,217–221) These approaches also lead to a $t^{-d/2}$ decay for the correlation function, and for $\rho_D^{(3)}(t)$

*But less divergent than the ring events.

a result identical to Eq. (254) is obtained, except that D_0 and η_0 are replaced by their full values D and η, respectively. The precise relation of these methods to that of kinetic theory has not yet been settled.

5. Discussion

We conclude this survey by mentioning a number of problems whose solution would lead to a deeper understanding of the kinetic theory of gases.

5.1. Problems Connected with the Boltzmann Equation

1. What are the mathematical properties of the nonlinear Boltzmann equation?

In particular, for what class of intermolecular potentials does this equation have a unique solution for given initial and boundary conditions? Is the normal solution of Chapman and Enskog in any sense a good approximation to the actual solution, at least sufficiently far from the boundaries, and for sufficiently long times? Can one find explicit solutions to the nonlinear Boltzmann equation with sufficient accuracy so that the specifically nonlinear features of the Boltzmann equation can be tested, in shock wave or sound wave or sound propagation experiments,[70,71] for example?

2. What are the general features of the approach to equilibrium of a gas when there are boundaries present?

In particular, (a) can one clarify the role of the Burnett and higher-order hydrodynamic equations in describing the gas flow in the case that $l_c/l_h \lesssim 1$? That is, can one find boundary conditions to be used in solving these higher-order equations, and can one show that the higher-order hydrodynamic equations give a better description of the system than the Navier–Stokes equations? If not, then (b) What is the physical significance of the μ-expansion and why do Navier–Stokes equations work so well?

In this connection it is worth pointing out that for the Kramers problem, an exact solution to the linearized Boltzmann equation for a BGK model gas has been constructed by Cercignani.[6,7] Even for this case the answers to questions (a) and (b) are not known.[81]

(c) Can one connect the theory used to describe the hydrodynamic region for which the Knudsen number l_c/l_h is small to that used for the rarefied gas region where $l_c/l_h \gg 1$?

For the case of a gas flow around a simple convex object such as a sphere or cylinder, where l_h characterizes the size of the object, some progress has been made on this problem by demonstrating the relation between the various microscopic processes that are important in each region.[44,45,77–80,101–103] The solution of this problem is closely related to the solution of:

(d) Can one construct solutions to the Boltzmann equation valid when the Knudsen number is of order unity?

3. Can one determine the spectrum of the linearized Boltzmann collision operator L for the class of interaction potentials illustrated in Fig. 1?

So far, the complete spectrum of L is only known for Maxwell molecules. The main features of the spectrum but not all of its detailed properties are known for the hard-sphere potential,[65-68] and for r^{-s} potentials both with[65-68] and without[69] an angular cutoff. However, for the type of potentials illustrated in Fig. 1, almost nothing is known, yet this is the potential of most physical interest. For applications to problems in spatially inhomogeneous systems and in sound propagation, one would like to know the spectrum of the inhomogeneous Boltzmann collision operator $(i\mathbf{k}\cdot\mathbf{v}-L)$. However, very little is known about the spectrum of this operator either, for any potential, except that in special cases like hard spheres one can prove the existence of five discrete eigenvalues, for small k, generated by perturbing the zero eigenvalues of L with the perturbation $i\mathbf{k}\cdot\mathbf{v}$, and describe the shape of the continuum for small k.[72,81]

4. Is it possible to use the H-theorem and the Chapman–Enskog solution to suggest a generalization of the laws of irreversible thermodynamics so that the latter would apply even when higher-order gradients or nonlinear terms in the hydrodynamic fluxes are taken into account?

5.2. Problems Connected with the Kinetic Theory of Dense Gases

All of the problems discussed above in connection with the Boltzmann equation carry over to the kinetic theory of dense gases. In addition to the obvious analogs of the problems mentioned above, the following problems need to be solved.

1. Can one specify precisely the entire class of intermolecular potentials for which the four-body term in the density expansion of the kinetic equation is divergent in three dimensions, or the three-body term in two dimensions?

2. Can one show that the summation of the most divergent, or ring terms in the density expansion of the kinetic equation completely determines the coefficient of the $n^2 \log n$ term in the density expansion of the transport coefficients, i.e., that this coefficient is not affected by other, less divergent, terms?

3. Can one determine the behavior of the time correlation functions for all times, over a wide range of densities, and for the class of intermolecular potentials considered here, as illustrated in Fig. 1?

4. What is the connection between the various kinetic and hydrodynamic theories of the long-time tails of the time correlation functions?

5. In view of the apparent divergence of Navier–Stokes transport coefficients for two-dimensional systems, can one find the correct form of the hydrodynamic equations in two dimensions?

6. What is the structure of the linear and of the nonlinear hydrodynamic equations for three-dimensional systems? Do the apparent divergences of the Burnett and higher-order transport coefficients in the linearized equations mean simply that nonlocal—but linear—effects must be taken into account, or must nonlinear effects be taken into account before one gets a completely well-behaved theory? Moreover, almost nothing is known about the theory of nonlinear hydrodynamic equations for dense gases. This is an area that certainly needs to be explored both from a theoretical viewpoint,[192–195] and also with an eye toward suggesting experiments by means of which one can test the theories.

7. What are the effects of the boundaries on a dense gas? In particular, do the kinetic boundary layers have a qualitatively different structure for a dense gas than for a dilute gas?

Some progress has been made along these lines by the recent work of Wolynes,[222] who has shown how one can compute boundary effects for dense systems by means of the mode-coupling theory.

8. Is there a generalization of the H-theorem to dense gases that can be used to prove that a dense gas not in equilibrium approaches equilibrium after a sufficiently long time?

9. How can one explain the efficacy of hydrodynamic equations in treating nonequilibrium phenomena on a microscopic scale? Here we refer to the work of Zwanzig and Bixon[202] among others, who have used hydrodynamic equations with slip boundary conditions to discuss the motion of a sphere in a fluid of other identical spheres, by regarding the other spheres as forming a continuum fluid. The close agreement between the results of this calculation and the results of real and computer experiments is remarkable and certainly deserves to be investigated further.

ACKNOWLEDGMENTS

The authors would like to thank Professor J. V. Sengers for his many valuable comments on this paper as it was being written, and for preparing Fig. 18. They would also like to thank Mrs. Nancy King for her generous help in the preparation of this manuscript, and Drs. Elizabeth Johnson and Mary Dorfman for their careful reading of it and their comments.

References

1. S. Brush, *Kinetic Theory*, Vols. 1–3, Pergamon Press, New York (1966, 1972).
2. J. C. Maxwell, *The Scientific Papers of J. C. Maxwell*, Dover, New York (1965).

3. L. Boltzmann, *Lectures on Gas Theory* (S. Brush, transl.), University of California Press, Berkeley, California (1964).
4. M. Klein *in*: *The Boltzmann Equation* (E. G. D. Cohen and W. Thirring, eds.), Springer Verlag, Vienna (1973).
5. S. Chapman and T. G. Cowling, *The Mathematical Theory of Non Uniform Gases*, 3rd ed., Cambridge University Press, London (1970).
6. J. H. Ferziger and H. G. Kaper, *Mathematical Theory of Transport Processes in Gases*, North Holland Publishing Co., Amsterdam (1972).
7. C. Cercignani, *Mathematical Methods in Kinetic Theory*, Plenum Press, New York (1969).
8. J. O. Hirschfelder, C. F. Curtiss, and R. B. Bird, *Molecular Theory of Gases and Liquids*, Wiley, New York (1954).
9. M. N. Kogan, *Rarefied Gas Dynamics* (L. Trilling, transl.), Plenum Press, New York (1969).
10. H. Grad, Principles of the kinetic theory of gases, *in*: *Handbuch der Physik* (S. Flügge, ed.), Vol. 12, Springer Verlag, Berlin (1958).
11. S. Harris, *An Introduction to the Theory of the Boltzmann Equation*, Holt, Rinehart and Winston, New York (1971).
12. R. Liboff, *Theory of Kinetic Equations*, Wiley, New York (1968).
13. L. Waldmann, Transporterscheinungen in Gasen von mittlerem Druck, *in*: *Handbuch der Physik* (S. Flügge, ed.), Vol. 12, Springer Verlag, Berlin (1958).
14. P. Ehrenfest and T. Ehrenfest, *The Conceptual Foundations of the Statistical Approach in Mechanics* (M. Moravesik, transl.), Cornell Univ. Press, Ithaca (1959).
15. N. N. Bogoliubov, Problems of a dynamical theory in statistical physics (E. K. Gora, transl.), *in*: *Studies in Statistical Mechanics* (G. E. Uhlenbeck and J. de Boer, eds.), Vol. I, North Holland Publishing Co., Amsterdam (1962).
16. E. G. D. Cohen, *Physica* **28**, 1025, 1045, 1060 (1962).
17. E. G. D. Cohen, *J. Math. Phys.* **4**, 143 (1963).
18. E. G. D. Cohen, *in*: *Lectures in Theoretical Physics* (*Boulder*) (W. E. Brittin, ed.), Vol. 8A, p. 145, Univ. of Colorado Press, Boulder, Colorado (1966).
19. E. G. D. Cohen, *in*: *Lectures in Theoretical Physics* (*Boulder*) (W. E. Brittin, ed.), Vol. 9C, p. 279, Gordon & Breach, New York (1967).
20. M. S. Green, *J. Chem. Phys.* **25**, 836 (1956).
21. M. S. Green, *Physica* **24**, 393 (1958).
22. M. S. Green and R. A. Piccerelli, *Phys. Rev.* **132**, 1388 (1963).
23. J. R. Dorfman and E. G. D. Cohen, *J. Math. Phys.* **8**, 282 (1967).
24. M. H. Ernst, L. K. Haines, and J. R. Dorfman, *Rev. Mod. Phys.* **41**, 296 (1969).
25. Y. Pomeau and A. Gervois, *Phys. Rev.* **A9**, 2196 (1974).
26. A. Gervois, C. Normand-Alle, and Y. Pomeau, *Phys. Rev.* **A12**, 1570 (1975).
27. R. Zwanzig, *Ann. Rev. Phys. Chem.* **16**, 67 (1965).
28. W. A. Steele, *in*: *Transport Phenomena in Fluids* (H. J. M. Hanley, ed.), Chapter 8, Marcel Dekker, New York (1970).
29. M. H. Ernst, J. R. Dorfman, and E. G. D. Cohen, *Physica*, **31**, 493 (1965).
30. M. H. Ernst, *Physica* **32**, 209 (1966).
31. B. Chu, *Laser Light Scattering*, Academic Press, New York (1974).
32. B. J. Berne and G. D. Harp, *in*: *Advances in Chemical Physics* (I. Prigogine and S. A. Rice, eds.), Vol. XVII, p. 63 and references contained therein, Wiley (Interscience), New York (1970).
33. W. W. Wood, *in*: *Fundamental Problems in Statistical Mechanics* (E. G. D. Cohen, ed.), Vol. III, p. 331, North Holland Publishing Co., Amsterdam (1975).
34. J. R. Dorfman and E. G. D. Cohen, *Phys. Rev. Lett.* **25**, 1257 (1970).
35. J. R. Dorfman and E. G. D. Cohen, *Phys. Rev.* **A6**, 776 (1972); **A12**, 292 (1975).
36. J. Dufty, *Phys. Rev.* **A5**, 2247 (1972).
37. Y. Pomeau and P. Resibois, *Phys. Rep.* **19C**, 63 (1975), and references contained therein.
38. B. J. Alder and T. E. Wainwright, *Phys. Rev. Lett.* **18**, 988 (1967).
39. B. J. Alder and T. E. Wainwright, *J. Phys. Soc. Jap.*, **Suppl. 26**, 267 (1969).
40. B. J. Alder and T. E. Wainwright, *Phys. Rev.* **A1**, 18 (1970).

41. B. J. Alder, D. M. Gass, and T. E. Wainwright, *J. Chem. Phys.* **53**, 3813 (1970).
42. B. J. Alder, D. M. Gass, and T. E. Wainwright, *Phys. Rev.* **A4**, 233 (1971).
43. E. Dahlberg, *J. Phys.* **A6**, 1800 (1973).
44. C. F. McClure, Ph.D. thesis, Univ. of Maryland, College Park, Maryland (1972).
45. H. van Beijeren, J. R. Dorfman, and C. F. McClure, *Arch. Mech.* (Warsaw) **28**, 333 (1976).
46. J. Lebowitz and P. Bergmann, *Phys. Rev.* **99**, 578 (1955).
47. J. Lebowitz and P. Bergmann, *Ann. Phys.* **1**, 1 (1957).
48. C. Cercignani and M. Lampis, *Transp. Theory Stat. Phys.* **1**, 101 (1971).
49. C. Cercignani, *Transp. Theory Stat. Phys.* **2**, 27 (1972).
50. J. Lebowitz and P. Bergmann, *Phys. Rev.* **99**, 578 (1955).
51. G. E. Kelly and J. V. Sengers, *J. Chem. Phys.* **57**, 1441 (1972); **61**, 2800 (1974).
52. G. E. Uhlenbeck and G. W. Ford, *Lectures on Statistical Mechanics*, American Mathematical Society, Providence, Rhode Island (1963).
53. J. Darrozés and J. P. Guiraud, *CR Acad. Sci.* **A262**, 1368 (1966).
54. S. Chandrasekhar *in*: *Noise and Stochastic Processes* (N. Wax, ed.), Dover, New York (1954).
55. T. Carleman, *Problèmes Mathématiques dans la Théorie Cinétique des Gaz*, Almquist and Wiksells, Uppsala (1957).
56. E. Wild, *Proc. Camb. Phil. Soc.* **47**, 602 (1951).
57. D. Morgenstern, *Proc. Nat. Acad. Sci.* **40**, 719 (1954).
58. D. Morgenstern, *J. Rational Mech. Anal.* **4**, 533 (1955).
59. E. G. D. Cohen, *in*: *Transport Phenomena in Fluids* (H. J. M. Hanley, ed.), Marcel Dekker, New York (1970).
60. L. D. Landau and E. M. Lifshitz, *Fluid Mechanics* (J. B. Sykes and W. S. Reid, transls.), Pergamon, Oxford (1959).
61. G. W. Ford and J. Foch, *in*: *Studies in Statistical Mechanics* (G. E. Uhlenbeck and J. de Boer, eds.), Vol. 5, p. 103, North Holland Publishing Co., Amsterdam (1970).
62. C. S. Wang Chang and G. E. Uhlenbeck *in*: *Studies in Statistical Mechanics* (G. E. Uhlenbeck and J. de Boer, eds.), Vol. 5, North Holland Publishing Co., Amsterdam (1970).
63. G. W. Ford, *in*: *The Boltzmann Equation* (E. G. D. Cohen and W. Thirring, eds.), p. 141, Springer Verlag, Vienna (1973).
64. J. Foch and M. Losa, *Phys. Rev. Lett.* **28**, 1315 (1972).
65. H. Grad, *in*: *Rarefied Gas Dynamics* (J. A. Laurmann, ed.), Academic Press, New York (1963).
66. [J.] R. Dorfman, *Proc. Nat. Acad. Sci.* **50**, 804 (1963).
67. I. Kuščer and N. M. R. Williams, *Phys. Fluids* **10**, 1922 (1967).
68. C. C. Yan and G. H. Wannier, *Bull. Am. Phys. Soc.* **13**, 899 (1968).
69. Y. Pao, *Comm. Pure Appl. Math.* **27**, 407, 559 (1974).
70. J. Foch, *in*: *The Boltzmann Equation* (E. G. D. Cohen and W. Thirring, eds.), Springer Verlag, Vienna (1973).
71. J. Foch, *Arch. Mech. (Poland)* **26**, 369 (1974).
72. J. A. McLennan, *Phys. Fluids* **8**, 1580 (1965).
73. H. Grad, *Phys. Fluids* **6**, 147 (1963).
74. H. Grad, *in*: *Transport Theory*, SIAM-AMS Proceedings (R. Bellman, G. Birkhoff, and I. Abu-Shumays, eds.), Vol. 1, American Mathematical Society, Providence, Rhode Island (1969).
75. H. Grad, *in*: *Rarefied Gas Dynamics* (K. Karamcheti, ed.), Academic Press, New York (1974).
76. G. Scharf, *Helv. Phys. Acta* **40**, 929 (1967); **42**, 5 (1969).
77. H. van Beijeren and J. R. Dorfman, *in*: *Proc. Fourth Conf. on Transport Theory, Blacksburg (1975)* (to appear).
78. H. van Beijeren, J. R. Dorfman, and C. F. McClure (to be published).
79. G. Scharf, *Phys. Fluids* **13**, 848 (1970).
80. J. A. McLennan and S. C. Chiu, *Phys. Fluids* **17**, 1146 (1974).
81. C. Cercignani, *in*: *Rarefied Gas Dynamics* (K. Karamcheti, ed.), Academic Press, New York (1969).

82. E. H. Kennard, *Kinetic Theory of Gases*, McGraw-Hill Book Co., New York (1938).
83. Y. Sone, *in*: *Rarefied Gas Dynamics* (L. Trilling and H. Y. Wachman, eds.), p. 243, Academic Press, New York (1969).
84. M. de Wit, Approximate Solutions to Boundary Layer Problems in Linear Kinetic Theory, Ph.D. thesis, Technische Hogeschool, Eindhoven (1975).
85. C. Cercignani, *in*: *Transport Theory* (R. Bellman, G. Birkhoff, and I. Abu-Shumays, eds.), Vol. 1, American Mathematical Society, Providence, Rhode Island (1969).
86. I. Prigogine, *Physica* **15**, 272 (1949).
87. S. R. de Groot and P. Mazur, *Non Equilibrium Thermodynamics*, North Holland Publ. Co., Amsterdam (1962).
88. J. W. Haarman, *in*: *Transport Phenomena* (J. Kestin, ed.), AIP Conf. Proc. No. 11, p. 193, American Institute of Physics, New York (1973).
89. J. W. Haarman, *Physica* **52**, 605 (1971).
90. J. Kestin and E. A. Mason, *in*: *Transport Phenomena* (J. Kestin, ed.), AIP Conf. Proc. No. 11, American Institute of Physics, New York (1973).
91. J. Kestin, S. T. Ro, and W. Wakeham, *J. Chem. Phys.* **56**, 4036, 4086, 4114, 4119, 5837 (1972).
92. J. M. Hellemans, J. Kestin, and S. T. Ro, *J. Chem. Phys.* **57**, 4038 (1972).
93. E. A. Mason, *in*: *Kinetic Processes in Gases and Plasmas* (A. R. Hochstim, ed.), p. 57, Academic Press, New York (1969).
94. M. Klein and H. J. M. Hanley, *J. Chem. Phys.* **53**, 4722 (1970); **76**, 1743 (1972).
95. H. J. M. Hanley, *J. Phys. Chem. Ref. Data* **2**, 619)1973).
96. H. J. M. Hanley and J. F. Ely, *J. Phys. Chem. Ref. Data* **2**, 735 (1973).
97. J. C. Maxwell, *The Scientific Papers of S. C. Maxwell*, Dover, New York (1965),
98. E. Ikenberry and C. Truesdell, *J. Ratl. Mech. Anal.* **5**, 1 (1956).
99. H.Grad, *Comm. Pure Appl. Math.* **2**, 231 (1949).
100. R. Peralta–Fabi and R. Zwanzig (to be published).
101. C. Cercignani and C. D. Pagani, *Phys. Fluids* **11**, 1395 (1968).
102. C. Cercignani, C. D. Pagani, and P. Bassanini, *Phys. Fluids* **11**, 1399 (1968).
103. C. Cercignani and C. D. Pagani, *in*: *Rarefied Gas Dynamics* (C. L. Brunin, ed.), p. 555, Academic Press, New York (1967).
104. S. A. Schaaf, *in*: *Handbuch der Physik* (S. Flügge, ed.), Vol. VIII, Pt. 2, p. 591, Springer Verlag, Berlin (1958).
105. M. Lunc and J. Lubonski, *Arch. Mech. Stosow.* (*Warsaw*) **8**, 597 (1956).
106. R. M. L. Baker and A. F. Charwat, *Phys. Fluids* **1**, 73 (1958).
107. D. R. Willis, *in*: *Rarefied Gas Dynamics* (F. M. Devienne, ed.), p. 246, Pergamon, New York (1960).
108. M. H. Rose, *Phys. Fluids* **7**, 1262 (1964).
109. G. J. Maslach, D. R. Willis, S. Tang, and D. Ko, *in*: *Rarefied Gas Dynamics* (J. H. De Leeuw, ed.), p. 433, Academic Press, New York (1965).
110. Y. Pao and D. R. Willis, *Phys. Fluids* **12**, 435 (1969).
111. A. L. Cooper and B. B. Hamel, *Phys. Fluids* **16**, 35 (1973).
112. J. R. Dorfman, W. A. Kuperman, J. V. Sengers, and C. F. McClure, *Phys. Fluids* **16**, 2347 (1973).
113. Y. Pomeau, *Phys. Fluids* **18**, 277 (1975).
114. A. Gervois and Y. Pomeau, *Phys. Fluids* **17**, 2292 (1974).
115. Y. H. Kan, Ph.D. dissertation, University of Maryland (1975).
116. Y. H. Kan and J. R. Dorfman (to be published).
117. P. L. Bhatnagar, E. P. Gross, and M. Krook, *Phys. Rev.* **94**, 511 (1954).
118. J. Kestin and E. A. Mason, *in*: *Transport Phenomena* (J. Kestin, ed.), AIP Conf. Proc. No. 11, American Institute of Physics, New York (1973).
119. E. A. Mason, *in*: *Kinetic Processes in Gases and Plasmas* (A. R. Hochstim, ed.), p. 57, Academic Press, New York (1969).
120. K. E. Grew, *in*: *Transport Phenomena in Fluids* (H. J. M. Hanley, ed.), Marcel Dekker, New York (1970).

121. J. M. Burgers, *Flow Equations for Composite Gases*, Academic Press, New York (1969).
122. C. S. Wang Chang, G. E. Uhlenbeck, and J. de Boer, *in*: *Studies in Statistical Mechanics* (J. de Boer and G. E. Uhlenbeck, eds.), Vol. II, North Holland Publ. Co., Amsterdam (1964).
123. L. Waldmann, *in*: *The Boltzmann Equation* (E. G. D. Cohen and W. Thirring, eds.), Springer Verlag, Vienna (1973).
124. L. Waldmann, *Fundamental Problems in Statistical Mechanics* (E. G. D. Cohen, ed.), Vol. II, North Holland Publ. Co., Amsterdam (1968).
125. L. Waldmann, *Z. Naturforsch.* **12A**, 660 (1957).
126. R. F. Snider, *J. Chem. Phys* **32**, 1051 (1960).
127. R. F. Snider, *in*: *Transport Phenomena* (G. Kirczenow and J. Marro, eds.), p. 470, Springer Verlag, Berlin (1974).
128. J. J. M. Beenakker, H. F. P. Knaap, and B. E. Sanctuary, *in*: *Transport Phenomena* (J. Kestin, ed.), AIP Conf. Proc. No. 11, American Institute of Physics, New York (1973).
129. R. Gordon, *in*: *Transport Phenomena* (J. Kestin, ed.), AIP Conf. Proc. No. 11, American Institute of Physics, New York (1973).
130. H. Moraal, *Phys. Rep.* **17C**, 225 (1975).
131. C. S. Wang Chang and G. E. Uhlenbeck, *in*: *Studies in Statistical Mechanics* (J. de Boer and G. E. Uhlenbeck, eds.), Vol. V, Chap. V, North Holland Publ. Co., Amsterdam (1970).
132. M. B. Lewis, *in*: *Kinetic Processes in Gases and Plasmas* (A. R. Hochstim, ed.), Academic Press, New York (1969).
133. I. Kuščer, *in*: *The Boltzmann Equation* (E. G. D. Cohen and W. Thirring, eds.), p. 491, Springer Verlag, Vienna (1973).
134. S. I. Sandler, *in*: *Transport Phenomena* (J. Kestin, ed.), AIP Conf. Proc. No. 11, p. 203, American Institute of Physics, New York (1973).
135. L. G. H. Huxley and R. W. Crompton, *The Diffusion and Drift of Electrons in Gases*, Wiley, New York (1974).
136. E. W. McDaniel and E. A. Mason, *The Mobility and Diffusion of Ions in Gases*, Wiley, New York (1973).
137. D. A. Montgomery and D. Tidman, *Plasma Kinetic Theory*, McGraw-Hill, New York (1964).
138. C. M. Tchen and by A. R. Hochstim, *in*: *Kinetic Processes in Gases and Plasmas* (A. R. Hochstim, ed.), Academic Press, New York (1969).
139. G. A. Massel, *in*: *Kinetic Processes in Gases and Plasmas* (A. R. Hochstim, ed.), Academic Press, New York (1969).
140. N. G. van Kampen and B. U. Felderhof, *Theoretical Methods in Plasma Physics*, North Holland Publ. Co., Amsterdam (1967).
141. L. Klimontovich, *The Statistical Theory of Non Equilibrium Processes in a Plasma*, MIT Press, Cambridge, Massachusetts (1967).
142. F. T. Smith, *in*: *Kinetic Processes in Gases and Plasmas* (A. R. Hochstim, ed.), Academic Press, New York (1969).
143. J. Ross, J. C. Light, and K. E. Schuler, *in*: *Kinetic Processes in Gases and Plasmas* (A. R. Hochstim, ed.), Academic Press, New York (1969).
144. G. E. Uhlenbeck and G. W. Ford, *in*: *Studies in Statistical Mechanics* (G. E. Uhlenbeck and J. de Boer, eds.), Vol. I, p. 119, North Holland Publ. Co., Amsterdam (1961).
145. F. H. Ree and W. G. Hoover, *J. Chem. Phys.* **40**, 939 (1964).
146. J. V. Sengers, *Int. J. Heat Mass Transfer* **8**, 1103 (1965).
147. J. V. Sengers, *in*: *Recent Advances in Engineering Science* (A. C. Eringen, ed.), p. 153, Gordon & Breach, New York (1968).
148. H. J. M. Hanley, R. D. McCarty, and E. G. D. Cohen, *Physica* **60**, 322 (1972); **83A**, 215 (1976).
149. L. Barajas, L. S. Garcia Colin, and B. Pina, *J. Stat. Phys.* **7**, 161 (1973).
150. H. van Beijeren and M. H. Ernst, *Physica* **68**, 437 (1973); **70**, 225 (1973).
151. J. K. Percus, *in*: *Classical Fluids* (H. L. Frisch and J. L. Lebowitz, eds.), Benjamin, New York (1964).
152. J. Lebowitz, J. Percus, and J. Sykes, *Phys. Rev.* **188**, 487 (1969).
153. G. F. Mazenko, T. Y. C. Wei, and S. Yip, *Phys. Rev.* **A6**, 1981 (1972).
154. H. H. U. Konijnendijk and J. M. J. van Leeuwen, *Physica* **64**, 342 (1973).

155. H. van Beijeren and M. H. Ernst, *Phys. Lett.* **43A**, 367 (1973).
156. R. Tolman, *The Principles of Statistical Mechanics*, Oxford University Press, London (1955).
157. M. Born and H. S. Green, *A General Kinetic Theory of Fluids*, Cambridge University Press, Cambridge (1949).
158. J. G. Kirkwood, *J. Chem. Phys.* **14**, 180 (1946); **15**, 72 (1947).
159. J. Yvon, La théorie statistique des fluids, *in*: *Actualités Scientifiques et Industrielles*, No. 203, Herman, Paris (1935).
160. E. G. D. Cohen, *in*: *Fundamental Problems in Statistical Mechanics* (E. G. D. Cohen, ed.), Vol. I, p. 110, North Holland Publ. Co., Amsterdam (1962).
161. E. G. D. Cohen, *in*: *Boulder Lectures in Theoretical Physics*, Vol. 8A, p. 170, Univ. of Colorado Press, Boulder, Colorado (1966).
162. S. T. Choh and G. E. Uhlenbeck, *The Kinetic Theory of Dense Gases*, University of Michigan Report (1958).
163. G. E. Uhlenbeck and G. W. Ford, *Lectures on Statistical Mechanics*, Chap. 6, American Mathematical Society, Providence, Rhode Island (1963).
164. J. V. Sengers, *in*: *The Boltzmann Equation* (E. G. D. Cohen and W. Thirring, eds.), p. 177, Springer Verlag, Vienna (1973).
165. W. R. Hoegy and J. V. Sengers, *Phys. Rev.* **A2**, 2461 (1970).
166. D. T. Gillespie and J. V. Sengers, *in*: *Proc. Fifth Symp. Thermophysical Properties* (C. F. Bonilla, ed.), American Society of Mechanical Engineers, New York (1970).
167. J. V. Sengers, *in*: *Kinetic Theory* (R. Liboff and N. Rostoker, eds.), Gordon & Breach, New York (1970).
168. Y. Pomeau, *J. Math. Phys.* **12**, 2286 (1971).
169. C. Alle and Y. Pomeau, *Physica* **66**, 145 (1973).
170. J. V. Sengers, *Phys. Fluids* **9**, 1685 (1966).
171. L. K. Haines, J. R. Dorfman, and M. H. Ernst, *Phys. Rev.* **144**, 207 (1966).
172. K. Kawasaki and I. Oppenheim, *Phys. Rev.* **139**, 1763A (1965).
173. K. Kawasaki and I. Oppenheim, *in*: *Statistical Mechanics* (T. Bak, ed.), p. 313, Benjamin, New York (1967).
174. A. Weijland and J. M. J. van Leeuwen, *Physica* **36**, 457 (1967); **38**, 3 (1968).
175. E. H. Hauge and E. G. D. Cohen, *Phys. Lett.* **25A**, 78 (1967).
176. E. H. Hauge and E. G. D. Cohen, *J. Math. Phys.* **10**, 397 (1969).
177. J. Weinstock, *Phys. Rev. Lett.* **17**, 130 (1966).
178. E. A. Frieman and R. Goldman, *J. Math. Phys.* **7**, 2153 (1966); **8**, 1410 (1967).
179. R. Goldman, *Phys. Rev. Lett.* **17**, 910 (1966).
180. E. G. D. Cohen, *in*: *Statistical Mechanics at the Turn of the Decade* (E. G. D. Cohen, ed.), Dekker, New York (1971).
181. J. R. Dorfman and E. G. D. Cohen (to be published).
182. R. Zwanzig, *Phys. Rev.* **129**, 486 (1963).
183. J. R. Dorfman, *in*: *Boulder Lectures in Theoretical Physics* (W. E. Britten, ed.), Vol. 9C, p. 443, Gordon & Breach, New York (1967).
184. M. H. Ernst, J. R. Dorfman, W. R. Hoegy, and J. M. J. van Leeuwen, *Physica* **45**, 129 (1969).
185. J. Kestin, *in*: *Proc. Fourth Int. Conf. High Pressure*, p. 518, Kyoto (1974).
186. J. Kestin, E. Paykoc, and J. V. Sengers, *Physica* **54**, 1 (1971).
187. H. J. M. Hanley and W. M. Haynes, *J. Chem. Phys.* **63**, 358 (1975).
188. H. J. M. Hanley, R. D. McCarty, and J. V. Sengers, *J. Chem. Phys.* **50**, 857 (1969).
189. C. Bruin, *Physica* **72**, 261 (1974).
190. J. R. Dorfman, *in*: *Fundamental Problems in Statistical Mechanics* (E. G. D. Cohen, ed.), Vol. III, North Holland Publ. Co., Amsterdam (1975).
191. J. A. McLennan, *Phys. Rev.* **A8**, 1479 (1973).
192. H. S. Green, *J. Math. Phys.* **2**, 344 (1961).
193. H. Mori, I. Oppenheim, and J. Ross, *in*: *Studies in Statistical Mechanics* (G. E. Uhlenbeck and J. de Boer, eds.), Vol. I, p. 213, North Holland Publ. Co. Amsterdam (1962).
194. M. H. Ernst, Ph.D. thesis, University of Amsterdam (1965).
195. J. H. Weare and I. Oppenheim, *Physica* **72**, 1, 40 (1974).

196. D. Levesque and W. T. Ashurst, *Phys. Rev. Lett.* **33**, 277 (1974).
197. H. J. M. Hanley and R. O. Watts, *Physica* **79A**, 351 (1975).
198. L. Verlet, *Phys. Rev.* **2A**, 2514 (1971).
199. M. S. Jhon and D. Forster, *Phys. Rev.* **A12**, 254 (1975).
200. P. M. Furtado, G. F. Mazenko, and S. Yip, *Phys. Rev.* **A12**, 1653 (1975).
201. I. M. de Schepper and E. G. D. Cohen, *Phys. Lett.* **55A**, 385 (1976).
202. R. Zwanzig and M. Bixon, *Phys. Rev.* **A2**, 2005 (1970).
203. J. Lebowitz, J. Percus, and J. Sykes, *Phys. Rev.* **188**, 487 (1969).
204. Y. Pomeau, *Phys. Rev.* **A3**, 1174 (1971).
205. Y. Pomeau, *Phys. Lett.* **27A**, 602 (1968).
206. D. Gass, *J. Chem. Phys.* **54**, 1898 (1971).
207. P. Resibois and Y. Pomeau, *Physica* **72**, 493 (1974).
208. M. Theodosopulu and P. Resibois, *Physica* **82A**, 47 (1975).
209. M. H. Ernst and J. R. Dorfman, *Physica* **61**, 157 (1972).
210. M. H. Ernst and J. R. Dorfman, *J. Stat. Phys.* **12**, 311 (1975) and references contained therein.
211. I. de Schepper, H. van Beijeren, and M. H. Ernst, *Physica* **75**, 1 (1974).
212. I. de Schepper and M. H. Ernst, Ph.D. thesis, University of Nijmegen (1975).
213. J. Dufty and J. A. McLennan, *Phys. Rev.* **A9**, 1266 (1974).
214. B. J. Alder, T. E. Wainwright, W. W. Wood, and J. J. Erpenbeck (to be published).
215. Y. Pomeau, *Phys. Lett.* **38A**, 245 (1972).
216. Y. Pomeau, *Phys. Rev.* **A5**, 2569 (1972); **A7**, 1134 (1973).
217. M. H. Ernst, E. H. Hauge, and J. M. J. van Leeuwen, *Phys. Rev. Lett.* **25**, 1254 (1970).
218. M. H. Ernst, E. H. Hauge, and J. M. J. van Leeuwen, *Phys. Rev.* **A4**, 2055 (1971).
219. T. Keyes and I. Oppenheim, *Phys. Rev.* **A7**, 1384 (1973).
220. J. Deutch, *in*: *Transport Phenomena* (J. Kestin, ed.), AIP Conf. Proc. No. 11, p. 71, American Institute of Physics, New York (1973).
221. P. Mazur, *in*: *Transport Phenomena* (G. Kirczenow and J. Morro, eds.), p. 126, Springer Verlag, Berlin (1974).
222. P. Wolynes, *Phys. Rev.* **A13**, 1235 (1976).

4

Renormalized Kinetic Theory of Dense Fluids

Gene F. Mazenko
and
Sidney Yip

1. Introduction

In this chapter we describe a new microscopic method of analyzing the dynamical properties of classical fluids. Our primary interest lies in the study of those time correlation functions which describe the fluctuations in fluids and which can be directly measured by neutron and light scattering or obtained from computer molecular dynamics experiments. While the discussions here deal specifically with simple fluids, such as argon, the techniques developed are also applicable to more complicated systems (see Section 7).

It has been recognized for some time that a fully microscopic treatment of the dynamics of fluids is feasible at low densities.[1,12] For dilute gases methods based on density expansion prove to be useful and a Boltzmann-like equation can be used to calculate time correlation functions.[2–5] The notion that microscopic or kinetic theory calculations could be formulated for dense fluids is much more recent.[6–9] The difficulty with dense systems is that there is no small parameter in the problem. Consequently a direct density expansion where one computes the first few corrections to the low-density result is not likely to give useful results. At first sight things are worse than this because the first few density corrections are not only mathematically exhausting, but the

Gene F. Mazenko • The James Franck Institute and Department of Physics, The University of Chicago, Chicago, Illinois, and *Sidney Yip* • Department of Nuclear Engineering, Massachusetts Institute of Technology, Cambridge, Massachusetts

expansion breaks down because the second-order correction turns out to be infinite. Fortunately a series of theoretical and molecular dynamics studies[10,11] have shown that many of the complexities (full 3-body, 4-body, etc., collisions) are not very important. On the other hand it is essential in treating dense fluids to take into account structural effects by including full static correlation functions. These studies also found that the infinity in the density expansion, associated with the so-called "divergence of transport coefficients,"[12] is symptomatic of a new set of dynamical correlations that are not amenable to a density expansion but understandable in terms of the coupling of single-particle and collective motions on a time scale of several collision times.

The purpose of this chapter is to explain how a tractable microscopic theory of dense fluids can be constructed that takes into account the exact static correlation functions and the effects of correlated dynamical processes. There are two separate stages in the development of such a theory. The first stage is to construct a flexible formalism that is useful for making approximations applicable to dense fluids. In the second stage, which follows after an appropriate collision kernel for a kinetic equation has been derived, one is concerned with solving this kinetic equation to find the time correlation functions of interest. We will consider both stages in our discussions.

In Sections 2 and 3, we set up a formalism for dealing with the dynamics of dense fluids at the molecular level. We begin in Section 2 by focusing attention on the phase space density correlation function from which the space–time correlation functions of interest in scattering experiments and computer simulations can be obtained. The phase space correlation function obeys a kinetic equation that is characterized by a memory function, or generalized collision kernel, that describes all the effects of particle interactions. The memory function plays the role of an effective one-body potential and one can regard its presence as a renormalization of the motions of the particles.

The main objective of our theory is to develop useful methods for approximating the memory function. Two basic motivations characterize our analysis in Section 3, the systematic inclusion of the exact correlation functions and the expression of the memory function in terms of an effective two-body problem, which constitutes a second renormalization. The second motivation is particularly important in setting up approximation schemes since we are conditioned to think in terms of two-body collision processes. In Section 4 we use these ideas to arrive at an approximate expression for the memory function. The memory function is the sum of a Boltzmann–Enskog term and a recollision term that describes the correlated dynamical processes mentioned above. We will call this theory the fully renormalized kinetic theory because the medium effects involving particles other than the interacting binary cluster are taken into account and also because "bare" interactions are replaced by effective interactions, which depend on the equilibrium correlations in the fluid.

We illustrate how the present formalism leads to practical microscopic calculations by considering two applications. In Section 5 we study the density correlation function in the Boltzmann–Enskog approximation, and in Section 6 the recollision effects on the velocity autocorrelation function and the self-diffusion coefficient are analyzed. In both cases, we assume, for simplicity that the fluid is a system of hard spheres. Although the calculations themselves are relatively crude and the numerical results have limited significance, these problems are of interest because they serve to indicate the level of complexity of current microscopic calculations and lead to a discussion of the areas where further calculations are needed. The chapter then closes with summary and discussions in Section 7.

2. Phase Space Time Correlation Functions

We consider a simple classical fluid system containing N particles confined in a volume Ω at an equilibrium temperature $T = (k_B\beta)^{-1}$. It is assumed that the particles interact via two-particle central potentials and that the interaction part of the Hamiltonian is the sum of pairwise contributions. The Hamiltonian of the system is therefore of the form

$$H = \sum_{i=1}^{N} \frac{\mathbf{P}_i^2}{2m} + \frac{1}{2} \sum_{i,j=1}^{N}{}' V(|\mathbf{R}_i - \mathbf{R}_j|) \tag{1}$$

where m is the particle mass, and $\mathbf{R}_i$ and $\mathbf{P}_i$ denote the position and momentum of the ith particle. We adopt the convention that a prime on the summation symbol means omitting those terms for which the summation indices are equal.

The dynamical variable fundamental in our study is the classical phase space density

$$f(1) = \sum_{i=1}^{N} \delta(1 - q_i) = \sum_{i=1}^{N} \delta(\mathbf{r}_1 - \mathbf{R}_i)\, \delta(\mathbf{p}_1 - \mathbf{P}_i) \tag{2}$$

where $1 = (\mathbf{r}_1, \mathbf{p}_1)$ denotes the phase space coordinates of a field variable and $q_i = (\mathbf{R}_i, \mathbf{P}_i)$ denotes the phase space coordinates of an internal variable. Internal variables are particle coordinates over which we perform the thermodynamic average. Any physical variable of interest can be constructed from a single f or a product of f's. We note, for example, that the density, current, and Hamiltonian can be written as

$$n(r_1) = \int d^3p_1 f(1) = \sum_{i=1}^{N} \delta(\mathbf{r}_1 - \mathbf{R}_i) \tag{3}$$

$$\mathbf{J}(r_1) = \int d^3p_1\, \mathbf{p}_1 f(1) = \sum_{i=1}^{N} \mathbf{P}_i\, \delta(\mathbf{r}_1 - \mathbf{R}_i) \tag{4}$$

$$H=\int d^3r_1\, d^3p_1 \frac{p_1^2}{2m} f(1)+\frac{1}{2}\int d^3r_1\, d^3r_1'\, d^3p_1\, d^3p_1'\, V(|\mathbf{r}_1-\mathbf{r}_{1'}|)g(11') \quad (5)$$

where $g(11')=f(1)f(1')-\delta(1-1')f(1)$.

The equilibrium averages of products of f's at equal times give a complete description of the static correlations in the system. If we use one of the canonical ensembles we can show that

$$\langle f(1)\rangle \equiv \omega_0(1)=n\Phi(p_1) \quad (6)$$

and*

$$\begin{aligned}\tilde{C}(12) &\equiv \langle \delta f(1)\, \delta f(2)\rangle \\ &= \omega_0(1)\, \delta(12)+\omega_0(1)\omega_0(2)h(\mathbf{r}_1-\mathbf{r}_2)\end{aligned} \quad (7)$$

where

$$\delta f(1)=f(1)-\langle f(1)\rangle \quad (8)$$

$$\delta(12)=\delta(\mathbf{r}_1-\mathbf{r}_2)\, \delta(\mathbf{p}_1-\mathbf{p}_2) \quad (9)$$

n is the equilibrium number density, and $\Phi(p)$ the normalized Maxwellian,

$$\Phi(p)=(\beta/2\pi m)^{3/2}\exp(-\beta p^2/2m) \quad (10)$$

In (7)

$$h(\mathbf{r}_1-\mathbf{r}_2)=g(\mathbf{r}_1-\mathbf{r}_2)-1 \quad (11)$$

and $g(\mathbf{r}_1-\mathbf{r}_2)$ is the equilibrium pair distribution function

$$n^2 g(\mathbf{r}_1-\mathbf{r}_2)=\left\langle \sum_{i,j=1}^{N}{}' \delta(\mathbf{r}_1-\mathbf{R}_i)\, \delta(\mathbf{r}_2-\mathbf{R}_j)\right\rangle \quad (12)$$

We see, for example, that by combining these results one has, for the total energy,

$$\langle H\rangle=\frac{3}{2}nk_BT+\frac{n^2}{2}\int d^3r\, V(r)g(r) \quad (13)$$

One can go on to define higher-order static correlation functions beyond $g(r)$,[8] but we will introduce them only when they are needed. It is important to note that in dense fluids static correlation functions play an important role in determining the dynamical properties. Consequently, detailed knowledge of these functions is essential in developing any dynamical theory. In our discussions we will regard static correlation functions as known quantities and rely on other theories for their calculations. This means that $g(r)$ is understood to be

*Throughout this chapter we use the tilde to denote that a correlation function is a static (equal-time) correlation function.

the exact pair distribution function wherever it appears. In an actual calculation, the precise representation of a static correlation function like $g(r)$ will depend on what is available in the literature.[13]

Since we are interested in dynamical phenomena, we will be concerned with the time dependence of f. Using the notation $f(1,t)=\sum_i \delta(1-q_i(t))$, we have the equation of motion

$$df(1,t)/dt = iLf(1,t) \tag{14}$$

where L, the Liouville operator, is defined as the sum

$$L = L_0 + L_I \tag{15}$$

where the noninteracting and interacting parts are defined, respectively, by

$$L_0 = -i \sum_{i=1}^{N} \frac{1}{m} \mathbf{P}_i \cdot \nabla_{\mathbf{R}_i} \tag{16a}$$

$$L_I = i \sum_{i,j=1}^{N}{}' \nabla_{\mathbf{R}i} V(|\mathbf{R}_i - \mathbf{R}_j|) \cdot \nabla_i^{\mathbf{P}} \tag{16b}$$

Equation (14) is just another way of expressing Newton's equations, It can be formally integrated to give

$$f(1,t) = \exp(itL) f(1) \tag{17}$$

To discuss the dynamics of a system near equilibrium one need only consider the equilibrium average of products of f's at different times. The average of $f(1,t)$ itself contains no dynamical information, since due to time translational invariance,

$$\langle f(1,t)\rangle = \langle f(1)\rangle \tag{18}$$

The basic quantity that we wish to calculate is the time-dependent phase space density correlation function defined by

$$C(12; t-t') = \langle \delta f(2,t')\, \delta f(1,t)\rangle \tag{19}$$

Because the equilibrium ensemble is time translationally invariant, C is a function of $t-t'$. Since we will be more interested in the Fourier and Laplace transforms of (19), we introduce the definitions

$$C(\mathbf{k}t, p_1 p_2) = \frac{1}{\Omega} \int d^3 r_1\, d^3 r_2\, e^{-i\mathbf{k}\cdot(\mathbf{r}_1 - \mathbf{r}_2)} C(12; t) \tag{20a}$$

$$C(12; \omega) = \int_{-\infty}^{\infty} dt\, e^{i\omega t} C(12; t) \tag{20b}$$

$$C(12) = -i \int_0^{\infty} dt\, e^{izt} C(12; t) = \int_{-\infty}^{\infty} \frac{d\omega}{2\pi} \frac{C(12; \omega)}{z-\omega} \tag{20c}$$

Notice that frequency ω will always be real, while z is complex and lies on the upper half-plane. Using (17) and noting that the Liouville operator commutes with a canonical distribution function, we can write

$$\begin{aligned} C(12) &= -i\int_0^\infty d(t_1-t_2)e^{iz(t_1-t_2)}\langle \delta f(2)e^{i(t_1-t_2)L}\,\delta f(1)\rangle \\ &= \langle \delta f(2)(z+L)^{-1}\,\delta f(1)\rangle \end{aligned} \tag{21}$$

We will construct our theory in terms of $C(12)$. Once we find $C(12)$ all the space–time correlation functions of interest can be calculated. For example, the density–density correlation function,[14–16] also known as the dynamic structure factor, is given by

$$S(k_1\omega) = -\frac{2}{n}\,\mathrm{Im}\int d^3p_1\,d^3p_2\int \frac{d^3r_1\,d^3r_2}{\Omega}e^{-i\mathbf{k}\cdot(\mathbf{r}_1-\mathbf{r}_2)}C(12)\bigg|_{z=\omega+i0^+} \tag{22}$$

In a similar way, quantities like the longitudinal and transverse current correlation functions can be extracted.

The phase space correlation function $C(12;t)$ is not the appropriate quantity for the discussion of single-particle motions in a fluid. For the study of particle diffusion one should consider the van Hove self-correlation function $S_s(k,\omega)$, which is measured by incoherent neutron scattering experiments.[14–16] We can calculate S_s using phase space correlation functions by introducing the single-particle density

$$f_s(1) = \sqrt{N}\,\delta(1-q_1) \tag{23}$$

Essentially all of the definitions given above carry over to this case if we simply replace f by f_s. The van Hove function is then obtained from the phase space correlation function

$$C_s(12) = \langle \delta f_s(2)(z+L)^{-1}\,\delta f_s(1)\rangle \tag{24}$$

which has the same form as (21). The self-diffusion problem is simpler than the case of density fluctuations because the associated static correlation functions are simpler. For example, in the thermodynamic limit,

$$\langle f_s(1)\rangle = 0 \tag{25}$$

$$\tilde{C}_s(12) = \langle \delta f_s(2)\,\delta f_s(1)\rangle = \omega_0(1)\,\delta(12) \tag{26}$$

We will take up the discussion of single-particle motions again in Section 6.

3. The Memory Function Formulation

Our calculation of $C(12)$ is based on a kinetic equation, which in turn is derived from the equation of motion for $C(12)$. The derivation consists of a rather involved analysis of the collision kernel in the kinetic equation, which we will call the memory function. In this section we formulate the memory function calculation in terms of an appropriate two-body correlation function. [In contrast, $C(12)$ is considered a one-body correlation function.] The calculation will be completed in the following section.

To find the equation of motion satisfied by $C(12)$, we use the operator identity

$$(z+L)^{-1} = z^{-1} - z^{-1}(z+L)^{-1}L \tag{27}$$

in (21) to find

$$zC(12) = \tilde{C}(12) - \langle \delta f(2)(z+L)^{-1}L\, \delta f(1)\rangle \tag{28}$$

The operator L acts on the space of internal coordinates. Using the δ-function properties of f, one finds that the action of L on the phase space densities can be expressed in terms of operators that act upon the external field variables,

$$Lf(1) = -L_0(1)f(1) - \int d\bar{1}\, L_I(1\bar{1})f(1)f(\bar{1}) \tag{29}$$

where, in contrast to (16a) and (16b),

$$L_0(1) = -\frac{i}{m}\,\mathbf{p}_1 \cdot \nabla_{\mathbf{r}_1} \tag{30}$$

$$L_I(12) = L_I(21) = i\nabla_{\mathbf{r}_1} V(|\mathbf{r}_1 - \mathbf{r}_2|) \cdot (\nabla_{\mathbf{p}_1} - \nabla_{\mathbf{p}_2}) \tag{31}$$

are operators that act on external field variables. In (29) we have introduced the abbreviation

$$\int d\bar{1} = \int d^3 r_{\bar{1}}\, d^3 p_{\bar{1}} \tag{32}$$

Using (29) one can rewrite (28) as

$$[z - L_0(1)]C(12) - \int d\bar{1}\, L_I(1\bar{1})C(1\bar{1};2) = \tilde{C}(12) \tag{33}$$

where

$$C(11';2) = \langle \delta f(2)(z+L)^{-1} f(1) f(1')\rangle \tag{34}$$

is a new, higher-order correlation function.

The term $L_0(1)$ on the left-hand side of (33) describes the free streaming of particles. It is an important advantage of working with phase space correlation functions that free streaming can be treated exactly. This is so because the noninteracting part of the Liouville operator, L_0, is diagonalized by $f(1)$.* The right-hand side of (33) gives the initial condition. Unlike transport theories[17]

*This is not true in generalized hydrodynamics.

that deal with nonequilibrium distribution functions with unknown initial conditions, the problem of time correlation function calculation is completely specified since one deals with equilibrium-averaged quantities. Just as in transport equations, an integral term representing interaction effects appears in (33). This term couples $C(12)$ to a higher-order correlation function $C(11'; 2)$. Since $C(11'; 2)$ satisfies an equation of motion that couples it to a still higher-order correlation function, we have in (33) the first equation in an infinite (as $N \to \infty$) hierarchy of coupled equations. (For a discussion of the BBGKY hierarchy, see Uhlenbeck and Ford[18] and Liboff.[17]) Our aim is to develop a method of truncating the hierarchy that avoids difficulties associated with the hierarchical approach.

In the formalism of fully renormalized kinetic theory (FRKT) one directs one's attention not to the phase space correlation function $C(12)$ but to the associated memory function $\phi(12)$, which is defined by[5-8]

$$[z - L_0(1)]C(12) - \int d\bar{1}\phi(1\bar{1})C(\bar{1}2) = \tilde{C}(12) \tag{35}$$

The memory function is of interest because we expect it to be a "less varied" function of its arguments than the correlation function. The Fourier transform of the correlation function is resonant (has a pole structure) in the hydrodynamical regime (small k and ω). The memory function, however, has a simple behavior in this regime and gives the placement and width of the hydrodynamical poles.[19] Since $z - L_0(1)$ is just the operator that describes the streaming of a single free particle, it is clear that the memory function describes the effects of the other $N-1$ particles on this free-streaming particle. One may say that the memory function serves to renormalize the free-particle motion in the system. Since the memory function has to describe the effects of many-particle dynamics, it is not a simple quantity to calculate.

Equation (35) can be regarded as a generalized kinetic equation. The relationship of this equation to the Boltzmann equation* is well understood.[5,22] Using linear response theory, one can show that the linear deviation of the singlet distribution function from equilibrium,[23] for a system initially held in constrained equilibrium, is proportional to

$$\int d\bar{1}C(1\bar{1})U(\bar{1})$$

where U is a weak adiabatic external potential that sets up the constrained initial state. Consequently, under the conditions of low densities, long times, and distances where the singlet distribution satisfies the linearized Boltzmann equation, $C(11')$ will also obey the linearized Boltzmann equation. It has been

*Besides Liboff[17] and Uhlenbeck and Ford,[18] interested readers should see Cohen[20] and Brittin.[21]

shown[5,22] explicitly that the memory function reduces to the linearized Boltzmann collision operator K_B in the limit of low densities, low frequencies, and small wavenumbers,

$$\lim_{z\to i0^+}\lim_{k\to 0}\lim_{n\to 0}\frac{1}{n}\phi(k\mathbf{pp}'z)=\frac{1}{n}K_B(\mathbf{pp}') \tag{36}$$

In this context one should keep in mind that an approximation to lowest order (in some parameter) for ϕ corresponds to keeping an infinite set of terms in a direct expansion for C.

Equation (35) is simply a definition of ϕ. We need to obtain an explicit expression for ϕ that can be used for making approximations. Comparing (33) and (35), we see that

$$\int d\bar{1}\phi(1\bar{1})C(\bar{1}2)=\int d\bar{1}L_I(1\bar{1})C(1\bar{1};2) \tag{37}$$

This is an integral equation for ϕ in terms of $C(12)$ and $C(11;2)$. One can extract an explicit expression for ϕ in terms of correlation functions after a few manipulations. By rewriting (21) as

$$C(12)=\langle\delta f(1)(z-L)^{-1}\,\delta f(2)\rangle \tag{38}$$

one finds the "transposed" equation of motion

$$[z+L_0(2)]C(12)+\int d\bar{2}\,L_I(2\bar{2})C(1;2\bar{2})=\tilde{C}(12) \tag{39}$$

Applying the operator $(z+L_0(2))$ to Eq. (37) and using (39), one then obtains

$$\int d\bar{1}\phi(1\bar{1})\Big[\tilde{C}(\bar{1}2)-\int d\bar{2}\,L_I(2\bar{2})C(\bar{1};2\bar{2})\Big]$$
$$=\int d\bar{1}\,L_I(1\bar{1})\tilde{C}(1\bar{1};2)-\int d\bar{1}\,d\bar{2}\,L_I(1\bar{1})L_I(2\bar{2})C(1\bar{1};2\bar{2}) \tag{40}$$

where

$$C(11';22')=\langle\delta(f(2)f(2'))(z+L)^{-1}\,\delta(f(1)f(1'))\rangle \tag{41}$$

The form of (40) suggests that the memory function can be split into a "static" (z-independent) part and a "collisional" part,

$$\phi(12)=\phi^{(s)}(12)+\phi^{(c)}(12) \tag{42}$$

where, from (40),

$$\int d\bar{1}\,\phi^{(s)}(1\bar{1})\tilde{C}(\bar{1}2)=\int d\bar{1}\,L_I(1\bar{1})\,\tilde{C}(1\bar{1};2) \tag{43}$$

We are left with an equation for collisional part of the memory function,

$$\begin{aligned}\int d\bar{1}\phi^{(c)}(1\bar{1})C(\bar{1}2) = &-\int d\bar{1}\, d\bar{2}\, L_I(1\bar{1})L_I(2\bar{2})C(1\bar{1};2\bar{2}) \\ &+\int d\bar{1}\,\phi(1\bar{1})\int d\bar{2}\, L_I(2\bar{2})C(\bar{1};2\bar{2})\end{aligned} \tag{44}$$

To write this as an equation for $\phi^{(c)}$ we note from (3.9) that*

$$\phi(12) = \int d\bar{1}\, d\bar{2}\, L_I(1\bar{1})C(1\bar{1};\bar{2})C^{-1}(\bar{2}2) \tag{45}$$

Therefore

$$\int d\bar{1}\phi^{(c)}(1\bar{1})\tilde{C}(\bar{1}2) = -\int d\bar{1}\, d\bar{2}\, L_I(1\bar{1})L_I(2\bar{2})G(1\bar{1};2\bar{2}) \tag{46}$$

where

$$G(11';22') = C(11';22') - \int d\bar{2}\, d\bar{3}\, C(11';\bar{2})C^{-1}(\bar{2}\bar{3})C(\bar{3};22') \tag{47}$$

Equations (43) and (46) are still integral equations for $\phi^{(s)}$ and $\phi^{(c)}$. In Appendix A we show, using the properties of $\tilde{C}(12)$ and a few simple manipulations valid for a canonical ensemble, that

$$\phi^{(s)}(kp) = -\frac{1}{m}\mathbf{k}\cdot\mathbf{p}\omega_0(p)C_D(k) \tag{48}$$

where C_D is the direct correlation function defined by

$$C_D(k) = h(k)/[1+nh(k)] \tag{49}$$

We also show in Appendix B that (46) can be reduced to

$$\phi^{(c)}(12)\omega_o(2) = -\int d\bar{1}\, d\bar{2}\, L_I(1\bar{1})L_I(2\bar{2})G(1\bar{1};2\bar{2}) \tag{50}$$

The approximation where we ignore $\phi^{(c)}$ and keep $\phi^{(s)}$ can be solved exactly.[24,25] This approximation corresponds to a generalized random-phase approximation or a linearized Vlasov approximation where the bare potential $V(r)$ is replaced by $-\beta^{-1}C_D(r)$. Since ϕ does not take into account any damping processes it does not lead to a correct description of the hydrodynamical region.

The correlation function $G(11';22')$ is the central quantity in our calculation of the memory function. It describes the dynamical correlations of two different particles in the medium. Given its definition (47), G does not appear

*The inverse $C^{-1}(12)$ is defined as

$$\int d\bar{1}C^{-1}(1\bar{1})C(\bar{1}2) = \int d\bar{1}C(1\bar{1})C^{-1}(\bar{1}2) = \delta(12)$$

to be a simple function. Further examination shows, however, that G has many desirable properties. Suppose in (47) we introduce

$$f(i)f(j) = g(ij) + \delta(ij)f(i) \tag{51}$$

which defines $g(ij)$; we then find that all of the δ-functional terms cancel.* This indicates that $G(11'; 22')$ possesses the property that there are no terms proportional to $\delta(11')$ or $\delta(22')$. The field labels in the argument of G therefore must correspond to *different* particles.

For purposes of making approximations it is important to note the clustering properties of G. If we denote by the disconnected part that portion of a correlation function which is nonzero when a field variable becomes statistically independent of the other field variables, the correlation function $C(11'; 22')$ is seen to have many disconnected pieces. Looking at $G(11'; 22')$ we see that it has the simple connectedness property

$$G(11'; 22') = G_D(11'; 22') + G_C(11'; 22') \tag{52}$$

where the inverse Laplace transform of the disconnected part G_D is given in time space by

$$G_D(11'; 22'; t) = C(12, t)C(1'2', t) + C(12', t)C(1'2, t) \tag{53}$$

G_D is disconnected in that if we let particles 1 and 2 go off to infinity and keep 1' and 2' fixed, G_D does not vanish. This follows because $C(12, t)$ and $C(1'2', t)$ do not vanish. Notice that G_D is singly connected since it vanishes if we remove any one index to infinity. On the other hand, G_c is completely connected since it vanishes unless all four indices 11'22' remain statistically correlated. As a result of these connectedness properties, one can define the matrix inverse

$$\begin{aligned}\int d3\, d3'\, G^{-1}(11'; 33')G(33'; 22') &= \langle 11'|22'\rangle \\ &= \tfrac{1}{2}[\delta(12)\,\delta(1'2') + \delta(12')\,\delta(1'2)]\end{aligned} \tag{54}$$

In choosing to deal with G instead of $C(11'; 22')$ and $C(11'; 2)$, we are able to avoid a rather subtle technical point known as the "plateau-value" problem.[6] Since there exist disconnected terms that *cancel* between $C(11'; 22')$ and $\int d3\, d3' C(11'; 3)C^{-1}(33')C(3'; 22')$, it is essential to treat G as a single entity and not deal separately with the two terms in (47).[26]

We can gain some feeling for G by investigating its low-density limit. First one can derive equations of motion for $C(11'; 2)$ *and* $C(11'; 22')$ in basically

*The function $G(11'; 22')$ is the same as that introduced in Mazenko.[8]

the same way as the derivation of (33). Then combining these equations with (47) one finds that G satisfies the equation of motion[8]

$$[z-L(11')]G(11';22')+\int d3\,d4\,d\bar{4}\,\tilde{C}(11';3)\tilde{C}^{-1}(34)L_{\mathrm{I}}(4\bar{4})G(4\bar{4};22')$$
$$-\int d\bar{1}'\,\sigma(11'\bar{1}')\Big[C(11'\bar{1}';22')-\iint d3\,d4\,C(11'\bar{1}';3)C^{-1}(34)C(4;22')\Big]$$
$$=\tilde{G}(11';22')$$

where

$$\sigma(123)=L_{\mathrm{I}}(13)+L_{\mathrm{I}}(23) \tag{56}$$

$$C(11'\bar{1}';22')=\langle\delta g(22')(z+L)^{-1}\,\delta h(11'\bar{1}')\rangle \tag{57}$$

$$C(11'\bar{1}';3)=\langle\delta f(3)(z+L)^{-1}\,\delta h(11'\bar{1})\rangle \tag{58}$$

$$h(11'\bar{1}')=\sum_{i\neq j\neq k}\delta(1-q_i)\,\delta(1'-q_j)\,\delta(\bar{1}'-q_k) \tag{59}$$

The quantity h enters the analysis because of the identity

$$Lg(11')=-L(11')g(11')-\int d\bar{1}\,\sigma(11'\bar{1})h(11'\bar{1}) \tag{60}$$

In (55) $\tilde{G}(11';22')$ is the equal time expression of G and is given by

$$\tilde{G}(11';22')=\tilde{C}(11';22')-\int d3\,d4\,\tilde{C}(11';3)\tilde{C}^{-1}(34)\tilde{C}(4;22') \tag{61}$$

This equation is exact.

The low-density limit can be extracted directly from (55). At low densities $G(11';22')$ is of order n^2 since the correlation between two *different* particles has a density expansion starting at order n^2.* Similarly, $C(11'\bar{1}';3)$ begins at order n^3 because $h(11'\bar{1}')$ requires treating three different particles. On the other hand, the quantity

$$\int d3\,d4\,\tilde{C}(11';3)\tilde{C}^{-1}(34)L_{\mathrm{I}}(4\bar{4})$$

is of first order in the density. We see therefore that to lowest order in the density (55) reduces to

$$[z-L(11')]G(11';22')=\tilde{G}(11';22') \tag{62}$$

To be consistent, only the contribution from $G(11';22')$ to lowest order in the density should appear on the right-hand side of (62). The low-density limit for $\tilde{G}$ is given by those pieces where $1=2$ and $1'=2'$ or $1'=2$ and $1=2'$, and

*From cluster expansions one can see that the lowest power of the density is equal to the minimum number of particles.

these pieces are of order n^2. If these conditions are not satisfied then we will have an average over a product of δ-functions corresponding to three distinct particles. Such an average must necessarily be of order n^3. We then see explicitly that, to lowest order in the density,

$$\begin{aligned}\tilde{G}(11';22') &= [\delta(12)\,\delta(1'2') + \delta(12')\,\delta(1'2)]\langle g(11')\rangle \\ &= [\delta(12)\,\delta(1'2') + \delta(12')\,\delta(1'2)]\omega_0(1)\omega_0(1')g(\mathbf{r}_1 - \mathbf{r}_{1'}) \qquad (63)\end{aligned}$$

where g is the pair distribution function. Expressing g in its low-density form, this can be written as

$$\tilde{G}_0(11';22') = [\delta(12)\,\delta(1'2') + \delta(12')\,\delta(1'2)]n^2\left(\frac{\beta}{2\pi m}\right)^3 \exp[-\beta H_2(11')] \qquad (64)$$

where

$$H_2(11') = \frac{p_1^2}{2m} + \frac{p_{1'}^2}{2m} + V(|\mathbf{r}_1 - \mathbf{r}_{1'}|) \qquad (65)$$

is just the Hamiltonian for a two-particle system. The static correlation function $\tilde{G}_0(11';22')$ is related to the probability of finding particles at 2 and 2′, given that there are two distinct particles at 1 and 1′. To lowest order in the density these must be the same particles; consequently we have the δ-functions. The $\exp(-\beta H_2)$ gives the thermal distribution for the two particles. The probability of finding the particles very close together should be small since this corresponds to a system with a higher energy.

Going back to the frequency-dependent G evaluated to lowest order in the density, we obtain from (62) and (63)

$$G_0(11';22') = [z - L(11')]^{-1}[\delta(12)\,\delta(1'2') + \delta(12')\,\delta(1'2)]\omega_0(11') \qquad (66)$$

or in time space,

$$G_o(11';22';t) = e^{-itL(11')}[\delta(12)\,\delta(1'2') + \delta(12')\,\delta(1'2)]\omega_0(11') \qquad (67)$$

Since $L(11')\omega_0(11') = \omega_0(11')L(11')$, this can be written in the form

$$G_0(11';22;t) = \omega_0(11')[\delta(12(t))\,\delta(1'2'(t)) + \delta(12'(t))\,\delta(1'2(t))] \qquad (68)$$

where we have used the property that $\exp(-itL)$ propagates the two particles backward in time in the usual way.[1,12] Therefore, $G_0(11';22';t)$ has the physical interpretation as the probability of finding two particles at the points 2 and 2′ at time t given that two particles are at 1 and 1′ at $t = 0$. The low-density expression for the memory function is given by putting G_0 back into (50). We discuss the further reduction of this equation in Appendix C.

4. Symmetrized Two-Particle Correlations with Renormalized Interactions

According to (50) the calculation of the memory function $\phi^{(c)}$ for dense fluids is reduced to finding a suitable approximation for the two-particle correlation function G. While the discussion of the low-density limit has given us some feeling for this quantity, the task of developing a tractable approximation appropriate to dense fluids is still a formidable problem. There are three essential considerations in our analysis of G. First, we want to replace the interaction operators L_I in (50) by effective interactions and we can do this by symmetrically extracting factors of $\tilde{G}$ from G to combine with L_I. We therefore define a new two-particle correlation function $\bar{G}$,

$$G(11'; 22') = \int d3\, d3'\, d4\, d4'\, \tilde{G}(11'; 33')\bar{G}(33'; 44')\tilde{G}(44'; 22') \quad (69)$$

Second, we know that the Boltzmann–Enskog transport theory gives a reasonable description of transport coefficients[27–29] and the short-time properties of correlation functions,[30–32] so the approximation for $\bar{G}$ should be formulated in such a way that the leading term corresponds to the Boltzmann–Enskog approximation. Third, in deriving the "correction" to the leading term we must consider the effects of recollision processes.[33,34]

4.1. Formal Solution for G

To carry out the program outlined above, we need to develop some formal techniques for treating G. If we regard (35) as a definition of the memory function for $C(12)$, we can introduce the corresponding memory function for G,

$$[z - L(11')]G(11'; 22') - \int d3\, d3'\, M(11'; 33')G(33'; 22') = \tilde{G}(11'; 22') \quad (70)$$

A formal expression for M can be found by a development similar to that used in treating $\phi(12)$.[8] Using an obvious matrix notation, we formally solve (70) to obtain

$$G = (z - L - M)^{-1}\tilde{G} \quad (71)$$

This solution is not completely satisfactory since it does not manifestly satisfy the exact symmetry relation,* which follows from (47),

$$G(11'; 22; z) = -G(22'; 11'; -z) \tag{72}$$

On the other hand, this important property can be built into our formal solution by defining an operator D,

$$G = \tilde{G}D \tag{73}$$

and substituting (73) in (70) to obtain

$$[(z-L)\tilde{G} - M\tilde{G}]D = \tilde{G} \tag{74}$$

If we now solve (74) f or D and put the result back into (73) the result can be written in the form of (69),

$$G = \tilde{G}\bar{G}\tilde{G} \tag{75}$$

with

$$\bar{G} = [z\tilde{G} - W - \Gamma(z)]^{-1} \tag{76}$$

In (76) we have introduced the quantities

$$W = L\tilde{G} + M^{(s)}\tilde{G} \tag{77}$$

$$\Gamma(z) = M^{(c)}(z)\tilde{G} \tag{78}$$

where $M^{(s)}$ and $M^{(c)}$ are the static and dynamical (z-dependent) parts of the memory function M. Notice that W and $\Gamma(z)$ are, respectively, the renormalized static and dynamical interactions in the equation of motion for $\bar{G}$.

With the introduction of $\bar{G}$ the memory function expression (50) can now be written in the form

$$\phi^{(c)}(12)\omega_0(2) = -\int d3\, d3'\, d4\, d4' \mathscr{V}(1; 33')\bar{G}(33; 44')\mathscr{V}(2; 44') \tag{79}$$

where

$$\mathscr{V}(1; 22') = \int d\bar{1} L_I(1\bar{1})\tilde{G}(1\bar{1}; 22') \tag{80}$$

is a renormalized interaction, which will be called an end point vertex. Upon comparing (79) with (50) one sees that the calculation of $\phi^{(c)}$ has been formally

*To obtain (72) apply the identity

$$\langle A(z+L)^{-1}\rangle = \langle B(z-L)^{-1}A\rangle = -\langle B[-z+L]^{-1}A\rangle$$

to the various correlation functions in (47). Equation (72) leads to the "detailed balance" condition[36]

$$\phi^{(c)}(12)\omega_0(2) = -\phi^{(c)}(21; -z)\omega_0(1)$$

which imposes a strong constraint on the approximations one can make for G.

transformed to the problem of analyzing a symmetrized two-particle correlation function $\bar{G}$.

4.2. Properties of Renormalized Interactions

We have introduced three renormalized interactions $\mathcal{V}$, W, and $\Gamma(z)$. An understanding of their properties is essential for deriving useful approximations to $\bar{G}$. We first consider the purely static quantity $\mathcal{V}$. By using (51) we have

$$\mathcal{V}(1;22') = -\langle \delta g(22') L_I f(1) \rangle + \int d4\, d4' \langle \delta f(4) L_I f(1) \rangle \tilde{C}^{-1}(44') \tilde{C}(4;22') \tag{81}$$

A general technique for dealing with quantities of the form $\langle (L_I f(1)) f(2) f(3) \cdots f(n) \rangle$ has been discussed elsewhere.[8] Using these results one can show that

$$\mathcal{V}(1;22') = \mathcal{V}_1(1;22') + \mathcal{V}_2(1;22') \tag{82a}$$

$$\mathcal{V}_1(1;22') = -\omega_0(22') \tilde{L}_I(22')[\delta(12) + \delta(12')] \tag{82b}$$

$$\tilde{L}_I(22') = -i\beta^{-1} [\nabla_{\mathbf{r}_2} \ln g(\mathbf{r}_2 - \mathbf{r}_{2'})] \cdot (\nabla_{\mathbf{p}_2} - \nabla_{\mathbf{p}_{2'}}) \tag{82c}$$

$$\mathcal{V}_2(1;22') = \frac{i}{m} p_1 \cdot \nabla_{\mathbf{r}_1} \int d^3 r_3\, \omega_0(1) \omega_0(2) \omega_0(2') \tilde{S}^{-1}(\mathbf{r}_1 - \mathbf{r}_3) B(\mathbf{r}_3; \mathbf{r}_2 \mathbf{r}_{2'}) \tag{82d}$$

$$B(\mathbf{r}_3; \mathbf{r}_2 \mathbf{r}_{2'}) = g_3(\mathbf{r}_3 \mathbf{r}_2 \mathbf{r}_{2'}) - g(\mathbf{r}_2 - \mathbf{r}_{2'})[h(\mathbf{r}_2 - \mathbf{r}_3) + h(\mathbf{r}_{2'} - \mathbf{r}_3) + 1] \tag{82e}$$

where g_3 is the three-particle static correlation function, and $\tilde{S}(r) = \delta(r) + nh(r)$.

We note that the quantity $\tilde{L}_I(11')$ is the same as $L_I(11')$ except the bare interaction potential is replaced by the average potential $-\beta^{-1} \ln g(\mathbf{r}_1 - \mathbf{r}_{1'})$.* This replacement involves a summation of an entire class of diagrams in a direct density expansion of the memory function. The Fourier transform of $\mathcal{V}_2$ over the variable $\mathbf{r}_1$ can be written in the following form:

$$\mathcal{V}_2(kp_1;22') = -\frac{1}{m} \mathbf{k} \cdot p_1 \omega_0(1) \omega_0(2) \omega_0(2') \tilde{S}^{-1}(k) B_k(\mathbf{r}_2 \mathbf{r}_{2'}) \tag{83}$$

$$B_k(\mathbf{r}_2 \mathbf{r}_{2'}) = \int d^3 r_3\, e^{-i\mathbf{k}\cdot\mathbf{r}_3} B(\mathbf{r}_3; \mathbf{r}_2 \mathbf{r}_{2'}) \tag{83a}$$

*The average potential is frequently called the potential of mean force.

where $\tilde{S}(k)=1+nh(k)$ is the static structure factor. $\mathcal{V}_2$ vanishes as $k\to 0$ and couples only to the longitudinal current matrix element of $\phi^{(c)}$ since it is proportional to $\mathbf{k}\cdot\mathbf{p}_1\Phi(p_1)$.* It is interesting to note that in the long-wavelength limit, B, which appears to be an intrinsic three-particle quantity, can be written as[37]

$$\lim_{k\to 0} B_k(\mathbf{r}_2\mathbf{r}_{2'}) = \left[\frac{\partial}{\partial n} h(\mathbf{r}_2-\mathbf{r}_{2'})\right]_\beta \tag{84}$$

We next consider the static interaction W, which is similar to $\mathcal{V}$, but is considerably more complicated.[38–40] We see that W plays essentially the same role in the equation for $\bar{G}$ that $\phi^{(s)}$ plays in the memory function equation for C. An analysis of the connectedness property of W leads to the result[8]

$$W = W_D + W_c \tag{85}$$

where the disconnected (in the same sense as G_D) piece is given by

$$W_D(11';22') = (1+P_{11'})(1+P_{22'})\tilde{C}(1'2')\omega_0(2)L_0(1)\,\delta(12) \tag{86}$$

and the connected piece is

$$W_c(11';22') = (1+P_{11'})(1+P_{22'}) \times\left\{\frac{-i}{\beta}[\nabla_{\mathbf{r}_1}F(1;1'2')\cdot\nabla_{\mathbf{p}_1} + F(1;1'2')L_0(1)]\,\delta(12)\right\} \tag{87}$$

The operator P_{ij} interchanges i and j when it operates on a function of i and j, and in (87)

$$F(1;1'2') = \omega_0(1)\omega_0(1')\{\delta(1'2')h(\mathbf{r}_1-\mathbf{r}_{1'}) + \omega_0(2')[\sigma(\mathbf{r}_1\mathbf{r}_{1'}\mathbf{r}_2) - h(\mathbf{r}_1-\mathbf{r}_{1'})h(\mathbf{r}_1-\mathbf{r}_{2'})]\} \tag{88}$$

where

$$\sigma(\mathbf{r}_1\mathbf{r}_{1'}\mathbf{r}_{2'}) = g_3(\mathbf{r}_1\mathbf{r}_{1'}\mathbf{r}_{2'}) - h(\mathbf{r}_1-\mathbf{r}_{1'}) - h(\mathbf{r}_1-\mathbf{r}_{2'}) - h(\mathbf{r}_{1'}-\mathbf{r}_{2'}) - 1 \tag{89}$$

and

$$n^3 g_3(\mathbf{r}_1\mathbf{r}_{1'}\mathbf{r}_2) = \left\langle \sum_{i\neq j\neq k} \delta(\mathbf{r}_1-\mathbf{R}_i)\,\delta(\mathbf{r}_{1'}-\mathbf{R}_j)\,\delta(\mathbf{r}_2-\mathbf{R}_k)\right\rangle \tag{90}$$

Notice that W satisfies the symmetry relation

$$W(11';22') = -W(22';11') \tag{91}$$

which is closely related to (72). The expression for W is formidable as it stands.

*One should note that $\mathcal{V}_1$ and $\mathcal{V}_2$ describe very different types of coupling. While $\mathcal{V}_1$ is like the original bare coupling with a renormalized potential, $\mathcal{V}_2$ represents a longer-range coupling between particles. Thus far, the effects of $\mathcal{V}_2$ have not been thoroughly investigated. At low densities $\mathcal{V}_2$ is of order n^3, whereas $\mathcal{V}_1$ is of order n^2.

It is useful to note that, for moderate densities, a reasonable approximation is

$$W(11';22') \simeq \omega_0(11')[L_0(11') + \tilde{L}_I(11')]]2\langle 11'|22'\rangle \tag{92}$$

where $\langle 11'|22'\rangle$ is defined by (54).

We now consider the interaction $\Gamma(z)$, which represents the "collisional" contributions to the inverse of $\bar{G}$. Although a detailed analysis of Γ is extremely complicated, the basic physics is fairly direct. One finds that Γ separates naturally into the sum of two pieces, $\Gamma = \Gamma_D + \Gamma_c$, where Γ_D and Γ_c are the disconnected and connected pieces, respectively. The disconnected piece Γ_D serves as the collisional part of the inverse for $\bar{G}_D$, since one can show that

$$[z\tilde{G}_D - W_D - \Gamma_D(z)]\bar{G}_D = 1 \tag{93}$$

Since $\bar{G}_D$ can be expressed in terms of two-point correlation functions, Γ_D can also be expressed in terms of two-point quantities. In the simple approximation where the memory function $\phi^{(c)}$ is approximately Markovian,

$$\phi^{(c)}(12;t-t') \simeq i\phi_0^{(c)}(12)\,\delta(t-t') \tag{94}$$

where ϕ_0 is the low-density expression of ϕ, one finds that

$$\Gamma_D(11';22') = (1+P_{11'})(1+P_{22'})\phi_0^{(c)}(12)\omega_0(2)\tilde{C}(1'2') \tag{95}$$

Thus Γ_D has the physical interpretation of the memory function describing the effects of the $(N-2)$-particle medium on two independent streaming particles.

The connected part of Γ is related to fully connected three-body collisions. This corresponds to a collision where three particles are simultaneously in each others' force fields. The analysis of these types of collisions is very difficult; at the present time we know very little about these processes except that at low densities their effects are small.[10,11] We will assume that even for dense fluids the effects of $\Gamma_c(z)$ can be taken to be small.

4.3. Mode-Coupling Approximations

We now show that the elaborate formalism we have developed is a convenient starting point for approximations. The type of approximations one makes depends, of course, on the nature of the problem. In our case, which is a classical fluid with short-range forces, we want to rewrite $\bar{G}$ as a sum of pieces, one of which leads to the Boltzmann–Enskog memory function. We can do this by applying the identity

$$\begin{aligned}(A-B)^{-1} &= A^{-1} + A^{-1}B(A-B)^{-1}\\ &= A^{-1} + (A-B)^{-1}BA^{-1}\end{aligned} \tag{96}$$

twice to (76) to obtain

$$\bar{G} = \bar{G}_s + \bar{G}_s K \bar{G}_s \tag{97}$$

where

$$\bar{G}_s = (z\tilde{G} - W)^{-1} \tag{98}$$

$$K = \Gamma + \Gamma\bar{G}\Gamma \tag{99}$$

The term $\bar{G}_s$ will lead to a generalized Boltzmann–Enskog approximation. This approximation will be discussed in some detail in Section 5.

Our next task is to deduce a tractable representation of the $\bar{G}_s K \bar{G}_s$ term. Notice that when we put $\bar{G}_s K \bar{G}_s$ into the memory function expression (79), at each end of the interaction K one has

$$\int d3\, d3'\, \mathcal{V}(1; 33')\bar{G}_s(33'; 44')$$

One can argue that this quantity has its largest values in the region near $\mathbf{r}_4 \simeq \mathbf{r}_{4'}$. Since this quantity multiplies K, we need to treat $K(44'; 22')$ accurately for $\mathbf{r}_4 \simeq \mathbf{r}_{4'}$. Since K is proportional to Γ, an accurate treatment of K requires an understanding of the behavior of $\Gamma(44'; 22')$ for $\mathbf{r}_4$ near $\mathbf{r}_{4'}$. An analysis of Γ in the low-density limit shows that Γ can be written in the form

$$\Gamma(44'; 22') = \int d5\, d5'\, d6\, d6'\, \tilde{G}(44'; 55')\gamma(55'; 66')\tilde{G}(66'; 22') \tag{100}$$

so that for $\mathbf{r}_4 \simeq \mathbf{r}_{4'}$, the main variation of $\Gamma(44'; 22')$ with $\mathbf{r}_4$ and $\mathbf{r}_4$ is given by $\tilde{G}(44'; 22')$. From our discussion in Section 3 we know that a basic property of $\tilde{G}(44'; 22')$ is that it restricts particles 4 and 4' from getting too close together.

Equation (100) can be taken as a definition of γ and we will assume that Γ has this structure for all densities. Returning to (99) we now want to extract factors of $\tilde{G}$ in the same way as (75),

$$K = \tilde{G}\bar{K}\tilde{G} \tag{101}$$

which just defines another quantity $\bar{K}$. When we combine (101) with (79), (97), and (99), we find the memory function is symbolically of the form

$$\phi^{(c)}\omega_0 = -\mathcal{V}\bar{G}_s\mathcal{V} - \mathcal{V}\bar{G}_s\tilde{G}\bar{K}\tilde{G}\bar{G}_s\mathcal{V} \tag{102}$$

We can regard $\mathcal{V}\bar{G}_s\tilde{G}$ as representing a new endpoint vertex. Notice that while $\mathcal{V}\bar{G}_s\tilde{G}$ may not be simple to interpret, the quantity

$$T(1; 33') = \int d4\, d4'\, d5\, d5'\, d6\, d6'\, \mathcal{V}(1; 44')\bar{G}_s(44'; 55') \times \tilde{G}(55'; 66')G_0^{-1}(66'; 33') \tag{103}$$

where

$$G_0^{-1}(66';33')=[z-L_0(66')]\langle 66'|33'\rangle \tag{104}$$

can be identified as a generalization of the classical T-matrix.[1,8,41] We therefore rewrite the memory function expression once more:

$$\begin{aligned}\phi^{(c)}(12)\omega_0(2)= &-\int d3\,d3'\,d4\,d4'\,\mathscr{V}(1;33')\bar{G}_s(33';44')\mathscr{V}(2;44')\\ &-\int d3\,d3'\,d4\,d4'\,T(1;33')R(33';44')T(2;44';-z)\end{aligned} \tag{105}$$

where

$$R=G_0\bar{K}G_0=G_0\tilde{G}^{-1}(\Gamma+\Gamma\bar{G}\Gamma)\tilde{G}^{-1}G_0 \tag{106}$$

In arriving at this result we have used the result

$$\begin{aligned}T(2;44';-z)&=\mathscr{V}(2;55')G_0^{-1}(44';66')\tilde{G}(66';77')\bar{G}_s(77';55')\\ &=\mathscr{V}(2;55')(-1)G_0^{-1}(66';44';-z)\tilde{G}(77';66')\\ &\quad\times(-1)\bar{G}_s(55';77';-z)\\ &=\mathscr{V}(2;55')\bar{G}_s(55';77';-z)\tilde{G}(77';66')G_0^{-1}(66';44';-z)\end{aligned} \tag{107}$$

For an understanding of the physical meaning of the new endpoint interaction, it is instructive to analyze it in the "moderate" density approximation. In this case we neglect all higher-density effects except those which occur in the pair distribution function. We therefore make the approximations

$$\mathscr{V}(1;44')=-\omega_0(44')\tilde{L}_I(44')[\delta(14)+\delta(14')] \tag{108}$$

$$\bar{G}_s(44';55')=\frac{1}{\omega_0(44')}[z-\tilde{L}(44')]^{-1}2\langle 44'|55'\rangle \tag{109}$$

$$\tilde{G}(55';66')=\omega_0(55')2\langle 55'|66'\rangle \tag{110}$$

which give

$$\begin{aligned}T(1;33')=&-4\int d4\,d4'\,[\tilde{L}_I(44')(\delta(14)+\delta(14'))]\\ &\times[z-\tilde{L}(44')]^{-1}\omega_0(44')[z-L_0(44')]\langle 44'|33'\rangle\end{aligned} \tag{111}$$

After integration by parts, we find

$$\begin{aligned}T(1;33')=&-4[z+L_0(33')]\omega_0(33')\\ &\times[z+\tilde{L}(33')]^{-1}\tilde{L}_I(33')[\delta(13)+\delta(13')]\end{aligned} \tag{112}$$

In Section 6 we show that this quantity has a very simple form in the case of hard spheres.

Our expression for $\phi^{(c)}$ [Eq. (105)] is thus far still exact. The first term in (105) is seen to be the contribution from local two-body types of interaction including static correlation effects. This term represents a generalized Boltzmann–Enskog term, which controls the short-time behavior of $\phi^{(c)}$. We note that the first two terms in a power series expansion of $\phi^{(c)}$ in $(1/z)$ both come from $\mathscr{V}\bar{G}_s\mathscr{V}$. The second term in (105) contains recollision[33] and mode-coupling[34,35] effects. The T-matrices on the ends have a simple form in the case of hard spheres, and they describe close collisions between two particles. The quantity R describes the motion of these two particles after their collision.

To develop an approximation to $\phi^{(c)}$ we have to obtain a tractable representation for R. The simplest approximation that is consistent with our emphasis on recollision processes is to assume that after a collision the two particles propagate independently of each other but are both under the influence of the medium. Mathematically this corresponds to making the "disconnected" approximation for R,

$$R \simeq R_D \tag{113}$$

$$R_D = G_0\tilde{G}_D^{-1}[\Gamma_D + \Gamma_D\bar{G}_D\Gamma_D]\tilde{G}_D^{-1}G_0 \tag{114}$$

Notice that this approximation is closely connected with the assumption that $\Gamma_c(z)$ is small. Recalling the equation of motion for $\bar{G}_D$, (33), we have the identity

$$\Gamma_D\bar{G}_D = -1 + G_d^{-1}\bar{G}_D \tag{115}$$

where

$$G_d^{-1} = z\tilde{G}_D - W_D \tag{116}$$

We can rewrite R_D as

$$R_D = G_0\tilde{G}_D^{-1}G_d^{-1}[-G_d + \bar{G}_D]G_d^{-1}\tilde{G}_D^{-1}G_0 \tag{117}$$

Using the result

$$L_0(11')\tilde{G}_D^{-1}(11';33') = -L_0(33')\tilde{G}_D^{-1}(11';33') \tag{118}$$

we can show that

$$G_0\tilde{G}_D^{-1}G_d^{-1} = 1 \tag{119}$$

Thus R_D is simply of the form

$$R_D(11';22') = \langle 11'|(\bar{G}_D - G_d)|22'\rangle \tag{120}$$

Our approximation to the memory function is

$$\phi^{(c)}(12) = \phi_{BE}(12) + \delta\phi(12) \tag{121}$$

where, schematically, the Boltzmann–Enskog term is $\phi_{BE} = -\mathscr{V}\bar{G}_s\mathscr{V}$ and the recollision term is $\delta\phi = -T[\bar{G}_D - G_d]T$. We will examine the approximation where one keeps only ϕ_{BE} in (121) in the next section, and the effects of $\delta\phi$ in the case of self-diffusion will be investigated in Section 6.

5. The Boltzmann–Enskog Theory of Thermal Fluctuations

The approximation wherein one retains the first term in the memory function expression (105) is of special interest because it leads to a kinetic equation that is closely related to the Boltzmann–Enskog equation in transport theory. In this section we will investigate this particular approximation in some detail not only from the standpoint of further analytical analysis but also from the standpoint of practical calculations. We will see that within certain limitations the approximation results in a reasonably realistic description of dense gases and liquids, and in this sense represents the first step in a systematic microscopic calculation.

We consider the approximate memory function

$$\phi(12)\omega_0(2) \simeq \phi^{(s)}(12)\omega_0(2) - \mathscr{V}(1;33')\bar{G}_s(33';44')\mathscr{V}(1;44') \tag{122}$$

where the static contribution $\phi^{(s)}$ is given by (48), and $\mathscr{V}$ and $\bar{G}_s$ are given by (108) and (109), respectively. For a fluid of hard spheres the term $\mathscr{V}\bar{G}_s\mathscr{V}$ can be reduced to an explicit form suitable for practical calculations. This is discussed in Appendix C. Using (122) and the results of Appendix C we find that (35) becomes an explicit kinetic equation describing the Fourier–Laplace transformed phase space density correlation function $C(kpp'z)$, defined according to (20a) and (20c). This equation reads

$$\left(z - \frac{\mathbf{k}\cdot\mathbf{p}}{m}\right)C(k\mathbf{pp}'z) + n\Phi(p)\frac{\mathbf{k}\cdot\mathbf{p}}{m}[C(k) - g(\sigma)C_0(k)]\int d3\bar{p}\, C(k\bar{p}p'z)$$
$$= ing(\sigma)J[C] - \tilde{C}(k\mathbf{pp}') \tag{123}$$

where

$$J[C] = \sigma^2 \int d^3p_1\, d\Omega_r[\hat{\boldsymbol{r}}\cdot(\boldsymbol{p}-\boldsymbol{p}_1)/m]\theta[\hat{\boldsymbol{r}}\cdot(\boldsymbol{p}_1-\boldsymbol{p})]$$
$$\times[\Phi(p_1^*)C(k\mathbf{p}^*\mathbf{p}'z) - \Phi(p_1)C(k\mathbf{pp}'z)$$
$$+ e^{i\mathbf{k}\cdot\hat{\mathbf{r}}\sigma}\Phi(p^*)C(k\mathbf{p}_1^*\mathbf{p}'z) - e^{-\mathbf{k}\cdot\hat{\mathbf{r}}\sigma}\Phi(p)C(k\mathbf{p}_1\mathbf{p}'z)] \tag{124}$$

It should be noted that (124) differs from the well-known collision integral in the linearized Boltzmann equation[42,43] (see Chapter 3) for hard spheres only in the presence of the phase factors $\exp(\pm i\mathbf{k}\cdot\hat{\mathbf{r}}\sigma)$. Moreover, when one compares (123) with the linearized Enskog equation[42,43] one finds that (123) is identical to the Enskog equation if $C(k)$ is replaced by $[C(0)/C_0(0)-g(\sigma)]C_0(k/2)$.[31,32] The distinction between the two kinetic equations lies therefore in the static part of the memory function. Since we have made no approximation in evaluating $\phi^{(s)}$, this difference should be regarded as an inherent defect in the Boltzmann–Enskog theory as a description of thermal fluctuations at finite wavelengths and frequencies. Indeed it is known that the linearized Enskog equation does not give correctly the second frequency moment of the density correlation function $S(k, \omega)$,[44,45] whereas (123) does. Since the static memory function affects the sound speed the two kinetic equations do not give the same hydrodynamic limit. On the other hand, it is known that transport coefficients calculated in the two descriptions are the same.[30]

The aim of our calculation is to extract $S(k, \omega)$ from (123). From (22) we have the relation

$$S(k,\omega) = -\frac{2}{n}\,\mathrm{Im}\left[\int d^3p\, d^3p'\, C(k\mathbf{p}\mathbf{p}'z)\right]_{z=\omega+i0^+} \qquad (125)$$

From the standpoint of obtaining numerical results, the most effective method of solution appears to be the method of kinetic models.[46,47] This is an approach where one introduces a model memory function that retains many of the essential properties of the original memory function. The approximate memory function is designed so that $S(k, \omega)$ can be directly computed from the resulting kinetic model equation. The advantage of the kinetic model method is that the approximation can be systematically improved. Each model memory function is specified by one or more relaxation times which are the matrix elements of the collision operator in an appropriate set of basis functions. By including a larger set of these matrix elements, one obtains a better approximation to the original memory function, and in this manner one can even discuss the convergence of the model equations.

The kinetic model solutions are particularly useful for analyzing fluctuation spectra at finite wavenumbers. The solutions do not involve any expansion in k, and since they are constrained to give the correct limiting behavior at large and small k, the models prove to be quite successful means of interpolating between well-known limits.

It is sufficient to illustrate the essential techniques of kinetic model solution by considering a triple relaxation time model. This is the simplest kinetic equation capable of describing quantitatively the thermal fluctuations at long wavelengths. We will first summarize the kinetic model method and formulate the triple relaxation time model explicitly. Then we will examine some of the calculations in the context of experimental data on $S(k, \omega)$.

The basic ingredients in the formulation of kinetic models are the matrix elements of the memory function calculated using an appropriate set of momentum basis functions. In the analysis of the Boltzmann collision operator it has been found convenient to use the Sonine polynomials.[42,43,48] However, the $\mathbf{k}\cdot\hat{\mathbf{r}}$ dependence in (124) destroys the rotational symmetry present in the Boltzmann collision operator for a spherically symmetric potential; therefore it is equally appropriate to choose the function

$$\psi_\alpha(\boldsymbol{\xi}) = (l!m!n!)^{-1/2}\bar{H}_l(\xi_x)\bar{H}_m(\xi_y)\bar{H}_n(\xi_z) \tag{126}$$

where

$$\bar{H}_l(x) = 2^{-l/2}H_l(x/2)$$

with $H_l(x)$ being the standard Hermite polynomial.[49] In (126) the single index α is understood to denote three indices (l, m, n), and $\boldsymbol{\xi}$ is the dimensionless momentum variable $\boldsymbol{\xi} = \mathbf{p}/mv_0$, $v_0 = (m\beta)^{-1}$. The orthonormality condition for ψ_α is

$$\int d^3\xi\, \Phi(\xi)\psi_\alpha(\boldsymbol{\xi})\psi_\beta(\boldsymbol{\xi}) = \delta_{\alpha\beta} \tag{127}$$

where $\Phi(\xi) = (2\pi)^{-3/2}\exp(-\xi^2/2)$.

Consider the expansion

$$\phi(k\boldsymbol{\xi}\boldsymbol{\xi}'z) = \sum_{\alpha\beta} \phi(\alpha|\beta)\psi_\alpha(\boldsymbol{\xi})\psi_\beta(\boldsymbol{\xi}')\Phi(\xi) \tag{128}$$

with

$$\phi(\alpha|\beta) = \int d^3\xi\, d^3\xi'\Phi(\xi')\psi_\alpha(\boldsymbol{\xi})\phi(k\boldsymbol{\xi}\boldsymbol{\xi}'z)\psi_\beta(\boldsymbol{\xi}') \tag{129}$$

We will not be concerned with the general symmetry and other formal properties of the memory function ϕ or the matrix elements $\phi(\alpha|\beta)$.[47] The actual evaluation of $\phi(\alpha|\beta)$ in the case of (123) and (124) can be carried out by reducing the integrals to a series of sums. The resulting general expressions[50] are quite bulky and need not be reproduced here. For purposes of illustration it is sufficient to consider a few of the lowest-order matrix elements. The most important such elements are given in Table 1, which introduces further notation for labeling the matrix elements.[51]

It is useful to regard the momentum basis functions ψ_α as states onto which one can project the memory function or the phase space density correlation function. By referring to the notations established in Table 1 and (125), we see that ψ_1 is the number density state, ψ_2 the longitudinal current (z-component) state, ψ_3 the energy state, ψ_4 and ψ_5 the two transverse current (x- and y-component) states, etc. The elements given in Table 1, $\psi(\alpha|\beta)$, $\alpha, \beta \leq 4$, are those which determine the thermodynamics as well as the basic hydrodynamic structure of the fluid system.[8,52] Notice that in the limit of $k\sigma \to 0$ these

Table 1. Matrix Elements of Triple Relaxation Time Memory Function, $\phi(\alpha|\beta)[n\sigma^2 v_0\sqrt{\pi}g(r_0)]^{-1}$ [a]

ψ_α	ψ_β: $\psi_1=\|000\rangle$	$\psi_2=\|001\rangle$	$\psi_3=\|E\rangle$	$\psi_4=\|010\rangle$	$\psi_5=\|100\rangle$	$\psi_6=\|011\rangle$	$\psi_7=\|H\rangle$
$\psi_1=\|000\rangle$	0	0	0	0		0	0
$\psi_2=\|001\rangle$	$-\frac{kC(k)}{\sqrt{\pi}\sigma^2 g(\sigma)}$	$-8i\left[\frac{1}{3}+\frac{d^2}{dx^2}j_0(x)\right]$	$4\left(\frac{\pi}{6}\right)^{1/2} j_1(x)$	0		0	$-\frac{4i}{\sqrt{10}}\left[\frac{1}{3}+\frac{d^2}{dx^2}j_0(x)\right]$
$\psi_3=\|E\rangle$	0	$4\left(\frac{\pi}{6}\right)^{1/2} j_1(x)$	$-\frac{8i}{3}[1-j_0(x)]$	0		0	$6\left(\frac{\pi}{15}\right)^{1/2} j_1(x)$
$\psi_4=\|010\rangle$	0	0	0	$-4i\left[\frac{2}{3}-\left(1+\frac{d^2}{dx^2}\right)j_0(x)\right]$		$-4\sqrt{\pi}\frac{d}{dx}\left[\frac{1}{x}j_1(x)\right]$	0
$\psi_5=\|100\rangle$							
$\psi_6=\|011\rangle$	0	0	0	$-4\sqrt{\pi}\frac{d}{dx}\left[\frac{1}{x}j_1(x)\right]$		$-16i\left[\frac{4}{15}+\frac{d^2}{dx^2}\left(\frac{1}{x}\right)j_1(x)\right]$	0
$\psi_7=\|H\rangle$	0	$-\frac{4i}{\sqrt{10}}\left[\frac{1}{3}+\frac{d^2}{dx^2}j_0(x)\right]$	$6\left(\frac{\pi}{15}\right)^{1/2} j_1(x)$	0		0	$-\frac{i}{15}\left[59+81\frac{d^2}{dx^2}j_0(x)\right]$

[a] Note: $x=k\sigma$, $|E\rangle=\frac{1}{3}[|002\rangle+|020\rangle+|200\rangle]$, $|H\rangle=(1/\sqrt{5})[|201\rangle+|021\rangle+\sqrt{3}|003\rangle]$. $j_l(x)$ is the spherical Bessel function of order l. Elements involving ψ_5 are the same as those involving ψ_4.

elements vanish, and this is consistent with the relation between the present memory function and the linearized Boltzmann collision operator.[5] The nonvanishing of $\psi(001|001)$ and $\psi(E|E)$ indicates that at finite k one cannot speak of collisional invariants such as momentum and energy conservation.

The states ψ_α, $\alpha \leq 5$, may be called the hydrodynamic states since they are associated with the conserved variables of number density, longitudinal and transverse components of the current, and kinetic energy. The other two states, ψ_6, ψ_7, correspond to the stress tensor and heat current, respectively. Therefore, the diagonal matrix elements involving these states must be related to the transport coefficients of shear viscosity and thermal conductivity as is well known in conventional transport theory.[42,43] We will see below that these elements are important in formulating kinetic models. Besides the matrix elements shown in Table 1, we will include one additional element, namely,

$$\phi^{(c)}(M|M) = -in\sigma^2 v_0 g(\sigma)\sqrt{\pi}\left[\frac{62}{15} - 2j_0(k\sigma)\right] \tag{130}$$

where

$$\psi_M = \frac{1}{\sqrt{30}}\left(\frac{\xi^4}{2} - 5\xi^2 + \frac{15}{2}\right) \tag{131}$$

The choice of nonhydrodynamic states may seem somewhat arbitrary until one realizes that it is logical to choose those states which are most directly associated with the description of transport coefficients. It is obviously desirable to have kinetic models that give the proper values for transport coefficients and thus ensure the correct behavior of the fluctuation spectra at long wavelengths.

We are now in a position to formulate the kinetic model. Returning to the kinetic equation (35), we write

$$(z - kv_0\xi_3)F(k\boldsymbol{\xi}z) = -\tilde{F}(k\boldsymbol{\xi}) + \int d^3\bar{\xi}\,\phi(k\boldsymbol{\xi}\bar{\boldsymbol{\xi}})F(k\bar{\boldsymbol{\xi}}z) \tag{132}$$

where

$$F(k\boldsymbol{\xi}z) = \int d^3\xi' C(k\boldsymbol{\xi}\boldsymbol{\xi}'z) \tag{133}$$

We introduce an approximate memory function as follows[53,54]:

$$\begin{aligned}
\phi(k\boldsymbol{\xi}\boldsymbol{\xi}') &= \sum_{\alpha,\beta=1}^{\infty} \phi(\alpha|\beta)\psi_\alpha(\boldsymbol{\xi})\psi_\beta(\boldsymbol{\xi}')\Phi(\xi) \\
&\simeq \sum_{\alpha,\beta=1}^{N} \phi(\alpha|\beta)\psi_\alpha(\boldsymbol{\xi})\psi_\beta(\boldsymbol{\xi}')\Phi(\xi) - i\nu(k)\sum_{\alpha=N+1}^{\infty} \psi_\alpha(\boldsymbol{\xi})\psi_\alpha(\boldsymbol{\xi}')\Phi(\xi) \\
&= \sum_{\alpha,\beta=1}^{N} [\phi(\alpha|\beta) + i\nu(k)\,\delta_{\alpha\beta}]\psi_\alpha(\boldsymbol{\xi})\psi_\beta(\boldsymbol{\xi}')\Phi(\xi) \\
&\quad - i\nu(k)\sum_{\alpha=1}^{\infty} \psi_\alpha(\boldsymbol{\xi})\psi_\alpha(\boldsymbol{\xi}')\Phi(\xi)
\end{aligned} \tag{134}$$

where $\nu(k)$ is a wavelength-dependent collision frequency to be chosen appropriately. It seems reasonable to take as $\nu(k)$ one of the diagonal matrix elements in a state ψ_γ, $\gamma > N$. In our case we shall take $N = 7$ and choose $\nu(k)$ to be $\phi^{(c)}(M|M)$. Note that the last term in (134) is just $-i\nu(k)\,\delta(\boldsymbol{\xi}-\boldsymbol{\xi}')$. Inserting (134) into (132) and noting that $F(k\boldsymbol{\xi}z)$ can be also expanded in $\psi_\alpha(\boldsymbol{\xi})$, one finds

$$[z - kv_0\xi_3 + i\nu(k)]F(k\boldsymbol{\xi}z) = -\tilde{F}(k\boldsymbol{\xi}) + \sum_{\alpha,\beta=1}^{N} \gamma_{\alpha\beta}(k)\psi_\alpha(\boldsymbol{\xi})F_\beta(kz)\Phi(\xi) \quad (135)$$

where

$$\gamma_{\alpha\beta}(k) = \phi(\alpha|\beta) + i\nu(k)\,\delta_{\alpha\beta} \quad (136)$$

$$F_\alpha(kz) = \int d^3\xi\, \psi_\alpha(\boldsymbol{\xi})F(k\boldsymbol{\xi}z) \quad (137)$$

This is a set of coupled integral equations that can be readily converted into a set of coupled algebraic equations for the space–time correlation functions F_γ,

$$F_\gamma(kz) = -D_{\gamma 1}(kz) + \sum_{\alpha,\beta=1}^{N} D_{\gamma\alpha}(kz)\gamma_{\alpha\beta}(k)F_\beta(kz) \quad (138)$$

where

$$D_{\gamma\alpha}(kz) = \int d^3\xi \frac{\Phi(\xi)\psi_\gamma(\boldsymbol{\xi})\psi_\alpha(\boldsymbol{\xi})}{z - kv_0\xi_3 + i\nu(k)} \quad (139)$$

can be further expressed in terms of the plasma dispersion function.[55]

The kinetic model we will deal with specifically is (138) with $\psi(\alpha|\beta)$ given by Table 1 ($N = 7$) and $\nu(k)$ given by (130). Before we go on to the analysis of fluctuation spectra it is instructive first to examine the thermodynamic and hydrodynamic properties of this model. Although one expects that these properties must be fixed once the model memory function is determined, it is only recently that the explicit relations between specific properties and matrix elements of the memory function have been obtained.[8,52] It is beyond the scope of this chapter to derive such relations; we shall therefore be content to quote them and refer the reader to the literature for further details.

In discussing the thermodynamics of the kinetic model description we will concentrate only on those properties which directly affect the behavior of $S(k, \omega)$. These are the isothermal compressibility χ_T, the specific heats C_v and C_p, and the adiabatic sound speed c_s. The compressibility is the long-wavelength limit of the static structure factor $S(k)$,[14]

$$\lim_{k\to 0} S(k) = n\chi_T/\beta \quad (140)$$

Since $S(k)$ is considered an input function to the kinetic theory description, the compressibility is not a property specified by the memory function. In contrast, the specific heat at constant volume can be obtained from the expression[52]

$$mC_v = \frac{3}{2}k_B\left[1 - \lim_{z\to i0^+}\lim_{k\to 0}\frac{\partial}{\partial z}\langle E|\phi^{(c)}|E\rangle\right] \tag{141}$$

which gives in the present case $C_v = 3k_B/2m$. This result is correct for hard-sphere fluids since the system has no potential energy. The constant-pressure specific heat may be obtained indirectly. First we note that

$$c_s^2 = \frac{C_P}{C_v}\left(\frac{dP}{dmn}\right)_T = \frac{C_P}{C_v}v_0^2[1 - nC_D(k)]_{k\to 0} \tag{142}$$

and

$$c_s^2 = \left(\frac{dP}{dmn}\right)_T + \frac{T}{(mn)^2C_v}\left(\frac{dP}{dT}\right)_n^2 \tag{143}$$

where it is known[52] that

$$\left(\frac{dP}{dT}\right)_n = nk_B\left[1 + \frac{1}{v_0}\left(\frac{3}{2}\right)^{1/2}\lim_{z\to i0^+}\lim_{k\to 0}\frac{1}{k^2}\langle 001|\phi^{(c)}|E\rangle\right] \tag{144}$$

From Table 1 we have

$$(dP/dT)_n = nk_B[1 + 4\eta g(\sigma)] \tag{145}$$

where $\eta = \pi n\sigma^3/6$ is the packing fraction. Thus the sound speed is

$$c_s^2 = v_0^2\{1 - nC_D(0) + \tfrac{2}{3}[1 + 4\eta g(\sigma)]^2\} \tag{146}$$

and we also obtain

$$C_p = C_v + \frac{2}{3}C_v\frac{[1 + 4\eta g(\sigma)]^2}{1 - nC_D(0)} \tag{147}$$

We see that (146) is the correct sound speed for a hard-sphere system. We may conclude that our model memory function gives the same thermodynamic properties as the exact hard-sphere fluid.

The transport coefficients that we want to investigate are the shear viscosity, the thermal conductivity, and the sound attenuation coefficient. It has been shown recently that the shear viscosity can be expressed directly in terms of certain matrix elements of ϕ[8,52]:

$$\begin{aligned}\eta_s = imnv_0^2\Big\{&\lim_{k\to 0}(kv_0)^{-2}\langle 010|\phi^{(c)}|010\rangle - \lim_{k\to 0}(kv_0)^{-1}\langle 011|[Q\phi^{(c)}Q]^{-1}|011\rangle \\ &-2\lim_{k\to 0}(kv_0)^{-1}\langle 011|[Q\phi^{(c)}Q]^{-1}\phi^{(c)}|010\rangle \\ &+\lim_{k\to 0}(kv_0)^{-2}\langle 010|\phi^{(c)}[Q\phi^{(c)}Q]^{-1}\phi^{(c)}|010\rangle\Big\}\end{aligned} \tag{148}$$

where $Q = 1 - \sum_{\alpha=1}^{5} |\alpha\rangle\langle\alpha|$ is the projection operator for nonhydrodynamic states ψ_β ($\beta > 5$). From (134) we find

$$[Q\phi^{(c)}Q]^{-1} = \frac{1}{\phi(616)}|6\rangle\langle 6| + \frac{1}{\phi(7|7)}|7\rangle\langle 7| + \frac{i}{\nu(k)}\sum_{\alpha=8}^{\infty}|\alpha\rangle\langle\alpha| \tag{149}$$

Using (134) and (149) one finds that the matrix elements required for the calculation of η_s are $\phi(4|4)$, $\phi(4|6)$, and $\phi(6|6)$, which are all given in Table 1. After some algebra one obtains[51] from (148)

$$\eta_s = \eta_s^0(4\eta)\left\{\frac{1}{4\eta g(\sigma)} + 0.8 + 0.771[4\eta g(\sigma)]\right\} \tag{150}$$

where $\eta_s^0 = 5mv_0/16\sqrt{\pi}\sigma^2$ is the shear viscosity coefficient of a dilute gas. The Enskog result for the shear viscosity is identical to (150) except that the coefficient 0.771 is replaced by 0.7614.[42,43]

The calculation of the thermal conductivity is similar but more involved because the longitudinal hydrodynamic modes are coupled. For the present model memory function it can be shown that[51]

$$\lambda = \lambda^0(4\eta)\left\{\frac{1}{4\eta g(\sigma)} + 1.2 + 0.7674[4\eta g(\sigma)]\right\} \tag{151}$$

where $\lambda^0 = 75k_Bv_0/64\sqrt{\pi}\sigma^2$. Equation (151) also agrees with the corresponding Enskog expression except that the coefficient 0.7674 is replaced by 0.7574. In the case of the sound attenuation coefficient, one finds[51]

$$\Gamma = (\tfrac{4}{3}\eta_s + \eta_v)/mn + \lambda\left(\frac{C_p}{C_v} - 1\right)\Big/ mnC_p \tag{152}$$

where

$$\eta_v = 1.037\eta_s^0(4\eta)^2 g(\sigma) \tag{153}$$

is the bulk viscosity. Equation (153) differs from the first-order Enskog result only in that 1.037 is replaced by 1.0186.

In summary we see that the use of the triple relaxation time model gives the same thermodynamic properties and essentially the same hydrodynamic properties as the exact hard-sphere system. It therefore follows that the corresponding kinetic equation (135) should give the proper limit for $S(k, \omega)$ at long wavelengths and low frequencies. The hydrodynamic behavior of $S(k, \omega)$ can be calculated using the macroscopic equations of fluid dynamics. All the essential features are summarized in the expression[19]

$$S(k,\omega) = mnv_0^2\left(\frac{\partial n}{\partial P}\right)_T\left\{\left(1 - \frac{C_v}{C_p}\right)\frac{D_T k^2}{\omega^2 + (D_T k^2)^2} + \frac{C_v}{C_p}\frac{\omega c_s^2 k^4 \Gamma}{(\omega^2 - c_s^2k^2)^2 + (\omega\Gamma k^2)^2} - \left(1 - \frac{C_v}{C_p}\right)\frac{\omega k^2(\omega^2 - c_s^2k^2)}{(\omega^2 - c_s^2k^2)^2 + (\omega\Gamma k^2)^2}\right\} \tag{154}$$

where $D_T = \lambda/mnC_p$. The spectral distribution consists of a central peak (the Rayleigh component) and two symmetrically displaced peaks (the Brillouin doublet) at the adiabatic sound frequency $c_s k$. The respective widths depend on the thermal diffusivity D_T and the sound attenuation coefficient Γ, and both vary with wavelengths as k^2. Notice that the transport coefficients do not affect the integrated intensity of each component, the ratio of the central to sound peak areas being just $(C_p/C_v)-1$, the Landau–Placzek ratio.

Equation (154) represents a solution to our kinetic model equation when the explicit expressions given in this section are used to avaluate the thermodynamic properties and transport coefficients. The solution is known to be valid at long wavelengths, or more precisely when $\Gamma k \ll c_s$. At the opposite extreme of very short wavelengths all the collision effects become unimportant compared to free molecular flow. Then all kinetic model solutions will tend to the distribution

$$S(k, \omega) = \frac{1}{\sqrt{2\pi} k v_0} \exp\left[-\frac{1}{2}\left(\frac{\omega}{k v_0}\right)^2\right] \tag{155}$$

whose width varies linearly with k. Roughly speaking, (155) should be valid when $k > 40 n\sigma^2$.

At the intermediate wavelengths no useful analytic forms of solution are known. On the other hand, (138) yields readily to numerical solutions. Such results show a smooth interpolation between characteristic hydrodynamic and free-particle behavior. Notice, however, that in this region it is essential to treat the collisions and molecular flow on equal footing; for this reason it would be inappropriate to apply either (154) or (155). Since the intermediate k range is particularly relevant to neutron inelastic scattering studies of liquids [56] and related computer molecular dynamics simulations,[57,58] the validity of our kinetic model solutions is of interest.

Figure 1 shows a series of spectra calculated using the present model memory function as applied to liquid argon.[51] The neutron scattering data[59] are given for comparison and one can see that a reasonably satisfactory overall agreement is achieved. The model memory function is precisely that specified in this section except for one modification. It has been found that in analyzing neutron scattering and computer molecular dynamics data using a single relaxation time kinetic model based on the generalized Enskog equation, the factor $g(\sigma)$ in $\phi^{(c)}$ should be replaced by a k-dependent factor $\chi(k\sigma)$.[60] This factor, which was essentially obtained by an empirical fitting procedure, is shown in Fig. 2. We have found that the magnitude of $\chi(k\sigma)$ varies somewhat with kinetic model, but its structure always seems to have a close correlation with the behavior of the static structure factor $S(k)$.

The introduction of $\chi(k\sigma)$ has not been justified on theoretical grounds although it could arise from a number of effects such as the neglect of

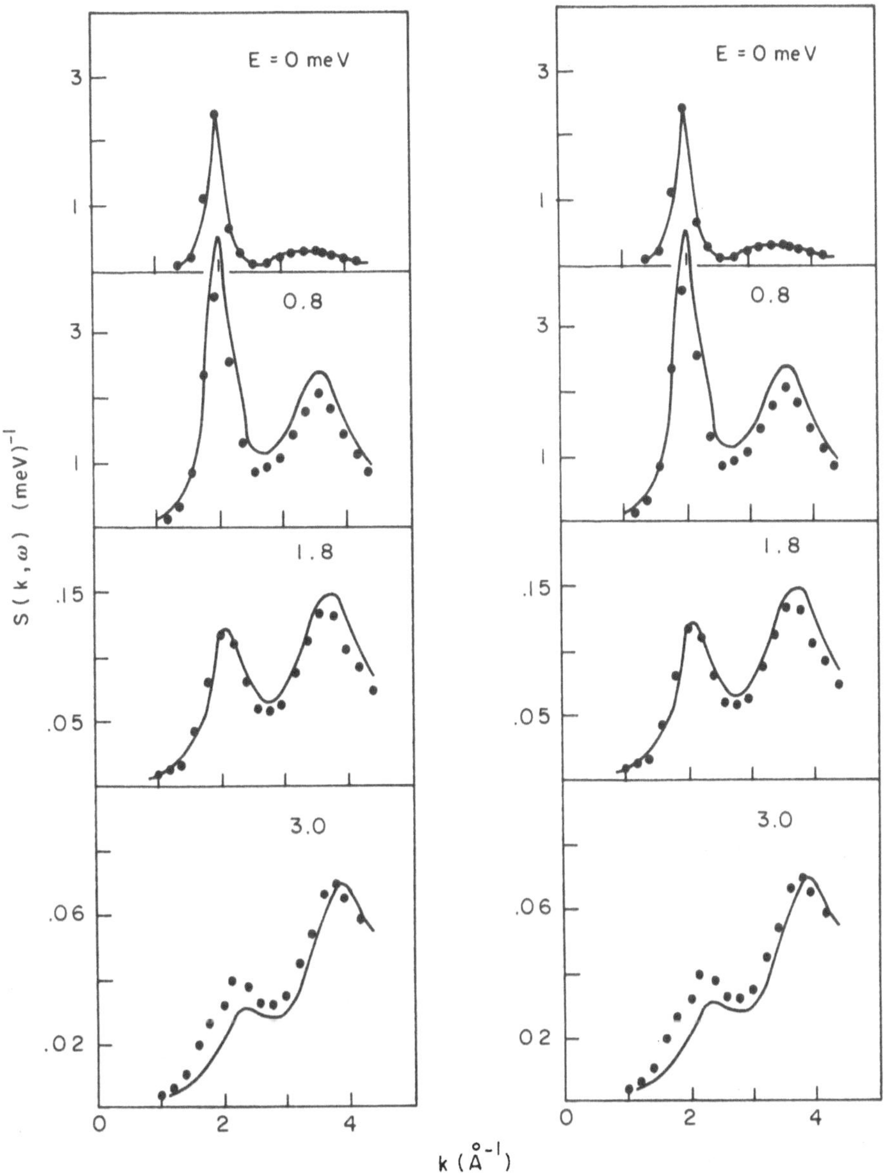

Fig. 1. Dynamic structure factor $S(k, \omega)$ of liquid ^{36}A at 85.2°K at various values of $E = \hbar\omega$. The solid curves are present kinetic model calculations; the points are coherent neutron inelastic scattering data.[59]

higher-order matrix elements in formulating the kinetic model approximation, the neglect of more complicated collision processes such as the ring terms discussed in Section 4, and the approximation of liquid argon as a hard-sphere fluid. Further analysis of this problem would be of considerable value.

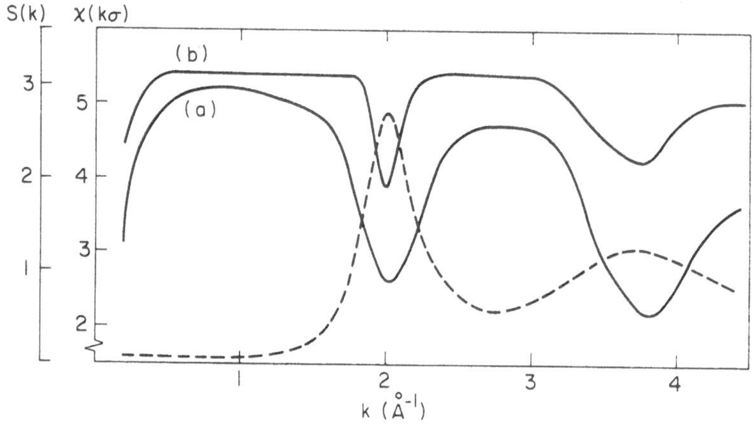

Fig. 2. Wave number dependence of the empirical function $\chi(k\sigma)$. Curve (b) corresponds to present kinetic model calculation, whereas curve (a) corresponds to a lower-order model calculation.[60] Dashed curve is the static structure factor $S(k)$ of liquid argon.

6. The Velocity Autocorrelation Function of a Dense Fluid

In our analysis of the memory function $\phi^{(c)}(12)$, the correction to the Boltzmann–Enskog approximation consists of recollision terms that describe correlated binary collisions. The effects of these processes on the behavior of time correlation functions have not yet been fully studied because the calculations involved are considerably more complicated. The problem where the effects of recollisions have been most extensively investigated is that of the velocity autocorrelation function, which is a simpler function than the dynamic structure factor. From this problem we can already see the kind of analysis involved in treating correlated collisions.[6,7]

We now consider the motions of an atom as it moves in a fluid environment. The memory function for this case can be written in a form like (105):

$$\phi_s^{(c)}(12) = \phi_{sE}^{(c)}(12) + \delta\phi_s(12) \tag{156}$$

where $\phi_{sE}^{(c)}$ is the Boltzmann–Enskog memory function appropriate for the self-diffusion problem.[30,47,61] The recollision term is $\delta\phi_s$ and it follows from (105) if we take into account the differences between the case of density fluctuations and single-particle motions. The first difference concerns the endpoint vertex. In the case of self-diffusion, one should replace $\mathscr{V}(1;22')$ by

$$\mathscr{V}_s(1;22') = -\omega_0(22')\tilde{L}_I(22')\,\delta(12) \tag{157}$$

and, for consistency, replace $2\langle 11'|22'\rangle$ by $\delta(12)\,\delta(1'2')$. This second replacement results from our tagging a particle so that only its motions are followed. In other words, there are no "exchange effects" as in the case of density fluctuations. As a consequence of this type of difference we see that the disconnected part of $G_s(11'; 22', t)$ in the tagged-particle problem is given by

$$G_D^s(11'; 22'; t) = C_s(12; t)C(1'2'; t) \tag{158}$$

which we can compare with (53). After we take into account these changes, we find

$$\delta\phi_s(12)\omega_0(2) = -\int d3\, d3'\, d4\, d4'\, T_s(1; 33') \times[\bar{G}_D^s(33'; 44') - \bar{G}_D^s(33'; 44']T_s(2; 44'; -z) \tag{159}$$

where $T_s(1; 33')$ is given by (112) with $[\delta(13) + \delta(13')]$ replaced by $\delta(13)$, and $\bar{G}_D^s(33'; 44')$ is the Laplace transform over t of

$$\bar{G}_D^s(33'; 44'; t) = \int d5\, d5'\, d6\, d6'\, \tilde{G}_s^{D^{-1}}(11'; 55')G_s^D(55; 66'; t)\tilde{G}_s^{D^{-1}}(66'; 44') \tag{160}$$

Since

$$\tilde{G}_s^{D^{-1}}(11'; 22') = \tilde{C}_s^{-1}(12)\tilde{C}^{-1}(1'2') \tag{161}$$

and

$$\tilde{C}_s^{-1}(12) = \delta(12)/\omega_0(1) \tag{162}$$

we directly obtain

$$\bar{G}_D^s(33'; 44'; t) = \omega_0^{-1}(3)C_s(34; t)\omega_0^{-1}(4)\int d5\, d5'\, \tilde{C}^{-1}(3'5)C(55'; t)\tilde{C}^{-1}(5'4') \tag{163}$$

In the present problem we make the approximation

$$\tilde{C}^{-1}(12) = \delta(12)\omega_0^{-1}(1) \tag{164}$$

and for consistency use the initial condition $C(55'; 0) = \delta(55')\omega_0(5)$. We then obtain

$$\bar{G}_D^s(33'; 44'; t) = \omega_0^{-1}(3)\omega_0^{-1}(3')G_s(34; t)G(3'4'; t)\omega_0^{-1}(4)\omega_0^{-1}(4') \tag{165}$$

where G_s and G are defined in the same way as C_s and C except for the initial conditions

$$G_s(12; 0) = G(12; 0) = \delta(12)\omega_0(1) \tag{166}$$

Note that the ω_0^{-1} factors in $\bar{G}_D^s$ will just cancel those that are included in the definition of T_s. If we carry out an analysis very similar to that given in

Appendix C we can show that, for hard spheres,

$$T_s(1;33') = \tilde{g}(\sigma)\omega_0(3)\omega_0(3')t_s(1;33') \tag{167}$$

$$t_s(1;33') = -\frac{2i}{m}\bar{\mathbf{p}}\cdot\hat{\mathbf{r}}\,\delta(|r|-\sigma)\theta(-\hat{\mathbf{r}}\cdot\bar{\mathbf{p}}) \times[\delta(\mathbf{p}_1-\boldsymbol{\alpha}-\bar{\mathbf{p}})-\delta(\mathbf{p}_1-\boldsymbol{\alpha}-\mathbf{p}^*)]\,\delta(\mathbf{r}_1-\mathbf{R}-\hat{r}\sigma) \tag{168}$$

$$\left.\begin{aligned}\bar{\mathbf{p}}=\tfrac{1}{2}(\mathbf{p}_3-\mathbf{p}_{3'}),\qquad \boldsymbol{\alpha}=\tfrac{1}{2}(\mathbf{p}_3+\mathbf{p}_{3'}),\qquad \mathbf{p}^*=\bar{\mathbf{p}}-2\hat{\mathbf{r}}(\hat{\mathbf{r}}\cdot\bar{\mathbf{p}})\\ \mathbf{r}=\mathbf{r}_3-\mathbf{r}_{3'},\qquad \mathbf{R}=\tfrac{1}{2}(\mathbf{r}_3+\mathbf{r}_{3'})\end{aligned}\right\} \tag{169}$$

$$T_s(1;44';-z) = -T_s(1;44') \tag{170}$$

Putting these results together and Fourier transforming over all spatial indices, we obtain

$$\begin{aligned}\delta\phi_s(k\mathbf{p}_1\mathbf{p}_2;t)\omega_0(2) = \tilde{g}^2(\sigma)\int d^3p_3\,d^3p_{3'}\,d^3p_4\,d^3p_{4'}\,d^3\bar{k}/(2\pi)^3\\ \times t(\bar{k}p_1p_3p_{3'})[G_s(\bar{k}+k/2,p_3p_4;t)G(-\bar{k}+k/2,p_{3'}p_{4'};t)\\ -G_s^0(\bar{k}+k/2,p_3p_4;t)G^0(-\bar{k}+k/2,p_{3'}p_{4'};t)]t(\bar{k}p_2p_4p_{4'})\end{aligned} \tag{171}$$

which is the form for the recollision term for self-diffusion. For the calculation of the velocity autocorrelation we need only deal with a reduced form of (171).

6.1 The Enskog Approximation

The velocity autocorrelation function

$$V(t) = \tfrac{1}{3}\langle\mathbf{v}_1(t)\cdot\mathbf{v}_1\rangle \tag{172}$$

where $\mathbf{v}_1$ is the velocity of the tagged particle, has been studied extensively using molecular dynamics techniques.[29,57,58] One can obtain $V(t)$ from the phase space self-correlation function through the relation

$$V(z) = \lim_{k\to 0}\frac{1}{n}\int d^3p\,d^3p'\,\frac{1}{3m^2}(\mathbf{p}\cdot\mathbf{p}')C_s(k\mathbf{p}\mathbf{p}'z) \tag{173}$$

where

$$V(z) = -i\int_0^\infty dt\,e^{izt}V(t) \tag{174}$$

and C_s is defined by (24) following the same transform definition as given in (20). It is clear that C_s also satisfies a kinetic equation like (35) with corresponding phase space memory function ϕ_s.* From (173) we see that to find $C_s(kz)$ one must solve an integral equation due to the coupling between the momentum indices. Strictly speaking, this is true, but we will proceed with an approximate calculation.[7] If we introduce a complete and orthonormal set of momentum states $|i>$ (as in Section 5), we can formally invert the kinetic equation for $C_s(k\mathbf{pp}'z)$ to obtain

$$C_{ij}^{s}(kz) = n\langle i|[z - \hat{\omega}(k) - \phi_s(kz)]^{-1}|j\rangle \tag{175}$$

where

$$\langle p|\hat{\omega}(k)|p'\rangle = \frac{1}{m}(\mathbf{k}\cdot\mathbf{p})\delta(\mathbf{p}-\mathbf{p}') \tag{176}$$

If the state $|2\rangle$ corresponds to the "momentum eigenstate," then

$$V(z) = v_0^2\langle 2|[z - \phi_s(0z)]^{-1}|2\rangle \tag{177}$$

The self-diffusion coefficient is related to the velocity autocorrelation function by

$$D = \int_0^\infty dt\, V(t) = \lim_{z\to i0+} iV(z) = -iv_0^2\langle 2|\phi_s^{-1}(0, i0^+)|2\rangle \tag{178}$$

To calculate the velocity autocorrelation function we therefore only need ϕ_s for $k = 0$. This simplifies matters considerably compared to a full calculation of the van Hove self-correlation function $S_s(k, \omega)$.

Before discussing the contribution of $\delta\phi_s$ to V, we first note the contribution of ϕ_{sE}. To a good approximation ϕ_{sE} is diagonal on the state $|2\rangle$ and the matrix element is

$$\langle 2|\phi_{sE}|2\rangle = -in\tfrac{8}{3}(\pi/m\beta)^{1/2}\sigma^2 g(\sigma) \tag{179}$$

Therefore,

$$V_E(z) \simeq v_0^2[z - \langle 2|\phi_{sE}|2\rangle]^{-1} \tag{180}$$

Since the Enskog expression for the self-diffusion coefficient is[42,43]

$$D_E = 3v_0[8\sqrt{\pi}nv_0^2 g(\sigma)]^{-1} \tag{181}$$

one has

$$V_E(z) = v_0^2[z + iv_0^2/D_E]^{-1} \tag{182}$$

Inverting the Laplace transform gives the simple exponential decay

$$V_E(t) = v_0^2 \exp(-tv_0^2/D_E) \tag{183}$$

*For a discussion of various model treatments of this problem, see Desai[62] and references therein.

6.2. Effects of Recollisions

We now want to investigate the effects of $\delta\phi_s$. A complete analysis of $V(t)$ has not yet been carried out; however, there exist approximate calculations that do give a qualitative understanding of the processes involved.[7,51,63,64] In such an analysis several approximations have to be introduced in order to make the calculations tractable. We will enumerate them as we go along and indicate later their justifications and where one can hope to do better.

The first approximation we will make is that $\delta\phi_s$ is diagonal on the momentum state $|2\rangle$,

$$\delta\phi_s \simeq |2\rangle\langle 2|\delta\phi_s|2\rangle\langle 2| \tag{184}$$

Next we must face up to the momentum integrations in (171). Clearly any attempt to do these integrals directly will be extremely difficult. We can formally carry out these integrations by introducing the complete set of momentum states as in Section 5. We can then carry out the integrations explicitly to obtain

$$\begin{aligned}\langle 2|\delta\phi_s(0,t)|\rangle &= \frac{1}{n} g^2(\sigma) \sum_{\alpha\beta\gamma\nu} \int \frac{d^3\bar{k}}{(2\pi)^3} T^{\mu}_{\alpha\beta}(\bar{k}) \\ &\quad \times [G^{s}_{\beta\nu}(\bar{k}t)G_{\alpha\gamma}(-\bar{k}t) - G^{s0}_{\beta\nu}(\bar{k}t)G^{0}_{\alpha\gamma}(-\bar{k}t)]T^{\mu}_{\nu\gamma}(\bar{k})\end{aligned} \tag{185}$$

where

$$\begin{aligned}T^{\mu}_{\alpha\beta}(k) &= \int d^3\xi_1\, d^3\xi_2 \psi_\alpha(\boldsymbol{\xi}_1)\psi_\beta(\boldsymbol{\xi}_2) \int d^3p_1 p_1^{\mu} t(kp_1\boldsymbol{\xi}_1\boldsymbol{\xi}_2)\Phi(\boldsymbol{\xi}_1)\Phi(\boldsymbol{\xi}_2) \\ &= \int d^3\xi_1\, d^3\xi_2 \Phi(\boldsymbol{\xi}_1)\Phi(\boldsymbol{\xi}_2)(-4iv_0) \\ &\quad \times \int d^3r\, e^{i\mathbf{k}\cdot\mathbf{r}}(\bar{\mathbf{p}}\cdot\hat{\mathbf{r}})^2\, \delta(|\mathbf{r}|-\sigma)\theta(-\hat{\mathbf{r}}\cdot\bar{\mathbf{p}})\hat{\mathbf{r}}_{\mu}\end{aligned} \tag{186}$$

and

$$G_{\alpha\beta}(kt) = \int d^3\xi_1\, d^3\xi_2 \psi_\alpha(\boldsymbol{\xi}_1)\psi_\beta(\boldsymbol{\xi}_2) G(k\boldsymbol{\xi}_1\boldsymbol{\xi}_2 t) \tag{187}$$

In (186) we have introduced the relative momentum $\bar{\mathbf{p}} = (\boldsymbol{\xi}_1 - \boldsymbol{\xi}_2)/2$.

Equation (185) is useful for extracting the long-time behavior of $\langle 2|\delta\phi_s|2\rangle$. One expects that only those contributions where all of the indices α, β, γ, and ν refer to hydrodynamical indices (1 for G_s and 1 to 5 for G) will dominate at long times. This is because the hydrodynamical modes decay like $e^{-k^2D_i t}$, where D_i is the associated transport coefficient, whereas the nonhydrodynamical states decay at least as fast as e^{-t/τ_i}, where τ_i is the relaxation time for the ith mode. Consequently for k small the hydrodynamical states persist for very long times compared to the nonhydrodynamical states. The observation that the hydrodynamical components give the largest contribution is essentially equivalent to the ideas used in mode–mode coupling theories.[34,35] A qualitative difference between the theory developed here and the mode–mode coupling theories is that the results here have been derived via a number of physically motivated microscopic approximations and one is free to go back and check the error made in each approximation.[8] Also, since the approximations have not been limited to any particular time or space region, one does not have to introduce wavenumber cutoffs and one can investigate the whole time region, not just long times.

The well-known nonanalytic structure for the density expansion of transport coefficients can be understood from our mode–mode coupling expression (185). Clearly one can not expand G_{11} and G_s in a power series in the density. This expansion would lead to singularities for small $\bar{k}$ and large t. To find the transport coefficients one has to integrate over $\bar{k}$ and t; then a density expansion of G and G_s in $\delta\phi_s$ will clearly lead to "divergences" term by term.

A detailed analysis shows that the coupling of the $G_{11}^{(s)}$ component to the transverse part of G_{22} dominates the long-time behavior.[7] As our second major approximation we assume that $\langle 2|\delta\phi_s|2\rangle$ can be approximated by keeping only the terms that dominate for long times. We write

$$\langle 2|\delta\phi_s(0,t)|2\rangle = \frac{2}{n}\bar{g}^2(\sigma)\int \frac{d^3\bar{k}}{(2\pi)^3}[T_{13}^{\mu}(\bar{k})]^2[G_{11}^{s}(\bar{k}t)G_{33}(-\bar{k}t) - \text{FP}] \tag{188}$$

where $|3\rangle$ is one of the two states (giving an overall factor of 2) transverse to $\bar{\mathbf{k}}$. The abbreviation FP means free particle expressions for the G functions. In this case (186) gives

$$T_{13}^{\mu}(k) = -i8\sqrt{\pi}v_0\sigma^2\frac{1}{x}\frac{d}{dx}j_0(x)\tilde{k}_{\mu} \tag{189}$$

and $x = |k|\sigma$. Inserting this into (185) and performing the angular integration, we obtain

$$\langle 2|\delta\phi_s(0,t)|2\rangle = -\frac{64}{3\pi n}\bar{g}^2(\sigma)v_0^2\sigma\int_0^{\infty} dx\, j_1^2(x)[G_s(\bar{k}t)G_t(-\bar{k}t) - \text{FP}] \tag{190}$$

Further calculation now depends on our knowledge of the correlation functions $G_s(\bar{k},t)$ and $G_t(\bar{k},t)$. For the purpose of illustration we will consider the simple approximation where G_s and G_t have a Gaussian wavenumber dependence,[65]

$$G_s(kt) = n\exp[-k^2W_D(t)/2] \tag{191}$$

$$G_t(kt) = n\exp[-k^2W_\nu(t)/2] \tag{192}$$

where the width functions W_ν and W_D have yet to be specified. One chooses the width functions to interpolate between the short-time free particle motion and the long-time hydrodynamical behavior. We will take[65]

$$W_D(t) = 2D_E\left[t + \frac{D_E}{v_0^2}(e^{-v_0^2 t/D_E} - 1)\right] \tag{193}$$

where D_E is the Enskog self-diffusion coefficient, and the same form for W_ν with D_E replaced by the Enskog expression for the kinematic shear viscosity $\nu_E = \eta_s/mn$, η_s being given by (150). After putting (190) and (191) into (189), we find

$$\langle 2|\delta\phi_s(0, t)|2\rangle = -\frac{3n}{\pi^2}\left(\frac{v_0}{D_E}\right)^2 \frac{1}{\sigma^3}\Phi_v(t) \tag{194}$$

where

$$\Phi_v(t) = \int_0^\infty dx\, j_1^2(x)[e^{-x^2\alpha(t)} - e^{-x^2\alpha_0(t)}] \tag{195}$$

$$\alpha(t) = \frac{1}{2\sigma^2}[W_D(t) + W_\nu(t)] \tag{196}$$

$$\alpha_0(t) = v_0^2 t^2/\sigma^2 \tag{197}$$

For short times, from (6.40) one finds

$$\langle 2|\delta\phi_s(0, t)|2\rangle \propto t^2 \exp(-2\sigma^2/v_0^2 t^2)$$

so $\delta\phi$ vanishes rapidly for small t.

For large times the major contribution to the integral $\Phi_v(t)$ comes from small x values. This is most easily seen by making the variable change $y = \alpha(t)^{1/2}x$ in the first integral and $y = \alpha_0(t)^{1/2}x$ in the second. We then obtain directly the asymptotic result

$$\Phi_v(t) \sim \frac{\sqrt{\pi}}{36}\{\alpha(t)^{-3/2} - \alpha_0(t)^{-3/2}\} \tag{198}$$

Since for long times

$$\alpha(t) \sim (D_E + \nu_E)t/\sigma^2 \tag{199}$$

it follows that

$$\Phi_v(t) \sim \frac{\sqrt{\pi}}{36}\sigma^3\{(D_E + \nu_E)t\}^{-3/2} + O(t^{-3}) \tag{200}$$

and

$$\begin{aligned}\langle 2|\delta\phi_s(0, t)|2\rangle &\sim -\frac{3n}{\pi^2}\left(\frac{v_0^2}{D_E}\right)^2 \frac{\sqrt{\pi}}{36}[(D_E + \nu_E)t]^{-3/2} + O(t^{-3}) \\ &= -\frac{2}{3n}\left(\frac{v_0^2}{D_E}\right)^2\left[\frac{1}{4\pi(D_E + \nu_E)t}\right]^{3/2} + O(t^{-3})\end{aligned} \tag{201}$$

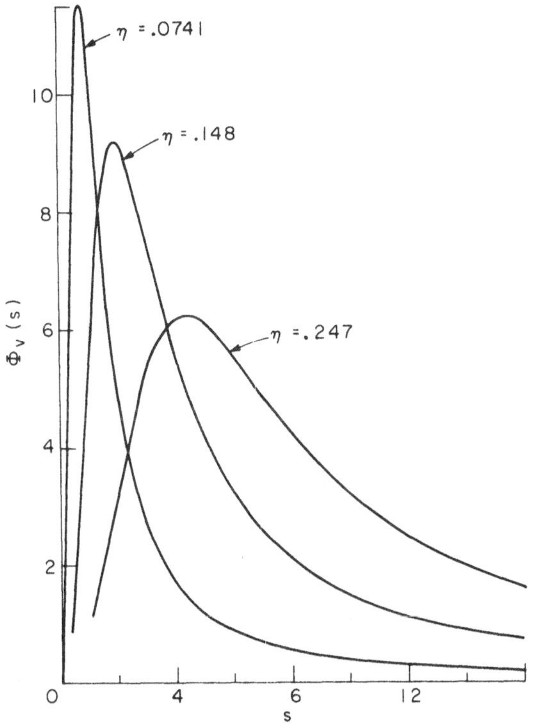

Fig. 3. Memory function $\Phi_V(s)$ of a hard-sphere fluid at several packing fractions η. Dimensionless time s is defined as $s = [4\sqrt{\pi} n\sigma^2 v_0 \tilde{g}(\sigma)]t$.

It is known that if the long-time behavior of ϕ is a power law decay, the asymptotic behavior of $V(t)$ is given by[7]

$$V(t) \sim -\left(\frac{D}{D_e}\right)^2 \left(\frac{D_E}{v_0^2}\right)^2 \Phi_v(t) \tag{202}$$

or

$$V(t) \sim \left(\frac{D}{D_E}\right)^2 \frac{2}{3n} \left[\frac{1}{4\pi(D_E + \nu_E)t}\right]^{3/2} \tag{203}$$

where D is the full diffusion coefficient including the contribution from ϕ^E and $\delta\phi_s$.* Using (178) we obtain the self-diffusion coefficient as

$$\begin{aligned} D/D_E &= [1 + \delta\phi_{22}(0, z=0)/\phi_{22}^E(0, z=0)]^{-1} \\ &= \left[1 - \frac{1}{6\pi\eta} \int_0^\infty ds\ \Phi_v(s)\right]^{-1} \end{aligned} \tag{204}$$

where $\eta = \frac{1}{6}\pi n\sigma^3$.

We have numerically evaluated $\Phi_v(t)$, $V(t)$, and D for a hard-sphere fluid at several densities. The results for $\varphi_v(t)$ are given in Fig. 3. In Fig. 4 we plot the

*Equation (203) differs from the results given by Dorfman and Cohen[33] by the factor $(D/D_E)^2$. See the discussion in Mazenko[7] and Desai.[62]

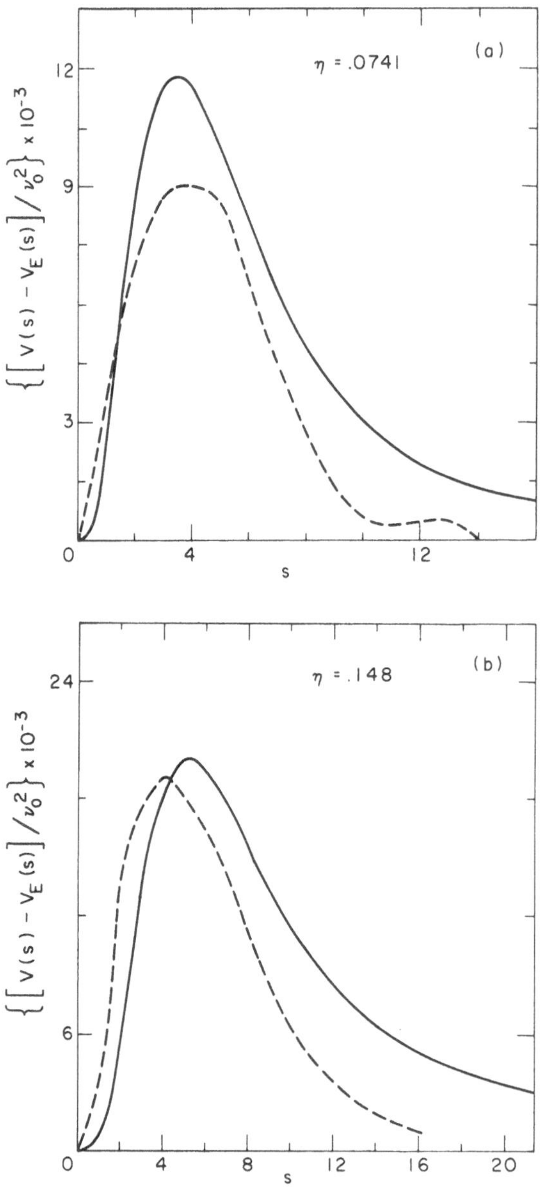

Fig. 4. Deviation of velocity autocorrelation function $V(s)$ from Enskog theory prediction at packing fractions (a) $\eta = 0.0741$ and (b) 0.148. Dashed curves are computer molecular dynamics results.[29]

deviation of $V(t)$ from the Enskog expression, and in Table 2 we give the ratio D/D_E. Comparing these results with the computer molecular dynamics calculations[29] one sees that our calculations are reasonable at low densities, while at high densities they break down rather badly. It is clear that the mode coupling terms do lead to correlations that persist for very long times. On the other hand, in order to produce the negative structure in $V(t)$ at high densities, further refinements of the calculation will be necessary.

Table 2. Calculated and Computer Values of Diffusion Coefficient Ratio D/D_E

η	D/D_E (calculated)	D/D_E (computer)[29]
0.074	1.07	1.07
0.148	1.204	1.16
0.247	1.498	1.34

Very recent calculations have taken into account other coupling terms that we have not considered in this discussion and have used more refined expressions for the G functions.[51,63,64] Although the results give further insight into the problem, further work is still needed. We believe that the memory function (170) is capable of describing $V(t)$ even in very dense systems. There is evidence that the mechanism responsible for the negative structure is taken into account in this expression, but this remains to be demonstrated explicitly. Significant improvements can be expected if one considers the other couplings that we have ignored here and more realistic expressions used in place of (191) and (192). Specifically, the coupling between G_s and the longitudinal current correlation function appears to be important for intermediate times, and the nonhydrodynamical modes also need to be taken into account in at least some approximate way.[51] The severe limitations of the Gaussian approximations (191) and (192) are now recognized.[63,64] The main defect here is the neglect of the very strong damping that can occur at large wavenumbers or short times, a process that is due to those matrix elements of $\delta\phi$ which also give rise to the phenomenon of collision transfer.

7. Summary and Discussions

We have studied a kinetic theory approach to the calculation of dynamical correlations in a dense fluid. By considering density correlations in phase space one can discuss the dynamics in terms of renormalized molecular interactions that take into account the existence of structural correlations in the fluid. The advantage of this approach is that through the properties of a single function, the phase space memory function, one can treat in a unified manner a variety of correlation functions without restriction to either wavelength–frequency domain or fluid density.

The central result of our analysis is a memory function expression in the form of (121). The memory function consists of two parts, a Boltzmann–Enskog term, which treats uncorrelated binary collisions, and a recollision term, which describes correlated binary collisions. We have shown that the Boltzmann–Enskog term alone already provides a first-order calculation of dense fluid correlations. The inclusion of recollision effects makes the problem nonlinear; this complexity is unavoidable since mode-coupling effects are known to be essential in describing the more subtle features of molecular dynamics in dense fluids. The present formalism allows even more complicated interactions to be included, but until the binary recollisions are fully analyzed the higher-order terms are not likely to be of much practical interest.

The applications discussed in Sections 5 and 6 serve to demonstrate that fully renormalized kinetic theory provides a practical method of calculating time correlation functions. In order to obtain numerical results, we have introduced rather crude approximations. Their effects should be investigated in future work. With regard to the Boltzmann–Enskog approximation one should try to deal directly with (221) rather than (223). We expect this refinement can be quite significant at high densities. Our calculation of the velocity autocorrelation function involved a number of simplifying assumptions. As discussed in Section 6, additional analysis would be very helpful in understanding the effects of mode coupling on transport processes.

The basic ingredients of the fully renormalized kinetic theory are such that they can be applied to a variety of dynamical problems. An analysis of the velocity autocorrelation function in a charged fluid has been carried out and the results shown to agree well with computer molecular dynamics experiments.[66] The Boltzmann–Enskog approximation has been used to study binary mixtures of hard spheres.[67] Other systems, such as quantum fluids and molecular fluids, appear to be appropriate problems for further application. It is important to recognize that the present formalism may also be used to investigate dynamical critical phenomena. To the extent that fluid behavior in the critical region is controlled largely by the static correlation functions, our kinetic theory description should be valid at least in a qualitative sense. A calculation of the thermal conductivity near the critical point lends support to this expectation.[8] It appears that the theory is capable of linking the normal and the critical regions; certainly more calculations along these lines would be of great interest.

Appendix A

Here we analyze the static part of the memory function. We start with (43):

$$\int d\bar{1}\, \phi^{(s)}(1\bar{1})\tilde{C}(\bar{1}2) = \int d\bar{1}\, L_{\mathrm{I}}(1\bar{1})\tilde{C}(1\bar{1};2) \tag{205}$$

If we note from (29) the identity

$$\int d\bar{1}\, L_{\mathrm{I}}(1\bar{1})f(1)f(\bar{1}) = -L_{\mathrm{I}}f(1) \tag{206}$$

then

$$\int d\bar{1}\, \phi^{(s)}(1\bar{1})\tilde{C}(\bar{1}2) = -\langle L_{\mathrm{I}}f(1)\,\delta f(2)\rangle \tag{207}$$

A correlation function like $\langle L_I f(1)\,\delta f(2)\rangle$ can be manipulated into the form

$$\begin{aligned}\langle L_I f(1)\,\delta f(2)\rangle &= -i\beta^{-1}\{-\nabla_{\mathbf{r}_1}\cdot\nabla_{\mathbf{p}_1}\tilde{C}(12)+\nabla_{\mathbf{p}_1}\cdot\nabla_{\mathbf{r}_1}\omega_0(1)\,\delta(12)\}\\ &= -i\beta^{-1}\nabla_{\mathbf{r}_1}\cdot\nabla_{\mathbf{p}_1}[-\tilde{C}(12)+\omega_0(1)\,\delta(12)]\end{aligned} \tag{208}$$

By making use of (7) we find the result

$$\int d\bar{1}\,\phi^{(s)}(1\bar{1})\tilde{C}(\bar{1}2) = -i\beta^{-1}\nabla_{\mathbf{p}_1}\cdot\nabla_{\mathbf{r}_1}\omega_0(1)\omega_0(2)h(r_1-r_2) \tag{209}$$

Equation (209) is still an integral equation for $\phi^{(s)}$. The first step in solving this equation is to Fourier transform over $\mathbf{r}_1$ to obtain

$$\int d^3\bar{p}\phi^{(s)}(k, p_1\bar{p})\tilde{C}(k, \bar{p}p_2) = -\frac{\mathbf{k}\cdot\mathbf{p}_1}{m}\omega_0(p_1)\omega_0(p_2)h(k) \tag{210}$$

Upon inserting the Fourier transform of (7) one finds

$$\begin{aligned}\phi^{(s)}(k, p_1p_2)\omega_0(p_2)+\int d^3\bar{p}\,\phi^{(s)}(k, p\bar{p})\omega_0(\bar{p})\omega_0(p_2)h(k)\\ = -\frac{\mathbf{k}\cdot\mathbf{p}_1}{m}\omega_0(p_1)\omega_0(p_2)h(k)\end{aligned} \tag{211}$$

Integrating over p_2 gives

$$\int d^3\bar{p}\,\phi^{(s)}(k, p_1\bar{p})\omega_o(\bar{p})[1+nh(k)] = -\frac{\mathbf{k}\cdot\mathbf{p}_1}{m}n\omega_0(p_1)h(k) \tag{212}$$

which we can then put back into (211) to obtain

$$\begin{aligned}\phi^{(s)}(k,\ p_1p_2)\omega_0(p_2) &= -\frac{\mathbf{k}\cdot\mathbf{p}_1}{m}\omega_0(p_1)\omega_0(p_2)h(k)\\ &\quad+\omega_0(p_2)h(k)\frac{\mathbf{k}\cdot\mathbf{p}_1}{m}\frac{n\omega_0(p_1)h(k)}{1+nh(k)}\\ &= -\frac{\mathbf{k}\cdot\mathbf{p}_1}{m}\omega_0(p_1)\omega_0(p_2)h(k)\left[1-\frac{nh(k)}{1+nh(k)}\right]\\ &= -\frac{\mathbf{k}\cdot\mathbf{p}_1}{m}\omega_0(p_1)\omega_0(p_2)\frac{h(k)}{1+nh(k)}\end{aligned} \tag{213}$$

The direct correlation function C_D is defined by

$$C_D(k) = \frac{h(k)}{1+nh(k)} \tag{214}$$

If we take the inverse Fourier transform of (213) and (214) we would find (48).

Appendix B

Here we solve the integral equation (46) to obtain $\phi^{(c)}$. Using (7) we write (46) as

$$\int d\bar{1}\,\phi^{(c)}(1\bar{1})\tilde{C}(\bar{1}2)=\phi^{(c)}(12)\omega_0(2)+\int d\bar{1}\,\phi^{(c)}(1\bar{1})\omega_0(\bar{1})\omega_0(2)h(r_{\bar{1}}-r_2)$$
$$=-\int d\bar{1}\,d\bar{2}\,L_{\mathrm{I}}(1\bar{1})L_{\mathrm{I}}(2\bar{2})G(1\bar{1};2\bar{2}) \tag{215}$$

This is not an explicit expression for $\phi^{(c)}$ because of the term proportional to

$$\int d^3\bar{p}\,\phi^{(c)}(r_1p_1;\,r_{\bar{1}}\bar{p})\omega_0(\bar{p})$$

We can evaluate this quantity if we integrate (215) over p_2. Simplification occurs because

$$\int d^3p_2\,d\bar{2}\,L_{\mathrm{I}}(2\bar{2})G(1\bar{1};2\bar{2})=0 \tag{216}$$

and

$$\int d^3\bar{p}\,\phi^{(c)}(r_1p_1;\,r_2\bar{p})\omega_0(\bar{p})+\int d^3\bar{r}\int d^3\bar{p}\,\phi^{(c)}(r_1p_1;\,\bar{r}\bar{p})\omega_0(\bar{p})nh(\bar{r}-r_2)=0 \tag{217}$$

Taking the Fourier transform over r_1 we find

$$\int d^3\bar{p}\,\phi^{(c)}(k,\,p_1\bar{p}z)\omega_0(\bar{p})[1+nh(k)]=0 \tag{218}$$

Since $1+h(k)\neq 0$, we conclude that

$$\int d^3\bar{p}\,\phi^{(c)}(1\bar{1})\omega_0(\bar{1})=0 \tag{219}$$

and therefore (215) reduces to (50). Notice that (219) is just a statement of conservation of particle number when viewed in the context of (35).

Appendix C. Reduction of the Memory Function in the Boltzmann–Enskog Approximation

We will show that in the case of hard-sphere interactions the second term in (122) can be expressed in the more familiar form of a collision integral. This reduction is rather instructive because it demonstrates a number of useful manipulations in working with the various expressions and operators in Sections 2–4.

The generalized Boltzmann–Enskog memory function is given by

$$\phi_{BE}(12)\omega_0(2) = -\int d3\, d3'\, d4\, d4'\, \mathscr{V}(1;33')\bar{G}_s(33';44')\mathscr{V}(2;44') \qquad (220)$$

where $\bar{G}_s$ is defined by (98). Since this quantity has not been fully investigated at present, we will analyze it only in the "moderate" density approximation in the sense of (108) and (109). In this case (220) reduces to

$$\phi_E^{(c)}(12)\omega_0(2) = -\int d3\, d3'\, \{\tilde{L}_I(33')[\delta(13)+\delta(13')] \times \tfrac{1}{2}[z-\tilde{L}(33')]^{-1}\omega_0(33')\tilde{L}_I(33')[\delta(23)+\delta(23')]\} \qquad (221)$$

where the subscript E indicates that (221) is related to the memory function corresponding to the Boltzmann–Enskog equation.

The observant reader will recognize this quantity as being identical with the low-density memory function[5] except the bare potential is replaced by $-\beta^{-1}\ln g(r)$. Consequently all of the general properties of the low-density memory function will hold also for $\phi_E^{(c)}$. Equation (222) is a reasonable generalization of the Enskog equation to smooth potentials. We note that the main approximation leading from (220) to (221) is the neglect of three-body static effects.

We can make contact with the usual Enskog collision kernel if we assume that $\tilde{L}_I$ is very sharply peaked near the point of collision. This leads to two approximations:

(i) $\omega_0(33')\tilde{L}_I(33') \simeq \omega_0(p_3)\omega_0(p_{3'})\tilde{g}(\sigma)\exp[-\beta V(r_3-r_{3'})]\tilde{L}_I(33')$

(ii) $\tilde{L}_I(33') \simeq L_I(33')$

Both approximations stem from writing the equilibrium pair distribution function as

$$g(r_3-r_{3'}) = \tilde{g}(r_3-r_{3'})\exp[-\beta V(r_3-r_{3'})] \qquad (222)$$

which is a definition of the function $\tilde{g}$. For short-ranged potentials $\tilde{g}$ behaves more smoothly than $\exp(-\beta V)$ so that $\nabla\tilde{g}$ is longer ranged and more slowly varying than the sharply peaked function $\nabla\exp(-\beta V)$. This means that for short-ranged interactions it is reasonable to neglect $\nabla\tilde{g}$ in comparison with $\nabla\exp(-\beta V)$, which is approximation (ii), and also replace the function $\tilde{g}$ by a constant, its value at interparticle separation σ, which is approximation (i). With these two approximations (221) becomes

$$\phi_E^{(c)}(12)\omega_0(2) = -\tfrac{1}{2}\tilde{g}(\sigma)\int d3\, d3'\, \{L_I(33')[\delta(13)+\delta(13')] \times [z-L(33')]^{-1}\omega_0(p_3)\omega_0(p_{3'})\exp[-\beta V(r_3-r_{3'})]L_I(33') \times [\delta(23)+\delta(23')]\} \qquad (223)$$

which is just $\tilde{g}(\sigma)$ times the low-density memory function. We want to emphasize that (i) and (ii) are introduced to make explicit contact with the conventional Enskog equation. One can, in fact, proceed without invoking such approximations, although the results of such calculations have not yet appeared in the literature.

The manipulation of (223) is best carried out in terms of the center-of-mass and relative coordinates of the colliding pair of particles. Writing out (223) explicitly and taking the Fourier transform using the convention of (16a), we find the result can be put in the form

$$\phi_{\mathrm{E}}^{(c)}(k\bar{p}p'z)n\Phi(p') = \Omega^{-1}\int d^3r_1\, d^3r_2\, e^{i\mathbf{k}\cdot(\mathbf{r}_1-\mathbf{r}_2)}\,\phi_{\mathrm{E}}^{(c)}(r_1r_2\bar{p}p'z)n\Phi(p')$$

$$= \tfrac{1}{2}n^2 g(\sigma)\int d^3r\, d^3\alpha\, M(\alpha) I_k(r\alpha) \tag{224}$$

where Ω is the system volume and

$$I_k(r\alpha) = \int d^3p\, [L_{\mathrm{I}}(rp)\Delta_k(rp|p')]M(p)g_0(r)[z+L_k(rp\alpha)]^{-1}L_{\mathrm{I}}(rp)\Delta_k^*(rp|\bar{p}) \tag{225}$$

$$M(\alpha) = \left(\frac{\beta}{\pi m}\right)^{3/2}\exp(-\beta\alpha^2/m) \tag{226}$$

$$\Delta_k(rp|p') = e^{i\mathbf{k}\cdot\mathbf{r}/2}\delta(\mathbf{p}'-\boldsymbol{\alpha}-\mathbf{p}) + e^{-i\mathbf{k}\cdot\mathbf{r}/2}\delta(\mathbf{p}'-\boldsymbol{\alpha}+\mathbf{p}) \tag{227}$$

$$L_k(rp\alpha) = -\frac{\boldsymbol{\alpha}\cdot\mathbf{k}}{m} + L_0(rp) + L_{\mathrm{I}}(rp) \tag{228}$$

$$L_0(rp) = -\frac{2i}{m}\mathbf{p}\cdot\boldsymbol{\nabla}_r \tag{229}$$

$$L_{\mathrm{I}}(rp) = i\boldsymbol{\nabla}V(r)\cdot\boldsymbol{\nabla}_p \tag{230}$$

In (225) $g_0(r)$ denotes $\exp[-\beta V(r)]$.

After an integration by parts, $I_k(r\alpha)$ can be written as

$$I_k(r\alpha) = -\int d^3p\, \Delta_k(rp|p')\{[L_{\mathrm{I}}(rp)M(p)g_0(r)] + M(p)g_0(r)L_{\mathrm{I}}(rp)\}$$

$$\times[z+L_k(rp\alpha)]^{-1}L_{\mathrm{I}}(rp)\Delta_k^*(rp|\bar{p}) \tag{231}$$

The integrands in (231) are rearranged by using two identities,

$$[L_{\mathrm{I}}(rp) + L_0(rp)]M(p)g_0(r) = 0 \tag{232}$$

$$L_{\mathrm{I}}(z+L_k)^{-1}L_{\mathrm{I}} = L_{\mathrm{I}} - \left(z - \frac{\boldsymbol{\alpha}\cdot\mathbf{k}}{m} + L_0\right)(z+L_k)^{-1}L_{\mathrm{I}} \tag{233}$$

One finds that $I_k(r\alpha)$ now consists of two parts, one of which is z independent. Inserting them into (221) one has

$$\phi_{\mathrm{E}}^{(c)}(k\bar{\mathbf{p}}\mathbf{p}'z) = \phi_1^{(c)}(k\bar{\mathbf{p}}\mathbf{p}'z) + \phi_2^{(c)}(k\bar{\mathbf{p}}\mathbf{p}') \tag{234}$$

where

$$\phi_1^{(c)}(k\bar{\mathbf{p}}\mathbf{p}'z)n\Phi(p') = -\tfrac{1}{2}n^2 g(\sigma)\int d^3r\, d^3\alpha\, d^3p\, M(\alpha)M(p)\Delta_k(rp|p') \times\left[z - \frac{\boldsymbol{\alpha}\cdot\mathbf{k}}{m} + L_0\right]g_0(r)(z+L_k)^{-1}L_1\Delta_k^*(rp|\bar{p}) \tag{235}$$

$$\phi_2^{(c)}(k\bar{\mathbf{p}}\mathbf{p}'z)n\Phi(p') = -\tfrac{1}{2}n^2 g(\sigma)\int d^3r\, d^3\alpha\, d^3p\, M(\alpha)M(p)\Delta_k \times(rp|p')g_0(r)L_1\Delta_k^*(rp|\bar{p}) \tag{236}$$

It should be noted that the low-density form of the T-matrix defined by (112) can be written in terms of center-of-mass variables, and upon taking Fourier transforms one finds

$$T(1;22') = -4\int d^3k\, e^{i\mathbf{k}\cdot[(\mathbf{r}_2+\mathbf{r}_{2'})/2-\mathbf{r}_1]}\left[z - \frac{\boldsymbol{\alpha}\cdot\mathbf{k}}{m} + L_0\right] g_0(r)(z+L_k)^{-1}L_1\Delta_k^*(rp|p_1)$$

$$= -4\int d^3k\, T_k(rp\alpha|p_1)e^{i\mathbf{k}\cdot[(\mathbf{r}_2+\mathbf{r}_{2'})/2-\mathbf{r}_1]} \tag{237}$$

Consequently many of the manipulations carried out in the rest of the appendix can be used to evaluate $T(1;22')$ explicitly.

The reduction of $\phi_2^{(c)}$ is fairly straightforward. Making use of

$$g_0(r)L_1(rp) = -\frac{i}{\beta}\boldsymbol{\nabla}_r[e^{-\beta V(r)}-1]\cdot\boldsymbol{\nabla}_p \tag{238}$$

we can carry out the r integration by defining the low-density direct correlation function

$$C_0(k) = \int d^3r\, e^{i\mathbf{k}\cdot\mathbf{r}}[e^{-\beta V(r)}-1] \tag{239}$$

The α and p integrals are then carried out with the aid of the delta functions in Δ_k and Δ_k^*. One obtains

$$\phi_2^{(c)}(k\bar{p}p')n\Phi(p') = n^2 g(\sigma)\frac{\mathbf{k}\cdot\bar{\mathbf{p}}}{m}C_0(k)\Phi(\bar{p})\Phi(p') \tag{240}$$

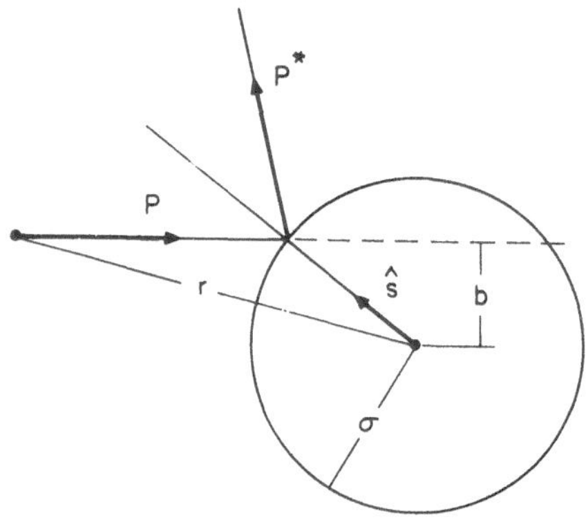

Fig. 5. Collision of two hard spheres.

In reducing $\phi_1^{(c)}$ we first replace $\Delta_k^*(rp|\bar{p})$ by $\exp(-i\mathbf{k}\cdot\mathbf{r}/2)\,\delta(\mathbf{p}'-\boldsymbol{\alpha}-\mathbf{p})$, since the integrand in (235) is invariant under the transformation $(\mathbf{r},\mathbf{p})\rightarrow(-\mathbf{r},-\mathbf{p})$. Then using (17) and (21) for the two-body case and noting that

$$e^{-i\mathbf{k}\cdot\mathbf{r}(t)/2}L_{\mathrm{I}}(rp)\,\delta(\mathbf{p}'-\boldsymbol{\alpha}-\mathbf{p}(t))=e^{-i\mathbf{k}\cdot\mathbf{r}(t)/2}\left(-i\frac{d}{dt}\right)\delta(\mathbf{p}'-\boldsymbol{\alpha}-\mathbf{p}(t)) \tag{241}$$

we can write (235) as

$$\begin{aligned}\phi_1^{(c)}(k\bar{p}p'z)n\Phi(p')=n^2g(\sigma)\int d^3r\,d^3\alpha\,d^3p\,M(\alpha)M(p)\Delta_k(rp|p')\\ \times\left[z-\frac{\boldsymbol{\alpha}\cdot\mathbf{k}}{m}+L_0(rp)\right]g_0(r)\int_0^\infty dt\exp\left[i\left(z-\frac{\boldsymbol{\alpha}\cdot\mathbf{k}}{m}\right)t\right]\\ \times\exp[-i\mathbf{k}\cdot\mathbf{r}(t)/2]\frac{d}{dt}\delta(\mathbf{p}'-\boldsymbol{\alpha}-\mathbf{p}(t))\end{aligned} \tag{242}$$

This is still a general expression valid for any central pair interaction potential. Further reduction will require the specification of $\mathbf{r}(t)$ and $\mathbf{p}(t)$, and the particle trajectories can only be determined if the intermolecular forces are known.

We now proceed to reduce (292) for the case of hard-core interaction. The dynamical description of a collision between two hard spheres of diameter σ is depicted in Fig. 5. The initial relative separation and momentum are $\mathbf{r}$ and $\mathbf{p}$. It is evident that there will be no collision unless $\hat{\mathbf{r}}\cdot\hat{\mathbf{p}}<0$ *and* $b\leq\sigma$, where b is the impact parameter. Given these conditions are satisfied, the time to collision is

$$\tau=\frac{m}{2p}[-(\mathbf{r}\cdot\hat{\mathbf{p}})-(\sigma^2-b^2)^{1/2}] \tag{243}$$

and postcollision relative momentum will be

$$\mathbf{p}^*=\mathbf{p}-\hat{\mathbf{s}}(\hat{\mathbf{s}}\cdot\mathbf{p}) \tag{244}$$

At the point of impact, the separation vector $\mathbf{r}(\tau)$ is

$$\sigma\hat{\mathbf{s}}=\mathbf{r}-(\mathbf{r}\cdot\hat{\mathbf{p}})\hat{\mathbf{p}}-\hat{\mathbf{p}}(\sigma^2-b^2)^{1/2} \tag{245}$$

where $b^2 = r^2 - (\mathbf{r} \cdot \hat{\mathbf{p}})^2$. With these quantities calculated we can express the time dependence of $\mathbf{p}(t)$ explicitly,

$$\partial(\mathbf{p}' - \boldsymbol{\alpha} - \mathbf{p}(t)) = \theta(\sigma^2 - b^2)\theta(-\hat{\mathbf{r}} \cdot \hat{\mathbf{p}}) \times [(\theta(\tau - t)\,\delta(\mathbf{p}' - \boldsymbol{\alpha} - \mathbf{p}) + \theta(t - \tau)\,\delta(\mathbf{p}' - \boldsymbol{\alpha} - \mathbf{p}^*)] + [\theta(b^2 - \sigma^2)\theta(-\hat{\mathbf{r}} \cdot \hat{\mathbf{p}}) + \theta(\hat{\mathbf{r}} \cdot \hat{\mathbf{p}})]\,\delta(\mathbf{p}' - \boldsymbol{\alpha} - \mathbf{p}) \quad (246)$$

where

$$\theta(t) = \begin{cases} 1, & t > 0 \\ 0, & t < 0 \end{cases} \quad (247)$$

Since

$$\frac{d}{dt}\theta(\pm(t - \tau)) = \pm\delta(t - \tau) \quad (248)$$

(242) becomes

$$\phi_1^{(c)}(k\bar{p}p'z)n\Phi(p') = n^2 g(\sigma) \int d^3r\, d^3\alpha\, d^3p\, M(\alpha)M(p)\Delta_k(rp|\bar{p}) \times \left[z - \frac{\boldsymbol{\alpha} \cdot \mathbf{k}}{m} + L_0\right] g_0(r) \exp\left[i\left(z - \frac{\boldsymbol{\alpha} \cdot \mathbf{k}}{m}\right)\tau\right] \exp[-i\mathbf{k} \cdot \hat{\mathbf{s}}\sigma/2] \times \theta(\sigma^2 - b^2)\theta(-\hat{\mathbf{r}} \cdot \hat{\mathbf{p}})[\delta(\mathbf{p}' - \boldsymbol{\alpha} - \mathbf{p}) - \delta(\mathbf{p}' - \boldsymbol{\alpha} - \mathbf{p}^*)] \quad (249)$$

It remains to eliminate the operator L_0. Explicit differentiation shows that L_0 acting on $\theta(\sigma^2 - b^2)$, $\hat{\mathbf{s}}$, or $\mathbf{p}^*$ gives a null result. Also $L_0\theta(-\hat{\mathbf{r}} \cdot \hat{\mathbf{p}})$ does not contribute because it results in a delta function $\delta(\hat{\mathbf{r}} \cdot \hat{\mathbf{p}})$ so $r = b$, $\mathbf{r} = \hat{\mathbf{s}}r$, $\hat{\mathbf{s}} \cdot \mathbf{p} = 0$, and $\mathbf{p}^* = p$. Thus we need only examine the effect of L_0 on $g_0(r) \exp i(z - \boldsymbol{\alpha} \cdot \mathbf{k}/m)\tau$. One finds that with $L_0\tau = i$ and $L_0 g_0(r) = -(2i/m)\hat{\mathbf{r}} \cdot \mathbf{p}\,\delta(r - \sigma^+)$, (249) reduces to

$$\phi_1^{(c)}(k\bar{p}p')n\Phi(p') = in^2 g(\sigma)\left(\frac{\beta}{\pi m}\right)^3 \int d^3r\, d^3\alpha\, d^3p\, e^{-\beta(\alpha^2 + p^2)/m} \times [\delta(\bar{\mathbf{p}} - \boldsymbol{\alpha} - \mathbf{p}) + e^{-i\mathbf{k} \cdot \hat{\mathbf{r}}\sigma}\delta(\bar{\mathbf{p}} - \boldsymbol{\alpha} + \mathbf{p})] \times 2\frac{\hat{\mathbf{r}} \cdot \mathbf{p}}{m}\delta(r - \sigma^+)\theta(-\hat{\mathbf{r}} \cdot \hat{\mathbf{p}})[\delta(\mathbf{p}' - \boldsymbol{\alpha} - \mathbf{p}) - \delta(\mathbf{p}' - \boldsymbol{\alpha} + \mathbf{p}^*)] \quad (250)$$

In arriving at (250) we have set $\tau = 0$ because of the factor $\delta(r - \sigma^+)$. This also means that in $\Delta_k(rp|\bar{p})$ we can put $\mathbf{r} = \sigma\hat{\mathbf{s}}$. Equation (250) shows that the memory function $\phi_1^{(c)}$ is explicitly frequency independent as one would expect for impulse interactions. Clearly this is true with hard-sphere potentials. For finite-range or continuous potentials one has a finite duration of collision, which then makes the memory function frequency dependent.

ACKNOWLEDGMENT

This work was supported by the National Science Foundation.

References

1. R. Zwanzig, *Phys. Rev.* **129**, 486 (1963).
2. M. Nelkin and A. Ghatak, *Phys. Rev.* **135**, A4 (1964).
3. S. Yip and M. Nelkin, *Phys. Rev.* **135**, A1241 (1964).
4. J. M. J. van Leeuwen and S. Yip, *Phys. Rev.* **139**, A1138 (1965).
5. G. F. Mazenko, *Phys. Rev.* **A3**, 2121 (1971); **A6**, 2545 (1972).
6. G. F. Mazenko, *Phys. Rev.* **A7**, 209 (1973).
7. G. F. Mazenko, *Phys. Rev.* **A7**, 222 (1973).
8. G. F. Mazenko, *Phys. Rev.* **A9**, 360 (1974).
9. G. F. Mazenko and S. Yip, *in*: *Molecular Motions in Liquids* (J. Lascomb, ed.), p. 79, D. Reidel, Dordrecht, Holland (1974).
10. J. V. Sengers, M. H. Ernst, and D. T. Gillespie, *J. Chem. Phys.* **56**, 5583 (1972).
11. J. V. Sengers, *Ber. Bunsenges. Phys. Chem.* **76**, 234 (1972).
12. M. H. Ernst, L. K. Haines, and J. R. Dorfman, *Rev. Mod. Phys.* **41**, 296 (1969).
13. H. C. Andersen, *Ann. Rev. Phys. Chem.* **145** (1975) and references therein.
14. P. A. Egelstaff, *An Introduction to the Liquid State*, Academic Press, London (1967).
15. P. C. Martin, *in*: *Many-Body Physics* (C. DeWitt and R. Balian, eds.), p. 39; Gordon & Breach, New York (1968).
16. B. Berne, *in*: *Physical Chemistry* (H. Eyring, ed.), Vol. 8B, Chap. 9, Academic Press, New York (1971).
17. R. J. Liboff, *Introduction to the Theory of Kinetic Equations*, Wiley, New York (1969).
18. G. E. Uhlenbeck and G. W. Ford, *Lectures in Statistical Mechanics*, Chap. 7, American Mathematical Society, Providence, Rhode Island (1963).
19. L. P. Kadanoff and P. C. Martin, *Ann. Phys.* (*N.Y.*) **24**, 419 (1963).
20. E. G. D. Cohen, *in*: *Statistical Mechanics and Non-Equilibrium* (J. Meixner, ed.), p. 140, North Holland, Amsterdam (1965).
21. W. E. Brittin, *Lectures in Theoretical Physics*, Vol. 9C, Gordon & Breach, New York (1967).
22. C. D. Boley, *Phys. Rev.* A*5*, 986 (1972).
23. P. Ortoleva and M. Nelkin, *Phys. Rev.* **181**, 429 (1969).
24. M. Nelkin and S. Ranganathan, *Phys. Rev.* **164**, 222 (1967).
25. R. Zwanzig, *Phys. Rev.* **144**, 170 (1966).
26. A. Z. Akcasu, *Phys. Rev.* **A7**, 182 (1973).
27. H. J. M. Hanley, R. D. McCarty, and E. G. D. Cohen, *Physica* **60**, 322 (1972).
28. H. L. Frisch and E. McLaughlin, *J. Chem. Phys.* **55**, 3706 (1971).
29. B. J. Alder, D. M. Gass, and T. E. Wainwright, *J. Chem. Phys.* **53**, 3813 (1970).
30. H. H. U. Konijnendijk and J. M. J. van Leeuwen, *Physica* **64**, 342 (1973).
31. J. Sykes, *J. Stat. Phys.* **8**, 279 (1973).
32. J. L. Lebowitz, J. K. Percus, and J. Sykes, *Phys. Rev.* **188**, 487 (1969).
33. J. R. Dorfman and E. G. D. Cohen, *Phys. Rev.* **A6**, 2247 (1972).
34. L. P. Kadanoff and J. Swift, *Phys. Rev.* **166**, 89 (1968).
35. K. Kawasaki, *Ann. Phys.* (*N.Y.*) **61**, 1 (1970).
36. D. Forster and P. C. Martin, *Phys. Rev.* **A2**, 1575 (1970).
37. P. Schofield, *Proc. Phys. Soc.* **88**, 149, Eq. (18) (1966).
38. C. D. Boley, *Phys. Rev.* **A11**, 328 (1975).
39. C. D. Boley, *Ann. Phys.* (*N.Y.*) **86**, 91 (1974).
40. E. P. Gross, *Ann. Phys.* (*N.Y.*) **69**, 42 (1972).
41. J. R. Dorfman and E. G. D. Cohen, *Phys. Lett.* **16**, 124 (1965).

42. S. Chapman and T. G. Cowling, *The Mathematical Theory of Non-Uniform Gases*, 3rd ed., Cambridge Univ. Press, London (1970).
43. J. O. Hirschfelder, C. F. Curtiss, and R. B. Bird, *Molecular Theory of Gases and Liquids*, Wiley, New York (1954).
44. S. Ranganathan and M. Nelkin, *J. Chem. Phys*, **47**, 4056 (1967).
45. E. P. Gross and D. Wisnivesky, *Phys. Fluids* **11**, 1387 (1968).
46. A. Sugawara, S. Yip, and L. Sirovich, *Phys. Fluids* **11**, 925 (1968).
47. G. F. Mazenko, T. Y. C. Wei, and S. Yip, *Phys. Rev.* **A6**, 1981 (1972).
48. J. D. Foch and G. W. Ford *in*: *Studies in Statistical Mechanics* (J. de Boer and G. E. Ulenbeck, eds.), North Holland, Amsterdam (1970).
49. P. M. Morse and H. Feshbach, *Methods of Theoretical Physics*, Vol. 1, McGraw-Hill, New York (1953).
50. T. Y. C. Wei, M.Sc. thesis, MIT (1972).
51. P. M. Furtado, Ph.D. thesis, MIT (1975).
52. D. Forster, *Phys. Rev.* **A9**, 943 (1974).
53. E. P. Gross and E. A. Jackson, *Phys. Fluids* **2**, 432 (1959).
54. L. Sirovich, *Phys. Fluids* **5**, 908 (1962).
55. B. D. Fried and S. D. Conte, *The Plasma Dispersion Function*, Academic Press, New York (1961).
56. J. R. D. Copley and S. W. Lovesey, Rep. *Progr. Phys.* **38**, 461 (1975) and references therein.
57. A. Rahman, *Phys. Rev.* **136**, A405 (1964).
58. D. Levesque, L. Verlet, and J. Kurkijarvi, *Phys. Rev.* **A7**, 1690 (1973).
59. K. Skold, M. J. Rowe, G. Ostrowsky, and P. D. Randolph, *Phys. Rev.* **A6**, 1107 (1972).
60. P. M. Furtado, G. F. Mazenko, and S. Yip, *Phys. Rev.* **A12**, 1653 (1975).
61. S. H. Chen, Y. Lefevre, and S. Yip, *Phys. Rev.* **A8**, 3163 (1973).
62. P. C. Desai, *Phys. Rev.* **A3**, 320 (1971).
63. P. Resibois, *J. Stat. Phys.* **13**, 393 (1975).
64. R. Resibois and J. L. Lebowitz, *J. Stat. Phys.* **12**, 483 (1975).
65. G. H. Vineyard, *Phys. Rev.* **110**, 999 (1958).
66. H. Gould and G. F. Mazenko, *Phys. Rev. Lett.* **35**, 1455 (1975).
67. J. C. Castresana, Ph.D. thesis, MIT (1974).

5

Projection Operator Techniques in the Theory of Fluctuations

Bruce J. Berne

1. Introduction

Projection techniques have become standard in the study of certain dynamic processes.* In this chapter some of the more important relationships are derived, and certain symmetry properties are discussed. No effort is made to be comprehensive.†

Because several other chapters in this volume require a certain background in the use of projection operators, it was felt that this chapter should provide some of this background. In particular, Chapter 6 on mode–mode coupling theory by Keyes and Chapter 4 on modern kinetic theory by Mazenko and Yip use some of the basic notions presented here. Thus we are constrained somewhat in our choice of topics. Nevertheless, some effort will be made to indicate simple applications of this technique.

*The review by Berne and Forster[1] covers most of the important applications prior to 1971. Particular attention should be given to the seminal papers of Zwanzig and Mori cited therein. Since the writing of this review, most of the important applications involve mode–mode coupling theories. These subjects are covered in Chapters 4 and 6 of this volume.

†The application of projection operators to kinetic equations is presented in an excellent didactic review article by Hynes and Deutch.[3] The excellent monograph by Forster[4] contains many novel applications of projection operators to symmetry-breaking phase transitions.

Bruce J. Berne • Department of Chemistry, Columbia University, New York, New York

2. Liouville Space

The *state* of a classical system of f degrees of freedom is completely specified by f generalized positions $(q_1, \ldots, q_f)$ and f conjugate momenta $(p_1, \ldots, p_f)$ or equivalently by a point or vector

$$\mathbf{\Gamma} = (q_1, \ldots, q_f, p_1, \ldots, p_f) \tag{1}$$

in *phase space*—a Cartesian space with orthogonal axes corresponding to the $2f$ positions and momenta. Once the state $\mathbf{\Gamma}$ is known no additional information is required to specify any of the *dynamical properties* A_i of the system, because these are prescribed functions of the state, that is,

$$A_i = A_i(\mathbf{\Gamma}) \tag{2}$$

The state $\mathbf{\Gamma}$ changes in time according to the canonical equations of motion, which can be expressed in vector form as

$$\dot{\mathbf{\Gamma}} = \{\mathbf{\Gamma}, H\} \equiv iL\mathbf{\Gamma} \tag{3}$$

where $\{\mathbf{\Gamma}, H\}$ is the Poisson bracket of $\mathbf{\Gamma}$ with the Hamiltonian H of the system and L is the Liouville operator or *Liouvillian*

$$iL \equiv \{\ldots, H\} \tag{4}$$

L is a linear Hermitian operator and Eq. (3) can be formally solved for the state at time t given the state at time 0,

$$\mathbf{\Gamma}(t) = e^{iLt}\mathbf{\Gamma}(0) \tag{5}$$

Clearly, since A_i is a function of $\mathbf{\Gamma}$ its value would be $A_i(0) = A_i(\mathbf{\Gamma}(0))$ at time 0 and $A_i(t) = A_i(e^{iLt}\mathbf{\Gamma}(0))$ at time t. Thus as the state of the system evolves in time, the value of each of the dynamical properties also varies except for those properties that are "constants of the motion." Because L is a linear operator it is clear that $A_i(t)$ can also be expressed as

$$A_i(t) = e^{iLt}A_i(0) \tag{6a}$$

or

$$\partial A_i(t)/\partial t = iLA_i(t) \tag{6b}$$

The Liouvillian is thus seen to be the generator of the natural motion of the system and the operator e^{iLt} is called the propagator. The propagator

$$G(t) \equiv e^{iLt} \tag{7a}$$

and its resolvent

$$\tilde{G}(s) = 1/(s - iL) \tag{7b}$$

are basic to the theory of time-dependent processes. The foregoing is depicted schematically in Fig. 1. The ensemble average of A at time t can be expressed in

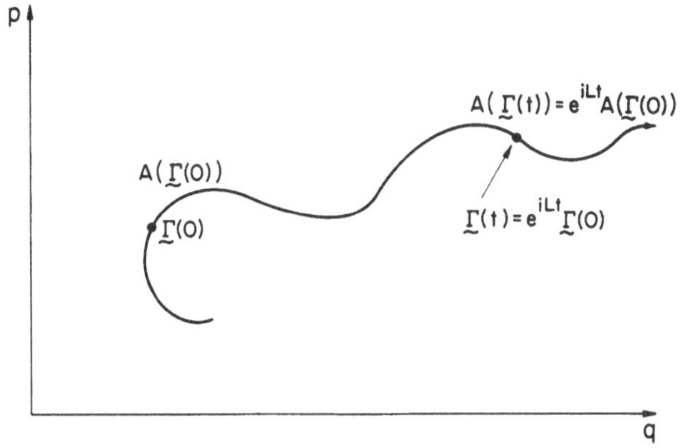

Fig. 1. A schematic diagram of phase space. The state of the system changes from $\mathbf{\Gamma}(0)$ to $\mathbf{\Gamma}(t)$ under the action of the propagator $G(t) = e^{iLt}$ with a concomitant change in the dynamical property from $A(0) \equiv A(\mathbf{\Gamma}(0))$ to $A(t) \equiv A(\mathbf{\Gamma}(t))$.

two equivalent forms:

$$\langle A(t)\rangle = \int d\mathbf{\Gamma}\, \rho(\mathbf{\Gamma}, t) A(\mathbf{\Gamma}) \tag{8a}$$

$$\langle A(t)\rangle = \int d\mathbf{\Gamma}\, \rho(\mathbf{\Gamma}, 0) e^{iLt} A(\mathbf{\Gamma}) \tag{8b}$$

where $\rho(\mathbf{\Gamma}, t)$ is the phase space distribution function at time t. In Eq. (8a), the value of A at the point $\mathbf{\Gamma}$ in phase space is weighted by $\rho(\mathbf{\Gamma}, t)\, d\mathbf{\Gamma}$, the probability of finding a system in the neighborhood $d\mathbf{\Gamma}$ of $\mathbf{\Gamma}$ at time t, whereas in Eq. (8b), $[e^{iLt}A(\mathbf{\Gamma})]$—the value of A at time t given that the system is in the state $\mathbf{\Gamma}$ at time $t = 0$—is weighted by $\rho(\mathbf{\Gamma}, 0)\, d\mathbf{\Gamma}$, the probability of finding the initial state in the neighborhood $d\mathbf{\Gamma}$ of $\mathbf{\Gamma}$. Naturally Eqs. (8a) and (8b) must give identical results. These are the analogs of the Schrödinger and Heisenberg representations in quantum mechanics. Now e^{iLt} is a unitary operator [see Eq. (17)] so that Eq. (8b) can also be expressed as

$$\langle A(t)\rangle = \int d\mathbf{\Gamma} A(\mathbf{\Gamma}) e^{-iLt} \rho(\mathbf{\Gamma}, 0) \tag{8c}$$

Equations (8a) and (8c) must be valid for all dynamical properties. It thus follows that

$$\rho(\mathbf{\Gamma}, t) = e^{-iLt} \rho(\mathbf{\Gamma}, 0) \tag{9a}$$

or

$$\partial \rho(\mathbf{\Gamma}, t)/\partial t + iL\rho(\mathbf{\Gamma}, t) = 0 \tag{9b}$$

The phase space distribution function is thus seen to evolve in time according to Eqs. (9). Note the difference in the sign accompanying L in Eqs. (6) and (9). Equation (9b) is called the *Liouville equation*. Substitution of the explicit form of L from Eq. (4) gives various forms for the Liouville equation:

$$\partial\rho/\partial t + \{\rho, H\} = 0 \tag{10a}$$

$$\partial\rho(\mathbf{\Gamma}, t)/\partial t + \nabla_{\Gamma} \cdot [\dot{\mathbf{\Gamma}}\rho(\mathbf{\Gamma}, t)] = 0 \tag{10b}$$

$$\frac{\partial\rho}{\partial t}(\mathbf{\Gamma}, t) + \sum_{j=1}^{N} [\mathbf{v}_j \cdot \nabla_{\mathbf{r}_j} + \mathbf{F}_j \cdot \nabla_{\mathbf{p}_j}]\rho(\mathbf{\Gamma}, t) = 0 \tag{10c}$$

In Eq. (10), $\dot{\mathbf{\Gamma}} \equiv (\dot{q}_1, \ldots, \dot{p}_f)$ is the time rate of change of $\mathbf{\Gamma}$ and $\nabla_{\Gamma} \equiv (\partial/\partial_{q_1}, \ldots, \partial/\partial_{p_f})$. Equation (10b) is an "equation of continuity" in phase space. Since the ensemble is represented by a cloud of points in phase space, each point giving the instantaneous state of a replica system, the change in state of the ensemble is represented by a "cloud" of points swarming in phase space. $\rho(\mathbf{\Gamma}, t)$ is proportional to the density of this cloud at the point $\mathbf{\Gamma}$ at time t. Equation (10b) governs the flow of these points. By analogy with the equation of continuity in hydrodynamics, Eq. (10b) tells us that this cloud moves as an incompressible fluid with flux $\dot{\mathbf{\Gamma}}\rho(\mathbf{\Gamma}, t)$. The trajectory of each point represents a streamline, and the number of points in a fixed volume element changes by virtue of an imbalance between the number of points flowing in versus the number flowing out. The integral $\int d\Gamma\, \rho(\mathbf{\Gamma}, t)$ is conserved. This implies the conservation of probability.

Equation (10c) is often used in calculations. Here Eq. (10a) has been specialized to N point particles whose position and velocities are expressed in a Cartesian coordinate system. Here $\mathbf{r}_j$, $\mathbf{v}_j$, $\mathbf{p}_j$, and $\mathbf{F}_j$ are vectors specifying, respectively, the position, velocity, momentum, and force on particle j.

That the Liouvillian is important has been amply demonstrated by the foregoing. In classical mechanics L plays a role analogous to the role played by the Hamiltonian operator $\hat{H}$ in quantum mechanics. this analogy has been extensively developed in connection with the calculation of time correlation functions of two arbitrary dynamical properties

$$C_{AB}(t) \equiv \langle A(t)B^*(0)\rangle \tag{11}$$

These functions play a central role in transport theory and in the theory of spectral line shapes.(5–7) They measure the correlation between $A(t)$ and $B^*(0)$. The brackets $\langle\cdots\rangle$ in Eq. (11) indicate an average over any of the equilibrium ensembles. To be specific we shall use the canonical ensemble so that

$$C_{AB}(t) \equiv \int d\mathbf{\Gamma}\, \rho_0(\mathbf{\Gamma})[e^{iLt}A(\mathbf{\Gamma})]B^*(\mathbf{\Gamma}) \tag{12}$$

where B^* is the complex conjugate of B and $\rho_0(\Gamma)$ is the equilibrium canonical distribution function

$$\rho_0(\Gamma) = Q^{-1} \exp -\beta H(\Gamma) \tag{13}$$

where $\beta = (k_B T)^{-1}$, and Q is the classical canonical partition function.

Correlation functions like Eq. (11) can be defined for any pair of dynamical properties for which the integral in Eq. (12) exists for all time. It is not difficult to prove that if the two properties have finite mean square values, such correlation functions do indeed exist. In the following we restrict our attention to such properties.

Equation (12) can be cast in the suggestive form

$$C_{AB}(t) = \int d\Gamma\, \psi_B^*(\Gamma) e^{iLt} \psi_A(\Gamma) \tag{14}$$

where $\psi_A(\Gamma) \equiv [\rho_0(\Gamma)]^{1/2} A(\Gamma)$. This follows from the fact that $\{Q^{-1}e^{-\beta H}, H\} = iL\rho_0(\Gamma) = 0$, with the concomitant result that $\rho_0(\Gamma)$ and e^{iLt} commute. These look very much like matrix elements of the operator e^{iLt}.

Consider now the set of all dynamical properties $A, B, C, \ldots$, and define the scalar product

$$(A, B^*) = \int d\,\Gamma \psi_A(\Gamma) \psi_B^*(\Gamma) = \int d\Gamma\, \rho_0(\Gamma) A(\Gamma) B^*(\Gamma) \tag{15}$$

between any pair of these properties. This set of properties plus the definition of the scalar product defines a function space called *Liouville space*. The properties themselves may be thought of as vectors in this space whose motion is given by Eq. (6). From Eq. (15) it is clear that $(A, A^*) = \langle |A|^2 \rangle \geq 0$, with the equality sign pertaining only to trivial vectors $A \equiv 0$. The length of a vector A in Liouville space is given by $\|A\| = (A, A^*)^{1/2} = \langle |A|^2 \rangle^{1/2}$. A property whose length (or norm) is unity is said to be normalized. Two different properties A and B are said to be orthogonal if $(A, B^*) = 0$. Thus all functions of Γ of finite norm, together with the definition of the scalar product [cf. Eq. (15)] define Liouville space.

Any operator on functions of Γ transforms one vector in Liouville space into another. Of particular importance to us are the operators L and e^{iLt}. It is easy to show by an integration by parts[(7)] that L is a linear Hermitian operator in Liouville space, that is,

$$L^+ = L$$

where L^+ is the Hermitian conjugate of L. It follows from this that the propagator $G(t)$ is unitary, that is, $G^+(t) = e^{-iLt} = G^{-1}(t)$. Thus for any pair of vectors A and B,

$$(LA, B^*)^* = (LB, A^*) \tag{16}$$

$$(e^{iLt}A, B^*)^* = (e^{-iLt}B, A^*) \tag{17}$$

The scalar products in Eq. (17) are the correlation functions $C_{AB}(t)$ and $C_{BA}(-t)$, respectively, and Eq. (17) gives

$$C^*_{AB}(t) = C_{BA}(-t) \tag{18}$$

This symmetry relation follows directly from the unitarity of e^{iLt}.

Because the operator $G(t) = e^{iLt}$ is unitary, or norm preserving, the vector $e^{iLt}A$ can be regarded as varying in time in such a way that its length (or norm) is preserved. The time evolution of $A(t)$ is thus represented by a rotation in Liouville space. The scalar product of $A(t)$ and $A(0)$ gives the time correlation function $C_{AA}(t)$, and leads to the interpretation of $C_{AA}(t)$ as the component (or projection) of $A(t)$ on $A(0)$. An operator that projects an arbitrary vector onto A is

$$P = (\ldots, A^*)(A, A^*)^{-1}A \tag{19}$$

When P acts on an arbitrary vector B it gives $PB = (B, A^*)(A, A^*)^{-1}A$, where $(B, A^*)(A, A^*)^{-1}$ is the component of B on the vector A.

The operator

$$Q \equiv 1 - P \tag{20}$$

is also a projection operator. When Q acts on an arbitrary property B it gives a vector $QB = B - (B, A^*)(A, A^*)^{-1}A$, which is orthogonal to A, that is, $(QB, A^*) = 0$. Thus Q projects onto a subspace of Liouville space that is orthogonal to A. It follows immediately from Eq. (20) that

$$Q + P = 1 \tag{21}$$

These projection operators will be useful in the next section.

The resolvent $G(s)$ given by Eq. (7b) can also be expressed as

$$\tilde{G}(s) = 1/[s - i(Q+P)L] \tag{22}$$

where we use the identity $iL = i(Q+P)L$, which follows directly from Eq. (21). Now using the well-known operator identity

$$\frac{1}{A} - \frac{1}{B} = \frac{1}{A}(B - A)\frac{1}{B} \tag{23}$$

with $A = s - i(Q+P)L$ and $B = s - iQL$, gives

$$\tilde{G}(s) = \frac{1}{s - iQL} + \frac{1}{s - iL}\, iPL \frac{1}{s - iQL} \tag{24}$$

Laplace inversion of the resolvent then gives

$$G(t) = e^{iQLt} + \int_0^t d\tau\, e^{iL(t-\tau)} iPL e^{iQL\tau} \tag{25}$$

This completes the background material required for the discussion of relaxation equations.

3. Relaxation Equations

The time dependence of the dynamical property A is given by Eq. (6a). Differentiation of this equation gives

$$dA(\mathbf{\Gamma})/dt = e^{iLt} iLA(\mathbf{\Gamma}) = e^{iLt}(P+Q)iLA(\mathbf{\Gamma}) \tag{26}$$

where the identity operator $P+Q$ has been inserted without altering the result. From Eq. (19) it follows that

$$e^{iLt}PiLA = (iLA, A^*)(A, A^*)^{-1} e^{iLt}A = i\Omega A(t) \tag{27}$$

where the *frequency* Ω is defined as

$$i\Omega \equiv (iLA, A^*)(A, A^*)^{-1} \tag{28}$$

Substitution of Eqs. (25) and (27) into Eq. (26) gives

$$\frac{dA(t)}{dt} = i\Omega A(t) + e^{iQLt}QiLA + \int_0^t d\tau\, e^{iL(t-\tau)} iPLe^{iQL\tau}QiLA \tag{29}$$

For simplicity we define the quantity f as

$$f \equiv QiLA \tag{30}$$

This property evolves in time according to the equation

$$f(\tau) = e^{iL\tau}f \tag{31}$$

Note, however, that one of the propagators that appear in Eq. (29) is not $e^{iL\tau}$ but $e^{iQL\tau}$. We thus define the quantity

$$F(\tau) \equiv e^{iQL\tau}f = e^{iQL\tau}QiLA \tag{32}$$

This quantity is called the *random force*. It should not be confused with $f(\tau)$. In general, $f(\tau) \neq F(\tau)$ except at $\tau = 0$, where $F(0) = f(0)$. The quantity $F(\tau)$ can also be written

$$F(\tau) = Qe^{iQL\tau}f = QF(\tau) \tag{33}$$

This means that $F(\tau)$ is a vector orthogonal to A, or

$$(F(\tau), A^*) = 0 \tag{34}$$

Thus there is no correlation between $A(0)$ and the random force. This is a very important formal conclusion.

The last integral in Eq. (29) involves the term $iPLF(\tau)$. From the preceding remarks and definitions it follows that this term can be expressed as

$$iPLF(\tau) = iPLQF(\tau) = (iLQF(\tau), A^*)(A, A^*)^{-1}A \tag{35}$$

Because Q and L are both Hermitian operators it follows that $(iLQF(\tau), A^*) = -(F(\tau), (QiLA)^*) = -(F(\tau), F^*(0))$. Consequently

$$iPLF(\tau) = -(F(\tau), F^*(0))(A, A^*)^{-1}A \tag{36}$$

We now define the *memory function* $K(\tau)$ as

$$K(\tau) \equiv (F(\tau), F^*(0))(A, A^*)^{-1} \tag{37}$$

Combining Eqs. (29), (36), and (37) we obtain the *generalized Langevin equation*,[8]

$$dA(t)/dt = i\Omega A(t) - \int_0^t d\tau\, K(\tau)A(t-\tau) + F(t) \tag{38}$$

where Ω, $K(\tau)$, and $F(t)$ are defined in Eqs. (28), (37), and (32), respectively.

From Eq. (37) we note that the memory function is proportional to the autocorrelation function of the random force. This is called the *second fluctuation–dissipation* theorem.[9]

Generally what is wanted is an equation for the time correlation function $C(t) = (A(t), A^*(0))$. Taking the scalar product of Eq. (38) with A^* and using Eq. (34) therefore gives

$$dC(t)/dt = i\Omega C(t) - \int_0^t d\tau\, K(\tau)C(t-\tau) \tag{39}$$

This is called the *memory function equation.*[10] In this equation the random force appears implicitly in $K(\tau)$.

If Eq. (38) is averaged over an initial nonequilibrium distribution function as in Eq. (8b), the resulting equation is

$$d\bar{A}(t)/dt = i\Omega \bar{A}(t) - \int_0^t d\tau\, K(\tau)\bar{A}(t-\tau) + \bar{F}(t) \tag{40}$$

where $\bar{A}(t)$ is the ensemble average of $A(t)$ and $\bar{F}(t)$ is the ensemble average of F, that is,

$$\bar{F}(t) = \int d\mathbf{\Gamma}\, \rho(\mathbf{\Gamma}, 0)e^{iQLt}QiLA(\mathbf{\Gamma}) \tag{41}$$

This function is not zero and is in general a very complicated function of time. Since Eq. (40) describes the relaxation of a macroscopic property of the system there should come a time beyond which this relaxation equation is identical with a phenomenological equation. If this happens, then at long times the decay is independent of the initial preparation of the system and we are at liberty to choose the most convenient initial distribution function $\rho(\mathbf{\Gamma}, 0)$. For simplicity we choose that $\rho(\mathbf{\Gamma}, 0)$ which describes a small deviation from equilibrium, that is, which gives $\bar{A}(0)$ very close to the equilibrium value (here assumed to be zero). Then it is easy to show that[8]

$$\rho(\mathbf{\Gamma}, 0) = \rho_0(\mathbf{\Gamma}) + A^*(\mathbf{\Gamma})F_A\rho_0(\mathbf{\Gamma}) \tag{42}$$

where $\rho_0(\mathbf{\Gamma})$ is the equilibrium distribution function and F_A the thermodynamic parameter conjugate to $\bar{A}$ in the free energy relation. F_A need not be specified here. Then $\bar{A}(t)$ becomes

$$\bar{A}(t) = (A(t), A^*)F_A \tag{43}$$

and $\bar{F}(t)$ becomes

$$\bar{F}(t) = (F(t), A^*)F_A = 0 \tag{44}$$

where the last equality follows from Eq. (34). Substitution of Eqs. (43) and (44) into Eq. (40) then gives Eq. (39).

The foregoing discussion was restricted to a consideration of a single variable A. Many circumstances arise in which we will be interested in the time evolution of many coupled variables. For example, in hydrodynamics the mass density, momentum density, and energy density are coupled. It is possible to extend the previous analysis to the case of many variables $\{A_1, \ldots, A_M\}$. These properties can be represented by the column matrix

$$\mathbf{A} = \begin{pmatrix} A_1 \\ \cdot \\ \cdot \\ \cdot \\ A_M \end{pmatrix} \tag{45}$$

For convenience these properties are chosen such that their equilibrium values are zero, $\langle \mathbf{A} \rangle = 0$. Moreover we demand that the set be linearly independent, that is, that none of the A_i are a linear combination of the others.

Let us now define the *correlation matrix*

$$\mathbf{C}(t) = (\mathbf{A}(t), \mathbf{A}^+(0)) \tag{46}$$

where $\mathbf{A}(t)$ is the column matrix $e^{iLt}\mathbf{A}$ and $\mathbf{A}^+$ is the Hermitian conjugate of $\mathbf{A}$, that is, the row matrix $\mathbf{A}^+ = (A_1^*, \ldots, A_M^*)$. $\mathbf{C}(t)$ is an $M \times M$ matrix whose ijth element is the correlation function $C_{ij}(t) = \langle A_i(t)A_j^*(0)\rangle = (A_i(t), A_j^*)$. The initial value of the correlation matrix $\mathbf{C}(0)$ will be denoted $\beta^{-1}\boldsymbol{\chi}$, where $\beta = k_B T)^{-1}$ and $\boldsymbol{\chi}$ is called the static *susceptibility matrix*,

$$\mathbf{C}(0) = (\mathbf{A}, \mathbf{A}^+) \equiv \beta^{-1}\boldsymbol{\chi} \tag{47}$$

Since $(A_i, A_j^*)^* = (A_j, A_i^*)$, the matrix $\boldsymbol{\chi}$ is a Hermitian matrix, so that $\boldsymbol{\chi}^+ = \boldsymbol{\chi}$.

The properties $\{A_1, \ldots, A_M\}$ define a subspace of Liouville space. This subspace is the set of all vectors that can be expressed as linear combinations of $\{A_1, \ldots, A_M\}$. Let us now determine the projection operator P that projects a vector onto this M-dimensional subspace. The projection operator must satisfy the conditions $PA = A$ and $P^2 = P$. Note that

$$P \equiv (\ldots, \mathbf{A}^+) \cdot (\mathbf{A}, \mathbf{A}^+)^{-1} \cdot \mathbf{A} = \beta(\ldots, \mathbf{A}^+) \cdot \boldsymbol{\chi}^{-1} \cdot \mathbf{A} \tag{48}$$

satisfies these requirements. As before, the operator $Q \equiv 1 - P$ is a projector onto the subspace orthogonal to $\{A_1, \ldots, A_M\}$. Following the same arguments used to derive the generalized Langevin equation for the single variable yields a generalized Langevin equation for the column vector $\mathbf{A}$:

$$d\mathbf{A}(t)/dt = i\mathbf{\Omega}\cdot\mathbf{A}(t) - \int_0^t d\tau\,\mathbf{K}(\tau)\cdot\mathbf{A}(t-\tau) + \mathbf{F}(t) \tag{49}$$

where $\mathbf{\Omega}$ is a matrix of *frequencies*

$$\mathbf{\Omega} \equiv (L\mathbf{A}, \mathbf{A}^+)\cdot(\mathbf{A}\mathbf{A}^+)^{-1} = \beta(L\mathbf{A}, \mathbf{A}^+)\cdot\boldsymbol{\chi}^{-1} \tag{50}$$

$\mathbf{K}(\tau)$ is a matrix of *memory functions,*

$$\mathbf{K}(\tau) \equiv (\mathbf{F}(\tau), \mathbf{F}^+(0))\cdot(\mathbf{A}, \mathbf{A}^+)^{-1} = \beta(\mathbf{F}(\tau), \mathbf{F}(0))\cdot\boldsymbol{\chi}^{-1} \tag{51}$$

$\mathbf{F}(\tau)$ is the random force, $\mathbf{F}(\tau) = e^{iQL\tau}\mathbf{f}$, and $\mathbf{f}$ is $iQL\mathbf{A}$. Equation (51) is the multidimensional *second-fluctuation dissipation theorem.* Again we note that $\mathbf{F}(\tau)$ is orthogonal to A, so that

$$(\mathbf{F}(\tau), \mathbf{A}^+) = 0 \tag{52}$$

Taking the scalar product of Eq. (49) with $\mathbf{A}^+$ gives the equation for the correlation function matrix,

$$d\mathbf{C}(t)/dt = i\mathbf{\Omega}\cdot\mathbf{C}(t) - \int_0^t d\tau\,\mathbf{K}(\tau)\cdot\mathbf{C}(t-\tau) \tag{53}$$

In terms of their components Eqs. (49) and (53) become

$$\frac{dA_\nu(t)}{dt} = \sum_\mu \left\{ i\Omega_{\nu\mu}A_\mu(t) - \int_0^t d\tau\,K_{\nu\mu}(\tau)A_\mu(t-\tau)\right\} + F_\nu(t) \tag{54}$$

$$\frac{dC_{\nu\mu}(t)}{dt} = \sum_\lambda \left\{ i\Omega_{\nu\lambda}C_{\lambda\mu}(t) - \int_0^t d\tau\,K_{\mu\lambda}(\tau)C_{\lambda\mu}(t-\tau)\right\} \tag{55}$$

where the components of the frequency matrix $\Omega_{\nu\lambda}$ and the memory matrix $K_{\nu\lambda}(\tau)$ are

$$\Omega_{\nu\lambda} = \beta\sum_\kappa (LA_\nu, A_\kappa^*)\chi_{\kappa\lambda}^{-1} \tag{56}$$

$$K_{\nu\lambda}(\tau) = \beta\sum_\kappa (F_\nu(\tau), F_\kappa^*(0)\,\chi_{\kappa\lambda}^{-1} \tag{57}$$

where $\chi_{\kappa\lambda}^{-1}$ is the $\kappa\lambda$th element of the inverse of the susceptibility matrix. The random force $F_\mu(t)$ is, of course,

$$F_\mu(t) = e^{iQLt}QiLA_\mu = e^{iQLt}f_\mu = QF_\mu(t) \tag{58}$$

In Eqs. (49) and (53) the dot denotes matrix multiplication. If $\mathbf{K}(t)$ and $\mathbf{F}(t)$ are known, these equations represent a set of closed equations from which the time evolution of the properties $\{A_1, \ldots, A_M\}$ can be computed. *Equations (49) and (51) are an exact consequence of the equations of motion.*

The *generalized Langevin equation* and the *memory function equation* simplify considerably when the set $\{A_1, \ldots, A_M\}$ relaxes much more slowly than all other properties. If *all* such slowly relaxing variables are included in the set $\{A_1, \ldots, A_M\}$, the set is called a *good set of variables.* At the outset it is important to note that there are no rules by which a good set of variables can be chosen. Generally this is a matter of one's intuition. It is, however, the crucial step in the application of the Zwanzig–Mori formalism to specific problems. There are several possible reasons for a given set of variables to be regarded as "slow" with respect to all other variables.

The Fourier component $\delta A(q, t)$ of the fluctuation of a conserved density* has a lifetime $\tau(q)$ such that $\tau(q) \to \infty$ as $q \to 0$, that is, $\delta A(q, t)$ varies slowly for small q. Thus we expect that the small ($q \to 0$) wave number Fourier components of the "densities" of all the conserved properties form a good set of variables. For example, in an isotropic monatomic fluid we surmise that a good set consists of the low-q Fourier components of the mass, linear momentum, and energy densities.

Another good example of a separation of time scales is Brownian motion. Because the Brownian particle is much more massive than the solvent particles, it moves much more slowly. Thus the position and velocity of the Brownian particle should constitute a good set of variables.

Highly anisotropic molecules reorient very slowly in dense fluids and liquid crystals. In these fluids the "conserved densities" do not by themselves constitute a good set. It is necessary to include "densities of orientational properties." (This is made more specific later.)

Let us assume that we can list the independent variables whose decay is slow, that is, those whose relaxation times τ_r satisfy

$$\tau_r \gg \tau_c$$

where τ_c typifies the longest relaxation times for all other variables. We choose these slowly relaxing variables as our set A, and thus the subspace of slowly relaxing variables is the subspace spanned by $\{A_1, \ldots, A_M\}$. Then the projection operator P projects onto a subspace of Liouville space containing all "slowly" relaxing properties of the system and the projector Q projects onto the orthogonal complement of this *slow subspace*, which by construction contains all of the rapidly decaying properties. Since the "random force" is always in this *fast subspace* it fluctuates rapidly, and its time correlation

*A conserved density $A(\mathbf{r}, t)$ must satisfy a conservation equation

$$\partial A(\mathbf{r}, t)/\partial t + \nabla_r \cdot \mathbf{J}_A(\mathbf{r}, t) = 0$$

where $\mathbf{J}_A(\mathbf{r}, t)$ is a corresponding current. It follows from a spatial Fourier transform of this conservation equation that

$$\partial \delta A(\mathbf{q}, \mathbf{t})/\partial t = i\mathbf{q} \cdot \mathbf{J}_A(\mathbf{q}, t)$$

Thus as $q \to 0$, $\delta \dot{A} \to 0$, indicating a vanishing rate of change or an infinite lifetime.

function (i.e., memory function) should decay on the time scale τ_c. Thus there will be a large separation in the time scales and **A** will decay much more slowly than **F**. Hence for times $t \gg \tau_c$ it is permissible to treat the memory function as a very rapidly decaying function so that

$$\mathbf{K}(t) = 2\,\boldsymbol{\Gamma}\sigma(t) \tag{59}$$

This is called the *Markov* approximation. Substitution of this into Eqs. (49) and (53) then yields the equations

$$d\mathbf{A}(t)/dt = i\boldsymbol{\Omega}\cdot\mathbf{A}(t) - \boldsymbol{\Gamma}\cdot\mathbf{A}(t) + \mathbf{F}(t) \tag{60}$$

$$d\mathbf{C}(t)/dt = i\boldsymbol{\Omega}\cdot\mathbf{C}(t) - \boldsymbol{\Gamma}\cdot\mathbf{C}(t) \tag{61}$$

where $\boldsymbol{\Gamma}$ is called the *relaxation matrix.*

Integration of Eq. (59) then gives the *relaxation matrix* $\boldsymbol{\Gamma}$ in terms of the *memory matrix* $\mathbf{K}(t)$ as

$$\boldsymbol{\Gamma} = \int_0^\infty d\tau\,\mathbf{K}(\tau) = \beta\int_0^\infty d\tau\,(\mathbf{F}(\tau), \mathbf{F}^+(0))\cdot\boldsymbol{\chi}^{-1} \tag{62}$$

where the second equality follows from Eq. (59). The relaxation matrix $\boldsymbol{\Gamma}$ can be regarded as the product of two parts: a *kinetic coefficient* $\mathbf{L}$ and an inverse susceptibility $\boldsymbol{\chi}^{-1}$, that is,

$$\boldsymbol{\Gamma} = \mathbf{L}\cdot\boldsymbol{\chi}^{-1} \tag{63}$$

where

$$\mathbf{L} \equiv \beta\int_0^\infty d\tau\,(\mathbf{F}(\tau), \mathbf{F}^+(0)) \tag{64}$$

The kinetic coefficients can be expressed in terms of ordinary time correlation functions. Such relations are called *Green–Kubo relations.*[11]

One consequence of Eq. (63) follows immediately. In the neighborhood of certain points of instability such as the gas–liquid critical point or order–disorder phase transitions, the susceptibilities corresponding to the fluctuations in the order parameters become very large. Thus if L does not increase as rapidly as χ, the corresponding relaxation rates Γ will become small. This phenomenon is called "critical slowing" of the fluctuations. There has been much recent work on this phenomenon.[12]

In the foregoing we assumed **A** was the only slowly varying set of properties in our system. This assumption is wrong because if **A** is slowly varying there is no reason to assume that products like A_iA_j, $A_iA_jA_k$, etc., are fast. Nevertheless, the parts of these products orthogonal to A have been included in the memory function and the random force. This does not contradict the exact equations, but certainly argues against the validity of the Markov approximation. Near a phase transition or at very long times, these nonlinear fluctuations must be included for a correct evaluation of the transport coefficients or the long-time tails. This is precisely what is considered in the chapter on mode–mode coupling. Nevertheless, the Markov approximation is still very useful.

4. *Symmetry Properties*

In this section we consider several different kinds of symmetry transformations on phase space.[2] For this discussion it is useful to choose the cartesian coordinate system for the positions and momenta of the N particle system. Then the state is given by $\mathbf{\Gamma}=(x_1, y_1, z_1, \ldots, p_{xN}, p_{yN}, p_{zN})$ and the various transformations considered here are

$\mathbf{\Gamma}\to(-x_1, -y_1, -z_1, \ldots, -p_{xN}, -p_{yN}, -p_{zN})$	parity	(65)
$\mathbf{\Gamma}\to(x_1, -y_1, z_1, \ldots, p_{xN}, -p_{yN}, p_{zN})$	reflection in xz plane	(66)
$\mathbf{\Gamma}\to(-x, y_1, z_1, \ldots, -p_{xN}, p_{yN}, p_{zN})$	reflection in yz plane	(67)
$\mathbf{\Gamma}\to(x_1, y_1, -z_1, \ldots, p_{xN}, p_{yN}, -p_{zN})$	reflection in xy plane	(68)
$\mathbf{\Gamma}\to(x_1+a_x, y_1+a_y, z_1+a_z, \ldots, p_{xN}, p_{yN}, p_{zN})$	translation by arbitrary vector $a=(a_x, a_y, a_z)$	(69)
$\mathbf{\Gamma}\to(x_1, y_1, z_1, \ldots, -p_{xN}, -p_{yN}, -p_{zN})$	time reversal	(70)

Although it is quite possible to discuss other transformations such as rotations,[6] these would take us too far afield here. To proceed it is important to observe the effects of these transformations on various quantities that come into the evaluation of time correlation function in Eq. (12). For example, the transformation might have an effect on the volume element $d\mathbf{\Gamma}$. This would be given by the Jacobian J of the transformation from $\mathbf{\Gamma}$ to $\mathbf{\Gamma}'$. In addition, the Hamiltonian, Liouvillian, and equilibrium distribution function might change under the various transformations. In Table 1 we summarize how these quantities are transformed. The primes on the headings of the columns indicate the values of the transformed quantities, whereas the unprimed quantities in the body of the table indicate the untransformed values.

In all of the following we assume that the Hamiltonian is invariant to these transformations. For example, in the absence of electromagnetic forces the Hamiltonian is a quadratic function of the momentum—hence for time reversal invariance $H(q, p)=H(q, -p)$. In addition, if the potential is only a function of the distances between particles, H is translationally invariant, reflection invariant, and has even parity. Because $\rho_0(\mathbf{\Gamma})\propto e^{-\beta H}$, $\rho_0(\mathbf{\Gamma})$ will be invariant to all of these transformations. In ordered states of matter (like solids or liquid crystals), H may be invariant but the symmetry of $\rho_0(\mathbf{\Gamma})$ is broken. Thus our table really applies only to homogeneous systems.

A glance at Table 1 shows that the only change is $iL \to -iL$ under time reversal. This gives rise to a change in the propagator

$$G(t) \to G(-t) = e^{-iLt} \tag{71}$$

under time reversal—an expected result. The transformation properties of L spring from the explicit form for L,

$$iL = \{\ , H\} = \sum_j \left\{ \frac{\partial H}{\partial p_{x_j}} \frac{\partial}{\partial x_j} - \frac{\partial H}{\partial x_j} \frac{\partial}{p_{x_j}} \right\} \tag{72}$$

and from the invariance of H under the various transformations.

Most of the applications of relaxation equations are to "hydrodynamic" variables, which have the form

$$A_\mu(q) = \sum_{j=1}^{N} a_\mu(\mathbf{r}_j, \mathbf{p}_j) \exp iqz_j \tag{73}$$

Here we have chosen a coordinate system in which the wave vector $\mathbf{q}$ is parallel to the $\hat{\mathbf{z}}$ axis. In this equation $a_\mu(\mathbf{r}_j, \mathbf{p}_j)$ is a single molecule property, that is, it depends only on the position and momentum (and internal degrees of freedom) of molecule j. z_j is the center of mass position of particle j, and q is to be regarded as a parameter. $A_\mu(q)$ is related to the spatial fourier transform of the "density" of molecular property a_μ, that is, $\sum_{j=1}^{N} a_\mu^j\, \delta(\mathbf{r}-\mathbf{r}_j)$.

For example, the properties

$$n(q) = \sum_{j=1}^{N} e^{iqz_j} \qquad \text{(number density)} \tag{74a}$$

$$\mathbf{g}(q) = \sum_{j=1}^{N} \mathbf{p}_j e^{iqz_j} \qquad \text{(momentum density)} \tag{74b}$$

are the Fourier transforms of the number and momentum density, respectively, and have the form of Eq. (73).

Table 1. The Behavior of Certain Properties under Certain Symmetry Transformations

Symmetry transformation	Volume element in phase space $d\mathbf{\Gamma}'$	H'	$\rho_0(\mathbf{\Gamma})'$	iL'
Parity (inversion)	$d\mathbf{\Gamma}$	H	$\rho_0(\mathbf{\Gamma})$	iL
Reflection	$d\mathbf{\Gamma}$	H	$\rho_0(\mathbf{\Gamma})$	iL
Translation	$d\mathbf{\Gamma}$	H	$\rho_0(\mathbf{\Gamma})$	iL
Time reversal	$d\mathbf{\Gamma}$	H	$\rho_0(\mathbf{\Gamma})$	$-iL$

Table 2. The Signatures of **g** *under Different Transformations*

Symmetry transformation	Signature	$g_x(q)$	$g_y(q)$	$g_z(q)$	$n(q)$
Parity	ε	-1	-1	-1	1
Reflection	α^{xz}	1	-1	1	1
	α^{yz}	-1	1	1	1
	α^{xy}	1	1	-1	1
Time reversal	γ	-1	-1	-1	1

The various single-molecule properties a_μ have different transformation properties, whereas the phase factors e^{iqz_j} are always invariant to reflection through the xz and yz planes but transform to e^{-iqz_j} under inversion and reflection through the xy plane and to $e^{iqa_z}e^{iqz_j}$ under translation **a**.

In the following the single-molecule properties a_μ^j are assumed to be invariant to translation and to have definite symmetry with respect to the other transformations, so that we can write

$$A_\mu(q) \to \varepsilon_\mu A_\mu(-q) \qquad \text{parity} \tag{75}$$

$$A_\mu(q) \to \alpha_\mu^{xz} A_\mu(q) \qquad \text{reflection through } xz \text{ plane} \tag{76}$$

$$A_\mu(q) \to \alpha_\mu^{yz} A_\mu(q) \qquad \text{reflection through } yz \text{ plane} \tag{77}$$

$$A_\mu(q) \to \alpha_\mu^{xy} A_\mu(-q) \qquad \text{reflection through } xy \text{ plane} \tag{78}$$

$$A_\mu(q) \to \gamma_\mu A_\mu(q) \qquad \text{time reversal symmetry} \tag{79}$$

$$A_\mu(q) = e^{iqa_z} A_\mu(q) \qquad \text{translation} \tag{80}$$

Here ε_μ, α_μ^{xz}, α_μ^{yz}, α_μ^{xy}, γ_μ are the signatures of A_μ under the indicated transformations. These signatures are either $+1$ or -1 depending on the symmetry of α_μ^j under the transformation. For $n(q)$, for example, all of these signatures are $+1$. Table 2 shows the various signatures for n, g_x, g_y, and g_z — the different components of the momentum density.

These symmetries allow us to make some definite statements about various correlation functions. For example, consider the two properties $A_\mu(q)$ and $A_\nu(q')$ corresponding to wave numbers q and q'. It is clear from the foregoing that the cross correlation function is

$$(A_\mu(q, t), A_\nu^*(q')) = \int d\mathbf{\Gamma}\, \rho_0(\mathbf{\Gamma})[e^{iLt}A_\mu(q, t)]A_\nu^*(q') \tag{81}$$

These properties transform under translation **a** to

$$A_\mu(q, t) \to e^{iqa_z} A_\mu(q, t), \qquad A_\nu(q') \to e^{+iq'a_z} A_\nu(q')$$

Since L, H, $\rho_0(\boldsymbol{\Gamma})$, and $d\boldsymbol{\Gamma}$ are unchanged, the transformed integral gives

$$(A_\mu(q, t), A_\nu^*(q')) = e^{i(q-q')a_z}(A_\mu(q, t), A_\nu^*(q')) \tag{82}$$

This relationship must be satisfied for arbitrary translations $\mathbf{a}$ of the coordinate system. Thus if $q \neq q'$ the correlation function is identically zero. This shows that $\langle A_\mu(q, t)A_\nu^*(q', 0)\rangle$ is zero unless $q = q'$. Translational symmetry therefore implies that, for homogeneous systems,

$$(A_\mu(\mathbf{q}, t), A_\nu^*(\mathbf{g}')) = (A_\mu(q, t), A_\nu(q'))\delta_{q',q} \tag{83}$$

This means that different wave vector components of the "fluctuations" are uncorrelated. In solids the theorem is valid if we restrict attention to vectors in the same Brillouin zone.

Let us consider the reflection through the yz plane. The correlation functions

$$(A_\mu(q, t), A_\nu^*(q)) = \int d\boldsymbol{\Gamma}\,[e^{iLt}A_\mu(q)]A_\nu^*(q)$$

transform to

$$(A_\mu(q, t), A_\nu^*(q)) = \alpha_\mu^{yz}\alpha_\nu^{yz}(A_\mu(q, t), A_\nu^*(q)) \tag{84}$$

under this reflection because $d\boldsymbol{\Gamma}$, $\rho_0(\boldsymbol{\Gamma})$, and e^{iLt} are invariant and A_μ and A_ν transform according to Eq. (77). Now if A_μ and A_ν have different reflection symmetries with respect to the yz plane, $\alpha_\mu^{yz}\alpha_\nu^{yz} = -1$ and Eq. (84) implies that the correlation function must be zero. Likewise for reflection through the xz plane. The important consequence of this is that dynamic properties with different reflection symmetries are totally uncorrelated. For example, g_x, g_y, and g_z are uncorrelated.

In like manner we can show that the frequencies defined in Eq. (50), the force correlation functions defined in Eq. (51), and the susceptibilities defined in Eq. (47) are all zero if the two properties have different reflection symmetry through either the xz or yz plane.

This means that our set of variables can be separated into uncoupled subsets according to whether they have even or odd reflection symmetry through the xz and yz planes.

The situation is more complicated with respect to symmetry through the xy plane. Then the correlation functions transform to

$$C_{\mu\nu}(q, t) = (A_\mu(q, t), A_\nu^*(q)) = \alpha_\mu^{xy}\alpha_\nu^{xy}C_{\mu\nu}(-q, t) \tag{85}$$

Then if A_μ and A_ν have different symmetries under reflection through this plane, $C_{\mu\nu}(q, t)$ is an odd function of q, whereas if they have the same symmetry, $C_{\mu\nu}(q, t)$ is an even function of q. In the former case, $\lim_{q\to 0}C_{\mu\nu}(q, t) = 0$. The same is true of $\Omega_{\mu\nu}$ and (F_μ, F_ν^*). From Eq. (85) we see that autocorrelation functions $C_{\mu\mu}(q, t)$ and $(F_\mu(t), F_\mu(0))$ and correspondingly $\chi_{\mu\mu}$ must be even functions of q. Considerations of parity give the same conclusions.

Time reversal symmetry plays a very important role in fluctuation dynamics. Under time reversal we note that

$$C_{\mu\nu}(t) = (e^{iLt}A_\mu(q), A_\nu^*(q))$$

becomes

$$C_{\mu\nu}(t) = \gamma_\mu\gamma_\nu(e^{-iLt}A_\mu(q), A_\nu^*(q)) = \gamma_\mu\gamma_\nu C_{\mu\nu}(-t) \tag{86}$$

This means that if A_μ and A_ν have different time reversal symmetry $C_{\mu\nu}(t)$ is an odd function of the time, whereas if they have the same symmetry $C_{\mu\nu}(t)$ is an even function of t. The same is true of the force matrix. In addition,

$$C_{\mu\nu}(0) = \langle A_\mu^* A_\nu\rangle = \gamma_\mu\gamma_\nu\langle A_\mu A_\nu^*\rangle \tag{87}$$

so that $\chi_{\mu\nu} = 0$ if A_μ and A_ν have different time reversal symmetries. From Eq. (86) we see that autocorrelation functions $C_{\mu\mu}(t)$ must be even functions of time. Likewise for $(F_\mu(t), F_\mu^*(0))$.

The quantities $(iLA_\mu(q), A_\nu^*(q))$ that appear in the frequencies [Eq. (50)] transform under time reversal to

$$(iLA_\mu(q), A_\nu^*(q)) = -\gamma_\mu\gamma_\nu(iLA_\mu(q), A_\nu^*(q)) \tag{88}$$

because $iL \to -iL$. These quantities are nonzero only if A_μ and A_ν have opposite time reversal symmetry. It follows from this and the fact that $\chi_{\mu\nu} = 0$ if $\gamma_\mu = \gamma_\nu$, that $\Omega_{\mu\nu}$ only couples properties of opposite time reversal symmetry.

These symmetry selection rules are summarized in Table 3, where the term nonzero or correlated is used if there is no symmetry restriction. Some of these functions could be zero for reasons having nothing to do with symmetry. The term zero or uncorrelated is used only if the functions are required by symmetry to be zero. In addition, we can state that for autocorrelation functions,

$$C_{\mu\mu}(-q, t) = C_{\mu\mu}(q, t) \tag{89a}$$

$$C_{\mu\mu}^*(q, t) = C_{\mu\mu}(q, t) = C_{\mu\mu}(q, -t) \tag{89b}$$

$$\chi_{\mu\mu}(q) = \chi_{\mu\mu}(-q) = \chi_{\mu\mu}^*(q) \tag{89c}$$

$$\Omega_{\mu\mu}(q) = 0 \tag{89d}$$

and

$$L_{\mu\mu}^*(q, t) = L_{\mu\mu}(-q, t) \tag{90a}$$

$$L_{\mu\mu}^*(q, t) = L_{\mu\mu}(q, t) = L_{\mu\mu}(q, -t) \tag{90b}$$

This means that $C_{\mu\mu}(q, t)$ and $L_{\mu\mu}(q, t)$ are real, even functions of q and t.

Table 3. Summary of Symmetry Selection Rules

Quantity	Relative symmetry of A_μ and A_ν	Reflection through xz or yz plane	Reflection through xy plane	Time reversal
$C_{\mu\nu}(q, t)$	Same	Correlated	$C_{\mu\nu}(q, t) = C_{\mu\nu}(-q, t)$	$C_{\mu\nu}(q, t) = C_{\mu\nu}(q, -t)$
	Different	Uncorrelated	$C_{\mu\nu}(q, t) = -C_{\mu\nu}(-q, t)$	$C_{\mu\nu}(q, t) = -C_{\mu\nu}(q, -t)$
$L_{\mu\nu}(q, \tau) = (F_\mu(q, t), F_\nu^*(q))_Q$	Same	Correlated	$L_{\mu\nu}(q, t) = L_{\mu\nu}(-q, t)$	$L_{\mu\nu}(q, t) = L_{\mu\nu}(q, -t)$
	Different	Uncorrelated	$L_{\mu\nu}(q, t) = -L_{\mu\nu}(-q, t)$	$L_{\mu\nu}(q, t) = -L_{\mu\nu}(q, -t)$
$\chi_{\mu\nu}(q)$	Same	Nonzero	$\chi_{\mu\nu}(q) = \psi_{\mu\nu}(-q)$	Correlated
	Different	Zero	$\chi_{\mu\nu}(q) = -\chi_{\mu\nu}(-q)$	Uncorrelated
$\omega_{\mu\nu}(q) = (iLA_\mu(q), A_\nu^*(q))$	Same	Nonzero	$\omega_{\mu\nu}(q) = \omega_{\mu\nu}(-q)$	Zero
	Different	Zero	$\omega_{\mu\nu}(q) = -\omega_{\mu\nu}(-q)$	Nonzero

5. Navier–Stokes Equations

The foregoing formalism is easily applied to the derivation of phenomenological equations and to the identification of the transport coefficients with well-defined statistical averages. In this section we consider the densities $n(g)$ and $g_x(q)$, $g_y(q)$, and $g_z(q)$ defined in Eqs. (74a) and (74b). A glance at Table 2 shows that with respect to reflections in the xz, yz, and xy planes these four variables can be subdivided into three sets $(n(g), g_z(q))$, $g_x(q)$, and $g_y(q)$ such that each set has the same reflection symmetry.

According to the arguments presented in Section 4 it follows that the relaxation equations for these four variables consist of two coupled equations for $n(g)$ and $g_z(q)$ and two totally uncoupled equations for $g_x(q)$ and $g_y(q)$. This breakup is a consequence of our choice of coordinate axes such that the wave vector $\mathbf{q}$ defines the z axis. $g_z(q)$ is the momentum density parallel to $\mathbf{q}$, namely the longitudinal momentum, and $g_x(q)$ and $g_y(q)$ are the momentum densities perpendicular to $\mathbf{q}$, namely the transverse components of the momentum density. By symmetry, $g_z(q)$, $g_x(q)$, and $g_y(q)$ are independent, and furthermore $g_x(q)$ and $g_y(q)$ should be dynamically equivalent. Let us therefore only treat $g_x(q)$.

The instantaneous rates of change of these variables are explicitly

$$iLn(q) = iqm^{-1} \sum_j p_{zj} e^{iqz_j} = iqm^{-1} g_z(q) \tag{91a}$$

$$iLg_x(q) = \sum_j [F_{xj} + iqm^{-1} p_{xj} p_{zj}] e^{iqz_j} \tag{91b}$$

$$iLg_y(q) = \sum_j [F_{yj} + iqm^{-1} p_{yj} p_{zj}] e^{iqz_j} \tag{91c}$$

$$iLg_z(q) = \sum_j (F_{zj} + iqm^{-1} p_{zj} p_{zj}) e^{iqz_j} \tag{91d}$$

where F_{xj}, etc., are the various components of the total force exerted on particle j by the rest of the fluid. It is clear that in the absence of external forces,

$$\sum_{j=1}^{N} F_{jx} = \sum_{j=1}^{N} F_{jy} = \sum_{j=1}^{N} F_{jz} = 0 \tag{92}$$

When Eq. (92) is combined with our expansion of Eq. (91) in powers of q, it follows that

$$iLG_x(q) = iq \sum_j (z_j F_{jx} + m^{-1} p_{xj} p_{zj}) + O(q^2) \tag{93a}$$

$$iLg_y(q) = iq \sum_j (z_j F_{jy} + m^{-1} p_{xj} p_{zj}) + O(q^2) \tag{93b}$$

$$iLg_z(q) = iq \sum_i (z_j F_{jz} + m^{-1} p_{zj} p_{zj}) + O(q^2) \tag{93c}$$

The rates of change $\dot{g}_x(q)$, $\dot{g}_y(q)$, and $\dot{g}_z(q)$ are thus seen to vary as the first power of q for small q and moreover to go to zero as $q \to 0$. One immediate consequence of this is that fluctuations in $\mathbf{g}(q)$ become longer and longer lived as $q \to 0$, and moreover that the set of variables (n, g_x, g_y, g_z) constitute a slow set of variables. Thus we surmise that we can use Eq. (60). Before applying this equation it is worthwhile to note that Eq. (25) can be rearranged to give

$$e^{iQLt} = e^{iLt} - \int_0^t d\tau\, e^{iL(t-\tau)} iPLe^{iQL\tau} \tag{94}$$

If P projects onto any of the variables in Eq. (74) then

$$PiLB = (iLB, A^*) = -(B, (iLA)^*) \tag{95}$$

where B is an arbitrary property, and the Hermitian property of L is used. According to Eqs. (91) and (93), iLA is at least first order in q so that $PiLB$ and consequently the integral in Eq. (94) are at least first order in q. It follows that

$$\lim_{q\to 0} e^{iQL\tau} = e^{iL\tau} \tag{96}$$

First we treat the transverse components and then we treat the longitudinal components. The evolution of the single variable $g_x(q)$ is described by Eq. (60) with $A \equiv g_x(q)$. It follows from Eq. (89d) that in this equation $\Omega = 0$. Substituting the explicit form of $g_x(q)$ from Eq. (74b) into the definition of the scalar product defined in Eq. (15) gives

$$(g_x(g), g_x^*(q)) = \left\langle \sum_{j,k=1}^{N} p_{xj} p_{xk} e^{iq(z_j - z_k)} \right\rangle = nm\beta^{-1} \tag{97}$$

where the last equality follows from the fact that $\langle p_{xj}^2 \rangle = mkT$, and from the statistical independence of the momenta on different molecules (in a canonical ensemble). Thus of the $N(N-1)$ terms in the double sum only N survive the averaging process. The random force is, to first order in q,

$$f_x = (1-P)iLg_x(q) = iq\tau_{zx} \tag{98}$$

where we have used the fact that $PiLg_x(q) = (iLg_x, g_x^*) = 0$, have substituted Eq. (93a) for iLg_x, and have defined

$$\tau_{zx} = \sum_{j=1}^{N} (z_j F_{xj} + m^{-1} p_{xj} p_{zj}) \tag{99}$$

It follows from Eqs. (96) and (98) that, to lowest order in q,

$$F_x(\tau) = e^{iQL\tau} f_x = iqe^{iL\tau}\tau_{xz} + O(q^2) \tag{100}$$

For the force correlation function, this gives

$$(F_x(\tau), F_x^*(0)) = q^2 \langle \tau_{xz}(0) \tau_{xz}(\tau) \rangle + O(q^2) \tag{101}$$

Substitution of Eqs. (97) and (101) into Eq. (37) then gives the memory function as

$$K_x(\tau) = \beta q^2 \frac{\langle \tau_{zx}(0)\tau_{xz}(\tau)\rangle}{Nm} + O(q^2) \tag{102}$$

The damping coefficient follows directly from Eq. (63),

$$\Gamma_x(q) = q^2\left[\beta\int_0^\infty d\tau \frac{\langle \tau_{xz}(0)\tau_{xz}(\tau)\rangle}{Nm}\right] \tag{103}$$

The term in the brackets has the dimensions of a diffusion coefficient (cm^2/sec) and describes the diffusion of transverse momentum.

The Langevin equation for the transverse momentum for small values of q is consequently

$$\partial g_x(q)/\partial t = -\Gamma_x(q)g_x(q) + F_x \tag{104}$$

Similarly the autocorrelation function of transverse momentum

$$C_\perp(\tau) = \langle g_x(\tau)g_x(0)\rangle \tag{105}$$

evolves in time according to Eq. (39),

$$\partial C_\perp(\tau)/\partial\tau = -\Gamma_x(q)C_\perp(\tau) \tag{106}$$

Solving this subject to the initial value $C_\perp(0) = \langle g_x^2\rangle = \beta^{-1}Nm$ gives

$$C_\perp(q, \tau) = \beta^{-1}Nme^{-\Gamma_x(q)\tau} \tag{107}$$

If the phenomenological linearized Navier–Stokes equations are used instead, one finds

$$C_\perp(q, \tau) = \beta^{-1}Nme^{-q^2(\eta_s/mn)t} \tag{108}$$

where η_s is the shear viscosity and mn the mass density. Comparing the decay rate Γ_x given by Eq. (107) with the hydrodynamic decay rate $q^2\eta_s/mn$ allows us to write the formula

$$\eta_s = \frac{mn\Gamma_x(g)}{q^2} = \frac{1}{Vk_BT}\int_0^\infty d\tau\langle \tau_{xz}(0)\tau_{xz}(\tau)\rangle \tag{109}$$

Thus we see that the shear viscosity of phenomenological hydrodynamics is directly proportional to the area under the ordinary time correlation function of τ_{xz} and inversely proportional to Vk_BT, where V is the volume of the system. Equation (109) is called a Green–Kubo relation. Such relations can always be derived for transport coefficients and are useful in conjunction with molecular dynamics calculations in the determination of these transport coefficients. Projection operator techniques can thus be used to derive Green–Kubo relations although these relations were first derived in other ways. We have thus far shown that in the small-q limit transverse momentum decays monotonically, that is, it is a diffusive mode with a diffusion coefficient $\Gamma_x(q)/q^2$, which can be determined by a Green–Kubo formula. This diffusion coefficient can be identified with the kinematic viscosity η_s/mn.

Because $n(q)$ and $g_z(q)$ have the same reflection symmetry it is necessary to use the coupled relaxation equations given by Eq. (49) with

$$\mathbf{A} = \begin{pmatrix} n(q) \\ g_z(q) \end{pmatrix} \tag{110}$$

The susceptibility matrix $\boldsymbol{\chi}$ is then

$$\boldsymbol{\chi} = \beta^{-1}\begin{pmatrix} S(q) & 0 \\ 0 & \langle |g_z|^2\rangle \end{pmatrix} \tag{111}$$

where

$$S(q) \equiv \langle |n(q)|^2\rangle \tag{112}$$

is the structure factor and $\langle |g_z|^2\rangle = \beta^{-1}Mn$. The off-diagonal terms are zero because $n(q)$ and $g_z(q)$ have different time reversal symmetries.

Now the frequency matrix can be computed from

$$(iL\mathbf{A}, \mathbf{A}^+) = \begin{pmatrix} 0 & (\dot{n}, g_z^*) \\ (\dot{g}_z, n^*) & 0 \end{pmatrix} \tag{113}$$

where the diagonal terms must be zero because of time reversal symmetry [cf. Eq. (88)].

Because L is Hermitian we find that

$$(\dot{g}_z, n^*) = (\dot{n}, g_n^*) = iq\langle |g_z|^2\rangle \tag{114}$$

and consequently Eq. (50) becomes

$$i\mathbf{\Omega} = \begin{pmatrix} 0 & iq \\ \dfrac{iq\langle |g_z|^2\rangle}{S(q)} & 0 \end{pmatrix} \tag{115}$$

Now note that

$$iL\mathbf{A} = \begin{pmatrix} \dot{n}(q) \\ \dot{g}_z(q) \end{pmatrix} = \begin{pmatrix} iqg_z(q) \\ \dot{g}_z(q) \end{pmatrix} \tag{116}$$

Since q projects onto the space orthogonal to $\mathbf{A}$, it follows that $Qg_z(q) = 0$ and $Q\dot{g}_z(q) = \dot{g}_z(q)$. Thus

$$\mathbf{f}_z = QiL\mathbf{A} = \begin{pmatrix} 0 \\ \dot{g}_z(q) \end{pmatrix} \tag{117}$$

This means that the memory function matrix is [cf. Eq. (51)]

$$K_z(\tau) = \begin{pmatrix} 0 & 0 \\ 0 & \dfrac{(e^{iQL\tau}\dot{g}_z(q), \dot{g}_z^*(q))}{\langle |g_z(q)|^2\rangle} \end{pmatrix} \tag{118}$$

Now in the low q limit we have already seen that $e^{iQL\tau} \to e^{iL\tau}$ and

$$\dot{g}_z(q) = iq\tau_{zz} + O(q^2) \tag{119}$$

where

$$\tau_{zz} \equiv \sum_{j=1}^{N} (z_j F_{zj} + m^{-1} p_{zj} p_{zj}) \tag{120}$$

Then to lowest order in q the memory function matrix has only one nonvanishing element,

$$K_{22}(\tau) = \beta q^2 \langle \tau_{zz}(0)\tau_{zz}(\tau)\rangle / Nm \tag{121}$$

and the damping matrix is

$$\mathbf{\Gamma} = \begin{pmatrix} 0 & 0 \\ 0 & \Gamma_{zz}(q) \end{pmatrix} \tag{122}$$

where

$$\Gamma_{zz}(q) = q^2 \left[\beta \int_0^\infty d\tau \frac{\langle \tau_{zz}(0)\tau_{zz}(\tau)\rangle}{Nm} \right] \tag{123}$$

To be consistent, all the elements in the matrices $\boldsymbol{\chi}$ and $\boldsymbol{\Omega}$ should be expanded to lowest order in q. The only element for which more need be said is $\chi_{\|}(q) = \beta^{-1} S(q)$. From thermodynamic fluctuation theory,

$$\lim_{q\to 0} S(q) \equiv S(0) = \beta^{-1} N n \chi_{\mathrm{T}} \tag{124}$$

where χ_{T} is the isothermal compressibility. Then combining Eqs. (60), (111), and (115) gives the two coupled equations

$$\begin{aligned} \frac{\partial n(qt)}{\partial t} &= iqg_z(qt) \\ \frac{\partial g_z(q,t)}{\partial t} &= -\Gamma_{zz}(q) g_z(q,t) + \frac{iq\langle |g_z|^2\rangle}{S(0)} n(q,t) + f_z(q,t) \end{aligned} \tag{125}$$

Equation (125) can be solved for

$$S(q,\tau) \equiv \langle n(q,\tau) n^*(q,0)\rangle \tag{126a}$$

and

$$C_{\|}(q,\tau) \equiv \langle g_z(q,\tau) g_z^*(q,0)\rangle \tag{126b}$$

To lowest order in q, the solution of Eq. (125) gives

$$S(q,\tau)=S(q)e^{-\frac{1}{2}\Gamma_{zz}(q)\tau}\left(\cos\omega_s(q)\tau+\frac{\Gamma_{zz}(q)}{2\omega_s(q)}\sin\omega_s(q)\tau\right) \tag{127a}$$

$$C_{\|}(q,\tau)=\beta^{-1}Nme^{-\frac{1}{2}\Gamma_{zz}(q)\tau}\left(\cos\omega_s(q,\tau)-\frac{\Gamma_{zz}(q)}{2\omega_s(q)}\sin\omega_s(q)\tau\right) \tag{127b}$$

where the frequency $\omega_s(q)$ is

$$\omega_s(q)\equiv q(m/n\chi_T)^{1/2} \tag{127c}$$

and the quantity $\Gamma_{zz}(q)/2\omega_s(q)$ is obviously first order in q.

Thus the longitudinal momentum propagates with propagation frequency $\omega_s(q)$ and damping constant $\Gamma_{zz}(q)/2$. Solution of the phenomenological Navier–Stokes equations gives the same behavior with

$$\omega_s(q)=c_T q \tag{128a}$$

$$\tfrac{1}{2}\Gamma_{zz}(q)=(\eta_v+\tfrac{4}{3}\eta_s)/mn \tag{128b}$$

where c_T is the isothermal speed of sound, η_v the volume viscosity, and η_s is the shear viscosity. Comparing Eqs. (123), (127), and (128) allows us to identify

$$c_T=(m/n\chi_T)^{1/2} \tag{129a}$$

$$\left(\eta_v+\frac{4}{3}\eta_s\right)=\frac{1}{2Vk_BT}\int_0^\infty d\tau\langle\tau_{zz}(\tau)\tau_{zz}(0)\rangle \tag{129b}$$

If energy conservation is included in this scheme, the full set of linearized fluid dynamic equations are obtained.

6. Conclusions

Clearly then, projection operator techniques are useful for the derivation of relaxation equations and for the derivation of formulas for transport coefficients in terms of microscopic properties. The various symmetry properties discussed here are useful for reducing a set of dynamical variables into subsets that are statistically independent of each other.

This formalism is particularly valuable for the analysis of fluctuations for which there is no existing phenomenology. Noteworthy in this regard is the treatment of the coupling between molecular reorientations and the transverse momentum of the fluid. Experiment has revealed that the depolarized component of scattered light has a well-diffused splitting that varies with scattering angle. Projection operator techniques have been very useful in describing this splitting.

The substance of this chapter has been covered elsewhere. Because Chapters 4 and 6 in this volume depend on this material, it was felt that a short chapter pointing out the main features of the formalism would be useful.

References

1. B. J. Berne and D. Forster, *Ann. Rev. Phys. Chem.* **22**, 563 (1971).
2. B. J. Berne and R. Pecora, *Dynamic Light Scattering*, Chaps. 11 and 12, Wiley, New York (1976).
3. J. Hynes and J. Deutch, *in*: *Physical Chemistry* (H. Eyring, D. Henderson, and W. Jost, eds.), Academic Press, New York (1975).
4. D. Forster, *Fluctuations, Broken Symmetries, and Time Correlation Functions*, Addison-Wesley, Reading, Massachusetts (1976).
5. R. Gordon, *Adv. Mag. Res.* **3**, 1 (1968).
6. W. Steele, *in*: *Transport Phenomena in Fluids* (H. J. M. Hanley, ed.), Chap. 8, Marcel Dekker, New York (1969).
7. B. J. Berne, *in*: *Physical Chemistry* (H. Eyring, D. Henderson, and W. Jost, eds.), Academic Press, New York (1971).
8. H. Mori, *Progr. Theor. Phys.* **33**, 423 (1965).
9. R. Kubo, *in*: *Many Body Theory* (R. Kubo, ed.), Benjamin, New York (1966).
10. R. Zwanzig, *Lectures on Theoretical Physics*, Vol. III, p. 135, Wiley (Interscience), New York (1961).
11. R. Zwanzig, *Ann. Rev. Phys. Chem.* **16**, 67 (1965).
12. H. L. Swinney, *in*: *Photon Correlation and Light Beating Spectroscopy* (H. F. Cummins and E. R. Pike, eds.), Plenum, New York (1973).
13. L. Landau and E. Lifshitz, *Fluid Mechanics*, Addison-Wesley, Reading, Massachusetts (1961).

6

Principles of Mode–Mode Coupling Theory

T. Keyes

1. Introduction

A very large class of experiments performed in large systems may be divided into two categories: the observation of nonequilibrium ensemble averages and the observation of equilibrium time-dependent correlation functions. That many experiments probe a nonequilibrium ensemble average is fairly obvious. Consider an arbitrary dynamical variable A, a function of the phase of the system. The equilibrium ensemble average, denoted by $\langle\ \rangle$, of all A discussed here is zero. If we perturb the system, the nonequilibrium ensemble average of A, denoted by $\bar{A}$, may become nonzero. We might then measure the steady state value of $\bar{A}$ as a function of the strength of the perturbation, as in an electrical conductivity measurement ($A \equiv$ current), or we might wait until $\bar{A}$ reaches its steady-state value, turn off the perturbation at a time defined as $t = 0$, and observe the way $\overline{A(t)}$ decays to equilibrium, $\langle A(t)\rangle = 0$. An example of the second type of experiment is the decay of turbulence, where $\overline{A(t)}$ is equivalent to a time-dependent velocity field.

The time correlation functions that we shall discuss are objects of the form $\langle A(t)B^*(0)\rangle$, where an asterisk indicates a complex conjugate; if $A = B$, we refer to a time autocorrelation function. It is often useful to express $A(t)$ in terms of the classical propagator,

$$A(t) = e^{i\mathscr{L}t}A \tag{1}$$

where $\mathscr{L}$ is the classical Liouville operator. The importance of time correlation functions in modern experiments is not immediately obvious. However, given

T. Keyes • Department of Chemistry, Yale University, New Haven, Connecticut

that all spectroscopy and scattering arises from processes that perturb an electromagnetic field, and given that instruments are sensitive to the squared amplitude of the perturbed field averaged over a time on the order of the instrument time constant, it is straightforward to show[1] that the results of all scattering and spectroscopy experiments may be formulated in terms of time correlation functions.

In 1931, Onsager[2] showed that an intimate relation exists between the time correlation functions and the dynamics of the nonequilibrium ensemble averages. Assume that some dynamical variable A has a nonzero average at time zero, at which time constraints are removed and the system begins to return to equilibrium; then

$$\overline{A(t)}/\overline{A(0)} = \langle A(t)A(0)\rangle / \langle A(0)A(0)\rangle \tag{2}$$

in the limit that $\overline{A(0)}$ is very small. Thus, in many discussions, time correlation functions and the nonequilibrium ensemble averages are interchangeable. A tremendous amount of work has been expended in attempts to find the dynamical laws that these quantities obey.

Various phenomenological, linear laws for the time dependence of the nonequilibrium ensemble averages existed before any systematic theory of such laws was developed, Fourier's heat flow law, Fick's law, and the hydrodynamic equations are examples.[3] By linear laws we mean laws of the form

$$\frac{\partial}{\partial t}\overline{\mathbf{A}(t)} = \mathsf{F} \cdot \bar{\mathbf{A}} \tag{3}$$

where F is a matrix of constants and $\mathbf{A}$ denotes a vector whose components are the several dynamical variables of interest. We shall not write vector and matrix symbols for this vector space of dynamical variables in most cases; the number of sub- and superscripts in some of our equations shall be formidable, and it seems best to omit tensor notation unless it is essential. It is surely reasonable that linear laws should hold close to equilibrium, and thus the linear laws should also apply to appropriate time correlation functions. The first systematic version of the linear laws was "irreversible thermodynamics,"[4] according to which

$$F = [i\Omega + L]\chi^{-1} \tag{4}$$

where $i\Omega$ and χ are matrices of equal-time correlation functions,

$$i\Omega_{ij} = \langle \dot{A}^i A^{j*}\rangle \tag{5}$$

$$\chi_{ij} = \langle A^i A^{j*}\rangle \tag{6}$$

and $\dot{A}^i$, the time derivative of A^i, is given by

$$\dot{A}^i = i\mathscr{L}A^i \tag{7}$$

The elements of the matrix L are the "Onsager coefficients." We introduce a matrix of decay rates Γ,

$$\Gamma \equiv L\chi^{-1} \tag{8}$$

and a matrix of frequencies $i\omega$,

$$i\omega \equiv i\Omega\chi^{-1} \tag{9}$$

The presence of $i\omega$ in Eq. (3) gives rise to oscillatory, nondissipative dynamics for $\overline{A(t)}$, while Γ gives rise to decay and dissipation. Since $i\Omega$, χ, and $i\omega$ are composed entirely of equal-time correlation functions, these quantities are, in principle, available from equilibrium statistical mechanics. On the other hand, irreversible thermodynamics gives no expression for L, aside from certain symmetry requirements, the "Onsager relations."[2]

Modern research has concentrated on elucidating the nature of L. If one has an equation representative of Eq. (3), questions abound concerning L, questions more subtle than the obvious "what is it?" Equation (3) is irreversible in time, while any exact equation for $\overline{A(t)}$ must be reversible. This dilemma could not be resolved with irreversible thermodynamics. Also, consider that Eq. (3) may represent a set of coupled, linear differential equations. Suppose a certain set of variables $\mathbf{A}$ must be taken together to obtain proper dynamics. Then, if one tried to apply the principles of irreversible thermodynamics with a set smaller than the true set, an incorrect dynamical law would result no matter what was used for L. How was one to know the correct set of $\mathbf{A}$? Finally, the obvious question does occur: Given that it is reasonable to write an irreversible equation and given that one has a proper set of variables, how does one find an expression for L?

Considerable progress in dealing with the questions just mentioned was made through projection operator methods, originally introduced by Green,[5] Kubo,[6] and Zwanzig.[7] Probably the most useful projection operator formalism for the writing of linear laws is that of Mori.[8] With Mori's method, one utilizes a formally exact rewriting of the exact equation of motion, Eq. (1), to obtain

$$\frac{\partial}{\partial t}A(t) = i\omega A(t) - \int_0^t K(t-\tau)\chi^{-1}A(\tau)\,d\tau + f(t) \tag{10}$$

where

$$K_{ij}(t) = \langle f^i(t) f^{j*}\rangle \tag{11}$$

$$f^i(t) = e^{(1-P)i\mathscr{L}t}(1-P)\dot{A}^i \tag{12}$$

and P is the Mori projection operator, defined such that for an arbitrary set of dynamical variables $\mathbf{g}$,

$$Pg = \langle gA^*\rangle\chi^{-1}A \tag{13}$$

The appearance of $(1-P)i\mathscr{L}$ in Eq. (12) ensures that, for all time, $\langle f(t)A^*\rangle = 0$. Equation (10) may be multiplied by A^* and averaged to obtain

$$\frac{\partial}{\partial t}\langle A(t)A^*\rangle = i\omega\langle A(t)A^*\rangle - \int_0^t K(t-\tau)\chi^{-1}\langle A(\tau)A^*\rangle d\tau \tag{14}$$

where we have used $\langle f(t)A^*\rangle = 0$. Equation (10) may also be averaged over a nonequilibrium ensemble. Near equilibrium, it is possible to show that $\overline{f(t)}$ vanishes, Thus,

$$\frac{\partial}{\partial t}\overline{A(t)} = i\omega\overline{A(t)} - \int_0^t K(t-\tau)\chi^{-1}\,\overline{A(\tau)}\,d\tau \tag{15}$$

Equations (14) and (15), derived by Mori's method, constitute a proof of Onsager's[2] contention that the dynamics of $\overline{A(t)}$ are identical to those of $\langle A(t)A^*\rangle$ near equilibrium.

The "orthogonality" of $f(t)$ to $A[\langle f(t)A^*\rangle = 0]$ has been[5-9] used as the guiding idea in the resolution of the problems concerning the linear laws. For, if $\langle f(t)A^*\rangle = 0$, we may in some sense state that $f(t)$ has none of the "character" of the time dependence of A. Suppose[5-9] that we can identify some dynamical variables that fluctuate in time much more slowly than any other conceivable dynamical variables. If we chose this set as the set **A** in the Mori formalism, then $\langle f(t)f^*\rangle$ may be thought of as having no "slowly fluctuating" character on the time scale of the variations in $\bar{A}$. This idea is embodied in the usual name for $f(t)$, the "random force." For the characteristic times of the variation in $\bar{A}$, $\langle f(t)f^*\rangle$ will have decayed to zero, and we may approximate

$$\int_0^t K(t-\tau)\bar{A}(\tau)\,d\tau \to \left[\int_0^\infty K(\tau)\,d\tau\right]\bar{A}(t)$$

If this fairly nebulous argument is correct, Eq. (15) yields the "linear laws," and the conditions for their validity. The laws are understood to be correct only for times longer than those over which $\langle f(t)f^*\rangle$ decays, in which case the reversibility of the true laws presents no conflict. It is seen that the linear laws can only be correct if all the "slowly fluctuating variables" are included in the set **A**; if a slowly varying variable is missing, $f(t)$ might have some projection on that variable, in which case the convolution form of Eq. (15) would have to be retained. Finally, an expression for L is obtained,[5-9]

$$L_{ij} = \int_0^\infty dt\,\langle f^i(t)f^{j*}\rangle = \int_0^\infty dt\,K_{ij}(t) \tag{16}$$

Most current attempts to evaluate decay rate coefficients [Eq. (8)] start with Eq. (16), which is often referred to as the Green–Kubo relation.

Thus, before the events that made mode–mode coupling of such interest, it was widely believed[5-9] that in order to obtain well-behaved linear laws for a given problem one need merely identify the "slowly varying variables," apply Eq. (15), and assume $K(t)$ is rapidly decaying. Of course, the choice of the slowly varying variables was not always obvious; indeed, there is no *a priori* reason that certain variables should fluctuate in a distinctly different way from all other dynamical variables for every system. However, a rigorous isolation of slowly varying variables may be made if[(5–11)] the system Hamiltonian possesses some constants of the motion. An α-particle constant of the motion is a molecular quantity $C_i(t)$ associated with the ith molecule, such that $\sum_{i=1}^{N_\alpha} C_i(t) = \text{const}$, where N_α is the number of particles of type α. The density of such a quantity is written

$$C(\mathbf{r}, t) = \sum_{i=1}^{N_\alpha} C_i(t)\, \delta(\mathbf{r} - \mathbf{r}_i(t)) \tag{17}$$

where $\mathbf{r}$ is a point in the system and $\mathbf{r}_i(t)$ is the position of the center of mass of the ith particle at time t. We use boldface to denote three-dimensional vectors. It is convenient to work with the spatial Fourier transform of $C(\mathbf{r}, t)$,

$$C_{\mathbf{k}}(t) = \sum_{i=1}^{N_\alpha} C_i(t) e^{i\mathbf{k}\cdot\mathbf{r}_i} \tag{18}$$

where $\mathbf{k}$ is the wave vector (the transform variable). Note that $C^*_{\mathbf{k}} = C_{-\mathbf{k}}$. For zero k, $C_{\mathbf{k}}(t) = \text{const}$ by the conservation law; thus, by making k as small as necessary, we may obtain a $C_{\mathbf{k}}$ that fluctuates as slowly as desired. This fact is embodied in the usual definition, $\dot{C}_{\mathbf{k}} = i\mathbf{k} \cdot \mathbf{j}_{\mathbf{k}}$, where $\mathbf{j}_{\mathbf{k}}$, the "flux," is nonconserved. For $k \neq 0$, $\langle C_{\mathbf{k}} \rangle = 0$, but $\langle C_{\mathbf{k}} \rangle = \sum_i C_i = \text{const}$ for $k = 0$. For $k = 0$, it is understood that $C_{\mathbf{k}}$ means $C_{\mathbf{k}} - \langle C_{\mathbf{k}} \rangle$.

Now there are an infinite number of $C_{\mathbf{k}}$ in an infinite system for each C_i, one for every possible $\mathbf{k}$. However, if one applies Eq. (15) using as variables all C's for all $\mathbf{k}$, one finds that, for a translationally invariant system in the thermodynamic limit, only C's with the same $\mathbf{k}$ "couple" to each other. For small enough k, the time scale of fluctuation of $C_{\mathbf{k}}$ may then be rigorously separated from that of any other variable.

In this article, we shall discuss fluid systems. The collective ($N_\alpha = N$, where N is the total number of particles) conserved variables are[(10)] the number density,

$$n_{\mathbf{k}}(t) = \sum_{i=1}^{N} e^{i\mathbf{k}\cdot\mathbf{r}_i(t)} \tag{19}$$

the momentum density,

$$\mathbf{g}_{\mathbf{k}}(t) = \sum_{i=1}^{N} \mathbf{p}_i(t) e^{i\mathbf{k}\cdot\mathbf{r}_i(t)} \tag{20}$$

where $\mathbf{p}_i(t)$ is the momentum of the ith particle at time t, and the energy density,

$$e_{\mathbf{k}}(t) = \sum_{i=1}^{N} e_i(t) e^{i\mathbf{k}\cdot\mathbf{r}_i(t)} \tag{21}$$

where $e_i(t)$ is the energy of the ith particle at time t, and it is assumed (short-range forces) that half of the potential energy of a pair of particles may be localized at the center of each particle. For multicomponent systems, additional conserved variables n^α arise corresponding to the conservation of the number of each type of particle,

$$n_{\mathbf{k}}^\alpha = \sum_{i=1}^{N_\alpha} e^{i\mathbf{k}\cdot\mathbf{r}_i} \tag{22}$$

Since $\mathbf{k}$ is adjustable, it is always possible to make k "small." Practically speaking, "small" means smaller than any other relevant inverse length in the system, such lengths are usually comparable to average interparticle spacings. Near a critical point, however, the correlation length ξ for order parameter fluctuations becomes very large, and $\xi^{-1} = \kappa$ becomes very small. Thus, as the critical point is approached, the criterion $k < \kappa$ for small k may reduce the small-k region to nothingness. Another way to see this is to note that the order parameter fluctuation rate vanishes as $\kappa \to 0$, and we need $k \ll \kappa$ if the conserved variables are to be more slowly varying than the order parameter. To avoid creating a theory that is of vanishing utility near a critical point, therefore, we stipulate[12,13] that relevant order parameters must be included with the conserved variables to form a complete set. The criterion for small k with the complete set is that k be much less than any relevant inverse length except κ.

Since the order parameter is to be included in our set of variables, the random force $f(t)$ will be orthogonal to the order parameter. Just as we have argued that the orthogonality of f to the slowly varying variables means that $f(t)$ has no slowly varying character, it is possible to argue that the orthogonality of f to the order parameter means that f has no critical anomaly. If the argument is correct, $K(t)$ will, in addition to being rapidly decaying, contain[12,13] no critical anomaly. We shall refer to a kernel that enjoys these properties as "well behaved" or "classical."

Once one has decided to formulate a dynamical theory for Fourier components conserved variables at long times and small k, extra simplifications occur. Note that the right-hand side of Mori's equation (10) may be a quite complicated function of k for arbitrary k. However, for "small" k, it is common[10,11] to expand $K(t)$, $i\Omega$, and χ in a power series in k. For the conserved-fluid variables, schemes that expand the equations of motion to order k, k^2, k^3, and k^4, respectively, are the Euler, Navier–Stokes, Burnett, and super-Burnett equations. Since $\dot{A}_{\mathbf{k}} = i\mathbf{k} \cdot \mathbf{j}_{\mathbf{k}}$, and since $i\Omega$ contains $\dot{A}$ once, K contains $\dot{A}$ twice, while χ does not contain $\dot{A}$ at all [10,11] [Eqs. (5), (6), and (11)], the leading term in $i\omega \equiv i\Omega\chi^{-1}$ is $O(ik)$, while the leading term in $K\chi^{-1}$ is $O(k^2)$. The leading term in $i\omega$ is imaginary while the leading term in Γ is real. The Euler equations are reversible, i.e., show no dissipation, while the introduction of the k^2 term in K at the Navier–Stokes level causes irreversibility.

Thus, the dynamical events of greatest interest to us first appear at $O(k^2)$, and we shall not attempt to consider higher orders in k in this article. Of course, as we have just discussed, functions of k may not be expanded if k/κ appears.

Since the coefficients in the linear laws for the conserved variables contain explicit factors of k, some extra notation is commonly introduced. We indicate that quantities can depend on k by attaching k as a subscript. For conserved variables, the decay rate Γ_k is proportional to k^2, and we let

$$\Gamma_k = k^2 \Lambda_k \tag{23}$$

where Λ_k is the "transport coefficient." Since $\dot{\mathbf{A}}_{\mathbf{k}} = i\mathbf{k} \cdot \mathbf{j}_{\mathbf{k}}$, it follows that

$$\Lambda_k = \int_0^\infty dt\, \langle j_{\mathbf{k}}^+(t) j_{-\mathbf{k}}^+ \rangle \chi_k^{-1} \tag{24}$$

for some appropriate component j^+ of $\mathbf{j}^+$, where

$$\mathbf{j}_{\mathbf{k}}^+(t) = \exp[(1-P)i\mathscr{L}t](1-P)\mathbf{j}_{\mathbf{k}} \tag{25}$$

i.e.,

$$\mathbf{f}(t) = i\mathbf{k} \cdot \mathbf{j}_{\mathbf{k}}^+(t) \tag{26}$$

and j^+ is referred to as[10,11] the "dissipative flux" to distinguish it from the flux j. Clearly, $j^+(t)$ enjoys the same properties as those discussed for $f(t)$. The Onsager coefficient for conserved variables $L_k^{(c)}$ is defined by the relation

$$\Lambda_k \equiv L_k^{(c)} \chi_k^{-1} \tag{27}$$

i.e.,

$$L_k^{(c)} = \int_0^\infty dt\, \langle j_{\mathbf{k}}^+(t) j_{-\mathbf{k}}^+ \rangle \tag{28}$$

In the future we shall need the generalized time-dependent transport coefficient,

$$\Lambda_k(t) = \langle j_{\mathbf{k}}^+(t) j_{-\mathbf{k}}^+ \rangle \chi_k^{-1} \tag{29}$$

its Laplace transform,

$$\Lambda_k(s) = \int_0^\infty dt\, e^{-st} \Lambda_k(t) \tag{30}$$

and the kernel for conserved variables $K^{(c)}(t)$,

$$\Lambda_k(t) = K_k^{(c)}(t) \chi_k^{-1} \tag{31}$$

In this notation, the "usual" transport coefficient Λ_k is

$$\Lambda_k = \Lambda_k(s=0) \tag{32}$$

and the convolution term in Mori's[8] equation is

$$\int_0^t d\tau\, K(t-\tau)\chi^{-1}\bar{A}(\tau) = k^2 \int_0^t d\tau\, \Lambda_k(t-\tau)\bar{A}(\tau) \tag{33}$$

From our discussion of the k expansion, the quantities that appear at the Navier–Stokes level are $\lim_{k\to 0}(\Lambda_k, L_k^{(c)}$, and $K_k^{(c)})$, which we shall just write as Λ, $L^{(c)}$, and $K^{(c)}$, respectively.

The popular viewpoint up to about the mid-1960s is summarized as follows. Find all the conserved variables and order parameters for a particular system, use these (for a particular k) as a set of variables in Eq. (10) or one of its many equivalent forms, expand the equations to at least order k^2, and linear laws, valid for the long times over which the variables of interest fluctuate, and for $k \ll$ all lengths but κ, will be obtained. The Onsager coefficients in the linear laws will be given by integrals over dissipative flux correlation functions that contain no slowly varying character and no "order parameter" character, i.e., no "critical anomaly."

The evaluation of the flux or random force correlation functions involves the many-body problem and is very difficult.[14,15] However, the assumption that $K^{(c)}(t)$ is rapidly decaying and insensitive to critical points allows much to be done. Many schemes for approximating $\Lambda(t)$ have been proposed, all based on the idea that $K^{(c)}(t)$ decays in a few molecular collision times. Furthermore, if $j_{\mathbf{k}}^{+}(t)$, and thus $L_k^{(c)}$, has no critical anomaly, a complete theory of the "critical slowing down" may[12,13] be constructed. All possible critical anomalies in the equations of motion of the slow variables will be contained in $i\Omega$ and in χ, which may be calculated from equilibrium theory (no dynamics). An equilibrium theory of critical phenomena will uniquely determine a dynamical theory. The solution to the linear law, Eq. (1), is

$$\bar{A}(t) = \exp[(i\Omega - L)\chi^{-1}t]\bar{A}(0) \tag{34}$$

where $L\chi^{-1} = \Gamma$ causes decay. If we assume that L has no critical anomaly, the critical-point properties of Γ vary as χ^{-1}. Suppose A represents a single variable, the order parameter. Then χ is the usual susceptibility associated with a critical point, which varies as $|T-T_c|^{-\gamma}$, and so Γ varies as $|T-T_c|^{\gamma}$.

During the 1960s, the classical view of dynamics described above was seen to be incorrect; both long-time[16] and critical-point[17] effects were observed in $K^{(c)}(t)$. To a good approximation, the sound absorption coefficient in a fluid is determined[11] by the shear and bulk viscosity coefficients. According to the classical theory, the viscosities should contain no critical anomalies at a binary critical point. However, it was observed[17] that sound absorption shows an anomalous increase at a binary critical point, which indicates that the appropriate elements of $L^{(c)}$ must also possess a critical anomaly. Critical sound absorption was first explained by Fixman,[19] and his work represents the first formulation of mode–mode coupling theory.

The Fixman theory was quite heuristic and bears little resemblance to later versions of mode–mode coupling. The theory was based on macroscopic hydrodynamic considerations of energy dissipation in a sound wave. The theory makes beautiful physical sense, but does not concern itself with the classical linear formalism that we have just discussed. Thus, the Fixman theory allows no determination of why our assumptions about $K^{(c)}(t)$ break down. Understanding of this point was achieved via the more formal theories of Kawasaki[20] and of Kadanoff and Swift.[21]

Alder and Wainwright[16] discovered that certain $K^{(c)}(t)$ behaved as $t^{-d/2}$ for long times, where d is the dimension. The existence of the now famous "long-time tails" was a second indication that, contrary to widespread prior assumption, slowly fluctuating character was entering $K^{(c)}(t)$. The presence of a critical anomaly in $K^{(c)}(t)$, of course, makes it perfectly reasonable that $K^{(c)}(t)$ should have slowly decaying character also and, in fact, Yamada and Kawasaki[22] made a little noticed remark to this effect before Alder's work. Alder and Wainwright[16] explained the "tails" with a hydrodynamic argument, and many subsequent explanations were formulated using the Kawasaki–Kadanoff and Swift mode–mode coupling theory.

A third difficulty,[23] which perhaps does not rank in the "breakdown" class but at least rates as a puzzle, perplexed certain investigators during the 1960s. The transport coefficients Λ characterize the decay of the slowly varying variables. Since classically $K^{(c)}(t)$ has no slowly varying character, one would not expect $K^{(c)}$ to contain any transport coefficients. Thus, one would not expect to find any expressions for transport coefficients of the form $\Lambda = f(\Lambda)$, i.e., equations for transport coefficients in terms of other transport coefficients. In fact, two expressions of this type are very well known, the Stokes–Einstein[24] law for the self-diffusion coefficient, and the Debye[25] law for the rotational diffusion coefficient. Both of these laws for the appropriate diffusion coefficient D are derived by hydrodynamics and have $D\alpha\eta^{-1}$, where η is the coefficient of shear viscosity, a transport coefficient. The Stokes–Einstein and Debye laws were reconciled [26] with formal theory with the use of mode–mode coupling theory.

A simple explanation of the breakdown of the old ideas about the linear laws, according to the Kawasaki–Kadanoff[20] and Swift[21] mode–mode coupling theory goes as follows. Consider some conserved variable $A_{\mathbf{k}}$; then $\langle j_{\mathbf{k}}^{+}(t)A_{-\mathbf{k}}\rangle = 0$. Even if $A_{\mathbf{k}}$ is the only slowly varying variable for small k, we may form products $A_{\mathbf{k}+\mathbf{k}'}A_{-\mathbf{k}'}$, $A_{\mathbf{k}+\mathbf{k}'+\mathbf{k}''}A_{-\mathbf{k}'}A_{-\mathbf{k}''}$, etc., characterized by wave vector $\mathbf{k}$ (the wave vector for a product is additive), which will be slowly varying if k', k'', etc., are also small. There is no reason to expect that, e.g., $\langle j_{\mathbf{k}}^{+}(t)A_{-\mathbf{k}-\mathbf{k}'}A_{\mathbf{k}'}\rangle = 0$, and $j_{\mathbf{k}}^{+}$ may attain slowly varying character by having a "part" that behaves as a product of A's. Almost all discussions of mode–mode coupling have considered only bilinear products of A's, so we shall limit this introductory discussion to that simplified case. For a single A, there are an infinite number of variables $A_{\mathbf{k}+\mathbf{k}'}A_{-\mathbf{k}'}$, corresponding to all the allowed $\mathbf{k}'$'s in the system. It is often useful to think of the allowed $\mathbf{k}'$'s as discrete during the early stages of these discussions. Since only slowly varying bilinear variables are of interest, it is usual to require[20,21] $k' \leq k_c$, where k_c is the "cutoff wave vector." For k' on the order of inverse interparticle spacings, $A_{\mathbf{k}'}$ is no longer expected to be slowly varying, so we write $k_c = (\pi/l)c$, where l is the average interparticle spacing and $c < 1$. The actual value of k_c does not enter many of our results. The variable(s) of interest, $A_{\mathbf{k}}$, obey Eq. (10). The quantity behaving strangely is $K^{(c)}(t)$ [Eq. (31)]. With the idea[8] of a Hilbert space of dynamical variables in mind, as is appropriate for the Mori formalism, we therefore write

$$j_{\mathbf{k}}^{+} = j_{\mathbf{k}}^{+\mathrm{NL}} + \sum_{\mathbf{k}'=0}^{k_c} \frac{\langle j_{\mathbf{k}}^{+} A_{-\mathbf{k}-\mathbf{k}'} A_{\mathbf{k}'} \rangle}{\langle A_{\mathbf{k}+\mathbf{k}'} A_{-\mathbf{k}-\mathbf{k}'} A_{\mathbf{k}'} \rangle} A_{\mathbf{k}+\mathbf{k}'} A_{-\mathbf{k}'} \tag{35}$$

where $j_{\mathbf{k}}^{+\mathrm{NL}}$, the part of $j_{\mathbf{k}}^{+}$ orthogonal to the bilinear variable, is assumed to have all the desirable properties that were formerly ascribed to $j_{\mathbf{k}}^{+}$, and the second term in Eq. (35) gives the projection of $j_{\mathbf{k}}^{+}$ onto the bilinear variable. If we write the autocorrelation function of $j_{\mathbf{k}}^{+}$ from Eq. (35), ignoring cross terms between $j_{\mathbf{k}}^{+\mathrm{NL}}$ and the bilinear variables, we find

$$K_k^{(c)}(t) = \langle j_{\mathbf{k}}^{+}(t) j_{-\mathbf{k}}^{+} \rangle = \langle j_{\mathbf{k}}^{+\mathrm{NL}}(t) j_{-\mathbf{k}}^{+\mathrm{NL}} \rangle$$
$$+ \sum_{\mathbf{k}', \mathbf{k}''}^{k_c} U_{\mathbf{k};\,\mathbf{k}+\mathbf{k}',\,-\mathbf{k}'} U_{-\mathbf{k};\,-\mathbf{k}-\mathbf{k}'',\,\mathbf{k}''} \langle A_{\mathbf{k}+\mathbf{k}'}(t^{+}) A_{-\mathbf{k}'}(t^{+})\, A_{-\mathbf{k}-\mathbf{k}''} A_{\mathbf{k}''} \rangle \tag{36}$$

where

$$U_{\mathbf{k};\,\mathbf{k}+\mathbf{k}',\,-\mathbf{k}'} \equiv \langle j_{\mathbf{k}}^{+} A_{-\mathbf{k}-\mathbf{k}'} A_{\mathbf{k}'} \rangle \langle A_{\mathbf{k}+\mathbf{k}'} A_{-\mathbf{k}'} A_{-\mathbf{k}-\mathbf{k}'} A_{\mathbf{k}'} \rangle^{-1} \tag{37}$$

and we write $A(t^{+})$ to indicate that time evolution is governed by the projected Liouville operator. For long times, we treat the microscopic part of Eq. (36) classically,

$$\langle j_{\mathbf{k}}^{+\mathrm{NL}}(t) j_{-\mathbf{k}}^{+\mathrm{NL}} \rangle \rightarrow L_k^0\, \delta(t) \tag{38}$$

where L_k^0 is the "bare" Onsager coefficient. The bilinear contribution to Eq. (36) is evaluated by the approximation

$$\langle A_{\mathbf{k}+\mathbf{k}'}(t^{+}) A_{-\mathbf{k}'}(t^{+}) A_{-\mathbf{k}-\mathbf{k}''} A_{\mathbf{k}''} \rangle \rightarrow \langle A_{\mathbf{k}+\mathbf{k}'}(t) A_{-\mathbf{k}-\mathbf{k}'}(t) \rangle \langle A_{-\mathbf{k}'}(t) A_{\mathbf{k}'}(t) \rangle\, \delta_{\mathbf{k}',\,\mathbf{k}''} \tag{39}$$

where δ is a Kroneker delta, i.e., we ignore the difference between t and t^{+} and we "factorize" the four-variable correlation function. With all the approximations just mentioned, Eq. (15) becomes

$$\frac{\partial}{\partial t}\langle A_{\mathbf{k}}(t)A_{-\mathbf{k}}\rangle = (i\omega_k - k^2\Lambda_k^0)\langle A_{\mathbf{k}}(t)A_{-\mathbf{k}}\rangle - k^2\sum_{\mathbf{k}'}^{k_c}\int_0^t (U_{\mathbf{k};\,\mathbf{k}+\mathbf{k}',\,-\mathbf{k}'})^2\langle A_{\mathbf{k}+\mathbf{k}'}(t-\tau)A_{-\mathbf{k}-\mathbf{k}'}\rangle \times\langle A_{-\mathbf{k}'}(t-\tau)A_{\mathbf{k}'}\rangle\langle A_{\mathbf{k}}A_{-\mathbf{k}}\rangle^{-1}\langle A_k(\tau)A_{-\mathbf{k}}\rangle \tag{40}$$

where $\Lambda_k^0 \equiv L_{kk}^0\chi_k^{-1}$ is the "bare" transport coefficient. For this brief introductory discussion of mode-mode coupling, we do not intend to discuss the validity of the approximations entering Eq. (40). In fact, Eq. (40) is a very good approximation to the true equation. The term in parentheses in Eq. (40) only differs from the complete linear laws by the presence of the bare transport coefficient Λ_k^0 instead of the true transporter coefficient Λ_k. If this term alone existed, the solution of Eq. (40) for long times would be trivial,

$$\langle A_{\mathbf{k}}(t)A_{-\mathbf{k}}\rangle = \langle A_{\mathbf{k}}A_{-\mathbf{k}}\rangle \exp[(i\omega_k - k^2\Lambda_k^0)t]$$

The presence of the second term, however, creates complications. The full convolution form of the equation must be kept, even at long times; furthermore, the kernel for the convolution involves (bilinearly) the correlation function itself, for all $k_c \geq k' \geq 0$. Thus, we may no longer deal with a single wave vector; to find the behavior of $\langle A_{\mathbf{k}}(t)A_{-\mathbf{k}}\rangle$ for a particular k it is necessary to know the correlation function for all $k' \leq k_c$. This point could grieviously complicate the making of a small wave vector expansion of the laws of motion, since no matter how small we make k, we still have to deal with the whole range of k'. We shall take the point of view,[27] however, that k_c is small enough that k' can be considered small, and thus we shall perform k' expansions just as we performed k expansions.

The exact solution of Eq. (40) is difficult to obtain. However, the three aspects of the breakdown of the classical theory are easily exhibited. The long-time form of $K^{r(c)}(t)$ is clearly

$$K_k^{(c)}(t) \xrightarrow[\text{long times}]{} \sum_{\mathbf{k}'}^{k_c} (U_{\mathbf{k};\,\mathbf{k}+\mathbf{k}',\,-\mathbf{k}'})^2\langle A_{\mathbf{k}+\mathbf{k}'}(t)A_{-\mathbf{k}-\mathbf{k}'}\rangle\langle A_{-\mathbf{k}'}(t)A_{\mathbf{k}'}\rangle \tag{41}$$

Assume that

$$\langle A_{\mathbf{k}'}(t)A_{-\mathbf{k}'}\rangle = \langle A_{\mathbf{k}'}A_{-\mathbf{k}'}\rangle \exp(-k'^2\Lambda t) \tag{42}$$

i.e., we ignore $i\omega_k$ for the moment, and ignore the change in the long-time form of $\langle A_{\mathbf{k}'}(t)A_{-\mathbf{k}'}\rangle$ due to the long-time nature of $K^{(c)}$. If $U_{\mathbf{k};\,\mathbf{k}+\mathbf{k}',\,-\mathbf{k}'}$ is independent of k, k' for $k' < k_c$, which is true away from a critical point in most cases, we find Alder and Wainwright's[16] result,

$$\lim_{k\to 0} K_k^{(c)}(t) \xrightarrow[\text{long times}]{} \sum_{\mathbf{k}'}^{k_c} \exp(-2k'^2\Lambda t)\alpha t^{-d/2}, \qquad t \gg (k_c^2\Lambda)^{-1} \tag{43}$$

where we have used the substitution, $\sum_{\mathbf{k}} \to [V/(2\pi)^3] \int d\mathbf{k}$, valid for an infinite system [in a finite cube the $t^{-d/2}$ behavior is "cut off" for $t \gg ((\pi/S)^2\Lambda)^{-1}$ where S is the length of an edge]. Near a critical point $U_{\mathbf{k};\mathbf{k}+\mathbf{k}',-\mathbf{k}'}$ may[20,21] (depending on the variable) cut off the $\sum_{\mathbf{k}'}$ at $k' \approx \kappa$, so the region of $t^{-d/2}$ behavior may shrink to $(\kappa^2\Lambda)^{-1} \gg t \gg ((\pi/S)^2\Lambda)^{-1}$.

The critical behavior of $K^{(c)}(t)$, unlike the long-time behavior, depends sensitively on the bilinear variable used in Eq. (40). In many cases[20,21] (e.g., binary critical solution), if A^i is the order parameter, the bilinear variable of interest for the dynamics of $A^i_{\mathbf{k}}$ is $A^i_{\mathbf{k}+\mathbf{k}'}A^j_{\mathbf{k}'}$, where A^j is relatively insensitive to the critical point. If we carry out all the above analyses for this case, we find, as $k \to 0$,

$$\frac{\partial}{\partial t}\langle A^i_{\mathbf{k}}(t)A^i_{-\mathbf{k}}\rangle = (i\omega_{k,\,ii} - k^2\Lambda^0_{k,\,ii})\langle A^i_{\mathbf{k}}(t)A^i_{\mathbf{k}}\rangle$$
$$-k^2\sum_{\mathbf{k}'}^{k_c} U^2_{\mathbf{k};\mathbf{k}',-\mathbf{k}'}\int_0^t d\tau\,\langle A^i_{\mathbf{k}'}(t-\tau)A^i_{-\mathbf{k}'}\rangle\langle A^j_{-\mathbf{k}'}(t-\tau)A^j_{\mathbf{k}'}\rangle$$
$$\times\langle A_{\mathbf{k}}A_{-\mathbf{k}}\rangle^{-1}\langle A_{\mathbf{k}}(\tau)A_{-\mathbf{k}}\rangle \tag{44}$$

If A^i is the order parameter, assuming L^0 has no critical anomaly, Ornstein–Zernicke theory gives $\Lambda^0_{ii}\alpha\xi^{-2}$. Typically, U has no critical anomaly, and we assume

$$\langle A^j_{\mathbf{k}}(t)A^j_{\mathbf{k}}\rangle = \langle A^j_{\mathbf{k}}A^j_{\mathbf{k}}\rangle \exp(-k^2\Lambda_{jj}t) \tag{45}$$

where $\langle A^j_{\mathbf{k}}A^j_{k}\rangle$ has no critical anomaly, nor to first approximation does Λ_{jj}, and as before we approximate

$$\langle A^i_{\mathbf{k}}(t)A^i_{-\mathbf{k}}\rangle = \langle A^i_{\mathbf{k}}A^i_{-\mathbf{k}}\rangle \exp(-k^2\Lambda_{ii}t) \tag{46}$$

For many typical cases, the dimensionless ratio $\langle A^i_{\mathbf{k}'}A^i_{-\mathbf{k}'}\rangle/\langle A^i_{\mathbf{k}}A^i_{-\mathbf{k}}\rangle$, which appears in Eq. (44) due to Eq. (46), simply serves to cut off the $\sum_{\mathbf{k}'}$ at $k' \approx \kappa$. Thus, changing the $\sum_{\mathbf{k}'}$ to an integral, we find that the bilinear contribution to Λ_{ii} varies as $(\Lambda_{ii}+\Lambda_{jj})^{-1}\kappa$. Near a critical point, $\Lambda_{ii} \ll \Lambda_{jj}$, so the bilinear part of Λ_{ii} goes as $(\Lambda_{jj})^{-1}\kappa$, $\kappa = \xi^{-1}$. This part of Λ_{ii} will dominate Λ_{ii} near a critical point, and the ξ^{-2} classical behavior of Λ^0_{ii} breaks down. We emphasize that the demonstrations given of both aspects of the breakdown of classical theory hold for $k \to 0$; the results for $k_c > k \geqslant \kappa$ require a somewhat more sophisticated analysis. The third difficulty[23] mentioned earlier, that of the appearance of transport coefficients in transport coefficients, is also explained by mode–mode coupling theory. Since $K^{(c)}(t)$ contains $\langle A_{\mathbf{k}'}(t)A_{-\mathbf{k}'}\rangle$, $\int_0^\infty K^{(c)}(t)\,dt$, which gives $L^{(c)}$ and thus Λ, will contain Λ^{-1}. The relation $\Lambda \propto \Lambda^{-1}$ is precisely that of the Stokes–Einstein and Debye laws mentioned earlier, $D \propto \eta^{-1}$.

The discussion of mode-mode coupling just given is not meant to be rigorous, but expository. Nonetheless, the results given above are in fact[20,21] those obtained, to a very good first approximation, from the rigorous theory.

Thus, practically speaking, if one were approaching a new problem in mode–mode coupling with a desire for preliminary results rather than a desire for refinement, the procedure outlined above would probably be quite adequate. All the fundamental ideas of mode–mode coupling and the breakdown of classical theory have been discussed above.

In the remainder of this article, we shall discuss the proper mode–mode coupling theory. This discussion naturally falls into two parts. The twofold nature of the theory has already been seen in the previous pages. First, it was necessary to make approximations to obtain Eq. (40), the mode–mode coupling equation. Second, it was necessary to make approximations to solve Eq. (40). Thus, the first part of the problem is deriving the equations, and the second part is solving them. To solve either part completely, for all cases, is probably impossible. In the past, more emphasis has been given to solving a given equation than to determining how valid that equation actually is. This is probably because the careful derivation of the equations, in many cases, seems hopelessly complicated, while the solution of typical mode–mode equations is amenable to well-known techniques.

2. Writing Mode–Mode Equations

It is now desirable to deal with the nonclassical behavior of the kernel in the linear laws in a precise, formal way. Of course, one could simply try to improve the crude method just discussed; such an approach is perfectly valid. However, we feel that an alternate[(20)] procedure, which has almost always been used in the literature, is preferable. Mori's method allows the writing of equations with well-behaved kernels if the proper set of variables is chosen. The kernel in the linear laws is badly behaved due to the influence of the nonlinear variable. If we include[(20)] the linear and nonlinear variables in the set of variables to which Mori's method is applied, the random forces f^{NL} and the dissipative fluxes $j^{+\mathrm{NL}}$ ($j^{+\mathrm{NL}}$ will be defined precisely in this section) will be projected orthogonal to all of these variables. The kernels K^{NL} in the resulting equations, the "nonlinear Langevin equations," should behave classically. Thus, convolutions involving K will be converted into scalar multiplication by the "classical" relation,

$$\int_0^t K^{\mathrm{NL}}(t-\tau)\gamma(\tau)\,d\tau \rightarrow L^0\gamma(t) \tag{47}$$

$$L^0 = \int_0^\infty dt\, K^{\mathrm{NL}}(t) \tag{48}$$

and γ is any variable.

All critical anomalies in the nonlinear equations will be contained[(20)] in the matrices $i\omega^{\mathrm{NL}}$ and χ^{NL} appearing in Eq. (10) for the new set of variables, which contain only equal-time correlation functions and are available from equilibrium statistical mechanics. The nonlinear Langevin equations are to be solved, with the use of classical assumptions, to find the dynamics of the linear variables of interest.

In short, the basic classical ideas are kept, but the rule given in the previous section for the writing of equations with classical kernels is extended as follows.[20] Suppose, for a given problem, all the conserved variables and order parameters have been found. If we choose these variables, for a given wave vector $\mathbf{k}$, plus *all* products of these variables characterized by wave vector $\mathbf{k}$, and apply Mori's methods, then the kernel $K^{\mathrm{NL}}(t)$ will have no critical, or long-time, behavior. No nonlinear variables where the wave vector of any single variable in the product is greater than the cutoff wave vector k_c need be included. We refer to the linear variables with wave vector $\mathbf{k}$ as $A_{\mathbf{k}}(t)$ (as before), the bilinear variables with wave vector $\mathbf{k}$ and "intermediate" wave vector $\mathbf{k}'$ $(k'<k_c)$ as $B_{\mathbf{k},\mathbf{k}'}(t)$ [i.e., $A_{\mathbf{k}+\mathbf{k}'}(t)A_{-\mathbf{k}'}(t)$ is $B_{\mathbf{k},\mathbf{k}'}(t)$], the trilinear variables as $T_{\mathbf{k},\mathbf{k}',\mathbf{k}''}$, and so on. A, B, T, etc., may refer to a single variable or to a set of variables. The extended set of variables is always infinite in the thermodynamic limit, since the number of $\mathbf{k}'$ between zero and k_c, $\sum_{\mathbf{k}=0}^{k_c} \equiv M$, is $O(N)$. There are M variables corresponding to a single bilinear product, M^2 variables for a single trilinear product, etc. The matrices that will appear in the Mori[8] formalism, χ^{NL}, $i\omega^{\mathrm{NL}}$, and $K^{\mathrm{NL}}(t)$, will all be of infinite rank. The manipulation of infinite-rank matrices is much harder than the manipulation of the small, finite-rank matrices that arise when writing linear laws. Thus, it is important to simplify mode–mode coupling problems as much as possible, at the very beginning, by eliminating superfluous nonlinear variables.

In practice, mode–mode coupling calculations are almost always performed with bilinear variables alone, and with one or at most two different bilinear products at that. We are aware of no attempt to justify such extreme simplification. Our feeling is that in many cases the use of bilinear variables alone is completely correct, but that in other cases the possible importance of other nonlinear variables deserves further study. All the problems that we discuss in detail will involve only bilinear variables.

The equations obtained by applying Mori's method to an extended set of variables are of two types. First, the time derivative of the variable of interest, $A_{\mathbf{k}}(t)$, is expressed as a sum of the values of all the linear variables and nonlinear variables at time t, with each variable multiplied by an appropriate coefficient. Second, a similar equation results for the time derivative of each nonlinear variable; such equations are of interest to us only insofar as the dynamics of the nonlinear variables are required to obtain the dynamics of the nonlinear variable.

In most applications of the Mori formalism, one simultaneously solves all the first-order differential equations for the behavior of the complete set of variables. This same approach has been used for mode–mode coupling problems by some[26–29] authors; the equations for the nonlinear variables are used to eliminate these variables from the equation for $A_{\mathbf{k}}(t)$, giving a closed equation for $A_{\mathbf{k}}(t)$. Recently, Michaels and Oppenheim[29] have shown that, in this approach, it is necessary to retain nonlinear variable to infinite order. In a truncation of the complete set of variables that only includes up to n-linear variables, the kernel that appears in the equation for the nth linear variable is badly behaved due to the influence of the $(n+1)$th linear variable. The malign influence of an $(n+1)$-linear variable on the kernel for an n-linear variable occurs in exactly the same way as a bilinear variable can affect the kernel for a linear variable.

In Kawasaki's approach to mode–mode coupling, one does not use the complete set of Mori's equations for the complete set of variables. Rather the procedure is to keep only the equation for $A_{\mathbf{k}}$, discard the equations for the nonlinear variables, and employ various alternative techniques to solve the equation for $A_{\mathbf{k}}$. This approach is the one we shall discuss, as otherwise simplification to a very few nonlinear variables is impossible.

Thus, we shall be writing and solving equations of the form

$$\frac{\partial}{\partial t}\bar{A}_{\mathbf{k}}(t) = (i\omega_k^{\mathrm{NL}} - k^2\Lambda_k^0)\bar{A}_{\mathbf{k}}(t) + \sum_{\mathbf{k}'}^{k_c} \mathscr{V}^{A,B}_{\mathbf{k};\mathbf{k}+\mathbf{k}',-\mathbf{k}'}\bar{B}_{\mathbf{k},\mathbf{k}'}(t)$$
$$+\sum_{\mathbf{k}'}^{k_c} \mathscr{V}^{A,T}_{\mathbf{k};\mathbf{k}+\mathbf{k}'+\mathbf{k}'',-\mathbf{k}',-\mathbf{k}''}\bar{T}_{\mathbf{k},\mathbf{k}',\mathbf{k}''}(t) + \cdots \qquad (49)$$

Equation (49) is merely schematic at this point, and the coefficients are yet to be determined. A well-behaved kernel $K^{(c)\mathrm{NL}}(t)$ will determine Λ_k^0. The vertices $\mathscr{V}$ differ from the vertices U in Eq. (40) by a factor ik. As $\dot{A}_{\mathbf{k}}$ must vanish (except for nonconserved order parameters) as $k \to 0$, all the $\mathscr{V}$'s are at least linear in k. It is convenient to discuss the properties of Eq. (49) in terms of the formally exact linear law,

$$\frac{\partial}{\partial t}\bar{A}_{\mathbf{k}}(t) = i\omega_k\bar{A}_{\mathbf{k}}(t) - k^2\int_0^t d\tau\, \Lambda_k(t-\tau)\bar{A}_{\mathbf{k}}(\tau) \qquad (50)$$

which is derived from Mori's equation. Equation (50) may be solved by Laplace transform to obtain

$$\bar{A}_{\mathbf{k}}(s) = (s + i\omega_k + k^2\Lambda_k(s))^{-1}\bar{A}_{\mathbf{k}}(t=0) \qquad (51)$$

Equations (49) and (50) are completely equivalent, the only differences being that nonlinear effects are explicit in Eq. (49) and "hidden" in Λ, through ill-behaved flux correlation functions, in Eq. (50). Instead of actually solving Eq. (49) for $\bar{A}_{\mathbf{k}}(t)$, we shall solve for $\bar{A}_{\mathbf{k}}(s)$, and by comparison with Eq. (51) deduce $\Lambda_k(s)$. It is natural to concentrate on $\Lambda_k(s)$, as this quantity is observed directly in many experiments. In all the cases we are aware of, $i\omega^{\mathrm{NL}} = i\omega$.

The importance of an n-linear variable in Eq. (49) is determined by the N dependence, wave-vector dependence, and correlation length (ξ) dependence of $\mathscr{V}^{A-n}$. Since there are M wave vectors between 0 and k_c, and M is $O(N)$, each sum in Eq. (49) may be regarded as $O(N)$. Thus, $\mathscr{V}^{A-B}$ must be $O(1/N)$ if the contribution of B is to be nonvanishing as $N\to\infty$, $\mathscr{V}^{A-T}$ must be $O(1/N^2)$ (two sums), and so on. Assuming that the contribution of a given variable is finite, the importance of that variable for the long-time behavior of the time correlation function of the dissipative flux for $A_{\mathbf{k}}$, $K_k^{(c)}(t)$ follows[30-32] from the wave-vector dependence of $\mathscr{V}^{A,n}$. The projection of the badly behaved dissipative flux for $A_{\mathbf{k}}$, $j_{\mathbf{k}}^{+}$ (not to be confused with the well-behaved $j_{\mathbf{k}}^{+\mathrm{NL}}$) onto the nonlinear variables is just given by the terms in Eq. (49) involving the nonlinear variables. That is, if we substitute the relation, $\dot{A}=i\mathbf{k}\cdot\mathbf{j}$, into Eq. (49), we immediately find the nonlinear parts of $i\mathbf{k}\cdot\mathbf{j}$, which are basically the nonlinear parts of j^{+}. Thus, in analogy to the argument given for the effect of a bilinear variable on $K^{(c)}(t)$ in Section 1, we may estimate the contribution $K_k^n(t)$ of an n-linear variable γ^n to $K_k^{(c)}(t)$, by the relation[30-32]

$$k^2K_k^n(t)\sim\sum_{\substack{\text{intermediate}\\ \text{wave vectors}}}^{k_c}(\mathscr{V}^{A-n}_{\mathbf{k},\mathbf{k}',\ldots})(\mathscr{V}^{A-n}_{\mathbf{k};\mathbf{k}'',\ldots})\langle\gamma^n_{\mathbf{k},\mathbf{k}',\ldots}(t)\gamma^n_{-\mathbf{k},\mathbf{k}'',\ldots}(0)\rangle \tag{52}$$

If we "factorize" the average in Eq. (52), i.e.,

$$\begin{aligned}\langle\gamma^n_{\mathbf{k},\mathbf{k}',\mathbf{k}'',\ldots}(t)\gamma^n_{-\mathbf{k},\mathbf{k}''',\mathbf{k}'''',\ldots}(0)\rangle\to&\langle A_{\mathbf{k}+\mathbf{k}'+\mathbf{k}''+\ldots}(t)A_{-\mathbf{k}-\mathbf{k}'-\mathbf{k}''-\ldots}\rangle\\&\times\langle A_{\mathbf{k}'}(t)A_{-\mathbf{k}'}\rangle\langle A_{\mathbf{k}''}(t)A_{-\mathbf{k}''}\rangle\cdots\\&\times\delta_{\mathbf{k}'-\mathbf{k}'''}\times\delta_{\mathbf{k}''-\mathbf{k}''''}-\cdots\end{aligned} \tag{53}$$

and if we use Eq. (42), where the value of Λ is immaterial, we find

$$\begin{aligned}k^2K_k^n(t)\sim&\sum_{\mathbf{k}',\mathbf{k}'',\ldots}^{k_c}(\mathscr{V}^{A-n}_{\mathbf{k};\mathbf{k}',\mathbf{k}'',\ldots})^2\\&\times\langle A_{\mathbf{k}+\mathbf{k}'+\mathbf{k}''+\ldots}A_{-\mathbf{k}-\mathbf{k}'-\mathbf{k}''}\rangle\langle A_{\mathbf{k}'}A_{-\mathbf{k}'}\rangle\langle A_{\mathbf{k}''}A_{-\mathbf{k}''}\rangle\cdots\\&\times\exp(-|\mathbf{k}+\mathbf{k}'+\mathbf{k}''+\cdots|^2\Lambda t-k'^2\Lambda t-k''^2\Lambda t-\cdots)\end{aligned} \tag{54}$$

For all $A_{\mathbf{k}}$ of interest, $\langle A_{\mathbf{k}}A_{-\mathbf{k}}\rangle$ is independent of k as $k\to 0$. Thus, all the averages in Eq. (54) may be brought outside the sum and ignored when estimating long-time behavior. If κ is small such that $\langle A_{\mathbf{k}}A_{-\mathbf{k}}\rangle$ becomes k dependent at $k\sim\kappa\ll k_c$, the small-k part of the sum will still dominate the long-time behavior of K^c. An obvious requirement on $\mathscr{V}^{A-n}$ is that this quantity be linear in k as $k\to 0$. If, say, $\mathscr{V}^{A-n}$ had a leading term $O(k^2)$ or higher, K^n would not contribute to the complete K^c (the $K^{(c)}$ in the exact linear law) to Navier–Stokes order. We may therefore investigate the $k\to 0$ limit of K_k^n,

$$\lim_{k\to 0}K_k^n(t)\propto\lim_{k\to 0}k^{-2}\sum_{\mathbf{k}',\mathbf{k}'',\ldots}^{k_c}(\mathscr{V}^{A-n}_{\mathbf{k};\mathbf{k}',\mathbf{k}'',\ldots})^2\exp(-|\mathbf{k}'+\mathbf{k}''+\cdots|^2\Lambda t-k'^2\Lambda t-k''^2\Lambda t) \tag{55}$$

The long time form of K^n depends only on the number of summations present in Eq. (55) and on the intermediate wave-vector dependence of $\mathcal{V}$. Furthermore,[30–32] only the total power of the intermediate wave-vector dependence of the leading term in the wave-vector expansion of $\mathcal{V}$ is of interest, i.e., if $\mathcal{V}^{A-n}$ varies as $(k')^i(k'')^j(k''')^l \cdots$ as $k', k'', k''', \ldots \to 0$, K^n only depends on $i+j+l+\cdots \equiv \mu$ for long times. Changing the sums in Eq. (55) to integrals, it is easy to see that the long-time behavior of K^n is[30–32]

$$\lim_{\substack{t\to\infty \\ k\to 0}} K_k^n(t) \to t^{-(n-1)d/2} t^{-\mu} \tag{56}$$

where d is the dimension, and $n = 2, 3, \ldots$ for bilinear, trilinear, etc. Equation (56) may be used to attempt an estimation of the importance of an n-linear variable.

In almost every case, $\mathcal{V}^{A-B}$ $(n = 2)$ is independent of k' as $k' \to 0$. If this is so, the bilinear variable will give a contribution to $K^{(c)} \propto t^{-3/2}$ at long times in three dimensions, which will dominate the contribution of any other variable. Thus, if one is only interested in the very long time behavior of $K^{(c)}(t)$, bilinear variables alone may be kept if $k^{-1}\mathcal{V}^{A,B}$ is independent of k and k', $k, k' \to 0$. If $\mathcal{V}^{A,B}$ behaves differently, the bilinear variable, in principle, need not dominate $K^{(c)}$ at long times. However, in the one case[30–32] we are aware of where this occurs, single-particle molecular reorientation, the bilinear variable also happens to win out.

We are not going to pay much attention to the very strange properties[16,27,33–35] of mode–mode coupling in two dimensions in this article, for obvious reasons. However, it seems worth mentioning that in two dimensions the arguments just given probably do establish the preeminence of the bilinear variable. In most cases, the bilinear variable will give a t^{-1} tail to $K^{(c)}(y)$ in two dimensions. Such a tail is not integrable at large times, and so $\lim_{k\to 0, s\to 0} K_k^2(s)$ is infinite in two dimensions; this point has been much discussed.[16,27,33–35] The contributions to $K_k^{(c)}(s)$, for small k and s, of any other nonlinear variable will be finite, and thus should be negligible. All the arguments just presented are, of course, quite crude: however, we feel that enough of the basic physics is contained in our crude arguments to render the conclusions correct.

A second way to isolate a dominant nonlinear variable would be to study how K^n behaves as a critical point is approached, i.e., $\xi \to \infty$ or $\kappa \to 0$. Referring to Eq. (54), the ξ dependence of K^n follows from the ξ dependence of $\langle A_{\mathbf{k}'}A_{-\mathbf{k}'}\rangle$, of Λ_k, and of $\mathcal{V}_{\mathbf{k};\mathbf{k}',\ldots}$. Bear in mind that A may represent a set of variables, so when we write $\gamma^{\mathrm{NL}} = A_{\mathbf{k}+\mathbf{k}'+\mathbf{k}''+\ldots}A_{-\mathbf{k}'}A_{-\mathbf{k}''}\cdots$, the different A may be different variables, and, similarly, different Λ may appear in Eq. (54). If one estimates that ξ dependence of Λ classically, i.e., $\Lambda_k \propto \Lambda_k^0$, a good first-order estimate of K^n should be possible. However, the estimation of $\mathcal{V}$ for variables of order greater than two is fairly complicated, and we have not attempted to pursue this point. Kadanoff and Swift[36] have shown that, in some cases, K^n, $n>2$, has the same ξ dependence as K^2. For problems where the preeminence of the bilinear variable is not established by the fairly ironclad principle to be discussed next, we suggest the estimation of $K^{(n)}(\xi)$ to the reader as a problem of interest.

For certain cases, it is possible to represent the time derivative of a linear variable exactly, or at least quite reasonably, in terms of a bilinear variable. No expansion or limiting process is involved; the bilinear variable just happens to be well suited to express $A_{\mathbf{k}}$ or $j_{\mathbf{k}}$. Under these circumstances, the nonlinear Langevin equation may be deduced from simple physical arguments. The cleanest examples of problems where bilinear variables arise in a fairly obvious way are diffusion problems. In any diffusion problem, self, mutual, or whatever, the linear variable of interest is a concentration, $n_{\mathbf{k}}^{\alpha}$ [see Eq. (22)]. The time derivative of a concentration is a momentum density/mass,

$$\frac{\partial}{\partial t} n_{\mathbf{k}}^{\alpha}(t) = \frac{i\mathbf{k}}{m} \cdot \sum_{i=1}^{N_\alpha} \mathbf{p}_i e^{i\mathbf{k}\cdot\mathbf{r}_i} \equiv \frac{i\mathbf{k}}{m} \cdot \mathbf{g}_{\mathbf{k}}^{\alpha} \tag{57}$$

where m is the mass of the particles of interest. Equation (57) may be obtained by expanding the exponential in Eq. (22) and differentiating term by term. Now the contribution $\mathbf{g}^{\alpha}$ to the momentum density of a certain set α of particles may be obtained from the total momentum density $\mathbf{g}$, if a way can be found to count only the contribution of the particles of interest. The total momentum density as a function of position $\mathbf{g}(\mathbf{r})$ is

$$\mathbf{g}(\mathbf{r}) = \sum_{i=1}^{N} \mathbf{p}_i \, \delta(\mathbf{r}-\mathbf{r}_i) \tag{58}$$

and $n^{\alpha}(\mathbf{r})$ is

$$n^{\alpha}(\mathbf{r}) = \sum_{i=1}^{N_\alpha} \delta(\mathbf{r}-\mathbf{r}_i) \tag{59}$$

If we compare the products $n^{\alpha}(\mathbf{r})\mathbf{g}(\mathbf{r})$,

$$n^{\alpha}(\mathbf{r})\mathbf{g}(\mathbf{r}) = \sum_{i=1}^{N_\alpha} \sum_{j=1}^{N} \mathbf{p}_j \, \delta(\mathbf{r}-\mathbf{r}_i)\, \delta(\mathbf{r}-\mathbf{r}_j) \tag{60}$$

and $n(\mathbf{r})\mathbf{g}^{\alpha}(\mathbf{r})$,

$$n(\mathbf{r})\mathbf{g}^{\alpha}(\mathbf{r}) = \sum_{i=1}^{N_\alpha} \sum_{j=1}^{N} \mathbf{p}_i \, \delta(\mathbf{r}-\mathbf{r}_i)\, \delta(\mathbf{r}-\mathbf{r}_j) \tag{61}$$

we note a close resemblance. This resemblance is seen to be an identity because $\mathbf{r}_i$ and $\mathbf{r}_j$ in Eqs. (60) and (61) are not arbitrary variables, but the positions of particles, in a real system, which are not allowed to interpenetrate, and

$\delta(\mathbf{r}-\mathbf{r}_i)\,\delta(\mathbf{r}-\mathbf{r}_j)$, $i \neq j$, must always be zero. With this point in mind, we have

$$\mathbf{g}^{\alpha}(\mathbf{r}) = n^{\alpha}(\mathbf{r})\mathbf{g}(\mathbf{r})/n(\mathbf{r}) \tag{62}$$

In phenomenological hydrodynamics, one is accustomed[3] to dealing with the velocity field $\mathbf{v}(\mathbf{r})$. In terms of the conserved densities, $\mathbf{v}(\mathbf{r}) = \mathbf{g}(\mathbf{r})/\rho(\mathbf{r})$, $\rho = mn$, or $g^{\alpha}(\mathbf{r}) = mn^{\alpha}(\mathbf{r})\mathbf{v}(\mathbf{r})$. If we inverse Fourier transform to find $\mathbf{g}_{\mathbf{k}}^{\alpha}$ and substitute the result in Eq. (57), we obtain

$$\frac{\partial}{\partial t} n_{\mathbf{k}}^{\alpha} = \frac{i\mathbf{k}}{V} \cdot \sum_{\mathbf{k}'=0}^{\infty} n_{\mathbf{k}+\mathbf{k}'}^{\alpha} \mathbf{v}_{-\mathbf{k}'} \tag{63}$$

Now Eq. (63) is not a bilinear equation. Writing $n(\mathbf{r}) = \langle n \rangle + \delta n(\mathbf{r})$, we find[24]

$$\mathbf{v}(\mathbf{r}) = \left(\frac{\mathbf{g}(\mathbf{r})}{m\langle n \rangle}\right)\left\{1 - \frac{\delta n(\mathbf{r})}{\langle n \rangle} + \cdots\right\}$$

i.e., $\mathbf{v}(\mathbf{r})$ contains nonlinear variables of all orders of form $\mathbf{g}\,\delta n\,\delta n\,\delta n \cdots$. However, it is possible to ignore variables containing δn, and we may make the approximation $\mathbf{v}(\mathbf{r}) = \mathbf{g}(\mathbf{r})/m\langle n \rangle$. There are two fairly good reasons for this. First,[16,20,21,27] to zero order (i.e., ignoring mode–mode effects),

$$\langle n_{\mathbf{k}}(t) n_{-k} \rangle = \langle n_{\mathbf{k}} n_{-\mathbf{k}} \rangle \exp(ikct - k^2\theta t) \tag{64}$$

where c is the sound velocity,

$$\theta = \tfrac{1}{2}(\tfrac{4}{3}\eta + \eta^{\mathrm{B}}) \tag{65}$$

and η and η^{B} are the coefficients of shear and bulk viscosity, respectively. The factor e^{ikct} in Eq. (64), which arises from $i\omega$ in Eq. (1), changes the long-time contribution of any bilinear variable containing δn from the usual $t^{-d/2}$. Consider a bilinear variable formed from a variable that decays as $e^{-k_2\theta' t}$ and δn; the long-time contribution of this variable to $\lim_{k \to 0} K_k^{(c)}(t)$ is

$$\lim_{k\to 0} K_k^{(c)}(t) \sim \sum_{\mathbf{k}'}^{k_c} \exp[ik'ct - k'^2(\theta + \theta')t] \sim \exp\left[\frac{-c^2 t}{4(\theta + \theta')}\right] t^{-d/2}(\theta + \theta')^{-1} \tag{66}$$

Due to the decaying exponential in Eq. (66), the bilinear variable containing δn may be ignored, at long times, with respect to bilinear variables producing a time dependence $t^{-d/2}$ in $K(t)$. A similar argument shows that the variables that arise in Eq. (63) via the expansion of $\mathbf{v}(\mathbf{r})$ are also negligible.

Bilinear, and higher, nonlinear variables containing δn are also negligible[20] near a critical point. In estimating critical point (ξ) behavior of various contributions to $K^{(c)}(t)$, we have been approximating various correlation functions and transport coefficients by their classical form. Classically, θ has *no* critical anomaly, since χ for viscosity is just $Nmk_{\mathrm{B}}T$, where k_{B} is Boltzmann's constant and T the absolute temperature. However, η^{B} and to a lesser extent η

take on critical anomalies through mode–mode effects. Usually when some transport coefficient with a critical anomaly has the form $\Lambda = L/\chi$, χ varies as ξ^2, L increases with ξ less strongly than ξ^2 due to mode–mode effects, and Λ consequently vanishes slower than the classical prediction as $\xi \to 0$; nonetheless, vanishes it does. If χ is independent of ξ, however, Λ becomes infinite as $\xi \to \infty$, if L increases with ξ; such is the case for η^B. Thus, as $\xi \to \infty$, δn need not even be considered[20] a "slowly varying" variable any more, and all nonlinear variables containing δn may be ignored. Formally, this may be seen from the right-hand side of Eq. (66), which varies as θ^{-1}. As ξ and $\theta \to \infty$, the contribution of the bilinear variable containing δn to $K(t)$ will vanish as θ^{-1}.

Therefore, we may write, with some confidence,

$$\frac{\partial}{\partial t} n_{\mathbf{k}}^{\alpha} = \frac{i\mathbf{k}}{mN} \cdot \sum_{\mathbf{k}'=0}^{\infty} n_{\mathbf{k}+\mathbf{k}'} \mathbf{g}_{-\mathbf{k}'} \tag{67}$$

Equation (67) may be averaged to yield a bilinear Langevin equation for $\bar{n}_{\mathbf{k}}$.

It is often desirable to treat diffusion of particles of variable size; however, all the definitions of densities given so far treat particles as points, e.g., according to Eq. (22), particle i only contributes to the density if $\mathbf{r}$ is infinitesimally close to its center of mass. To introduce particle size we redefine the concentration,[26]

$$n^{\alpha}(\mathbf{r}) = \frac{1}{\frac{4}{3}\pi R^3} \sum_{i=1}^{N_\alpha} S_R(|\mathbf{r} - \mathbf{r}_i|) \tag{68}$$

where R is the particle radius and S the step function,

$$S_R(X) = \begin{cases} 1, & X < R \\ 0, & X > R \end{cases} \tag{69}$$

Equation (68) reduces to Eq. (22) in the limit $R \to 0$. An identical redefinition is made for all other variables, such as $\mathbf{g}^{\alpha}$, containing contributions from α-type particles. The Fourier transform of Eq. (68) is

$$n_{\mathbf{k}}^{\alpha} = \chi(kR) \sum_{i=1}^{N_\alpha} e^{i\mathbf{k}\cdot\mathbf{r}_i} \tag{70}$$

where

$$\chi(kR) = \frac{1}{3} \frac{\sin kR - kR \cos kR}{(kR)^3} \tag{71}$$

All the arguments used for a point particle may be used for a finite particle to show that Eq. (67) holds with the new definitions (also use $S^2 = S$). Note that R is a new, variable, length that may appear in our equations, in addition to $\xi, k^{-1}, k'^{-1}, \ldots$. Just as $\xi \to \infty$ near a critical point, $R \to \infty$ for "Brownian" or large particles.

Only bilinear variables with intermediate wave vector $\leq k_c$ are to be included in the nonlinear Langevin equation, while the relations just derived contain sums over all wave vectors. Let us split up the sum in Eq. (67) and use the definition of the diffusion flux $\dot{n}_{\mathbf{k}}^{\alpha} \equiv i\mathbf{k} \cdot j_{\mathbf{k}}^{\alpha}$, to obtain

$$\mathbf{j}_{\mathbf{k}}^{\alpha} = \frac{1}{mN} \sum_{\mathbf{k}'=0}^{k_c} n_{\mathbf{k}+\mathbf{k}'}^{\alpha} \mathbf{g}_{-\mathbf{k}'} + \frac{1}{mN} \sum_{\mathbf{k}_c}^{\infty} n_{\mathbf{k}+\mathbf{k}'}^{\alpha} \mathbf{g}_{-\mathbf{k}'} \tag{72}$$

Now, suppose we choose $n_{\mathbf{k}}^{\alpha}$ and $n_{\mathbf{k}+\mathbf{k}'}^{\alpha} \mathbf{g}_{-\mathbf{k}'}$, $k' < k_c$, as a variable in the Mori formalism. The well-behaved nonlinear dissipative flux $\mathbf{j}_{\mathbf{k}}^{\alpha+\mathrm{NL}}$ is then simply

$$\mathbf{j}_{\mathbf{k}}^{\alpha+\mathrm{NL}} = \sum_{\mathbf{k}'=k_c}^{\infty} (n_{\mathbf{k}+\mathbf{k}'}^{\alpha} \mathbf{g}_{-\mathbf{k}'})^{+} \tag{73}$$

where + indicates the "linear" orthogonality to $n_{\mathbf{k}}^{\alpha}$ since the $(1-P)$ operator subtracts off the bilinear variable for $k' \leq k_c$. Equation (67) then becomes[26]

$$\frac{\partial}{\partial t} n_{\mathbf{k}}^{\alpha}(t) = -k^2 D_k^0 n_{\mathbf{k}}^{\alpha}(t) + \frac{i\mathbf{k}}{mN} \cdot \sum_{\mathbf{k}'}^{k_c} n_{\mathbf{k}+\mathbf{k}'}^{\alpha}(t) \mathbf{g}_{-\mathbf{k}'}(t) + f^{\mathrm{NL}}(t) \tag{74}$$

where

$$D_k^0 = \int_0^{\infty} dt \, \langle j_{\mathbf{k}}^{\alpha+\mathrm{NL}}(t) j_{-\mathbf{k}}^{\alpha+\mathrm{NL}} \rangle / \langle n_{\mathbf{k}}^{\alpha} n_{-\mathbf{k}}^{\alpha} \rangle \tag{75}$$

Equations like Eq. (74) have been used[37] to study turbulent diffusion. We emphasize that the *only* reason D^0 appears is that we have chosen to treat the "rapidly" varying bilinear variables for $k' > k_c$ implicitly by hiding them in D^0, because we wish to treat explicitly only slowly varying quantities such as the bilinear variable for $k' < k_c$. The function $\chi(kR)$, which appears in Eqs. (70)–(73), is equal to unity for $kR \ll 1$, and falls to zero in damped oscillatory fashion for $kR > \pi$. If $\pi/R < k_c$, $j^{\alpha+\mathrm{NL}}$ and D^0 vanish, while D^0 must be kept for $\pi/R > k_c$. For a pure fluid and for binary mixtures of ordinary fluids, we expect $\pi/R > k_c$; as discussed earlier, $k_c/(\pi/R) = c(R/l)$, with c, $R/l < 1$. For a single large tagged particle moving in a dense solvent of small particles, it is possible to make $\pi/R < k_c$ by increasing R. Thus, D^0 is necessary for mutual diffusion (binary critical point), but may be ignored[26] for self-diffusion by a large particle. Self- and mutual diffusion are two of the most important processes where mode–mode coupling is of interest. For these problems, writing the nonlinear Langevin equation presents no difficulty whatsoever, provided we believe the arguments given for the neglect of δn in the velocity fields.

Sometimes only a part of the nonlinear Langevin equation can be rigorously expressed in bilinear form. Consider an arbitrary conserved variable,

$$C_{\mathbf{k}}^{\alpha} = \sum_{i=1}^{N_\alpha} C_i e^{i\mathbf{k} \cdot \mathbf{r}_i} \tag{76}$$

The time derivative of $C^\alpha_{\mathbf{k}}$ naturally divides into two parts, as $\partial/\partial t$ acts on C_i or $e^{i\mathbf{k}\cdot\mathbf{r}_i}$. The arguments given previously allow one to write

$$\sum_{i=1}^{N_\alpha} C_i \frac{\partial}{\partial t} e^{i\mathbf{k}\cdot\mathbf{r}_i} = \sum_{\mathbf{k}'}^{\infty} C^\alpha_{\mathbf{k}+\mathbf{k}'} i\mathbf{k}\cdot\mathbf{v}_{-\mathbf{k}'} \tag{77}$$

and, as before, $\mathbf{v}$ may be replaced by $\mathbf{g}/m\langle n\rangle$. One then obtains an equation

$$\frac{\partial}{\partial t} C_{\mathbf{k}}(t) = -k^2\lambda_{\mathbf{k}} C_{\mathbf{k}}(t) + \frac{1}{N}\sum_{\mathbf{k}'}^{k_c} C_{\mathbf{k}'+\mathbf{k}'}(t)\frac{i\mathbf{k}}{m}\cdot \mathbf{g}_{-\mathbf{k}'}(t) + f^{\mathrm{NL}}(t), \tag{78}$$

where λ is determined by the flux,

$$j^{+\mathrm{NL}}_{\mathbf{k}} = \sum_{i=1}^{N_\alpha}\left[\left(\frac{\partial}{\partial t}C_i\right)e^{i\mathbf{k}\cdot\mathbf{r}_i}\right]^{+c\mathbf{g}} \tag{79}$$

which does not contain the variable $c\mathbf{g}$. In this way, for example, one may obtain[27,38] the nonlinear Navier–Stokes equation. The transport coefficient λ may still contain nonlinear effects in this case, with variables other than $c\mathbf{g}$, even for an incompressible fluid. There is no way to make a "clean" argument that bilinear variables exhaust the nonlinearities in λ, i.e., λ is not the true well-behaved transport coefficient Λ^0.

In the preceding paragraphs we have tried to explain just how one might argue that bilinear variables alone may be used in mode–mode coupling, a procedure that has almost always been used in the literature. We have seen that this is possible for diffusion, but that less solid arguments are available for other cases. From here on, we shall consider how to write equations for bilinear variables.

The mode–mode coupling problems of greatest interest to chemists involve treatment of pure fluids, where[10,11] the $\mathbf{A}_{\mathbf{k}}$ are $n_{\mathbf{k}}$, $\mathbf{g}_{\mathbf{k}}$, and $e_{\mathbf{k}}$, binary fluids, where the concentration $n^\alpha_{\mathbf{k}}$ is added to the above variables, and self-diffusion,[26,28,29] where the tagged particle density $n^1_{\mathbf{k}}$ replaces the concentration. Of all the $A_{\mathbf{k}}$, $n^1_{\mathbf{k}}$ alone does not involve a sum over $O(N)$ particles. Let[26] $A^c_{\mathbf{k}}$ schematically represent collective linear variables, $A^s_{\mathbf{k}}$ represent single-particle linear variables like n^1, and B^{cc}, B^{cs}, and B^{ss} represent the possible different types of bilinear variables. It is easy to see[26] that the following properties hold:

$$\langle A^s_{\mathbf{k}} A^s_{-\mathbf{k}}\rangle \sim O(1), \qquad \langle A^s_{\mathbf{k}} A^c_{-\mathbf{k}}\rangle \sim O(1)$$

$$\langle A^c_{\mathbf{k}} A^c_{-\mathbf{k}}\rangle \sim O(N), \qquad \langle A^s_{\mathbf{k}} B^{sc}_{-\mathbf{k},\mathbf{k}'}\rangle \sim O(1)$$

$$\langle A^s_{\mathbf{k}} B^{cc}_{-\mathbf{k},\mathbf{k}'} \sim O(1), \qquad \langle A^c_{\mathbf{k}} B^{cc}_{-\mathbf{k},\mathbf{k}'}\rangle \sim O(N)$$

$$\langle B^{sc}_{\mathbf{k},\mathbf{k}'} B^{sc}_{-\mathbf{k},\mathbf{k}''}\rangle \sim \begin{cases} O(N), & \mathbf{k}'=\mathbf{k}'' \\ O(1), & \mathbf{k}'\neq\mathbf{k}'' \end{cases}$$

$$\langle B^{cc}_{\mathbf{k},\mathbf{k}'} B^{cc}_{-\mathbf{k},\mathbf{k}''}\rangle \sim \begin{cases} O(N^2), & \mathbf{k}'=\mathbf{k}'' \\ O(N), & \mathbf{k}'\neq\mathbf{k}'' \end{cases} \tag{80}$$

We ignore B^{ss}, since A^s is almost always n^1, and $n^1_{\mathbf{k}+\mathbf{k}'}n^1_{-\mathbf{k}'} = n^1_{\mathbf{k}}$, i.e., $B^{ss} = A^s$ in this case. Certain variables may now be decoupled by $O(N)$ arguments. The flux for self-diffusion, j^1, is also an "A^s." As in Eq. (35), we have schematically

$$j^{1+}_{\mathbf{k}} = j^{1+,NL}_{\mathbf{k}} + \frac{\langle j^{1+}_{\mathbf{k}}\, A^c_{-\mathbf{k}}\rangle}{\langle A^c_{\mathbf{k}}\, A^c_{-\mathbf{k}}\rangle} A^c_{\mathbf{k}} + \sum_{\mathbf{k}'}^{k_c} \frac{\langle j^{1+}_{\mathbf{k}}\, B^{sc}_{-\mathbf{k},\mathbf{k}'}\rangle}{\langle B^{sc}_{\mathbf{k},\mathbf{k}'}\, B^{sc}_{-\mathbf{k},\mathbf{k}'}\rangle} B^{sc}_{\mathbf{k},\mathbf{k}'}$$
$$+ \sum_{\mathbf{k}'}^{k_c} \frac{\langle j^{1+}_{\mathbf{k}} B^{cc}_{\mathbf{k},\mathbf{k}'}\rangle}{\langle B^{cc}_{\mathbf{k},\mathbf{k}'} B^{cc}_{-k,k'}\rangle} B^{cc}_{\mathbf{k},\mathbf{k}'} \tag{81}$$

the coefficients in the second, third, and fourth terms in Eq. (81) are $O(1/N)$, $O(1/N)$, and $O(1/N^2)$, respectively. Since a sum counts as $O(N)$, it follows that the only bilinear variables to which A^s couples is B^{sc}. An analogous argument shows that A^c only couples to B^{cc}. We will therefore[26] be considering the coupled sets of variables (A^s, B^{sc}) and (A^c, B^{cc}).

We have now eliminated as many variables from consideration as is possible with the use of general agreements; it is time to start writing the nonlinear Langevin equation for the remaining variables. To obtain the equations, we must form the product $(i\Omega^{NL} + L^0)(\chi^{NL})^{-1}$. The only quantity in this product that is not at least formally available is χ^{-1}; the inversion of an infinite rank matrix can be a formidable task. Thus, we first attempt to find out if χ might be inverted in some simpler manner.

We order the variables for matrix multiplication within a coupled set, as $A_{\mathbf{k}}$, then $B_{\mathbf{k},\mathbf{k}'}$ for a given $\mathbf{k}'$, then $B_{\mathbf{k},\mathbf{k}'}$ for the next $\mathbf{k}'$, and so on. Note that a bilinear variable $A_{\mathbf{k}+\mathbf{k}_i}A_{-\mathbf{k}_i}$ is identical for $\mathbf{k}_i = \mathbf{k}'$ and $\mathbf{k}_i = -\mathbf{k} - \mathbf{k}'$. Thus, each variable of this type will be counted twice when we sum over all $\mathbf{k}_i$. On the other hand, a variable $A_{\mathbf{k}+\mathbf{k}_i}A'_{-\mathbf{k}_i}$ is different for all $\mathbf{k}_i$. Thus, as our bilinear variables for each $\mathbf{k}_i$, we keep both $A_{\mathbf{k}+\mathbf{k}_i}A'_{-\mathbf{k}_i}$ and $A'_{\mathbf{k}+\mathbf{k}_i}A_{-\mathbf{k}_i}$ for all $A \neq A'$. In this way, every bilinear variable will be counted twice when we sum over $\mathbf{k}_i$, and we shall simply multiply each sum by $\frac{1}{2}$ by incorporating a $\frac{1}{2}$ into every vertex. For manipulative purposes, it is probably most convenient[27] to introduce modified bilinear variables that are orthogonalized to the linear variables, i.e.,

$$B_{\mathbf{k},\mathbf{k}'} \to B_{\mathbf{k},\mathbf{k}'} - \frac{\langle B_{\mathbf{k},\mathbf{k}'}A_{-\mathbf{k}}\rangle}{\langle A_{\mathbf{k}}A_{-\mathbf{k}}\rangle} A_{\mathbf{k}} \equiv B'_{\mathbf{k},\mathbf{k}'} \tag{82}$$

where $\langle B'_{\mathbf{k},\mathbf{k}'}A_{-\mathbf{k}}\rangle = 0$. From Eq. (80), it is seen[26,27] that, in the collective-variable problem, χ^{NL} has a block in the upper left-hand corner, which is $O(N)$ ($A - A$ terms), successive blocks, $O(N^2)$, M in number, along the rest of the diagonal ($B'_{\mathbf{k},\mathbf{k}'} - B'_{\mathbf{k},\mathbf{k}'}$ terms), zeros by construction in the off-diagonal elements of the first row and column ($B - A$ terms), and off-diagonal block elements in the rest of the matrix, which are $O(N)$ ($B'_{\mathbf{k},\mathbf{k}'} - B_{\mathbf{k},\mathbf{k}''}$ terms). Of course, any particular matrix element may be zero. The order of the matrix χ^{NL} for the single-particle problem is obtained from that given above by dividing everything by N.

Near a critical point, or if large particles are involved, the elements of χ^{NL} may have important ξ or R dependence. Suppose, however, that we ignore such complexities for the moment. The inverse of χ^{NL} may be developed in a perturbation series about the inverse of the block diagonal χ^{NL} (i.e., no $B'_{\mathbf{k},\mathbf{k}'}-B'_{\mathbf{k},\mathbf{k}''}$ elements, $\mathbf{k}'\neq\mathbf{k}''$). In this way, we find that the corrections to the block diagonal inverse χ^{NL} are $B'_{\mathbf{k},\mathbf{k}'}-B'_{\mathbf{k},\mathbf{k}''}$, $\mathbf{k}'\neq\mathbf{k}''$, terms that are $O(1/N)$ with respect to the diagonal terms.

The N dependence of the matrices L^0 and $i\Omega^{\mathrm{NL}}$ is the same as that of χ, with the addition of off-diagonal $(A-B')$ elements of order N [or $O(1)$] for the collective (single-particle) problem. The N dependence of the equations of motion may be now exhibited,[26,27]

$$\frac{\partial}{\partial t}A_{\mathbf{k}}\sim O\left[A_{\mathbf{k}}+\left(\frac{1}{N}+\frac{M}{N^2}\right)\sum_{\mathbf{k}'}B'_{\mathbf{k},\mathbf{k}'}\right] \tag{83}$$

where we have used the $M\times M$ dimensionality of the bilinear part of χ^{-1} The coefficient of order M/N^2 in Eq. (83) is the only coefficient where the $B'_{\mathbf{k},\mathbf{k}'}-B'_{\mathbf{k},\mathbf{k}''}$ terms in χ^{-1} contribute. According to the definition of M, we expect[27] $M/N\approx(c)^3\ll 1$. Thus, $M/N^2\ll 1/N$, and the off-block diagonal elements of χ and χ^{-1} make a negligible contribution to the equations of motion and may be ignored. With the simple nature of χ^{-1} in hand, we may now write the linear-bilinear coupling coefficient, $\mathscr{V}^{A-B}$, which belongs in Eq. (49). We shall now drop the $A-B$ superscripts, as only bilinear variables are being considered:

$$\begin{aligned}\mathscr{V}_{\mathbf{k};\,\mathbf{k}+\mathbf{k}',\,-\mathbf{k}'}&\equiv\tfrac{1}{2}[i\Omega^{\mathrm{NL}}(A_{\mathbf{k}}-B'_{\mathbf{k},\mathbf{k}'})+L^0(A_{\mathbf{k}}-B'_{\mathbf{k},\mathbf{k}'})]\chi^{\mathrm{NL}-1}(B'_{\mathbf{k},\mathbf{k}'}-B'_{\mathbf{k},\mathbf{k}'})\\&\to\tfrac{1}{2}[i\Omega^{\mathrm{NL}}(A_{\mathbf{k}}-B'_{\mathbf{k},\mathbf{k}'})+L^0(A_{\mathbf{k}}-B'_{\mathbf{k},\mathbf{k}'})]\langle B'_{\mathbf{k},\mathbf{k}'}B'_{-\mathbf{k},\mathbf{k}'}\rangle^{-1}\end{aligned} \tag{84}$$

Of course, B' now appears in Eq. (49) instead of B. From the definition of Ω^{NL} and L^0, from Eqs. (5) and (16), and from $\dot{A}_{\mathbf{k}}=i\mathbf{k}\cdot\mathbf{j}_k$, it is easy to see that the leading term in the wave-vector expansion of Ω^{NL} is $O(k)$, while the leading term in L^0 is $O(k,k')$. We have previously explained that a fundamental idea for mode–mode coupling is that intermediate wave vectors k' may be treated[26,27] as small, just as k is treated as small for the ordinary linear laws; thus, we shall ignore L^0 in Eq. (84). To be consistent in expanding the equations of motion to second order in the wave vector, this term should really be kept; however, such terms are almost impossible to evaluate. Furthermore, since the quadratic L^0 term is added to an $i\Omega^{\mathrm{NL}}$ term linear in the wave vector, it is on a somewhat different footing from a quadratic term that is the leading term of a certain matrix element. For example, according to Eq. (56), the inclusion of L^0 in Eq. (84) would add a $t^{-5/2}$ tail to the $t^{-3/2}$ term arising from $i\Omega$. Whenever any term is the leading term in a matrix element, we shall retain it. We therefore obtain

$$\mathscr{V}_{\mathbf{k};\,\mathbf{k}+\mathbf{k}',\,-\mathbf{k}}=\tfrac{1}{2}\lim_{\substack{k\to 0\\ k'\to 0}} i\mathbf{k}\cdot\langle\mathbf{j}_{\mathbf{k}}B'_{-\mathbf{k},\mathbf{k}'}\rangle\langle B'_{\mathbf{k},\mathbf{k}'}B'_{-\mathbf{k},\mathbf{k}'}\rangle^{-1} \tag{85}$$

with the restriction that kR, $k\xi$, $k'R$, and $k'\xi$ may not, as the occasion demands, be small. Equation (85) may be further simplified by noting that, as $N \to \infty$ (Let $B_{\mathbf{k},\mathbf{k}'} = A^i_{\mathbf{k}+\mathbf{k}'}A^j_{-\mathbf{k}'}$)

$$\langle B'_{\mathbf{k},\mathbf{k}'}B'_{-\mathbf{k},\mathbf{k}'}\rangle = \langle A^i_{\mathbf{k}+\mathbf{k}'}A^i_{-\mathbf{k}-\mathbf{k}'}\rangle\langle A^j_{\mathbf{k}'}A^j_{-\mathbf{k}'}\rangle \tag{86}$$

It is easy to verify that corrections to the right-hand side of Eq. (86) are $O(1/N)$ smaller than the term kept.

The coefficient of $\bar{A}_{\mathbf{k}}$ on the right-hand side of Eq. (49) is easily obtained, using the block diagonality of χ^{-1}:

$$\frac{\partial}{\partial t}\bar{A}_{\mathbf{k}}(t) = [i\Omega^{\mathrm{NL}}(A_{\mathbf{k}} - A_{\mathbf{k}}) + L^0(A_{\mathbf{k}} - A_{\mathbf{k}})]\langle A_{\mathbf{k}}A_{-\mathbf{k}}\rangle^{-1}\overline{A_{\mathbf{k}}(t)} + \text{bilinear terms} \tag{87}$$

and $[L^0(A_{\mathbf{k}} - A_{\mathbf{k}})$ is $O(k^2)]$

$$\Lambda^0_k = k^{-2}L^0(A_{\mathbf{k}} - A_{\mathbf{k}})\langle A_{\mathbf{k}}A_{-\mathbf{k}}\rangle^{-1} = L^{(c)0}(A_{\mathbf{k}} - A_{\mathbf{k}})\langle A_{\mathbf{k}}A_{-\mathbf{k}}\rangle^{-1} \tag{88}$$

Also, by definition,

$$i\Omega^{\mathrm{NL}}(A_{\mathbf{k}} - A_{\mathbf{k}})\langle A_{\mathbf{k}}A_{-\mathbf{k}}\rangle^{-1} \equiv i\omega^{\mathrm{NL}}_k = i\omega_k \tag{89}$$

The complete mode–mode equation is thus Eq. (49), with $\omega^{\mathrm{NL}}_k = \omega_k$, Λ^0_k given by Eq. (88), with $\mathcal{V}$ given by Eq. (85), and with $\mathcal{V}^{A-n} = 0$, $n > 2$.

For any problem, once we have isolated the correct set of variables, we now have a rule for writing the equation of motion for the nonequilibrium ensemble average of the linear variables. Of course, the same equation holds for the time correlation functions of the linear variables, and with the addition of the random force f^{NL}, for the fluctuating linear variables [Eqs. (14) and (10)]. The equation contains $i\omega_k$, $\langle A_{\mathbf{k}}A_{-k}\rangle$, $\mathcal{V}$, and $\langle B'_{\mathbf{k},\mathbf{k}'}B'_{-\mathbf{k},\mathbf{k}'}\rangle$, which we have to evaluate in most cases, and L^0. Hopefully, all we need to know about L^0 is that it behaves classically. The equation we have just obtained after considerable argument is in fact written immediately by most authors. However, we feel that this section has not been a waste of time. We have dwelt at length on the mechanics of writing mode–mode equations because there exists relatively little literature on the subject. It is important to realize, when manipulating mode–mode equations, just what assumptions those equations are based upon. In the next section, we shall see that considerable effort has been expended in attempts to obtain better and better solutions to our final version of Eq. (49). Since in many cases the equations being solved are based upon the numerous approximations just discussed, one questions the utility of extreme refinement of the solutions.

In the last stages of our recent argument, we assumed that the matrix elements of χ did not contain R or ξ, which may become large for large particles or near a critical point. For large R or ξ, we must be careful, in taking the limit in Eq. (85), that kR, $k'R$, $k\xi$, and $k'\xi$ are not small. In addition, our argument that χ may be inverted by iteration can break down[26] for large ξ, R. For example, if the magnitude of the off-block diagonal elements in the bilinear part of χ increases as some $f(\xi)$ or $f(R)$, the argument given previously now indicates that the coefficient $(1/N+M/N^2)$ in Eq. (83) becomes $(1/N+Mf/N^2)$; although $M/N \ll 1$, there is no assurance that $Mf/N \ll 1$ as $f(\xi)$ or $f(R)$ becomes infinite with $\xi, R \to \infty$. In these cases, χ must be inverted by other means; one must simply check from problem to problem to see if this difficulty arises. An example of this point is[26] self-diffusion by a large particle.

Before continuing, it seems worthwhile to comment on a potentially confusing point. We have decided that it is necessary to write a nonlinear Langevin equation [Eq. (49)] to describe the motion of $\bar{A}_{\mathbf{k}}(t)$ arbitrarily near equilibrium. However, one might expect that $\overline{A_{\mathbf{k}+\mathbf{k}'}(t)A_{-\mathbf{k}'}} \propto \bar{A}_{\mathbf{k}+\mathbf{k}'}\bar{A}_{-\mathbf{k}}$, in which case the bilinear term should be negligible with respect to the linear term close enough to equilibrium $(\bar{A} \to 0)$. Why then must the bilinear term be kept in Eq. (49)? The reason[39,40] is that $\overline{A_{\mathbf{k}+\mathbf{k}'}A_{-\mathbf{k}'}}$ may be thought to have a "part" that behaves as $\bar{A}_{\mathbf{k}+\mathbf{k}'}\bar{A}_{-\mathbf{k}'}$, but also a part that behaves as $\bar{A}_{\mathbf{k}}$; the $\bar{A}_{\mathbf{k}}$ term does not vanish near equilibrium. Indeed, it is possible to show[40] that the true bilinear equation (in a mathematical sense), i.e., an equation where $\dot{\bar{A}}_{\mathbf{k}}$ is expressed in terms of $\bar{A}_{\mathbf{k}}$ and $\bar{A}_{\mathbf{k}+\mathbf{k}'}\,\bar{A}_{-\mathbf{k}'}$, has precisely the same badly behaved kernel as the linear laws. Thus, as intuition insists, a true bilinear term, a product of averages, in an equation of motion may be neglected near equilibrium; average of products may not be ignored.

3. Solving Mode–Mode Equations

A method of solving the nonlinear Langevin equation has been given[20] by Kawasaki. Kawasaki's method is based upon the pioneering work of Wyld[41] on turbulence theory, which is itself based upon standard techniques[42] of quantum field theory. These techniques are familiar to physicists, but usually appear fairly esoteric to chemists. A different method has been given by Bedeaux and Mazur.[43] Both methods give formally exact solutions to the nonlinear Langevin equation. Of course, approximations must be made to these formally exact solutions by mortal man. The two methods give the same zero-order solution, but differ in the way successive approximations naturally arise. In both cases, there is no obvious small parameter associated with neglected terms. However, Lo and Kawasaki[44] has shown *a posteriori* that his "vertex correction" is small, while Bedeaux and Mazur[43] argue concerning the asymptotic time dependence of neglected terms. Since both methods give the same zero-order solution, and since we are not going to discuss refinements of that solution here, we shall just discuss the method of Kawasaki. Most published work on mode–mode coupling employs this method. Of course, Kawasaki himself has presented[20] his own ideas in great detail, and we do not wish simply to repeat his expositions. Therefore, this section is basically a "primer," which will hopefully aid the nonspecialist in reading the original papers.

In attempting to solve the nonlinear Langevin equation [Eq. (49)], one is immediately struck that a solution is not obtainable by ordinary mathematical techniques. Since the nonlinear variables $\overline{A_{\mathbf{k}+\mathbf{k}'}A_{-\mathbf{k}'}}$ are not known as a function of $\bar{A}_{\mathbf{k}}$ ($\overline{AA} \neq \bar{A}\bar{A}$), Eq. (49) is simply not a closed equation for the variable of interest, $\bar{A}_n$ To circumvent this difficulty, it is necessary[20] to start with the fluctuating form of the nonlinear Langevin equation. The fluctuating form of Eq. (49) is

$$\frac{\partial}{\partial t}A_{\mathbf{k}}(t) = (i\omega_k - k^2\Lambda_k^0)A_{\mathbf{k}}(t) + \sum_{\mathbf{k}'}^{k_c} \mathcal{V}_{\mathbf{k};\,\mathbf{k}+\mathbf{k}',\,-\mathbf{k}'}A_{\mathbf{k}+\mathbf{k}'}(t)A_{-\mathbf{k}'}(t) + f^{\mathrm{NL}}(t) \quad (90)$$

In Eq. (90), the bilinear term is truly a product of linear terms and may be manipulated by standard techniques. The resulting expression for $A_{\mathbf{k}}(t)$ may then be averaged over a nonequilibrium ensemble to obtain $\bar{A}_{\mathbf{k}}(t)$, or the expression may be multiplied on the left by $A_{-\mathbf{k}}(0)$ and averaged over an equilibrium ensemble to obtain $\langle A_{\mathbf{k}}(t)A_{-\mathbf{k}}\rangle$. In the following, we shall concentrate on the calculation of the time correlation function, which is, of course, identical to the calculation of $\overline{A_{\mathbf{k}}(t)}$ in the limit of small deviations from equilibrium.

A problem arises when trying to write Eq. (90) in vector form in the space of dynamical variables. If n linear variables are needed to form a complete set, then there exist n^2 possible bilinear variables. Thus, $\mathcal{V}$ must be an $n \times n^2$ matrix, and AA is a vector in an n^2-dimensional space; as it turns out, these considerations would complicate the manipulations we are about to perform. Kawasaki has given a very simple alternate way to treat the vectorial structure of Eq. (90). This is to extend the definition of the wave vector of a variable to include an index, which specifies the identity of the variable itself. Thus, instead of $A_{\mathbf{k}}^i$, we now just write $A_{\mathbf{k}}$, where $\mathbf{k}$ is understood to contain both the Fourier transform variable and the index i. Sums over $\mathbf{k}$ are similarly reinterpreted. Now, combinations of the linear variables can always be chosen such that the matrix $(i\omega_k - k^2\Lambda_k^0)$ is diagonal in the space of dynamical variables. Henceforth, we shall always assume that such a choice has been made. Then Eq. (90) holds as a vector equation with the new interpretation of wave vector. The wave-vector subscripts $\mathbf{k}$, $\mathbf{k}+\mathbf{k}'$, and $-\mathbf{k}'$ on $\mathcal{V}$ now index the identity of $A_{\mathbf{k}}$, $A_{\mathbf{k}+\mathbf{k}'}$, and $A_{-\mathbf{k}'}$ in Eq. (90).

The starting point of Kawasaki's[20] method is a well-known identity. If

$$\partial\gamma(t)/\partial t = \beta\gamma(t) + \varepsilon(t) \quad (91)$$

and if $\gamma^0(t)$ is available, where

$$\partial\gamma^0(t)/\partial t = \beta\gamma^0(t), \qquad \gamma^0(t=0) = \gamma(t=0) \quad (92)$$

then it may be useful to write the exact expression,

$$\gamma(t)=\gamma^0(t)+\int_0^t d\tau\frac{\gamma^0(t-\tau)}{\gamma^0(t=0)}\varepsilon(\tau) \tag{93}$$

Equation (93) may be iterated to obtain $\gamma(t)$ as an infinite series if ε contains $\gamma(t)$. The series may converge quickly, if ε is small in some sense, or an infinite order resummation of selected terms may be carried out. If we partition the fluctuating Langevin equation according to Eq. (91), with

$$\beta=i\omega_k-k^2\Lambda_k^0 \tag{94}$$

then

$$A_{\mathbf{k}}^0(t)=A_{\mathbf{k}}(t=0)\exp(i\omega_k t-k^2\Lambda_k^0 t) \tag{95}$$

and one might be tempted to iterate the solution of Eq. (90) starting with $A_{\mathbf{k}}^0$. The choice of $A_{\mathbf{k}}^0$ as given by Eq. (95) is[20] a poor one, because this $A_{\mathbf{k}}^0$ is badly behaved at negative times, and as we shall see, $A_{\mathbf{k}}^0$ does take on negative times in the calculation of time correlation functions. We are not particularly interested in negative times, but simply want to assure that $A_{\mathbf{k}}^0$, given by Eq. (95), does not appear anywhere in the mathematics with a negative time argument. It therefore appears natural to define[20] a zero-order propagator,

$$G_k^0(t)=\theta(t)\exp(i\omega_k t-k^2\Lambda_k^0 t) \tag{96}$$

where

$$\theta(t)=\begin{cases}1, & t>0\\ 0, & t<0\end{cases} \tag{97}$$

and a full propagator,

$$G_k(t-t')=\theta(t-t')\langle A_{\mathbf{k}}(t)A_{-\mathbf{k}}(t')\rangle/\langle A_{\mathbf{k}}A_{-\mathbf{k}}\rangle \tag{98}$$

where we have noted that the exact equilibrium time correlation function only depends on the difference of the two time arguments. We may now use Eq. (93) to write a solution of Eq. (90) for $t>0$,

$$A_{\mathbf{k}}(t)=A_{\mathbf{k}}(t=0)G_k^0(t)+\int_0^t d\tau\, G_k^0(t-\tau)f_{\mathbf{k}}^{\mathrm{NL}}(\tau)$$
$$+\int_0^t d\tau\, G_k^0(t-\tau)\sum_{\mathbf{k}'}^{k_c}\mathcal{V}_{\mathbf{k};\,\mathbf{k}+\mathbf{k}',-\mathbf{k}'}A_{\mathbf{k}+\mathbf{k}'}(\tau)A_{-\mathbf{k}'}(\tau) \tag{99}$$

which may be used to obtain the full propagator.

For a zero-order solution of Eq. (99) we do not take A^0 as given by Eq. (95), but $(t>0)$

$$A_{\mathbf{k}}^0(t)=A_{\mathbf{k}}(t=0)G_k^0(t)+\int_0^t d\tau\, G_k^0(t-\tau)f_{\mathbf{k}}^{\mathrm{NL}}(\tau) \tag{100}$$

i.e.,

$$\frac{\partial}{\partial t}A_{\mathbf{k}}^0(t)=(i\omega_k-k^2\Lambda_k^0)A_{\mathbf{k}}^0(t)+f_{\mathbf{k}}^{\mathrm{NL}}(t) \tag{101}$$

Thus, $A^0_{\mathbf{k}}$ is a fluctuating quantity that obeys an equation similar to the fluctuating linear law, but containing the well-behaved random force. The motivation[20] for our final choice of A^0 is that the random force in Eq. (101) maintains the time reversibility of the equation for A^0. The equation for the earlier mentioned A^0, Eq. (95), was irreversible, giving rise to unphysical behavior at negative time. We now have

$$\langle A^0_{\mathbf{k}}(t')A^0_{-\mathbf{k}}(t'')\rangle/\langle A_{\mathbf{k}}A_{-\mathbf{k}}\rangle = G^0_k(t'-t'') + G^0_k(t''-t) \tag{102}$$

which is perfectly well behaved for all times.

Iteration of Eq. (99) about A^0 now gives

$$\begin{aligned} A_{\mathbf{k}}(t) &= A^0_{\mathbf{k}}(t) + \int_0^t G^0_k(t-\tau)\sum_{\mathbf{k}'}^{k_c} \mathcal{V}_{\mathbf{k};\,\mathbf{k}+\mathbf{k}',-\mathbf{k}'} A^0_{\mathbf{k}+\mathbf{k}'}(\tau)A^0_{-\mathbf{k}'}(\tau) + \cdots \\ &\equiv A^0_{\mathbf{k}}(t) + A^1_{\mathbf{k}}(t) + A^2_{\mathbf{k}}(t) + \cdots \end{aligned} \tag{103}$$

Due to the structure of Eq. (99), the products of $\mathcal{V}$, A^0, and G^0 that appear under sum and integral signs on the right-hand side of Eq. (103) have the following form. These are n vertices in a contribution to A^n. Immediately to the left of a given vertex is a G^0 with time argument, say, $t_j - t_i$, $t_j > t_i$; the vertex is assigned time t_i. Immediately to the right of the vertex are two A^0, and A^0, and a G^0, or two G^0. The time argument of any A^0 is t_i, and the time argument of any G^0 is $t_i - t_h$, where $t_i > t_h$. If the wave vectors of the two quantities to the right of the vertex are $\mathbf{k}'$ and $\mathbf{k}''$, then the wave vector of the propagator to the left is $\mathbf{k}'+\mathbf{k}''$, and the vertex is $\mathcal{V}_{\mathbf{k}'+\mathbf{k}'';\mathbf{k}',\mathbf{k}''}$.

The entity just described is represented graphically by three lines meeting at a dot representing the vertex. A solid line runs off to the left, representing a propagator. Two lines, each of which may be solid or dashed, representing a propagator or an A^0, respectively, run off to the right. Each contribution to A^n is built up by combining n of the fundamental units. Graphically, this is done by joining the propagator running to the left of one dot to a propagator running to the right of the dot to its immediate left (they then become the same propagator). Time increases from right to left along the graph. The time argument of a propagator running from the left of a vertex with time t_i to a vertex with time t_j, $t_j > t_i$, is $t_j - t_i$; the argument of the propagator on the extreme left of a graph is $t - t_l$, where t_l is the time assigned to the leftmost vertex. All topologically different graphs through fourth order are shown in Fig. 1.

The rules for writing the contribution to $A(t)$ of a graph, which have been given by Kawasaki[20] and follow from the discussion just given, are now summarized. Write down all the A^0, G^0, and $\mathcal{V}$ indicated in the graph. Assign time arguments as indicated, with $t_1, t_2, t_3, \ldots, t_l$ denoting increasing time from right to left. Assign wave vectors to each A^0 and G^0, subject to the conditions that the wave vectors $\mathbf{k}', \mathbf{k}''$ of the lines joining at the right of a vertex $\mathcal{V}_{\mathbf{k}'+\mathbf{k}'',\,\mathbf{k}',\,\mathbf{k}''}$

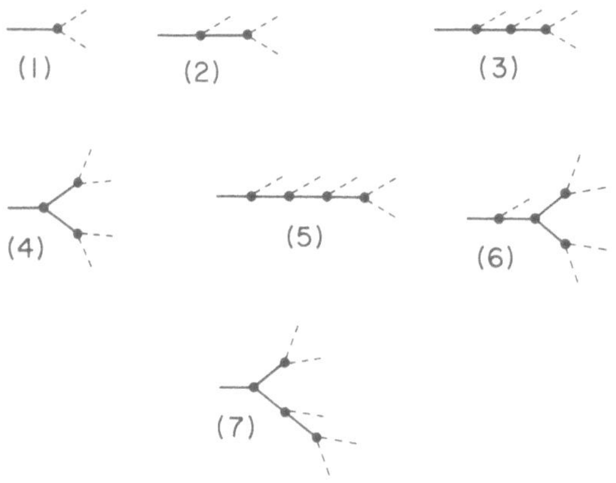

Fig. 1. Graphs for $A_k(t)$, $t>0$, to fourth order in $\mathscr{V}$.

add to determine that the wave vector of the propagator leaving from the left is $\mathbf{k}'+\mathbf{k}''$, and that the wave vector of the propagator farthest to the left is $\mathbf{k}$. All independent wave vectors are summed from zero to k_c; t_n is integrated from zero to t_{n+1}, and t_l from zero to t.

The combinatorial factors associated with the graphs are[20] very simple. Consider the two "trees" that run left to right from the right to each dot. If both trees are the same, the graph will only be generated once by Eq. (103). However, if the trees are different, the graph will be generated twice. Thus, if we agree to incorporate a factor of two into each vertex with asymmetric branching, A^n is represented by the sum of all topologically different graphs with n vertices.

The skeleton of the graphs in Fig. 1 is composed of solid propagator lines; the dashed A^0 lines just "hang" from the vertices. Thus, there is no need to carry the A^0 if one simply remembers that A^0 are always present such that each vertex has two lines entering on the right-hand side. With this understanding, the dashed lines in Fig. 1 may be erased, giving rise to the graphs[20] of Kawasaki.

Having studied the infinite series for $A(t)$, $t>0$, we now multiply each term by $A_{-\mathbf{k}}(0)$, average, and divide by $\langle A_{\mathbf{k}}A_{-\mathbf{k}}\rangle$ to form an infinite series for $G_k(t)$. Immediately we are forced to make an approximation, since correlation functions of large numbers of A^0 arise whose time dependence is unknown. (Note that the averaging process does not act on G^0, which has no fluctuating character.) To escape this dilemma, Kawasaki[20] assumes that time correlation functions of A^0 factorize,

$$\langle A^0_{\mathbf{k}_1}(t_1)A^0_{\mathbf{k}_2}(t_2)\cdots A^0_{\mathbf{k}_n}(t_n)\rangle = \begin{cases} 0, & n \text{ odd} \\ \text{(the sum of all products of } n/2 \text{ pair correlation functions formed by dividing the } n\, A^0 \text{ into } n/2 \text{ distinct pairs)}, & n \text{ even} \end{cases} \qquad (104)$$

Since

$$\langle A^0_{\mathbf{k}_i}(t_i) A^0_{\mathbf{k}_j}(t_j)\rangle = \langle A^0_{\mathbf{k}_i}(t_i) A^0_{\mathbf{k}_i}(t_j)\rangle \delta_{\mathbf{k}_i+\mathbf{k}_j} \quad (105)$$

the various decompositions of higher-order correlation functions into pair correlation functions give rise to important constraints on wave-vector sums. Equation (104) represents the first approximation involved in the solution of mode–mode equations. It is very hard to judge the validity of Eq. (104), since a time-dependent correlation function is involved, and equilibrium statistical mechanics cannot be used as a check. As has been seen in Eq. (86), it is possible to factor equal time correlation functions in many cases; typically, corrections to the factorization approximation for equal time correlation functions are $O(1/N)$. If this holds for time-dependent quantities, the corrections to Eq. (103) would be $O(M/N)$, since the nonfactorized part of the higher order correlation function is summed over wave vectors. We have argued that M/N is small, although any ξ or R dependence could destroy this smallness. Equation (104) seems reasonable, but in no way may be considered firmly established.

Let us now consider[20] multiplying a term in Eq. (103), represented by a graph, by $A_{-\mathbf{k}}(0)\langle A_{\mathbf{k}}A_{-\mathbf{k}}\rangle^{-1}$ and averaging. For odd n, $A^{(n)}$ will not contribute since such $A^{(n)}$ have even numbers of A^0, and thus, by Eq. (104), $\langle A^{(n)}A_{-\mathbf{k}}(0)\rangle$ vanishes, n odd. For $A^{(n)}$, n even, consider the graphs in Fig. 1. These graphs will have an odd number, $n+1$, of dashed A^0 lines. Of these, according to Eq. (104), n will be partitioned into $n/2$ pairs, and one will be paired with $A_{-\mathbf{k}}(0)$, to form $n/2+1$ pair correlation functions; all possible pairings will be formed. The pairing of two A^0 is indicated graphically by joining the corresponding dashed lines and writing the resulting single line as a wiggly line. The wave vectors of the dashed line that comprise each "half" of the wiggly line must sum to zero. The wiggly line is the graphical symbol for $\langle A^0_{\mathbf{k}}(t_1)A^0_{-\mathbf{k}}(t_2)\rangle$, given by Eq. (102). The sequence of propagators running from the extreme left of the graph to the vertex where the A^0, say $A^0_{\mathbf{k}_r}(t_r)$, with which $A_{-\mathbf{k}}(0)$ is paired, is attached, is called[20] the trunk of the graph. The factor of $\langle A_{\mathbf{k}}A_{-\mathbf{k}}\rangle^{-1}$ is incorporated into this pairing to form a propagator, $G^0_k(t_r)$. Thus, a straight propagator line extends to the right of the last vertex on the trunk; the trunk begins and ends with factors of G_k running to the left and right of the vertices with times t_ℓ and t_r, respectively.

If a wiggly line runs between two vertices on the same "branch" of the tree, the time argument of the correlation function represented cannot change sign, and according to Eq. (102) the line will be equivalent to a single propagator. However, the time argument of wiggly lines connecting vertices on different branches may change sign, and so the full representation of the time correlation function in terms of two propagators [Eq. (102)] is necessary.

The diagrams contributing to $G_k(t)$ derived from the diagrams in Fig. 1 are shown in Fig. 2. Kawasaki[20] has pointed out that diagrams where a subgraph may be cut off from the trunk by cutting a single propagator line, examples of

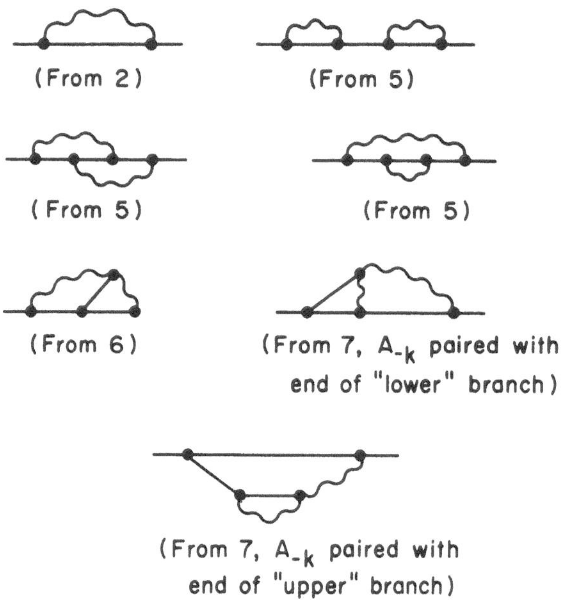

Fig. 2. Graphs contributing to $G_k(t)$, derived from the graphs in Fig. 1.

which are shown in Fig. 3, must be excluded. The reason for this is as follows. The trunk begins and ends with propagator lines with wave vector k. Wave vector is conserved at every vertex. thus, any propagator lines that "feed into" the trunk as it proceeds from right to left must carry zero wave vector. A zero wave-vector propagator contains $A_{k=0}$, which is an average quantity, not a fluctuating quantity. As discussed in Section 1, there is no need to keep $A_{k=0}$ as a dynamical variable. Thus, G_0 is effectively zero, as is the case of diagrams in question.

Kawasaki[20] also gives the rule, "In a diagram obtained by decorating another diagram which may be merely a set of vertices, if the decoration contains a vertex whose time coordinate is smaller than the smallest of the time coordinates of the vertices connecting the decoration to the original diagram, such a decoration must be omitted." An example of the type of diagram excluded by this rule is given in Fig. 4. Kawasaki[20] gives a very general argument to justify this rule for which we now make a "plausibility" argument.

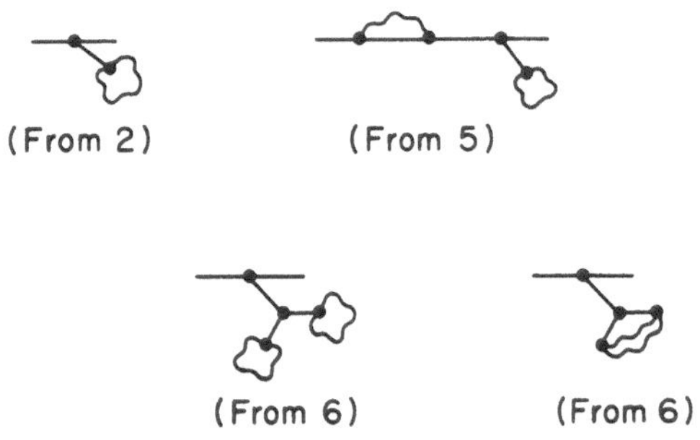

Fig. 3. Graphs not contributing to $G_k(t)$, derived from the graphs in Fig. 1, excluded due to the appearance of $A_{k=0}$.

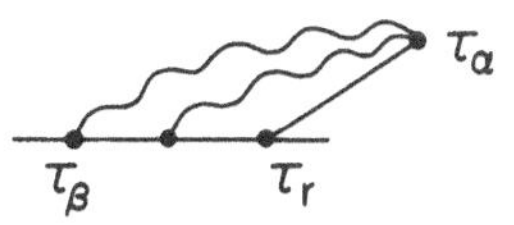

Fig. 4. Graph not contributing to $G_k(t)$, derived from the graph labeled (a) in Fig. 1, excluded due to the "trapped" vertex at t_α.

The diagrams for $G(t)$ involve time integrations for all of the vertices. However, for many configurations of the vertices, the integrand vanishes. This is because many vertices are connected by propagator or correlation function lines for some wave vector $\mathbf{k}'$, which is to be summed over. Although the wave-vector sum includes low wave vectors, for all practical purposes a wave vector that is summed over acts as a high wave vector. Propagator and correlation functions of high wave vectors are rapidly decreasing functions of their time argument. We are interested in $G(t)$ at large times, where t may be considered a large ordering parameter. In order to make estimates of the magnitude of the various diagrams for large t, it is possible to consider vertices connected by summed-over wave-vector lines as constrained to remain close together in the time integration. The magnitude of the diagram will then depend on how many free integrations remain.

If a diagram has a decoration with a vertex at an earlier time than any of the vertices of the main diagram in question, then that vertex will be connected, by a high-wave-vector line, to a vertex of the main diagram with later time than the time τ_r of the earliest vertex of the main diagram. For example, in Fig. 4 the vertex of the decoration at t_α is connected to the vertex of the diagram at t_β, later than the earliest vertex of the diagram at t_r. The effect of this connection is to "trap" all vertices with times between the connected vertices so that they must be close together. Thus, instead of just connecting two vertices, the line connecting the decoration and the diagram effectively amalgamates the two connected vertices plus the "trapped" vertices. The result is fewer free time integrations than present in the diagrams that are kept, and the consequent justification of Kawasaki's rule.

There is no reason to believe that the infinite series for G represented by the diagrams converges quickly. In fact, since the true transport coefficients may be infinitely greater than those in G^0 near a critical point, it is clear that the perturbation may dominate G^0 near a critical point. Thus, the equation for G is resummed by the method[42] of Dyson. First, it is recognized that all the diagrams for $G_k(t)$ begin and end with a propagator G_k^0. There is no reason to "carry" these propagators and we define the improper self-energy Σ^* by the relation

$$G_k(t) = G_k^0(t) + \int_0^t dt_l \int_0^{t_l} dt_r\, G_k^0(t-t_l)\Sigma_k^*(t_l - t_r)G_k^0(t_r) \tag{106}$$

The diagrams for the self-energy Σ^* are clearly obtained from the diagrams

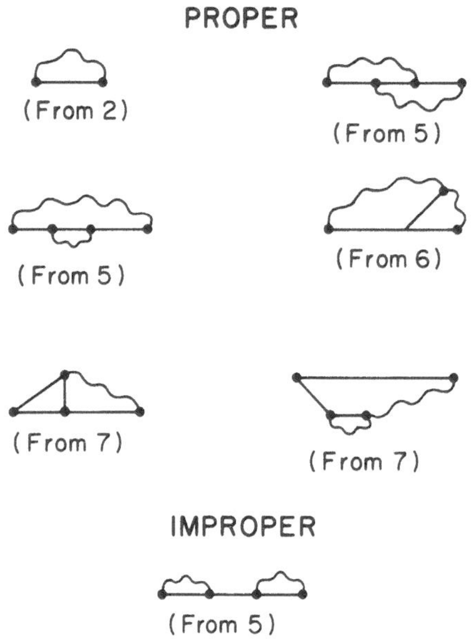

Fig. 5. Graphs contributing to the improper and proper self-energy, respectively, derived from the graphs in Fig. 1.

given for G by removing the propagator lines at the far left and far right, and by dropping the integration over the time coordinates t_l and t_r of the vertices at the far left and right. It is now noted that Σ^* contains diagrams that consist of chains of subdiagrams joined only by zero-order propagator lines G_k^0; such propagators must have wave vector k by wave-vector conservation. If we define the subclass of diagrams for Σ^* that may not be disconnected by cutting a single G_k^0 line as the proper self-energy Σ, then[42] graphically

$$\Sigma^* = \Sigma + \Sigma\text{——}\Sigma + \Sigma\text{——}\Sigma\text{——}\Sigma + \cdots \tag{107}$$

where Σ^* and Σ represent the improper and proper self-energies, respectively. The proper and improper self-energy diagrams to fourth order are shown in Fig. 5. Let us represent the full G by a thick line. Combining Eqs. (106) and (107), we find

$$\text{———} = \text{——} + \text{——}\Sigma\text{——} + \text{——}\Sigma\text{——}\Sigma\text{——} + \cdots \tag{108}$$

which is identical to [42]

$$\text{———} = \text{——} + \text{———}\Sigma\text{——} \tag{109}$$

or the Dyson equation,

$$G_k(t) = G_k^0(t) + \int_0^t dt_l \int_0^{t_l} dt_r\, G_k(t - t_l)\Sigma_k(t_l - t_r)G_k^0(t_r) \tag{110}$$

Equation (110) may be solved by Laplace transform to obtain

$$G_k(s) = G_k^0(s) + G_k(s)\Sigma_k(s)G_k^0(s) \tag{111}$$

or

$$G_k(s) = \{(G_k^0(s))^{-1} - \Sigma_k(s)\}^{-1} \tag{112}$$

The Laplace transform of Eq. (96) for $G_k^0(t)$ is easily taken:

$$G_k^0(s) = \{s - i\omega_k + k^2\Lambda_k^0\}^{-1} \tag{113}$$

Substituting into Eq. (112), we find

$$G_k(s) = \{s - i\omega_k + k^2\Lambda_k^0 + \Sigma_k(s)\}^{-1} \tag{114}$$

Following the procedure discussed early in Section 2, we compare Eq. (114) with the exact linear law, Eq. (51), to obtain

$$\Lambda_k(s) = \Lambda_k^0 + k^{-2}\Sigma_k(s) \tag{115}$$

The contribution of Σ to Λ gives rise to nonclassical behavior. If we just keep the first term in the infinite series for Σ, Eq. (115) becomes

$\Lambda_k(s) = \Lambda_k^0 + k^{-2}$ 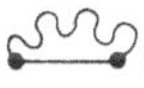 $_k(s)$ (116)

where Laplace transformation of the graph is understood.

As an exercise in interpreting Kawasaki's graphs, we shall now evaluate 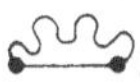 $_k(s)$ for the case of a single variable A. This graph is originally formed from Graph 2 for $A_{\mathbf{k}}(t)$ in Fig. 1. Corresponding to the asymmetric branching at the vertex on the left, Graph 2 has two contributions,

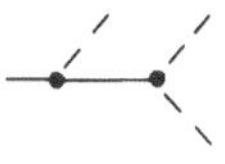

$$= \sum_{\mathbf{k}',\mathbf{k}''}^{k_c}\int_0^t d\tau_l \int_0^{\tau_l} d\tau_r\, G_k^0(t-\tau_l)\mathcal{V}_{\mathbf{k};\mathbf{k}+\mathbf{k}',-\mathbf{k}'}A^0_{\mathbf{k}+\mathbf{k}'}(\tau_l)$$
$$\times G^0_{-k'}(\tau_l-\tau_r)\mathcal{V}_{-\mathbf{k}';-\mathbf{k}'+\mathbf{k}'',-\mathbf{k}''}A^0_{-\mathbf{k}'+\mathbf{k}''}(\tau_r)A^0_{-\mathbf{k}''}(\tau_r)$$
$$+ \sum_{\mathbf{k}',\mathbf{k}''}^{k_c}\int_0^t d\tau_l \int_0^{\tau_l} d\tau_r\, G_k^0(t-\tau_l\,)_{\mathbf{k};\mathbf{k}+\mathbf{k}',-\mathbf{k}'}\,A^0_{-\mathbf{k}'}(\tau_l)$$
$$\times G^0_{k+k'}(\tau_l-\tau_r)\mathcal{V}_{\mathbf{k}+\mathbf{k}';\mathbf{k}+\mathbf{k}'+\mathbf{k}'',-\mathbf{k}''}A^0_{\mathbf{k}+\mathbf{k}'+\mathbf{k}''}(\tau_r)A^0_{-\mathbf{k}''}(\tau_r) \tag{117}$$

When Eq. (117) is multiplied on the right by $A_{-\mathbf{k}}$ and averaged, four contributions to the graph arise, corresponding to the successive pairing of $A_{-\mathbf{k}}$ with each of the four $A(\tau_r)$ in Eq. (117), according to Eq. (104) [pairing

$A_{-\mathbf{k}}$ with $A(\tau_l)$ forms a forbidden graph]. The factor of $\langle A_{\mathbf{k}}A_{-\mathbf{k}}\rangle$ in $\langle A_{\mathbf{k}}(\tau_r)A_{-\mathbf{k}}\rangle$ is divided out to turn the series for $\langle A_{\mathbf{k}}(t)A_{-\mathbf{k}}\rangle$ into one for $G_k(t)$:

$$
\begin{aligned}
= & \sum_{\mathbf{k}'}^{k_c}\int_0^t d\tau_l \int_0^{\tau_l} d\tau_r\, G_k^0(t-\tau_l)\mathcal{V}_{\mathbf{k};\,\mathbf{k}+\mathbf{k}',\,-\mathbf{k}'}\langle A_{\mathbf{k}+\mathbf{k}'}A_{-\mathbf{k}-\mathbf{k}'}\rangle \\
& \times G_{k+k'}^0(\tau_l-\tau_r)G_{-k'}^0(\tau_l-\tau_r)\mathcal{V}_{-\mathbf{k}';\,\mathbf{k},\,-\mathbf{k}-\mathbf{k}'}G_k^0(\tau_r) \\
& +\sum_{\mathbf{k}'}^{k_c}\int_0^t d\tau_l \int_0^{\tau_l} d\tau_r\, G_k^0(t-\tau_l)\mathcal{V}_{\mathbf{k};\,\mathbf{k}+\mathbf{k}',\,-\mathbf{k}'}\langle A_{\mathbf{k}+\mathbf{k}'}A_{-\mathbf{k}-\mathbf{k}'}\rangle \\
& \times G_{k+k'}^0(\tau_l-\tau_r)G_{-k'}^0(\tau_l-\tau_r)\mathcal{V}_{-\mathbf{k}';\,-\mathbf{k}-\mathbf{k}',\,\mathbf{k}}G_k^0(\tau_r) \\
& +\sum_{\mathbf{k}'}^{k_c}\int_0^t d\tau_l \int_0^{\tau_l} d\tau_r\, G_k^0(\tau-\tau_l)\mathcal{V}_{\mathbf{k};\,\mathbf{k}+\mathbf{k}',\,-\mathbf{k}'}A_{-\mathbf{k}'}\langle A_{\mathbf{k}'}\rangle \\
& \times G_{-k'}^0(\tau_l-\tau_r)G_{k+k'}^0(\tau_l-\tau_r)\mathcal{V}_{\mathbf{k}+\mathbf{k}';\,\mathbf{k},\,\mathbf{k}'}G_k^0(\tau_r) \\
& +\sum_{\mathbf{k}'}^{k_c}\int_0^t d\tau_l \int_0^{\tau_l} d\tau_r\, G_k^0(\tau-\tau_l)\mathcal{V}_{\mathbf{k};\mathbf{k}+\mathbf{k}',-\mathbf{k}'}\langle A_{-\mathbf{k}'}A_{\mathbf{k}'}\rangle \\
& \times G_{-k'}^0(\tau_l-\tau_r)G_{k+k'}^0(\tau_l-\tau_r)\mathcal{V}_{\mathbf{k}+\mathbf{k}';\,\mathbf{k}',\,\mathbf{k}}G_k^0(\tau_r)
\end{aligned}
\tag{118}
$$

The G_k^0 are removed to form the self-energy, and Laplace transformation leads to

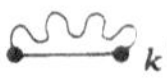

$$
\begin{aligned}
{}_k(s) = & \int_0^\infty e^{-st}\, dt \sum_{\mathbf{k}'}^{k_c} \mathcal{V}_{\mathbf{k};\,\mathbf{k}+\mathbf{k}',-\mathbf{k}'}G_{k+k'}^0(t) \\
& \times G_{-\mathbf{k}'}^0(t)\{\langle A_{\mathbf{k}+\mathbf{k}'}A_{-\mathbf{k}-\mathbf{k}'}\rangle(\mathcal{V}_{-\mathbf{k}';\mathbf{k},-\mathbf{k}-\mathbf{k}'}+\mathcal{V}_{-\mathbf{k}';-\mathbf{k}-\mathbf{k}',\mathbf{k}}) \\
& +\langle A_{-\mathbf{k}'}A_{\mathbf{k}'}\rangle(\mathcal{V}_{\mathbf{k}+\mathbf{k}';\,\mathbf{k},\,\mathbf{k}'}+\mathcal{V}_{\mathbf{k}+\mathbf{k}';\,\mathbf{k}',\,\mathbf{k}})\}
\end{aligned}
\tag{119}
$$

If we now rewrite the definition of $\mathcal{V}$ [Eq. (85)] with the use of "dot switching,"

$$\langle \dot{A}B^*\rangle = -\langle A\dot{B}^*\rangle \tag{120}$$

it is easily seen that the quantity in the brackets in Eq. (119) equals $\langle A_{\mathbf{k}+\mathbf{k}'}A_{-\mathbf{k}-\mathbf{k}'}\rangle\langle A_{\mathbf{k}'}A_{-\mathbf{k}'}\rangle/\langle A_{\mathbf{k}}A_{-\mathbf{k}}\rangle\mathcal{V}^*_{\mathbf{k};\mathbf{k}+\mathbf{k}',-\mathbf{k}'}$,

and

$$
\begin{aligned}
{}_k(s) = 2 & \int_0^\infty e^{-st}\, dt \sum_{\mathbf{k}'=0}^{\infty} |\mathcal{V}_{\mathbf{k};\,\mathbf{k}+\mathbf{k}',\,-\mathbf{k}'}|^2 \\
& \times\langle A_{\mathbf{k}+\mathbf{k}'}A_{-\mathbf{k}-\mathbf{k}'}\rangle G_{k+k'}^0(t)\langle A_{-\mathbf{k}'}A_{\mathbf{k}'}\rangle G_{-k'}^0(t)\times\langle A_{\mathbf{k}}A_{-\mathbf{k}}\rangle^{-1}
\end{aligned}
\tag{121}
$$

Now compare Eq. (121) to Eq. (40), which resulted from a crude argument for the mode–mode contribution to $K^{(c)}(t)$, and which gives Σ by implication. Noting that $\mathcal{V}$ contains an extra ik with respect to U, it is seen that the two are very similar. The differences are that zero-order propagators and correlation functions appear in Eq. (121), while exact propagators and correlation functions appear in Eq. (40), and a numerical factor in Eq. (121) of $2\times\frac{1}{4}=\frac{1}{2}$ [$\frac{1}{2}$ for each vertex, Eq. (85)] with respect to Eq. (40). The numerical factor appears simply because we were not careful about overcounting bilinear variables in Section 1. For bilinear variables A^iA^j, $i \neq j$, where overcounting is not a problem, it is easily seen that  is given by two terms, identical in form to Eq. (121), one containing $\mathscr{V}_{\mathbf{k};(i,\mathbf{k}+\mathbf{k}'),(j,-\mathbf{k}')}G^{i,0}_{k+k'}\ G^{j,0}_{-k'}$, the other containing $\mathscr{V}_{\mathbf{k};(j,\mathbf{k}+\mathbf{k}'),(i,-\mathbf{k}')}G^{j,0}_{k+k'}\ G^{i,0}_{k'}$.

When the solution of the nonlinear Langevin equation was first formulated by Kawasaki, the approximation $\Sigma \approx$  was always made. The presence of full correlation functions in "phenomenological" results for Λ, as contrasted to zero-order correlation functions in the results of Kawasaki, remained a puzzling point for a few years. Experimental evidence indicated that 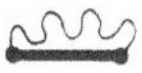 was a better approximation to Σ than 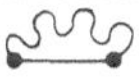. The problem was resolved[20] by Kawasaki by introducing an alternate form[42] of Dyson's equation. The first form of Dyson's equation [Eqs. (110) and (111)] expresses G in terms of G^0 and •. It is natural to ask if one might express G in terms of G itself, i.e., if Σ might be written in terms of G rather than G^0; a closed expression for G would result. A little study of the diagrams in Fig. 5 shows that this idea is quite sensible. Consider the diagram 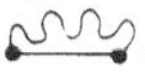. If the zero-order correlation function and propagator were replaced by exact quantities, these exact quantities could be expanded according to the method just described, with the result

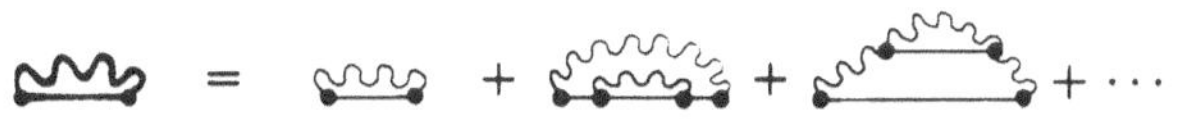 (122)

The diagrams in Eq. (122) all appear in the exact expression for Σ. Thus, a single diagram containing exact propagators represents an infinite subclass of diagrams containing bare propagators. The zero-order diagrams that 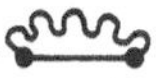 cannot represent are those in which there exist extra connections between the two basic parallel propagator and correlation function lines over and above those made by the vertices at the left and right; an example is . This diagram cannot be generated from 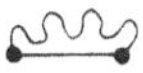 by expanding either the propagator or correlation function line, as such an expansion will not serve to place an extra connection between these two lines. The ideas just discussed are incorporated[42] into Dyson's second equation,

$\Sigma =$  (123)

where a heavy dot ● represents the "renormalized vertex." Equation (123) may be regarded as defining the renormalized vertex, whose role is to provide the connections between the propagator and correlation function line necessary to generate the complete Σ. A typical term in ● has three "terminals," one that connects to the propagator line, one to the correlation line, and one that provides the left-hand vertex of Σ. The first few contributions to ● are[20]

 (124)

For example, the diagram 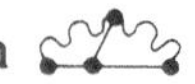 is formed by using bare propagators and  from Eq. (124).

Approximations are made to Dyson's second equation [Eq. (123)] by truncating the infinite series for the renormalized vertex. The zero-order approximation

 (125)

yields

$\Lambda_k(s) = \Lambda_k^0 +$  $_k(s)$ (126)

This approximation is almost always made. Lo and Kawasaki[44] have investigated the effect of the first vertex correction, shown in Eq. (124), on the critical behavior of the mutual diffusion coefficient; the effect is seen to be tiny. Bedeaux and Mazur,[43] who do not use Kawasaki's method, have shown that corrections to Eq. (126) give rise to a $t^{-5/2}$ tail on $K^{(c)}(t)$, which is dominated at long times by the $t^{-3/2}$ terms already included in Eq. (126). thus, the approximation Eq. (126) seems quite reasonable; we shall use this approximation from here on. Our final results are identical (except for counting problems) to those obtained by rough arguments in Section 1. Of course, we were guided in obtaining the result in Section 1 by a knowledge of the correct answer. In any event, we consider that a knowledge of the foundations upon which Eq. (126) rests is essential. As a final point, note that the solution of Eq. (126) is by no means trivial, and further approximations must be made, which will vary from case to case, actually to obtain $\Lambda_k(s)$.

4. Examples

To solve a specific mode–mode coupling problem, having now laid the general framework, we must simply find the variables of interest, evaluate $\mathcal{V}$, and solve Eq. (126). We shall consider three specific problems. The first, mutual diffusion near a binary critical point, is probably the best known[20,21] mode–mode coupling problem and is presented as a "standard" use of the technique. Self-diffusion is studied as an example[26] of how mode–mode coupling gives rise to the coupling of single particle and collective modes. Finally, molecular reorientation[30] is considered as an example of "long-time tails," where, in addition, nonconserved variables are of interest.

4.1. Mutual Diffusion in a Binary Critical Mixture

At a critical mixing point for a two-component liquid composed of species α and β, the order parameter is the concentration. For an incompressible fluid, the concentration is equivalent to the number density of species α (or $\hat{\beta}$), $n_{\mathbf{k}}^{\alpha}$, characterized by the mutual diffusion coefficient $D_k(s)$. We will now discuss the critical behavior of D. For $n_{\mathbf{k}}^{\alpha}$, the discussion on writing mode–mode equations for number densities in Section 2 holds [Eq. (74)], and we immediately obtain

$$\frac{\partial}{\partial t} n_{\mathbf{k}}^{\alpha} = -k^2 D_k^0 n_{\mathbf{k}}^{\alpha} + \frac{i\mathbf{k}}{2mN} \cdot \left(\sum_{\mathbf{k}'}^{k_c} n_{\mathbf{k}+\mathbf{k}'}^{\alpha} \mathbf{g}_{-\mathbf{k}'} + \sum_{\mathbf{k}'}^{k_c} \mathbf{g}_{\mathbf{k}+\mathbf{k}'} n_{-\mathbf{k}'}^{\alpha} \right) + f^{\mathrm{NL}} \tag{127}$$

where we have incorporated the symmetrization discussed in the paragraph following Eq. (81). In Eq. (127) we have left $\mathbf{k}'$ as a wave vector index alone; formally, Eq. (127) is just a representation of the vector Eq. (49) ($\mathcal{V}_{n^{\alpha};n^{\alpha},n^{\alpha}}, \mathcal{V}_{n^{\alpha};g,g} = 0$). Appearances aside, Eq. (127) does not involve just a single bilinear variable, since $\mathbf{g}_{\mathbf{k}}$ behaves differently for $\mathbf{g} \| \mathbf{k}$ (longitudinal mode) and $\mathbf{g} \perp \mathbf{k}$ (transverse modes). The transport coefficient for $\mathbf{g} \| \mathbf{k}$ is θ [Eq. (65)], and so $\mathbf{g} \| \mathbf{k}$ need not be kept near a critical point. In the solution of Eq. (127), we shall need (let $k = k_z$)$\langle g_{\mathbf{k}''}^z(s) g_{-\mathbf{k}''}^z \rangle$. For a given $\mathbf{k}''$, g^z is broken up into components parallel and perpendicular to $\mathbf{k}''$. The component parallel to $\mathbf{k}''$ is thrown away, and so

$$\langle g_{\mathbf{k}''}^z(s) g_{-\mathbf{k}''}^z \rangle \to \frac{\langle g_{\mathbf{k}''}^z g_{-\mathbf{k}''}^z \rangle (1 - (\hat{\mathbf{e}}_{\mathbf{k}''} \cdot \hat{\mathbf{e}}_z)^2)}{s + k''^2 \eta_{k''}(s)} \tag{128}$$

where $\hat{\mathbf{e}}_{\mathbf{k}''}$ is a unit vector parallel to $\mathbf{k}''$, and we have used the fact that different components of $\mathbf{g}$ are independent. Of course, $\eta_{k''}(s)$ is the full transport coefficient for $\mathbf{g}_{\mathbf{k}''} \perp \mathbf{k}''$, the shear viscosity. By reducing g^z to two independent components with identical behavior, $n\mathbf{g}$ and $\mathbf{g}n$ become equivalent to single bilinear variables.

The vertex in Eq. (127) is

$$\mathcal{V}_{\mathbf{k};\mathbf{k}+\mathbf{k}',-\mathbf{k}'} = ik_z/2mN \tag{129}$$

This relation does *not* depend on Eq. (85), although the two equations agree in this case. Equation (126) immediately yields

$$\begin{aligned} D_k(s) = D_k^0 + \frac{k_B T}{2(mN)} \sum_{\mathbf{k}'}^{k_c} \Bigg\{ & \frac{\langle n_{\mathbf{k}+\mathbf{k}'}^{\alpha} n_{-\mathbf{k}-\mathbf{k}'}^{\alpha} \rangle}{\langle n_{\mathbf{k}}^{\alpha} n_{-\mathbf{k}}^{\alpha} \rangle} [1 - (\hat{\mathbf{e}}_{\mathbf{k}+\mathbf{k}'} \cdot \hat{\mathbf{e}}_z)^2] \\ & \times \frac{1}{s + |\mathbf{k}+\mathbf{k}'|^2 D_{k+k'}(s) + k'^2 \eta_{k'}(s)} \\ & + \frac{\langle n_{\mathbf{k}'}^{\alpha} n_{-\mathbf{k}'}^{\alpha} \rangle}{\langle n_{\mathbf{k}}^{\alpha} n_{-\mathbf{k}}^{\alpha} \rangle} [1 - (\hat{\mathbf{e}}_{\mathbf{k}'} \cdot \hat{\mathbf{e}}_z)^2] \; \frac{1}{s + k'^2 D_{k'}(s) + |\mathbf{k}+\mathbf{k}'|^2 \eta_{k+k'}(s)} \Bigg\} \end{aligned} \tag{130}$$

where we have used the relation

$$\langle g_{\mathbf{k}}^z g_{-\mathbf{k}}^z \rangle = Nmk_{\mathrm{B}}T \tag{131}$$

which follows from the simple property of the momenta,

$$\langle p_i^\alpha p_j^\beta \rangle = mk_B T \delta_{ij} \delta_{\alpha\beta} \tag{132}$$

Away from the critical point, typically $D/\eta < 10^{-2}$ $[\Lambda \equiv \Lambda_{k=0}(s=0)]$; D decreases and η increases as the critical point is approached. Thus, $D_k(s)$ may always be neglected with respect to $\eta_k(s)$ on the right-hand side of Eq. (130). In the absence of mode–mode effects, $\eta_k(s)$ would be independent of k and s for small k, s, even near a critical point. Furthermore, the actual critical anomaly of η is weak. Thus, we expect[17,18] that the substitution $\eta_k(s) \to \eta_0(0) \equiv \eta$ in Eq. (130) will lead to a good approximation for $D_k(s)$. With this approximation, $D_k(s)$ is determined by D_k^0, η, and $\langle n_{\mathbf{k}}^\alpha n_{-\mathbf{k}}^\alpha \rangle$ alone. To proceed further, we must find an expression for $\langle n_{\mathbf{k}}^\alpha n_{-\mathbf{k}}^\alpha \rangle$ near the critical point. If we assume an Ornstein–Zernicke form,

$$\langle n_{\mathbf{k}}^\alpha n_{-\mathbf{k}}^\alpha \rangle \equiv \chi_{\alpha\alpha,k} = N_\alpha (\xi/w)^2 (1 + k^2 \xi^2)^{-1} \tag{133}$$

where w has no critical anomaly, and if we assume that we are very close to the critical point, with $\xi k_c \gg 1$, we may immediately calculate D_k $[\equiv D_k(0)]$ from Eq. (130),

$$D_k = D_k^0 + k^{-2} \frac{k_B T}{6\pi\eta\rho\xi^3} K_0(k\xi) \tag{134}$$

where

$$K_0(x) = \tfrac{3}{4}[1 + x^2 + (x^3 - x^{-1}) \tan^{-1} x] \tag{135}$$

s dependence is somewhat harder to obtain. From Eq. (135), it is easily seen that

$$\lim_{k\xi \to 0} D_k = \frac{w^2 L_{\alpha\alpha}^{(c),0}}{N_\alpha \xi^2} + \frac{k_B T}{6\pi\eta\rho\xi} \tag{136a}$$

and

$$\lim_{k\xi \to \infty} D_k = \frac{w^2 d^2 L_{\alpha\alpha}^{(c),0}}{N_\alpha} + \frac{k k_B T}{6\pi\eta\rho} \tag{136b}$$

where we have used the relation

$$D_k^0 = L_{\alpha\alpha}^{(c),0} / \chi_{\alpha\alpha,k} \tag{137}$$

with no critical anomaly in $L_{\alpha\alpha}^{(c),0}(L_{\alpha\alpha}^{(c),0} \propto O(N_\alpha)$. We have also assumed that no k-dependence in $L_{\alpha\alpha}^{(c),0}$ need be considered, which is valid for large ξ. Suppose $L_{\alpha\alpha,k}^{(c),0} = N_\alpha(a + bk^2)$. Thus, $D_k^0 = w^2 a/\xi^2 + k^2 w^2 (b/\xi^2 + a)$. For large ξ, the region of interest, the k^2 contribution to D_k^0 containing b is always small compared to the terms kept in Eq. (136).

Equations (136) illustrate many of the salient points of mode-mode coupling. We are considering large ξ, with $\kappa(\equiv\xi^{-1}) \ll k_c$. D_k shows a major deviation from $D_{k=0}$ for $k \sim \kappa$, even if κ is extremely small. Nevertheless, the theory treats D_k correctly for k all the way up to k_c. This is a consequence, as discussed in Section 1, of the inclusion of the order parameter in the set of variables. Note also that k_c does not enter Eq. (136). For $\xi k_c \gg 1$, the integral in Eq. (130) is automatically cut off by the Ornstein–Zernicke function. Finally, for $k\xi \to 0$, Eq. (136) show the dominance of the mode-mode part of D over D^0 for large ξ, with the mode-mode part $\alpha\xi^{-1}$ and $D^0 \propto \xi^{-2}$. The small $k\xi$, large ξ region, of course, shrinks as ξ increases. For large $k\xi$, but with k still "small" ($k < k_c$), Eqs. (136) show that the ξ dependence drops out of both contributions to D. The relative magnitudes of these two contributions[15] then must simply be determined numerically from the coefficients involved.

A very simple interpretation[16] of Eq. (136) is as follows. Near a critical mixing point, say in the one-phase region, "droplets" of pure α or β phase, of dimension $\alpha\xi$, are constantly forming and breaking up. As ξ becomes large, the rate constant for ordinary diffusion slows down as ξ^{-2}. On the other hand, since the "droplets" are becoming longer lived and larger, more and more of the motion of a single particle is describable in terms of the over all motion of the droplets to which the particle attaches itself. The self-diffusion of droplets of radius α_ξ should then determine the contribution to D of "droplet diffusion," and the mode-mode part of Eq. (136a) is just the Stokes–Einstein law (we use kinematic viscosity), with stick boundary conditions, for a sphere with radius ξ.

Of course, one must always be careful not to read too much into explanations like that just given; nevertheless, we do feel that a considerable amount of truth is contained in the "droplet" picture. We hasten to point out that the 6 in Eq. (136a), which occurs in the usual Stokes–Einstein law with stick boundary conditions, only occur because we have used O–Z theory; this has been stressed by Swinney *et al.*[47,48] In addition, note that "droplet diffusion" is a process that is *always* operative; Eq. (130) holds even far from the critical point, where, of course, we would not use the O–Z form for $\langle n_{\mathbf{k}}^{\alpha} n_{-\mathbf{k}}^{\alpha} \rangle$, and k_c would enter the final result. Near a critical point, however, it just so happens that "droplet diffusion" becomes more effective than "ordinary" diffusion.

Beautiful and useful though they are, Eqs. (130) and (136) do not represent the complete solution of a mode-mode coupling problem. Recall that the aim of mode-mode coupling is to predict all dynamical critical phenomena and "long-time tails" from equilibrium properties and "bare" transport coefficients. Thus, we have not really completed our calculation of D until η is eliminated. To do this, we calculate η via mode-mode coupling theory.

Equation (128), with $\mathbf{k}'' \perp z$, defines $\eta_k(s)$. Thus, to find $\eta_k(s)$, we must write and solve the mode–mode equations for, say, $g^x_{\mathbf{k}}$, $k = k_z$, and compare the solution with Eq. (128). If we ignore energy density fluctuations, as is almost always done, and recall that both total number and longitudinal momenta fluctuations are no longer "slowly varying" near the critical point, the linear variables for this problem are $n^\alpha_{\mathbf{k}}$, $g^x_{\mathbf{k}}$, and $g^y_{\mathbf{k}}$. For these variables, the matrix $i\boldsymbol{\omega}_k - k^2\boldsymbol{\Lambda}^0_{\mathbf{k}}$ is already diagonal (see third paragraph, Section 3), i.e., the linear variables do not "couple" directly. The vertex coupling $g^x_{\mathbf{k}}$ to bilinear variables $\mathbf{gg}$, $n^\alpha\mathbf{g}$, and $\mathbf{g}n^\alpha$ may be seen to vanish. Dynamical variables are[10] either even or odd with respect to time renewal, positions being even and momenta being odd; application of the time derivative changes the parity. Equilibrium averages of odd variables vanish. When these considerations are applied to Eq. (85) for $\mathscr{V}$, it is seen that $g^x_{\mathbf{k}}$ may not couple to the bilinear variables $\mathbf{g}n^\alpha$ and $n^\alpha\mathbf{g}$. Furthermore,[10,11,49] the equilibrium average of any quantity that does not transform as a scalar under rotations must vanish. For $k = k_z$, this means that the only products of $\mathbf{g}$ to which g^x may couple are g^xg^z and g^zg^x; since we do not retain g^z, we need keep no products of form gg.

Thus, the calculation of η involves[20,21], to a very good approximation, the variable $n^\alpha_{\mathbf{k}+\mathbf{k}'}n^\alpha_{-\mathbf{k}'}$ alone. To use the formalism developed earlier, this variable should first be orthogonalized to the linear variables; however, it is easy to see that the orthogonalization has no effect in this case, so we shall just keep $n^\alpha n^\alpha$. The vertex coupling g^x to $n^\alpha n^\alpha$ is

$$\mathscr{V}_{\mathbf{k};\,\mathbf{k}+\mathbf{k}',-\mathbf{k}'} = \frac{1}{2}\frac{\langle \dot{g}^x_{\mathbf{k}}(n^\alpha_{-\mathbf{k}-\mathbf{k}'}n^\alpha_{\mathbf{k}'})\rangle}{\langle n^\alpha_{\mathbf{k}+\mathbf{k}'}n^\alpha_{-\mathbf{k}'}\rangle\langle n^\alpha_{\mathbf{k}'}n^\alpha_{-\mathbf{k}'}\rangle} \tag{138}$$

We need not take any $\lim_{k,\,k'} \to 0$ in Eq. (138), since the only k, k' dependence to be included near the critical point enters as $k\xi$ and $k'\xi$, which need not be small. If we combine Eqs. (138), (120), (57), (132), and (133), we find

$$\mathscr{V}_{\mathbf{k};\mathbf{k}+\mathbf{k}',-\mathbf{k}'} = \frac{ik'_x}{2}k_BT\frac{\chi_{\alpha\alpha,\,k+k'} - \chi_{\alpha\alpha,\,k'}}{\chi_{\alpha\alpha,\,k+k'}\chi_{\alpha\alpha,\,k'}} \tag{139}$$

which is combined with the Ornstein–Zernicke form for χ_k, Eq. (133) (recall that $k\|z$), to yield

$$\mathscr{V}_{\mathbf{k};\,\mathbf{k}+\mathbf{k}',-\mathbf{k}'} = -\frac{ikk'_xw^2}{2N_\alpha}k_BT(k+2k'_z) \tag{140}$$

From Eqs. (140) and (126), we immediately find

$$\eta_k(s) = \eta^0_k + \frac{\xi^4k_BT}{2mN}\sum_{\mathbf{k}'=0}^{k_c}\frac{k'^2_x[k+2k'_z]^2[1+k'^2\xi^2]^{-1}[1+|\mathbf{k}+\mathbf{k}'|^2\xi^2]^{-1}}{s+k'^2D_{k'}(s)+|k+k'|^2D_{k+k'}(s)} \tag{141}$$

Equation (141), when combined with Eq. (130), gives, in principle, two integral equations that may be solved for two unknowns, $\eta_k(s)$ and $D_k(s)$. For now, we shall be content to find a first approximation to η, obtained by substituting Eq. (134) into Eq. (141). Even this equation must[20,21,45,50,51] be solved numerically However, it is not hard to exhibit the qualitative behavior of η near a critical point. The reason Eq. (141) is difficult to evaluate is the presence of the function K_0 [Eq. (135)] in D_k. If we just want to estimate η, we may write $\sum_{\mathbf{k}'=0}^{k_c}$ as $\sum_{\mathbf{k}'=0}^{\xi^{-1}} + \sum_{|\mathbf{k}'|=\xi^{-1}}^{k_c}$, and use the small and large $k'\xi$ form of D_k in the two sums, respectively. Changing the sum to an integral, it is easily seen that the $\int_0^{\xi^{-1}}$ has no critical anomaly, and so any possible critical contribution to η, η^c, is given by the relation

$$\eta^c \propto \int_{\xi^{-1}}^{k_c} dk' \frac{1}{(k'k_BT/6\pi\eta\rho) + k'^2 w^2 L_{\alpha\alpha}^{(c),0} N_\alpha^{-1}} \tag{142}$$

where we have performed the angular integration. Upon performing the elementary integral in Eq. (142) in the limit $\xi \to \infty$, under the assumption $6\pi\eta\rho L_{\alpha\alpha}^{(c),0} k_c w^2/k_BTN_\alpha \gg 1$, we find[45,50,51]

$$\eta^c/\eta \sim \ln \xi \tag{143}$$

For quite large ξ, η^0/η still remains close to unity. In this region, Eq. (143) predicts that η^c varies as $\ln \xi$. Note that the cutoff wave vector k_c drops out of Eq. (143); this is so because, as was recently pointed out by Oxtoby and Gelbart,[45] the large-k' form of D^0 [Eq. (136)], which provides the k'^2 term in the denominator of Eq. (142), provides an "internal" cutoff.

The results just obtained immediately give the divergent Onsager coefficients $L = \chi\Lambda$, corresponding to the breakdown of classical irreversible thermodynamics, for D and η. Since χ for viscous processes is independent of ξ, the ξ dependence of η dependence of η is identical to the ξ dependence of the integrated viscous flux correlation function. Since χ for diffusion is $\langle n_{\mathbf{k}}^{\alpha} n_{-\mathbf{k}}^{\alpha}\rangle$, we find in the Ornstein–Zernicke approximation that $L_{\alpha\alpha,k}^{(c),0} \propto \xi$, $k\xi \to 0$.

Thus, we have seen how to calculate D and η near a binary critical point. The problem of calculating the thermal conductivity and η near a liquid–gas critical point is[52] completely isomorphic to the problem discussed. These two problems are the most-studied examples of mode–mode coupling calculations of the critical behavior of transport coefficients. Future work will probably focus on the testing of the many assumptions upon which the results just obtained are based. We feel that Eq. (127) is soundly based, as discussed in Section 2. thus, the validity of Eq. (130), relating $D_k(s)$ to $\eta_k(s)$, only depends on the approximations made when solving Eq. (127).

The mode–mode equation for $\mathbf{g}_\mathbf{k}$, involving only $n^\alpha n^\alpha$ with a vertex given by Eq. (140), is much less soundly based than Eq. (127). There is no compelling reason to expect that $\mathbf{g}_\mathbf{k}$ should only couple to bilinear modes. The importance of triple and higher order products of concentration fluctuations on η^c should be investigated. Furthermore, recall that Eq. (85) for $\mathcal{V}$ is based on the assumptions that χ may be treated as block diagonal and that the first term in

Eq. (84) for $\mathscr{V}$ may be neglected with respect to the second term. Both of these assumptions break down for the case at hand. Since χ contains ξ, the corrections to the block-diagonal χ^{-1}, which we argued were small, may contain ξ and become large as $\xi \to \infty$. The neglect of the first term in Eq. (84) with respect to the second term was based on the first and second terms being quadratic and linear in the wave vector (k or k'), respectively. However, in the vertex coupling $\mathbf{g}$ to $n^\alpha n^\alpha$, the term we keep in $\mathscr{V}$ is quadratic in the wave vector, i.e., the linear part of this term just happens to vanish. There is no *a priori* reason then not to keep both terms in Eq. (85) in this case. Thus, even given that $\mathbf{g}$ couples to $n^\alpha n^\alpha$ alone, many assumptions made in deriving the bilinear equation must be checked.

4.2. Self-Diffusion

We now wish to illustrate another of the three main consequences of mode–mode coupling mentioned in the introduction, that of the existence of expressions for transport coefficients in terms of other transport coefficients; such expressions could not exist if classical theory held. Of course, the results derived in the previous section are examples of such expressions. However, we wish to discuss cases that have nothing to do with critical phenomena. The best known example of the expressions of interest is the Stokes–Einstein law for the self-diffusion coefficient D^s of a large spherical particle,

$$D^s_{SE} = k_B T / q\pi\eta\rho R \tag{144}$$

where q varies from 4 to 6 as the boundary conditions at the surface of the sphere vary from slip to stick. We shall now show[26] how the Stokes–Einstein law follows from mode–mode coupling theory.

The self-diffusion is defined in terms of the tagged particle correlation function,

$$\langle n^1_{\mathbf{k}}(s) n^1_{-\mathbf{k}} \rangle = \{s + k^2 D^s_k(s)\}^{-1} \tag{145}$$

Since we want to include the possibility that the tagged particle is large, we define n^1 according to Eq. (68). As discussed in Section 2, Eq. (74) still holds for the new definitions, so we simply repeat the steps taken in Section 4.1. The modified definition of the conserved variables makes little sense unless $kR \ll 1$, so we shall assume this immediately. Thus ($k = k_z$)

$$D^s_k(s) = D^{0,s}_k + \frac{k_B T}{2mN} \sum_{\mathbf{k}'}^{k_c} \chi^2(k'R) \times \left\{ \frac{[1 - (\hat{\mathbf{e}}_{\mathbf{k}+\mathbf{k}'} \cdot \hat{\mathbf{e}}_z)^2]}{s + k'^2 \eta_{k'}(s)} + \frac{[1 - (\hat{\mathbf{e}}_{\mathbf{k}'} \cdot \hat{\mathbf{e}}_z)^2]}{s + |\mathbf{k} + \mathbf{k}'|^2 \eta_{\mathbf{k}+\mathbf{k}'}(s)} \right\} \quad (kR \ll 1) \tag{146}$$

where we have used the relation

$$\langle n^1_{\mathbf{k}''} n^1_{-\mathbf{k}''} \rangle = \chi^2(k''R) \tag{147}$$

and χ is given in Eq. (71).

The expression for $\mathscr{V}$ implied by Eq. (146), which is just Eq. (129), can also in fact be obtained from our approximate, general expression for $\mathscr{V}$ [Eq. (85)]. Recall that Eq. (85) assumes that the matrix χ may be treated as block-diagonal. However, for the self-diffusion of a large particle, if the matrix χ^{-1} is iterated about the block-diagonal inverse, all the terms in the series are seen to be of the same order of magnitude. Thus, it appears that Eq. (85) holds in this case, not because χ may be treated as block-diagonal, but because the non-block-diagonal contributions to χ^{-1} simply do not contribute to the equation of motion for n^1.

Equation (146) shows the consequences of the coupling of the tagged particle density to the collective momentum density $\mathbf{g}$. The introduction of the bilinear variable $n^1\mathbf{g}$ is essential; as discussed earlier, n^1 cannot couple directly to $\mathbf{g}$. Before the advent of mode–mode coupling theory, it was impossible to see how the macroscopic, hydrodynamic effects that obviously occur when a large tagged particle moves through a solvent manifest themselves in a formal theory for D^s. It is now clear that the macroscopic hydrodynamics of rigid bodies in fluids is contained, on a microscopic level, in mode–mode coupling theory.

Equation (146) may be used to treat many interesting problems. For small particles ($Rk_c \ll 1$), $\chi(k'R)$ may be replaced by unity. If we then make the approximation $\eta_k(s) \to \eta$, which should be even better for the case at hand (no critical point) than it was for Section 4.1, we find

$$D^s = D^{0,s} + \frac{k_B T k_c}{3\pi^2 \rho \eta} \tag{148}$$

The second term in Eq. (148) has the form of the Stokes–Einstein law ($k_c = c\pi/l$, $c < 1$, and $l \sim a$, whcrc a is "R" in a purc fluid). This contribution to D contains the effects of "droplet diffusion."

For large particles, D^s is given very accurately by D^s_{SE} with $q = 6$; even for small particles, however, D^s_{SE} given a good estimate of D. In fact,[53] D^s_{SE}, with $q = 4$, is amazingly accurate for D^s of a dense pure fluid. It follows that the mode–mode contribution to D^s must always be important in dense fluids, even for small particles. This state of affairs holds for other types of diffusion, but not for nondiffusive processes. Diffusion is the slowest of transport processes, since it may only proceed by actual physical transport of particles, while other transport utilizes this mechanism *plus* "collisional transport" of the quantity of interest from particle to particle. However, the mode–mode part of all transport coefficients is comparable to D^s_{SE}, i.e., to D. Thus, mode–mode coupling becomes important in nondiffusive transport only when some phenomenon, such as a critical point, slows down the processes that usually dominate the transport.

Since ordinary kinetic theory approximations to D^s, such as[54] Enskog's theory, do not include hydrodynamics effects at all, it is conceivable[55] that, with an appropriate choice of k_c, $D^{0,s}$ could be identified in Eq. (148) as Enskog's approximation D_E to D^s. The mode–mode part of D would then represent[53,56] the "enhancement" of D over D_E, as lately discussed in the literature, due to hydrodynamic effects. The proper tool for the study of this point is kinetic theory, not mode–mode coupling, but we mention the subject due to its extreme interest.

For a large particle ($Rk_c \ll 1$), the $\sum_{\mathbf{k}'=0}^{k_c}$ in Eq. (146) may be replaced by $\sum_{\mathbf{k}'=0}^{\infty}$, since the summand is zero for $k' \gg R^{-1}$. The sum is then converted into an elementary integral $[\eta_{k'}(s) \to \eta]$. We have argued previously that D^0 vanishes when $Rk_c \gg 1$. Thus, we find[26]

$$D^s = k_B T/5\pi\eta R, \qquad Rk_c \gg 1 \tag{149}$$

the Stokes–Einstein law, with $q = 5$. The reason why we have obtained $n = 5$ instead of any other between 4 or 6 is unclear. Irrespective of the value of q, however, Eq. (149) shows how the basic physics of the Stokes–Einstein law is contained in mode–mode coupling.

As a final aspect of self-diffusion, we note[57] that $D_k^s(s)$ can contain critical-point effects, through $\eta_k(s)$ in Eq. (146). When a tagged particle is diffusing in a solvent that is near a critical point, one need simply substitute the mode–mode equation for $\eta_{k'}$ in the solvent into Eq. (146) to obtain the critical behavior of $D_{k'}^s$.

4.3. Reorientation

We now consider a problem where the third principal aspect of mode–mode coupling, the presence[16] of "long-time tails" on flux correlation functions, is of primary interest. In the theory of magnetic resonance in fluids, one is often interested in the autocorrelation function of the single-particle variable $Q^{(2)}_{\mathbf{k};\alpha\beta}$,

$$Q^{(2)}_{\mathbf{k};\alpha\beta} = [\hat{\mu}\hat{\mu}]^{(2)}_{\alpha\beta} e^{i\mathbf{k}\cdot\mathbf{r}_1} \tag{150}$$

where $\hat{\mu}$ is a unit vector fixed in the tagged particle (particle 1), and $[\hat{\mu}\hat{\mu}]^{(2)}_{\alpha\beta}$ denotes[32] the $\alpha\beta$ component of the second-rank irreducible product of the $\hat{\mu}$,

$$[\hat{\mu}\hat{\mu}]^{(2)}_{\alpha\beta} = \mu_\alpha\mu_\beta - \tfrac{1}{3}\delta_{\alpha\beta} \tag{151}$$

It is instructive to extend the calculation to irreducible products of arbitrary rank, so we shall consider the variable $Q^{(l)}_{\mathbf{k};\alpha,\beta,\gamma\cdots}$, where

$$Q^{(l)}_{\mathbf{k};\alpha,\beta,\gamma\cdots} = [\hat{\mu}\hat{\mu}\cdots]^{(l)}_{\alpha\beta\gamma\cdots} e^{i\mathbf{k}\cdot\mathbf{r}_1} \tag{152}$$

As usual, we write

$$\langle Q_{\mathbf{k}}^{(l)}(s)Q_{-\mathbf{k}}^{(l)}\rangle \equiv \{s+\Gamma_k^{(l)}(s)\}^{-1} \tag{153}$$

where $\Gamma_k^{(l)}(s)$ is the l-rank reorientational rate. Now, $Q^{(l)}$ is a nonconserved variable, ($\Gamma^{(l)}$ is independent of k for small k). Nonetheless, $Q^{(l)}$ may couple to the conserved variables and products of conserved variables, thereby attaining long-time behavior for $\Gamma_k^{(l)}(t)$. For simplicity, we shall consider the case where $k \to 0$.

It is important to understand that $Q^{(l)}$, a nonconserved variable, is not considered a "slowly varying variable." The point is, if we choose $Q^{(l)}$ plus all the usual "slowly varying variables" as a complete set, then we must obtain the proper result for $Q^{(l)}$ at long times; the presence of $Q^{(l)}$ should not affect the long-time behavior of the slowly varying variables at all. Being a single-particle variable, $Q^{(l)}$ may couple only to single-particle variables and products of single-particle and collective variables. We shall treat the system as isothermal and incompressible. Thus, if[30,31] trilinear and higher nonlinear variables may be ignored, the complete set consists of $Q_{\mathbf{k}}^{(l)}$, $n_{\mathbf{k}}^1$, $n_{\mathbf{k}+\mathbf{k}'}^1\mathbf{g}_{-\mathbf{k}'}$, and $\mathbf{g}_{\mathbf{k}+\mathbf{k}'}n_{-\mathbf{k}'}^1$, with the understanding that only transverse components of $\mathbf{g}$ need be kept.

Different components of $Q^{(l)}$, for the same l, may couple to each other. This coupling gives rise[58] to "anisotropic rotational diffusion" for asymmetric molecules, which we shall ignore, since it does not change long-time properties. We shall ignore the coupling of $Q^{(l)}$ to n^1 for the same reason. The long-time behavior of $Q^{(l)}$ will then be determined by a simple mode–mode equation in which $Q^{(l)}$ is coupled to the transverse components of $n'g$ through the l-dependent vertex $\mathcal{V}^{(l)}$.

The standard approximation for the vertex, Eq. (85), vanishes for the case at hand. This is because $\dot{Q}^{(l)}$ is linear in the angular momenta $\mathbf{J}^1$ of particle 1. When $\dot{Q}^{(l)}$ is multiplied by $n^1\mathbf{g}$, products of momenta J^1P_i arise that may be averaged independently of all functions of position. However, linear and angular momenta are themselves independent, so these averages vanish. Thus, the term we have ignored in Eq. (84) must now be kept, and we find

$$\mathcal{V}^{(l)}_{Q_{\mathbf{k}}^{(l)};n_{\mathbf{k}+\mathbf{k}'}^1,\mathbf{g}_{-\mathbf{k}'}} = \frac{\int_0^\infty \mathrm{dt}\,\langle \dot{Q}_{\mathbf{k}}^{(l)+\mathrm{NL}}(t)(n_{-\mathbf{k}-\mathbf{k}'}^1\mathbf{g}_{\mathbf{k}'})^{+\mathrm{NL}}\rangle}{2Nmk_{\mathrm{B}}T} \tag{154}$$

It should be clear from the discussion that there is no compelling reason why the standard approximation for $\mathcal{V}$, Eq. (85), should always hold. As we have tried to emphasize, the important contributions to mode–mode equations should be reexamined for every new problem.

Since Eq. (154) contains a time-dependent correlation function, $\mathcal{V}^{(l)}$ may not be evaluated via equilibrium statistical mechanics. The consequent impossibility of making any practical calculation of $\mathcal{V}$ is probably the main reason that such contributions to $\mathcal{V}$ are usually ignored; it is the fond hope of most theorists that intractable quantities are always small. Fortunately, the long-time behavior of $\dot{Q}_{(t)}^{(l)+}$, which we wish to obtain, only depends on the wave-vector dependence of $\mathcal{V}^{(l)}$. As should now be clear, the solution to the mode–mode equation for $Q^{(l)}$ (or $\Gamma^{(l)}$) is ($k = k_z \to 0$)

$$\Gamma_{k=0}^{(l)}(s) = \Gamma_{k=0}^{(l),0} + Nmk_B T \sum_{\mathbf{k}'=0}^{k_c} |\mathcal{V}_{k=0;\mathbf{k}',-\mathbf{k}'}^{(l)}|^2 \frac{(1-(\hat{\mathbf{e}}_{\mathbf{k}'}\cdot\hat{\mathbf{e}}_z)^2)}{s+k'^2\eta_{k'}(s)} \tag{155}$$

where we have used the results $\langle Q_{\mathbf{k}}^{(l)} Q_{\mathbf{k}}^{(l)}\rangle = 1$ and $\eta \gg D$. The now familiar approximation $\eta_k(s) \to \eta$ may be used to find the long-time behavior of the dissipative flux correlation for $Q^{(l)}$,

$$\Gamma_{k=0}^{(l)}(t) \underset{\substack{\text{long}\\ \text{times}}}{\propto} \text{ILT} \int_0^{k_c} k'^{d-1}\, dk' \, |\mathcal{V}_{k=0;\mathbf{k}',-\mathbf{k}'}^{(l)}|^2 \frac{1}{s+k'^2\eta} \tag{156}$$

where the results of the angular integrations have been absorbed into the constant of proportionality, and ILT means inverse Laplace transform. Suppose now that the leading term in the k' expansion of $\mathcal{V}_{k=0;\,\mathbf{k}',-\mathbf{k}'}^{l}$ is of order $(k')^{c^{(l)}}$; Eq. (56) then yields

$$\Gamma_{k=0}^{(l)}(t) \underset{\substack{\text{long}\\ \text{times}}}{\propto} \text{ILT} \int_0^{k_c} dk' \frac{(k')^{[2c^{(l)}+(d-1)]}}{s+k'^2\eta} \tag{157}$$

which is an elementary ILT,

$$\Gamma_{k=0}^{(l)}(t) \underset{\substack{\text{long}\\ \text{times}}}{\propto} t^{-(2c^{(l)}+d)/2} \tag{158}$$

and reduces to the "usual" $t^{-d/2}$ for the "usual" $c^{(l)} = 0$.

To evaluate $c^{(l)}$, recall that an average of products $\langle AB^*\rangle$ may be nonzero only if A and B have the same tensorial rank, i.e., if A and B can combine to form a scalar. For $k \to 0$, the tensorial rank of $\dot{Q}^{(l)}$ is l. Thus,[30] when $(n^1 \cdot \mathbf{g})$ is expanded in powers of k' in Eq. (154), the lowest power of k' that multiplies a dynamical variable of rank l is equal to $c^{(l)}$.

The time derivative $(n^1 \cdot \mathbf{g})$, may be written ($k \to 0$)

$$\left(n\frac{1}{\mathbf{k}'}\cdot \mathbf{g}_{-\mathbf{k}'}\right) = i\frac{\mathbf{k}'}{m}\cdot \mathbf{g}\frac{1}{\mathbf{k}'}\mathbf{g}_{-\mathbf{k}'} - i\frac{\mathbf{k}'}{m}\cdot n\frac{1}{\mathbf{k}'}\boldsymbol{\sigma}_{-\mathbf{k}'} \tag{159}$$

where $\boldsymbol{\sigma}_{\mathbf{k}}$ is the microscopic stress tensor. It is well known that $\boldsymbol{\sigma}_{k'=0}$ may be divided into zero-, first-, and second-rank tensorial parts. The zero-rank part of $\boldsymbol{\sigma}$, which shall not enter our results, corresponds to the pressure. The first-rank part of $\boldsymbol{\sigma}$ is actually a pseudovector, which may not couple to the true vector $\dot{Q}^{(l)}$. Thus, for our purposes, $\boldsymbol{\sigma}_{k=0}$ may be treated as second rank irreducible. The quantity $\mathbf{g}_{\mathbf{k}'}^{1}\mathbf{g}_{-\mathbf{k}'}$ may also be treated as second rank irreducible for $k' \to 0$.

Expansions of the exponentials ($e^{i\mathbf{k}'\cdot\mathbf{r}_i}$) that enter $\mathbf{g}^1_{\mathbf{k}'}$, $\mathbf{g}_{\mathbf{k}'}$, and $n^1_{\mathbf{k}'}$ leads to a k' expansion of $(n^1_{\mathbf{k}'} \cdot \mathbf{g}_{-\mathbf{k}'})$, the nth term of which will contain products of $(ik')^{n+1}$ with dynamical variables that are themselves products of $n\mathbf{r}_i$ (or $\mathbf{r}_{ij}$) and $\mathbf{g}^1_{k'=0}\mathbf{g}_{-k'=0}$ or $n^1_{k'=0}\sigma_{-k'=0}$. The nth term in the expansion may therefore have rank such that $n+2 \geq (\text{rank}) \geq n-2$. It follows[30] that the zeroth term in the k' expansion of $(n^1 \cdot \mathbf{g})$ (linear in k') may form a nonvanishing average with $\dot{Q}^{(2)}$, the first term (quadratic in k') may form a nonvanishing average with $\dot{Q}^{(3)}$, $\dot{Q}^{(2)}$, and $\dot{Q}^{(1)}$, and so on, i.e., $c^{(1)}=2$, $c^{(2)}=1$, $c^{(3)}=2$, $c^{(4)}=3, \ldots$. Thus, $\Gamma_{k=0}(t)$ decays at long times as $t^{-7/2}$, $l=1$, and $t^{-5/2-(l-2)}$, $l \geq 2$.

Of course, it is impossible actually to determine the amplitudes of the "long-time tails" under discussion. Nevertheless, we feel that this example exhibits three interesting points: (a) the direct influence of hydrodynamic modes on a nonconserved variable, (b) the breakdown, in this particular case, of the usual expression, Eq. (85), for $\mathcal{V}$, and (c) the importance of the tensorial rank of a variable in determining its long-time behavior.

5. Conclusion

In conclusion we would like to mention some unsolved (1975) mode–mode coupling problems that should provide fruitful sources of future research projects. Predictions of the theory, on the level discussed in Section 4.1, for the critical behavior of transport coefficients, have been[20,21,45,47,48,50,51,59] extensively compared with experiment. Agreement is remarkable for $k > \mathcal{K}$, but discernible discrepancies exist for $k > \mathcal{K}$. It is of major interest that the source of error in the theory be discovered. As we have tried to emphasize, the current standard version of the theory may only be derived from the basic theory with the use of a huge number of approximations. Presumably, near a critical point, one or more of these approximations is breaking down.

The "long-time tails" have been extensively studied via molecular dynamics, but have proved almost impossible to observe in real experiments. It would be most desirable to find experiments in which the "tails" could be unambiguously observed. We have suggested elsewhere[50] that one such experiment might be light scattering from very large macromolecules. The "tails" already have been observed[61] in experiments on the applied magnetic field$^{(H^0)}$ dependence of NMR linewidths in liquid crystals. When the dominant spin relaxation mechanism in molecular reorientation, the linewidth $\Delta\omega(H_0)$ depends on $\langle Q^{(2)}(S_L)Q^{(2)}\rangle$, where S_L (H_0) is the Larmor frequency. However, the liquid crystal experiments have not been widely recognized as examples of mode–mode coupling and "long-time tails." The theory[61] of these experiments has been done using older, heuristic techniques. In our opinion, this problem should be attacked by simply generalizing the approach of Section 4.3 to treat the anisotropic liquid-crystal system.

If Eq. (155) for $\Gamma^{(2)}$ could actually be evaluated, $\Gamma^{(2)}$ would be divided into a part $\Gamma^{2,0}$, which should not be a function of any other transport coefficient, and a mode–mode part proportional to η^{-1}. Recently,[62] detailed experiments have indeed established that if η is varied and other parameters kept constant,

$$(\Gamma^{(2)})^{-1} = \tau^{0})^{-1} + \text{const} \times \eta \tag{160}$$

Equation (155) is *not* compatible with Eq. (160), which indicates that some hidden viscosity dependence lurks in Eq. (155). It would be of great interest if this point could be explained and if the resulting theory could be refined to the point where the constant in Eq. (160) could be estimated. This may be impossible, however, due to the presence of k_c.

Of course, a whole host of other problems exists. There are no doubt many critical points where mode–mode coupling has not yet been used to calculate the critical slowing down, including the laser transition and the liquid-crystal transition. The effect of critical phenomena on observable nonconserved variables, via mode coupling, has hardly been studied at all. The solution of even the standard mode coupling equations is extremely difficult in two dimensions, and should be pursued. On a much more fundamental level, the basic assumptions of mode coupling have never been tested, i.e., it has never been shown that a "bare" transport coefficient possesses all the classical attributes. Continuing interest in all these problems will no doubt maintain mode–mode coupling as an exciting field for many years to come.

References

1. Robert Zwanzig, *Ann. Rev. Phys. Chem.* **16**, 67 (1965).
2. L. Onsager, *Phys. Rev.* **37**, 405 (1931); **38**, 2265 (1931).
3. L. Landau and E. M. Lifshitz, *Fluid Mechanics*, Addison-Wesley, Reading, Massachusetts (1968).
4. S. R. de Groot and P. Mazur, *Non-Equilibrium Thermodynamics*, North Holland Publishing Co., Amsterdam (1962).
5. M. S. Green, *J. Chem. Phys.* **20**, 1281 (1952).
6. R. Kubo, *J. Phys. Soc. Japan* **12**, 570 (1957).
7. R. Zwanzig, *Lectures Theor. Phys.* (*Boulder*) **3**, 106 (1961).
8. H. Mori, *Progr. Theor. Phys.* **33**, 423 (1965).
9. L. P. Kadanoff and P. C. Martin, *Ann. Phys.* **24**, 419 (1963).
10. B. U. Felderhof and I. Oppenheim, *Physica* **31**, 1441 (1965).
11. P. A. Selwyn and I. Oppenheim, *Physica* **54**, 161 (1971).
12. L. van Hove, *Phys. Rev.* **95**, 1374 (1954).
13. P. Debye, *Phys. Rev. Lett.* **14**, 783 (1965).
14. R. Zwanzig, *Phys. Rev.* **129**, 486 (1963).
15. M. H. Ernst, J. R. Dorfman, and E. G. D. Cohen, *Physica* **31**, 493 (1965).
16. B. J. Alder and T. E. Wainwright, *Phys. Rev.* **A1**, 18 (1970).
17. J. V. Sengers *in Proc. Conf. Phenomena near the Critical Point, Washington, D.C., 1965* (M. S. Green and J. V. Sengers, eds.), Nat. Bur. Std. Mics. Publ. No. 273 (1965).
18. K. F. Herzfeld and T. A. Litovitz, *Absorption and Dispersion of Ultrasonic Waves*, Academic Press, New York (1957).

19. M. Fixman, *J. Chem. Phys.* **36**, 310 (1962).
20. K. Kawasaki, *Ann. Phys.* (*N.Y.*) **61**, 1 (1970).
21. L. P. Kadanoff and J. Swift, *Phys. Rev.* **165**, 310 (1968).
22. T. Yamada and K. Kawasaki, *Progr. Theor. Phys.* **38**, 1031 (1967).
23. D. Kivelson, I. Oppenheim, private communications.
24. L. D. Landau and E. M. Lifshitz, *Fluid Mechanics*, Addison-Wesley, Reading, Massachusetts (1968).
25. P. Debye, *Polar Molecules*, Dover, New York (1929).
26. T. Keyes and I. Oppenheim, *Phys. Rev.* **A8**, 937 (1973).
27. T. Keyes and I. Oppenheim, *Phys. Rev.* **A7**, 1384 (1973).
28. Raymond Kapral and Michael Weinberg, *Phys. Rev.* **A8**, 1008 (1973).
29. Ira A. Michaels and Irwin Oppenheim, *Physica*, **81A**, 454 (1975).
30. T. Keyes and I. Oppenheim, *Physica* **75**, 583 (1974).
31. F. Garisto and R. Kapral, *Phys. Rev.* **A10**, 309 (1974).
32. Y. Pomeau and J. Weber, *Phys. Rev.* **A8**, 1422 (1973).
33. M. H. Ernst, E. H. Hauge, and J. M. J. van Leeuwen, *Phys. Rev. Lett.* **25**, 1254 (1970).
34. Y. Pomeau, *Phys. Rev.* **A5**, 2569 (1972).
35. J. W. Dufty, *Phys. Rev.* **A5**, 2247 (1972).
36. L. P. Kadanoff and J. Swift, *Phys. Rev.* **166**, 89 (1968).
37. B. A. Fleishman, *Quart. Appl. Math.* **14**, 145 (1956).
38. J. H. Irving and John G. Kirkwood, *J. Chem. Phys.* **18**, 817 (1950).
39. Mordechai Bixon and Robert Zwanzig, *J. Statist. Phys.* **3**, 245 (1971).
40. J. H. Weare and I. Oppenheim, *Physica* **72**, 1 (1974).
41. H. W. Wyld, *Ann. Phys.* (*N.Y.*) **14**, 143 (1961).
42. Alexander L. Fetter and John Dirk Walecka, *Quantum Theory of Many Particle Systems*, McGraw-Hill, New York (1971).
43. D. Bedeaux and P. Mazur, *Physica* **73**, 431 (1974).
44. Shih-Min Lo and Kyozi Kawasaki, *Phys. Rev.* **A5**, 421 (1972).
45. David W. Oxtoby and William M. Gelbart, *J. Chem. Phys.* **61**, 2957 (1974).
46. Richard A. Ferrell, *Phys. Rev. Lett.* **24**, 1169 (1970).
47. H. L. Swinney and D. L. Henry, *Phys. Rev.* **A8**, 2586 (1973).
48. H. L. Swinney and B. E. A. Saleh, *Phys. Rev.* **A7**, 747 (1973).
49. W. A. Steele, *in*: *Transport Phenomena in Fluids* (H. J. M. Hanley, ed.), Marcel Dekker, New York (1969).
50. R. Perl and R. A. Ferrell, *Phys. Rev.* **A6**, 2358 (1972).
51. R. Perl and R. A. Ferrell, *Phys. Rev. Lett.* **29**, 51 (1972).
52. J. Swift, *Phys. Rev.* **173**, 257 (1968).
53. B. J. Alder, D. M. Gass, and T. E. Wainwright, *J. Chem. Phys.* **53**, 3813 (1970).
54. Sydney Chapman and T. G. Cowling, *The Mathematical Theory of Non-Uniform Gasses*, Cambridge Univ. Press, London (1970).
55. Gene F. Mazenko, *Phys. Rev.* **A7**, 222 (1973).
56. J. H. Dymond and B. J. Alder, *J. Chem. Phys.* **48**, 343 (1968); **52**, 923 (1970).
57. T. Keyes, *J. Chem. Phys.* **62**, 1691 (1975).
58. T. Keyes, *Mol. Phys.* **23**, 437 (1972).
59. B. Chu, S. P. Lee, and W. Tscharnuter, *Phys. Rev.* **A7**, 353 (1973).
60. T. Keyes and I. Oppenheim, unpublished.
61. P. Pincus, *J. Appl. Phys.* **41**, 974 (1970).
62. G. R. Alms, D. R. Bauer, J. I. Brauman, and R. Pecora, *J. Chem. Phys.* **59**, 5310 (1973).

7

Global Analysis of Nonlinear Chemical Kinetics

Leon Glass

1. Introduction

1.1. Structural Stability of Kinetic Equations

The 1920s was a time of active research in nonlinear phenomena in chemical systems.[1] Bray discovered that in a limited concentration range, if H_2O_2, KIO_3, and H_2SO_4 are heated to 60°C there is a periodic production and destruction of iodine,[2] and Lord Rayleigh (whose father was the famous mathematical physicist) observed traveling luminous waves of phosphorous oxidation at low pressures of oxygen and phosphorous in a cylindrical tube.[3] In the same decade approximately 200 papers were written about Liesegang rings, a phenomenon in which there is periodic precipitation.[4] On the theoretical side, Lotka discovered an autocatalytic mechanism that leads to sustained oscillations.[5]

In recent years there has been a revival of interest in these exotic phenomena. Several chemical oscillations of biological origin have been studied,[6] and a new oscillating chemical reaction, the Belousov–Zhabotinsky reaction, has been found that can display varied geometrical patterns as it evolves in time.[7–10] Two seminal theoretical papers that have had a great deal of influence on recent work are a study of the stability properties of reaction–diffusion equations by Turing[11] and an analysis of biochemical mechanisms that might underly oscillation and multistability by Monod and Jacob.[12] A common goal of recent work has been to develop a theoretical chemistry

Leon Glass • Department of Physiology, McGill University, Montreal, Quebec, Canada

sufficiently rich to accommodate the varied oscillatory and spatially dependent phenomena that are occasionally found in inorganic chemistry but that appear to be ubiquitous in biological systems. Several outstanding reviews reflecting this perspective have recently been published.[13–15]

This paper describes mathematical methods appropriate for the study of problems of chemistry, biology, and other fields in which complicated nonlinear equations appear to be required to represent the interactions and dynamics of the systems being studied. In these fields, the detailed dynamical equations are often unknown (or perhaps are intrinsically unknowable) and attention has focused on studying qualitative properties of the dynamics. An example of a qualitative property amenable to theoretical analysis is the number and types of dynamic behavior in the limit $t \to \infty$. In Fig. 1 are phase plane representations of dynamics in systems with two interacting chemicals. Starting from initial points in concentration space the asymptotic behavior as $t \to \infty$ is different for the three cases. In Fig. 1a, from any initial concentration, the system will evolve to a single stable steady state, and in Fig 1b, after a period of time all orbits approach a single closed trajectory, a *stable limit cycle*. Similar phase plane representations have appeared in large numbers of studies of oscillations, for example, in chemical oscillations in glycolysis,[16] in autocatalytic reactions.[17–19] and in models of biochemical oscillations regulating mitosis.[20] In Fig. 1c, there is a phase plane representation of dynamics that approach one of two stable steady states, depending on the initial conditions. Similar phase plane diagrams have appeared in a number of studies in which there is bistability, where the two stable steady states represent two possible biochemical pathways in a single cell,[21–23] two stable modes of oscillation of a laser,[24] or two possible asymptotic behaviors in a system with two competing biological species.[25] An outstanding mathematical result dating from 1937 is that the equations that generate the dynamics depicted in Fig. 1 are *structurally*

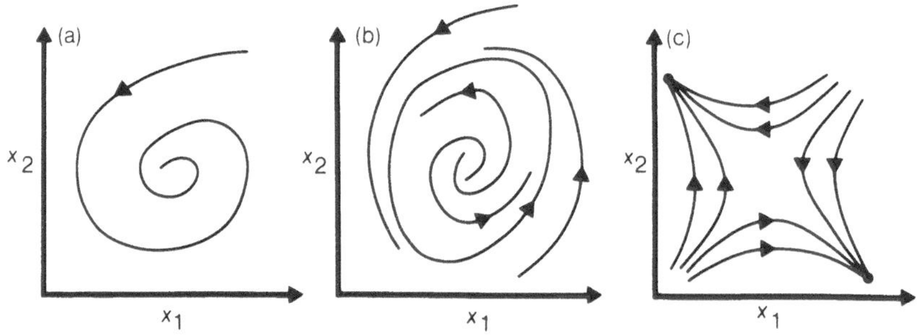

Fig. 1. Phase plane representations of structurally stable dynamics found in chemical and biological systems. (a) Single stable steady state, which is approached in oscillatory fashion. (b) Limit cycle attractor to which all trajectories tend in the limit $t \to \infty$. (*c*) Two stable steady states, which have different basins of attraction. The initial conditions determine which steady state is reached. There is also a saddle point.

stable, that is, the topological features of the dynamics remain invariant to small changes in the differential equations.[26–28]

Global analysis refers to the application of topological methods to the study of differential equations. Since I am not a topologist, anyone desiring a rigorous or formal development of the mathematics should consult texts on differential equations[29–32] and differential topology.[33,34] My interest here will be to show how some of the most basic topological results, the fixed-point and index theorems, have been used to study oscillations and multiple stability in chemical reaction systems. I shall also discuss topological and combinatorial methods that can be used to classify chemical networks on the basis of their qualitative dynamics. Before proceeding, it is necessary to introduce terminology dealing with qualitative analysis of ordinary differential equations.

1.2. Critical Points and Dynamics in the Linear Regime

Assume that a chemical reaction system of N interacting chemicals is described by the differential equations

$$dx_i/dt = R_i(\mathbf{x}), \qquad i = 1, \ldots, N \tag{1}$$

where x_i is the concentration of the ith species and R_i describes its reaction kinetics. Since concentrations are positive definite, only the dynamics in the positive orthant

$$0 \le x_i \le \infty, \qquad i = 1, \ldots, N \tag{2}$$

will be considered. In general, as a result of bimolecular collisions, R_i will be nonlinear. In writing (1) it is assumed that the system is deterministic and sufficiently small or well stirred that diffusion effects and spatial inhomogeneities can be ignored. Equation (1) can be written in vector notation,

$$\dot{\mathbf{x}} = R(\mathbf{x}) \tag{3}$$

A critical point of (1), $\mathbf{x}^0$ is defined by the expression

$$R(\mathbf{x}^0) = 0 \tag{4}$$

In the neighborhood of the critical point, the dynamics can be studied by computing the linear part of Eq. (3), which is given by

$$\dot{\mathbf{x}} = A(\mathbf{x} - \mathbf{x}^0) \tag{5}$$

where A is a matrix whose elements are defined by

$$a_{ij} = \partial R_i/\partial x_j|_{\mathbf{x}^0} \tag{6}$$

The *eigenvalues* of A, $p_1, p_2, \ldots, p_N$, which are found by solving the characteristic equation

$$\det|A - pI| = 0 \tag{7}$$

give information about the qualitative dynamics of (3) in the neighborhood of the critical point.[32] For example, if the real parts of all the eigenvalues are negative, the critical point is *stable* and all trajectories in the neighborhood of $\mathbf{x}^0$ approach it in the limit $t \to \infty$. If at least one eigenvalue is positive, the critical point is *unstable* and all trajectories (except perhaps a singular set of trajectories of measure zero) leave the neighborhood of the critical point in the limit $t \to \infty$. If

$$\operatorname{Re} p_i \neq 0, \qquad i = 1, \ldots, N \tag{8}$$

the critical point is *hyperbolic*.

1.3. Graphical Analysis of Stability Criteria

Several workers have recently been developing methods to analyze the local stability of a critical point on the basis of the signs of the elements of A.[17,35–40] It has been shown[35] that all the eigenvalues of a matrix have negative real parts independent of the magnitude of the nonzero elements if and only if the following five conditions apply:

(i) $a_{ii} \leq 0$ for all i.
(ii) $a_{ii} \neq 0$ for some i.
(iii) $a_{ij}a_{ji} \leq 0$ for all $i \neq j$.
(iv) $a_{ij}a_{jk} \ldots a_{pq}a_{qi} = 0$ for any sequence of three or more indices $i \neq j \neq k \neq \ldots \neq p \neq q$.
(v) $\det A \neq 0$.

If these five conditions apply, the matrix A is called *qualitatively stable*.

Clark[38,39] and Tyson[40] have discussed the chemical interpretation of the various destabilizing factors. To help visualize these conditions we use a graphical representation of the matrix A, which was introduced by Levins.[37] Each variable is considered a node of a graph where two nodes are connected if $a_{ij} \neq 0$, where the edge directed from node i to node j terminates in an arrow (circle) if a_{ij} is greater than (less than) 0. The graphs representing the linear part of the reaction kinetics for several kinetic equations are given in Fig. 2. Only Fig. 2a is qualitatively stable. If condition (i) is violated; this corresponds to autocatalysis. The autocatalysis in Fig. 2b can lead to oscillatory instabilities, for example, as in the limit cycle oscillation shown in Fig. 1b.[16–19] If condition (iii) is violated there is either mutual activation or mutual inhibition of synthesis, in which case there may be bistability as anticipated by Monod and

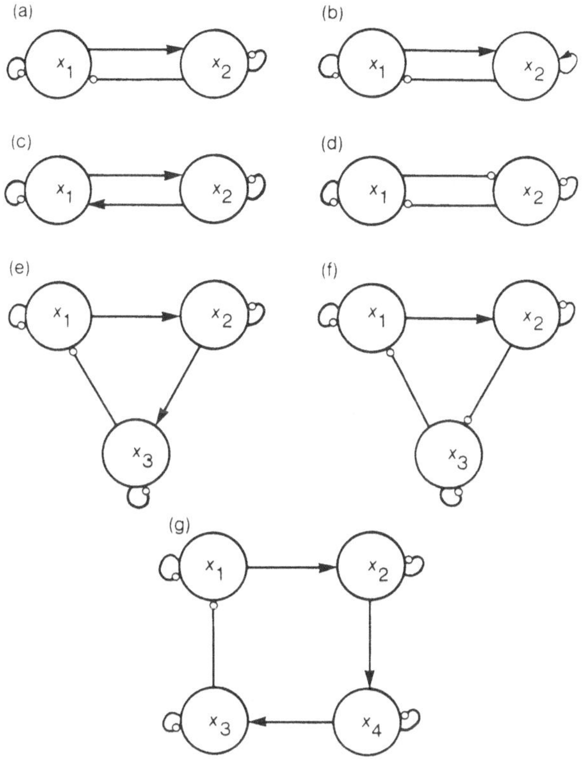

Fig. 2. Schematic representations of the linear parts of differential equations in the neighborhood of steady states. For the dynamical Eq. (1) the edge directed from node i to node j terminates in an arrow (circle) if $\partial R_i/\partial x_j|_{\mathbf{x}=\mathbf{x}^0}$ is greater than (less than) zero. (a) A feedback inhibition loop of two chemicals, where each chemical is self-damped. (b) A feedback inhibition loop of two chemicals, where x_2 undergoes autocatalysis. (c) Mutual activation between two chemicals. (d) Mutual inhibition between two chemicals. (e) A negative feedback loop involving three chemicals. (f) A positive feedback loop involving three chemicals. (g) A negative feedback loop involving four chemicals.

Jacob.[12] Kinetic schemes with both mutual inhibition[21,23,24] (Figs. 1c and 2d) and with mutual activation [22,23] (Fig. 2c) have been widely studied. If condition (iv) is violated and the product is less than 0, there is a *negative feedback loop* (Figs. 2e and 2g). There is a large literature dealing with dynamics of chemical networks with negative feedback loops,[42–47] which can be summarized by saying that there are usually parameter ranges in which there are globally stable limit cycle oscillations, but that such oscillations are easier to demonstrate using numerical rather than analytical means (see Section 4). Finally if condition (iv) is violated and the product is greater than 0 (Fig. 2f) there is a *positive feedback loop* and multiple steady states may arise.[48,49]

In general, the presence of a destabilizing factor is insufficient to ensure instability and there will be stable and unstable regimes that depend on the particular values of the coefficients in the kinetic equations. A set of parameters separating two structurally stable regimes with different qualitative dynamics is called a *bifurcation point.* At bifurcation points, differential equations are not structurally stable. There is an extensive literature on bifurcation theory,[50,51] but it will not be reviewed here.

The techniques described in this section are useful for studying chemical dynamics in the neighborhood of critical points. The remainder of this paper is devoted to the analysis of the global dynamics of nonlinear kinetic equations. In Section 2 a topological theorem is given, which can be used to place restrictions on the entire set of critical points of a chemical network. In Section 3 it is shown that many chemical networks can be classified on the basis of flows between volumes in concentration space. In Section 4, a number of techniques for establishing limit cycle oscillations in three and more dimensions are described. The topological methods are applied to analysis of compartmental chemical systems in Section 5. The results are discussed in Section 6. In the Appendix the principal mathematical results that have been used in the text are summarized.

2. An Index Theorem for Chemical Kinetics

2.1. Chemical Dynamics on Manifolds

In Section 1, the local analysis of the stability properties of nonlinear equations was discussed. In the 1880s Poincaré discovered a theorem that makes a statement about the entire set of critical points of dynamical systems on two-dimensional manifolds. This result was later extended by Hopf to arbitrary manifolds (see Appendix for greater detail and references). Here the theorem is applied to dynamics in chemical systems.

The index I if a hyperbolic critical point is defined by

$$I = (-1)^{\mu} \tag{9}$$

where μ is the number of eigenvalues for which the real part is negative. The Poincaré–Hopf index theorem is then

$$\sum_{\text{cps}} I = \chi(M) \tag{10}$$

where the sum is over all the critical points of a dynamical system defined on a manifold M, and $\chi(M)$ is the Euler–Poincaré characteristic of the manifold. For example, for an N-sphere (the set of points for which $\sum_{i=1}^{N+1} y_i^2 = \text{const}$) $\chi(M)$ is given by

$$\chi(M) = 1 + (-1)^N \tag{11}$$

By examining the physical constraints imposed on kinetic equations, it is possible to embed many chemical reaction systems on N-spheres. Assume that the reaction kinetics are given by Eqs. (1) and (2), where each of the N reactants is present initially and no new reactants are generated as time proceeds. Since chemical reactions are reversible, it is impossible for any reactant to disappear completely. Consequently, we assume that there exists a small number δ such that

$$\dot{x}_i > 0, \qquad \text{for } x_i = \delta, \quad i = 1, \ldots, N \tag{12}$$

Also, since the concentrations must remain finite we assume that, for some large number C,

$$\dot{x}_i < 0, \qquad \text{for } x_i = C, \quad i = 1, \ldots, N \tag{13}$$

Boundary conditions (12) and (13) constrain the dynamics to a box in N dimensions, where all trajectories on the boundary of the box enter the box. By associating the boundary of the box with a single unstable source at the South Pole, the chemical kinetic system defined by (1), (12), and (13) can be embedded on an N-sphere. By applying (9)–(11) we compute

$$\sum_{j=1}^{m} (-1)^{\pi_j} = 1 \tag{14}$$

where the sum is over the m critical points of the chemical system and π_j is the number of eigenvalues with positive real parts at the jth critical point.*[52–54]

Equation (14) places sharp restrictions on the sets of critical points in chemical reaction networks.[54] Consider the simplest cases that are possible for two reacting chemicals, $N = 2$.

(i) $m = 1, \pi_1 = 0$. Here there is a single stable steady state. An example of this critical point structure is given in Fig. 1a.

(ii) $m = 1, \pi_1 = 2$. Here there is a single critical point with two eigenvalues with positive real parts. By the Poincaré–Bendixson theorem (see Appendix) there must also be at least one closed trajectory. An example of (ii) is shown in Fig. 1b, in which the closed trajectory is a stable limit cycle.

(iii) $m = 3, \pi_1 = 1, \pi_2 = 0, \pi_3 = 0$. Here there are three critical points, one of which is unstable in one dimension, and two of which are stable. An example of this behavior is shown in Fig. 1c.

*A somewhat different derivation and statement of this theorem has been given by Gavalas.[52]

As discussed in Section 1 these three cases encompass many examples of kinetic systems in two dimensions that have been studied to date. Moreover, these three critical point structures are also possible in higher dimensions since the topological constraint (14) does not depend on the number of reacting chemicals.

One immediate and interesting consequence of (14) is that it places restrictions on bifurcations in the dynamics as a parameter changes. In particular, it is impossible for the real part of only one eigenvalue to change sign at a bifurcation point. One allowable bifurcation is the *Hopf bifurcation* in which the real parts of two complex eigenvalues cross the real axis simultaneously (see Section 4.1). Another allowable bifurcation is one in which a single stable steady state splits into three critical points to give (iii) above. This behavior has often been observed in mathematical models of competition and bistability in chemical and physical systems.[21,23,24]

2.2. Limitations of the Index Theorem

Although the topological theorem (14) is a global result, since it places restrictions on the entire set of critical points of chemical networks, the theorem does not give information about the dynamics outside the neighborhoods of the critical points of the system. It is consequently a question of some interest to analyze the global dynamics in kinetic equations away from the critical points. In dynamical systems of dimension three and higher, this is a problem of current mathematical interest since an enumeration of the topologically distinct attractors is not available. In addition to stable steady states and limit cycles, there are also found "strange attractors" in which the dynamics pass through complex nonperiodic behavior. There is speculation that hydrodynamic turbulence may be a manifestation of strange attractors in the hydrodynamic equations[55] and that strange attractors might underlie complicated dynamics in ecological systems.[56,57] There has been a suggestion that strange attractors might also be found in nonlinear chemical kinetics.[58] That would give rise to chaotic concentration changes in time, which most experimenters might incorrectly identify as noisy and spurious results and would not study further.

An additional limitation of the index theorem is that in higher dimensions, there may often be different time scales corresponding to fast and slow reactions, so that the dynamics rapidly relax to a manifold that is embedded in the N-sphere. In this case, the Euler–Poincaré characteristic of the submanifold may be different from that of the N-sphere and (14) would have to be modified appropriately. A very general mathematical approach in which the geometric and topological properties chemical kinetic equations is treated has recently been developed.[59,60] Although the formulation is in principle capable of treating some of these limitations in a rigorous way, the techniques described have not yet found extensive applications. In the next two sections we consider additional theoretical methods that can be used to analyze multiple stability and limit cycle oscillations in nonlinear kinetics.

3. Flow Box Analysis in Chemical Kinetics

3.1. Classification of Flows in Kinetic Systems

Assume that there is a chemical kinetic system described by Eqs. (1) and (2). We decompose the concentration space into 2^N volumes, which are arranged in space so that they are *homeomorphic* (topologically equivalent) to the 2^N orthants in the neighborhood of the origin in an N-dimensional coordinate space. There are $N \times 2^{N-1}$ boundaries between the contiguous volume elements. We assume that across each of the boundaries the flows are transverse and of unique orientation. With each of the volumes in concentration space it is possible to associate a vertex of an N-dimensional hypercube (N-cube). The transitions between contiguous volumes are represented by directed edges on the N-cube.[49,53] The representation in which each edge of an N-cube is oriented is called a *state transition diagram*. Two chemical systems are in the same *structural equivalence class* if their state transition diagrams can be superimposed under the symmetry group of the N-cube.

This construction takes continuous flows in N-dimensions and represents them in a finite way so that combinatorial methods can be employed to obtain a classification of the dynamics. (This should be compared with the methods of

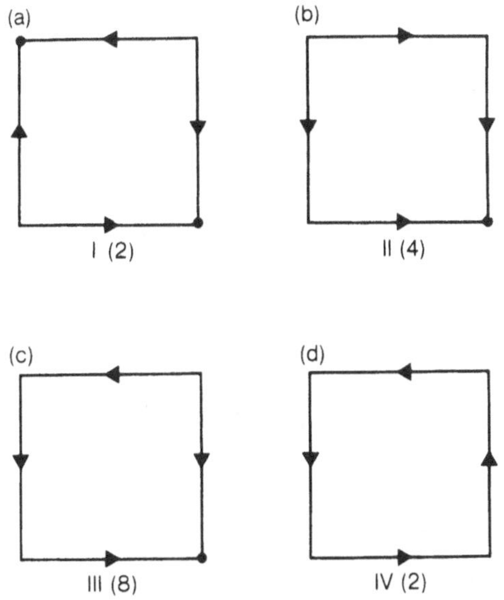

Fig. 3. The structural equivalence classes for $N = 2$. The number in parentheses under each structure gives the number of systems falling in each equivalence class.

symbolic dynamics, which have been used to study ergodic problems in mechanical systems.[61]) For example, for $N = 2$ there are only 4 structural equivalence classes (Fig. 3). By constructing vertical and horizontal axes through the central critical points and examining the flows between contiguous regions in Fig. 1, it is possible to identify Figs. 1a and 1b with class IV and Fig. 1c with class I.

For $N = 3$ the situation is more interesting. First, it is possible to compute the total number of structural equivalence classes without enumerating them. The combinatorial methods to do this were developed by Burnside and Polya, and have been extensively applied in the theory of switching networks.[62,63] For the present case, the number of structural equivalence classes is 112.[49] It is not a simple matter to obtain a representative assembly containing one member from each of the 112 classes. One way to approach the problem of generating a representative assembly is to recognize that there are restrictions on the local configurations of the edges at vertices and faces, which are imposed by a topological theorem closely related to the index theorems (see Appendix). For the present case, for any specification of edges on the cube,

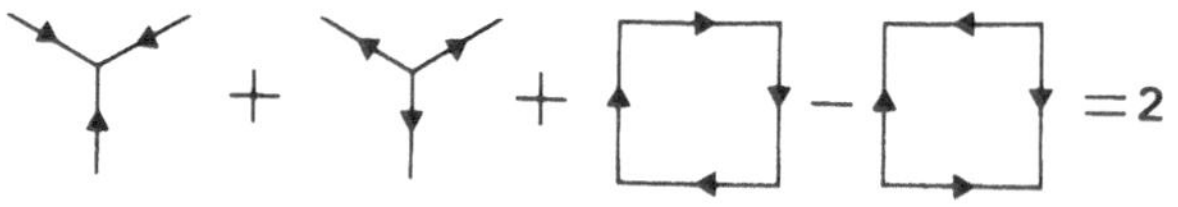

(15)

where each diagram in Eq. (15) represents the number of times each configuration appears in the state transition diagram. Two chemical networks are said to be in the same *surface equivalence class* if all the configurations in Eq. (15) can be superimposed under some symmetry operation of the cube. I have recently given a listing of the surface equivalence classes.[53] These results are summarized in Table 1.

One of the structural equivalence classes, shown in Fig. 4, has been of particular interest in the study of oscillations in chemical networks with three interacting chemicals. It has appeared[64] in an analysis of the Field–Noyes equations[65] for the Belousov–Zhabotinsky reaction, and in an analysis of

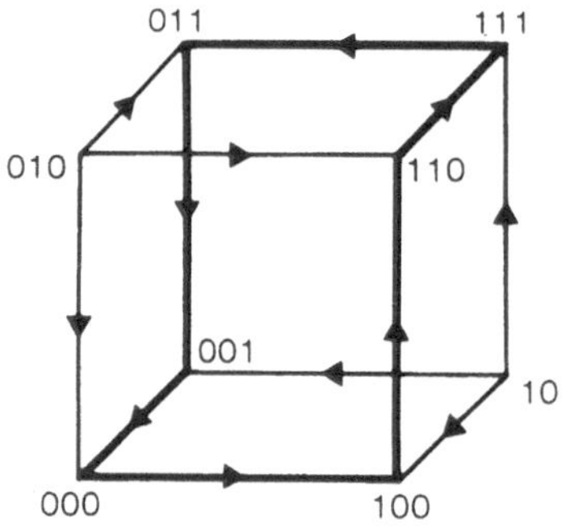

Fig. 4. The state transition diagram found for the Field–Noyes mechanism[65] for the Belousov–Zhabotinsky reaction by Hastings and Murray,[64] and for feedback inhibition.[47] The dynamics of an equation with this structure is discussed in Sections 4.1 and 4.2.

Table 1. A Summary of the Computation of the Surface Equivalence Classes Given by Glass[(53)a]

Number			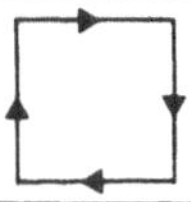	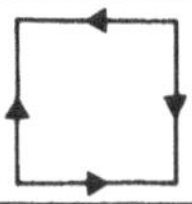	S_i	C_i
1	4	4	0	6	1	2
2	3	3	0	4	1	24
3	3	2	0	3	2	32
4	2	3	0	3	2	32
5	2	2	0	2	12	372
6	2	2	1	3	1	24
7	2	1	0	1	9	336
8	2	1	1	2	1	48
9	2	0	1	1	1	24
10	2	0	0	0	1	8
11	1	2	0	1	9	336
12	1	2	1	2	1	48
13	1	1	2	2	2	48
14	1	0	2	1	4	144
15	1	1	1	1	12	576
16	1	0	1	0	7	336
17	1	1	0	0	19	744
18	0	2	0	0	1	8
19	0	2	1	1	1	24
20	0	1	2	1	4	144
21	0	1	1	0	7	336
22	0	0	4	2	1	6
23	0	0	3	1	2	72
24	0	0	2	0	11	372

[a] The first four columns give the number of times each configuration appears on the state transition diagram. S_i the number of structural equivalence classes (out of 112) falling into each category, and C_i the number of distinct configurations (out of 4096) falling in each category.

feedback inhibition equations with three chemicals.[(47)] In Sections 4.1 and 4.2 the qualitative dynamics of some equations having this structure will be discussed.

3.2. Computation of State Transition Diagrams from the Nonlinear Equations

It is not a simple matter to determine whether the notion of state transition diagram introduced here provides a compact and useful method for representing flows of an arbitrary system of nonlinear kinetic equations. It is, however, possible to provide a list of some functional forms for $R_i(\mathbf{x})$ that when substituted in Eq. (1) give differential equations of interest in chemical kinetics that can be readily classified.

3.2.1. Volterra–Lotka Systems

One example is

$$R_i = (-1)^{m_i}\lambda_i x_i \prod_{j\neq i} (x_j - \theta_j), \qquad i = 1, \ldots, N \tag{16}$$

where λ_i and θ_i are production and threshold constants and m_i can either be 0 or 1. A famous example that was proposed by Volterra to account for population oscillation of fish in the Adriatic sea[66] and by Lotka to account for biochemical oscillation in systems with autocatalysis[5] can be written as Eq. (16):

$$\dot{x}_1 = \lambda_1 x_1(\theta_2 - x_2), \qquad \dot{x}_2 = \lambda_2 x_2(x_1 - \theta_1) \tag{17}$$

To determine the state transition diagram for Eqs. (1) and (16) it is convenient to define a Boolean variable $\tilde{x}_i$,

$$\tilde{x}_i = \begin{cases} 1, & \text{if } x_i \geq \theta_i \\ 0, & \text{if } x_i < \theta, \quad i = 1, \ldots, N \end{cases} \tag{18}$$

We shall name each of the 2^N volumes by a Boolean N-vector found from Eq. (18), which represents a vertex on the Boolean N-cube (the geometric dual of the concentration space). Consider the element $i = 1$. If $m_1 = 0$ in Eq. (16) then there will be a directed edge from $(011 \cdots 1)$ to $(111 \cdots 1)$ and if $m_1 = 1$, the orientation of this edge will be reversed. Similarly the edge joining $(001 \cdots 1)$ and $(101 \cdots 1)$ is directed from the first to the latter vertex if $m_1 = 1$, and if $m_1 = 0$ the orientation of this edge is reversed. The remainder of the edges are handled in similar fashion.

3.2.2. Boolean Kinetic Equations

Another functional form that generates kinetic equations classified using the same methods is[53]

$$R_i = \lambda_i B_i[\tilde{x}_1\tilde{x}_2 \cdots \tilde{x}_{i-1}\tilde{x}_{i+1} \cdots \tilde{x}_N] - \gamma_i x_i, \qquad i = 1, \ldots, N, \quad N+1 = 1 \tag{19}$$

where λ_i and γ_i are production and decay constants and B_i is a Boolean function* of the $N-1$ Boolean variables defined in Eq. (18). When Eq. (19) is substituted in Eq. (1) all trajectories are confined to the box

$$0 \leq x_i \leq \lambda_i/\gamma_i, \qquad i = 1, \ldots, N \tag{20}$$

in the limit $t \to \infty$. In the discussion that follows, only points in the volume defined in Eq. (20) are considered.

*A Boolean function of N variables assigns a value of 1 or 0 to each of the 2^N states of those variables.

Since B_i is discontinuous, Eq. (19) is not a realistic representation of nonlinear chemical networks. However, many control molecules in biochemical systems, for example, operator regions of the gene,[67] and enzymes[68] appear to undergo cooperative conformational changes so that synthetic rates vary from low basal levels to maximal rates over a comparatively small concentration range of control molecules. This "switchlike control has led to conjectures that switching networks might constitute a reasonable conceptualization for analysis of nonlinear kinetics in biological systems.[49,69–73] Previous studies (see also Section 4.2) have indicated that in at least some cases there is no difference in the qualitative dynamics of discontinuous equations [(1) and (19)] and equations that are mathematically "close" to it, in which the discontinuities are made smooth.[74,75]

The state transition diagrams for Eqs. (1) and (19) can be readily computed. A *truth table* gives the value of B_i for each of the 2^N states. In Table 2 we give the general form for the truth table, where each of the $b_{i,j}$ is either 1 or 0. Consider states S_1 and S_2. From the condition of no self-input in Eq. (19) we have $b_{1,1} = b_{2,1}$. If $b_{1,1} = b_{2,1} = 1$ then $\dot{x}_1 \geq 0$ in S_1 and S_2, and there can only be transitions from S_2 to S_1. If $b_{1,1} = b_{2,1} = 0$, then $\dot{x}_1 \leq 0$ in S_1 and S_2 and there can only be transitions from S_1 to S_2. These possible transitions are represented as oriented edges on the state transition diagram. From this construction, each of the columns in Table 2 will give 2^{N-1} directed edges in the state transition diagram. For example, in Table 3, we give the truth table for several chemical networks. The state transition diagrams for Table 3a, 3b, and 3c are given in Figs. 3a, 3d, and 4, respectively. These networks share the common feature that each B_i is a function of *only one* of the other Boolean variables of the system. Consequently, a simple verbal description of these networks is possible. For example, for Table 2b we say that x_1 activates the production of x_2, and x_2 inhibits the production of x_1.

Table 2. Truth Table for an N-Variable System[a]

	Boolean variable				Boolean function			
State	$\tilde{x}_1$	$\tilde{x}_2$	$\cdots$	$\tilde{x}_N$	B_1	B_2	$\cdots$	B_N
S_1	1	1	$\cdots$	1	$b_{1,1}$	$b_{1,2}$	$\cdots$	$b_{1,N}$
S_2	0	1	$\cdots$	1	$b_{2,1}$	$b_{2,2}$	$\cdots$	$b_{2,N}$
S_3	1	0	$\cdots$	1	$b_{3,1}$	$b_{3,2}$	$\cdots$	$b_{3,N}$
S_4	0	0	$\cdots$	1	$b_{4,1}$	$b_{4,2}$	$\cdots$	$b_{4,N}$
$\vdots$								
$S_{2^{N-1}}$	0	0	$\cdots$	1	$b_{2^{N-1},1}$	$b_{2^{N-1},2}$	$\cdots$	$b_{2^{N-1},N}$
$\vdots$								
S_{2^N}	0	0	$\cdots$	0	$b_{2^N,1}$	$b_{2^N,2}$	$\cdots$	$b_{2^N,N}$

[a] For each of the 2^N states, the truth table gives the values of B_i [Eq. (20)].

Table 3. Truth Tables for Some Simple Networks[a]

(a)				(b)				(c)					
$\tilde{x}_1$	$\tilde{x}_2$	B_1	B_2	$\tilde{x}_1$	$\tilde{x}_2$	B_1	B_2	$\tilde{x}_1$	$\tilde{x}_2$	$\tilde{x}_3$	B_1	B_2	B_3
1	1	0	1	1	1	0	0	1	1	1	0	1	1
0	1	0	0	0	1	0	1	0	1	1	0	0	1
1	0	1	1	1	0	1	0	1	0	1	0	1	0
0	0	1	0	0	0	1	1	0	0	1	0	0	0
								1	1	0	1	1	1
								0	1	0	1	0	1
								1	0	0	1	1	0
								0	0	0	1	0	0

[a] (a) Feedback inhibition with two chemicals. (b) Mutual inhibition with two chemicals. (c) Feedback inhibition with three chemicals. The state transitions diagrams for case (a) are given in structure IV, Fig. 3; for case (b) in structure I, Fig. 3, and for case (c) in Fig. 4.

Still another functional form that yields a classification using the state transition diagrams is

$$R_i = \lambda_i x_i\{2B_i[\tilde{x}_1\tilde{x}_2 \cdots \tilde{x}_{i-1}\tilde{x}_{i+1} \cdots \tilde{x}_N] - 1\} - \gamma_i x_i^2, \qquad i = 1, \ldots, N, \quad N+1 = 1 \tag{21}$$

If the set of B_i is the same as the B_i in Eq. (19) the state transition diagrams for both equations will be identical. For both Eqs. (19) and (21) it is possible to choose each B_i in $2^{2^{N-1}}$ different ways, so that the number of different networks possible is $2^{N\times 2^{N-1}}$. This corresponds to all the possible labelings of the $N\times 2^{N-1}$ edges of the N-cube with directed edges.[49]

3.3 Critical Points and Qualitative Dynamics from State Transition Diagrams

There have been only a very small number of structurally stable, topologically equivalent dynamics that arise in chemical examples that have identical state transition diagrams. For example, for structure IV, Fig. 3, the two dynamics in Figs. 1a and 1b have appeared widely in the chemical literature. Of course, dynamics having great complexity could also be consistent with the same flow box structure if, for example, there were nested limit cycles though all four quadrants, or limit cycles were entirely contained within a single quadrant. It would be of great interest to analyze the global features of the qualitative dynamics of nonlinear differential equations arising in chemical kinetics if only the state transition diagram is known, simply by learning to interpret the state transition diagrams properly.

As a first step in this direction, we shall study the global dynamics of a limited class of differential equations based on Eqs. (1), (2), and (19). If we assume $\gamma_1 = \gamma_2 = \cdots = \gamma_N = \gamma$ these equations can be rewritten

$$dX_i/d\tau = B_i[\tilde{X}_1\tilde{X}_2 \cdots \tilde{X}_{i-1}\tilde{X}_{i+i} \cdots \tilde{X}_N] - X_i, \qquad i = 1, \ldots, N, \quad N+1 = 1 \tag{22}$$

where

$$\tau = \gamma t, \qquad X_i = x_i/(\lambda_i/\gamma)$$

$$\tilde{X}_i = \begin{cases} 1 & \text{if } X_i \geq \theta_i/(\lambda_i/\gamma) \\ 0 & \text{if } X_i < \theta_i/(\lambda_i/\gamma) \end{cases} \tag{23}$$

Since all trajectories must tend to the box $0 \leq X_i \leq 1$, we consider only trajectories in this volume.

For Eq. (22) a geometrical construction of the trajectories in concentration space is possible. Consider any point $\mathbf{X}^0$, where $X_i^0 \neq \theta_i/(\lambda_i/\gamma)$. The trajectory through $\mathbf{X}^0$ is a straight line whose equation is

$$\frac{X_1 - B_1}{X_1^0 - B_1} = \frac{X_2 - B_2}{X_2^0 - B_2} = \cdots = \frac{X_N - B_N}{X_N^0 - B_N} \tag{24}$$

where the B_i are defined in the truth table for the initial state. Starting from any initial point, the dynamics is a piecewise linear flow that is iterated through the sequence of volumes determined by the equations. This construction has a number of consequences.

1. If $X_i^0 = B_i$, $i = 1, \ldots, N$, then all trajectories in $\tilde{\mathbf{X}}^0$ will asymptotically approach the concentration

$$X_i = B_i, \qquad i = 1, \ldots, N \tag{25}$$

This will be an asymptotically stable steady state of Eq. (22) called an *extremal steady state*. By construction, any vertex on the state transition diagram with only arrows directed in will be an extremal steady state.

2. If $X_i^0 \neq B_i$ for at least one i, then all trajectories in $\tilde{X}^0$ will leave this volume, and there will be no steady states in the volume.

3. In order for there to be a cycle in Eq. (22), there must be a cyclic path in the state transition diagram of the system.

By integrating particular examples, it is discovered that the cycles in Eq. (22) can be of two types, *unstable cycles*, in which the trajectories spiral into an intersection point of the threshold axes, and *stable cycles*, in which there is a stable limit cycle attractor in concentration space. Unstable oscillations are found for Eq. (22) in structure IV, Fig. 3, and stable limit cycle oscillations are found for Eq. (22) in Fig. 4 (see Section V.3). It is not known whether other types of asymptotic behavior besides extremal steady states and stable and unstable cycles are possible in Eq. (22).

The derivatives of Eq. (22) are discontinuous. It is therefore impossible to discuss the dynamics in terms of hyperbolic critical points. Let us assume, however, that there are equations "close" to Eq. (22), which are constructed by smoothing the discontinuities in the neighborhoods of the threshold axes, in which at least first derivations are continuous, and in which there are isolated hyperbolic critical points. For such systems it is possible to place restriction on the set of critical points by using the topological theorem, Eq. (14). In these systems, since smoothing only changes the equations in the neighborhoods of the threshold axes, the extremal steady states will still be stable steady states since they do not lie in the neighborhoods of the threshold axes. The smoothing operation will generate new critical points only in the neighborhoods of the threshold axes. Let us say that the smoothing operation generates ε critical points with an even number of positive real parts and ω critical points with an odd number of positive real parts. Then if s is the number of extremal steady states (the number of vertices in the state transition diagram with only edges directed in) we find[53]

$$\varepsilon - \omega = 1 - s \tag{26}$$

For example, in Fig. 3, structure I, $s = 2$. A smoothing of Eq. (22) generates one additional critical point, which is a saddle point, as in Fig. 1c. In Fig. 3, structure IV, $s = 0$. For this case, the smoothing operation generates a stable focus as is shown in Fig. 1a.

An additional example will be considered. I have stressed the importance of the structure in Fig. 4 in studies of oscillations in chemical systems with three interacting chemicals. A differential equation with this structure is[53]

$$\begin{aligned} \dot{X}_1 &= \frac{0.5^n}{0.5^n + X_N^n} - X_1 \\ \dot{X}_i &= \frac{X_{i-1}^n}{0.5^n + X_{i-1}^n} - X_i, \qquad i = 2, \ldots, N \end{aligned} \tag{27}$$

with $N = 3$. Here the dynamics are dominated by nonlinear terms of Hill function form.[67,68] In the limit $n \to \infty$, the equation can be given as in Eq. (22), where the set B_i are given in Table 2c. Equation (27) represents a feedback inhibition equation with three reactants.

There is only one critical point in Eq. (27), at the point $X_i = 0.5, i = 1, \ldots, N$. By choosing this point to define the threshold axes $\theta_i = 0.5$, the flow between volumes is easily seen to be given by Fig. 4. In Fig. 4, $s = 0$, so in the continuous system close to Eq. (22) having this structure [here Eq. (27), $N = 3$, n sufficiently large], we must verify Eq. (26). The eigenvalues of the critical point Eq. (27) can be readily computed and are[53]

$$p_1 = -1 - n/2, \qquad p_{2,3} = -1 + n/4 \pm 3^{1/2} ni/4 \tag{28}$$

For $n > 4$, there are only two positive eigenvalues and the topological theorem is satisfied.

In the next section we discuss some of the mathematical methods that can be used to study in more detail the global dynamics of equations that display nonlinear oscillations in more than two dimensions.

4. Some Limit Cycle Oscillations in Nonlinear Kinetic Equations

Here we consider the problem of determining whether or not there are limit cycle oscillations in kinetic equations with more than two interacting chemicals. To keep definite, we discuss the dynamics of Eq. (27), $N = 3$, in Sections 4.1 and 4.2, and Eq. (22) in Section 4.3.

4.1. The Hopf Bifurcation

In 1942 Hopf demonstrated a way to establish the presence of oscillations in limited regions of parameter space.[76] The methods developed by Hopf have recently had many applications to study of oscillations in chemical and biological systems.[46,55,56,58,75,77–79] After stating the theorem in general form, I will show how it can be applied to the equation for feedback inhibition. Let us assume that the dynamical system in Eq. (1) depends on the parameter a, so that

$$\dot{\mathbf{x}} = f(\mathbf{x}, a) \tag{29}$$

Assume also that $\mathbf{x}^0(a)$ is a critical point of Eq. (29). The eigenvalues $p_1(a), p_2(a), \ldots, p_N(a)$ will now also depend on the parameter a. If for some values of a, say $a < a_0$, the critical point is stable, and if in addition a pair of complex conjugate eigenvalues $p_1(a), p_2(a)$ cross the imaginary axes transversely ($d \operatorname{Re} p_1(a)/da|_{a=a_0} \neq 0$) then we say that a *Hopf bifurcation* takes place at the value $a = a_0$. If a Hopf bifurcation occurs in Eq. (29) then there exists a one-parameter family of periodic solutions for a in the neighborhood of a_0, with a period near $2\pi/|\operatorname{Im} p_1(a_0)|$. If the flow attracts to the critical point $\mathbf{x}^0(a_0)$ when $a = a_0$, then $\mathbf{x}^0(a_0)$ is called a "vague attractor." For this case the family of closed orbits is contained in $a > a_0$ and the orbits are of attracting type.[55]

The value $n = 4$ is a Hopf bifurcation point for the feedback inhibition in three dimensions, Eq. (26). Although algebraic methods are available to determine whether the critical point is a vague attractor[78] this can also be established by numerical integration. In Fig. 5, we give the amplitude of X_1 at maximum and minimum for a trajectory starting at (0.6, 0.7, 0.2) using a Runge–Kutta scheme to integrate the equations. Since the first-order terms of the real parts of the eigenvalues vanish, relaxation to the steady state is extremely slow as the system approaches the critical point. This numerical demonstration that the critical point is a vague attractor indicates there must be

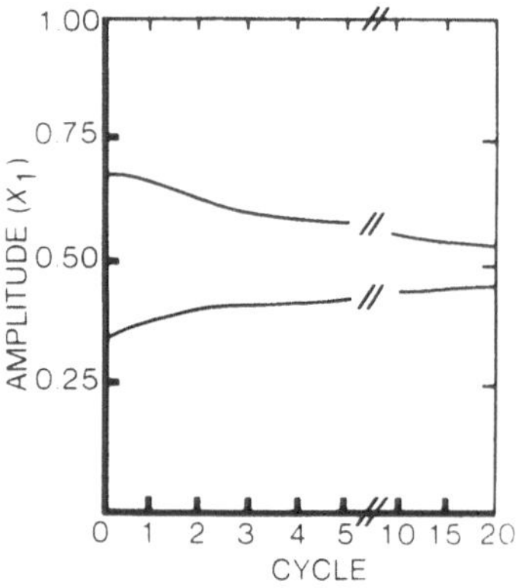

Fig. 5. The amplitude of x_1 at subsequent cycles in Eq. (27), $N = 3$, $n = 4$, starting from (0.6, 0.7, 0.2), using a Runge–Kutta integration scheme.

an attracting orbit for values of $n > 4$, *in the neighborhood of the value* $n = 4$, with a period near $2\pi/3^{1/2} = 3.63$. What is so interesting about the feedback inhibition equation is that numerical methods indicate there is a single unique limit cycle attractor for all values $n > 4$. In Fig. 6 we show the transient to a stable limit cycle oscillation for $n = 6$ starting from initial values (0.6, 0.7, 0.2). In Fig. 7a we plot the values of the maxima and minima of the limit cycles, and in Fig. 7b the period of the limit cycles, which are found by numerical integration as n is increased.

4.2. Global Analysis of Limit Cycle Oscillations

The Hopf bifurcation theorem is only valid in the neighborhood of the bifurcation value $n = 4$. However, it is possible to prove existence of a limit cycle for finite values $n > 4$, and uniqueness and stability for infinite n.

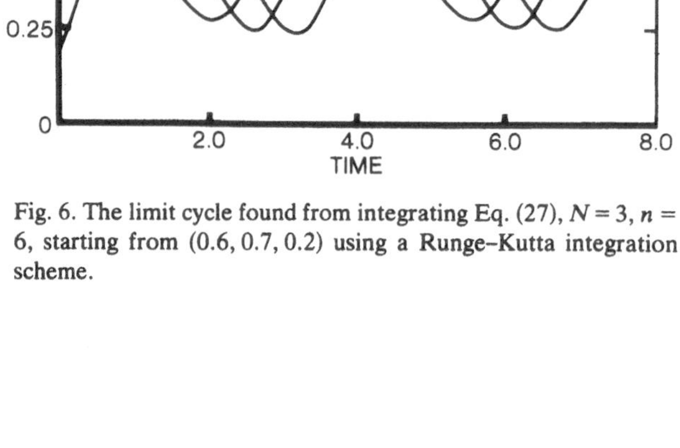

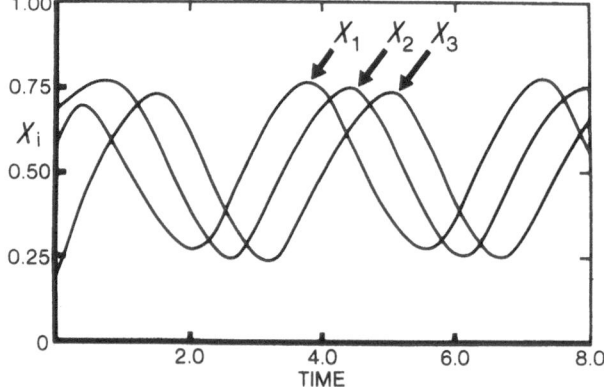

Fig. 6. The limit cycle found from integrating Eq. (27), $N = 3$, $n = 6$, starting from (0.6, 0.7, 0.2) using a Runge–Kutta integration scheme.

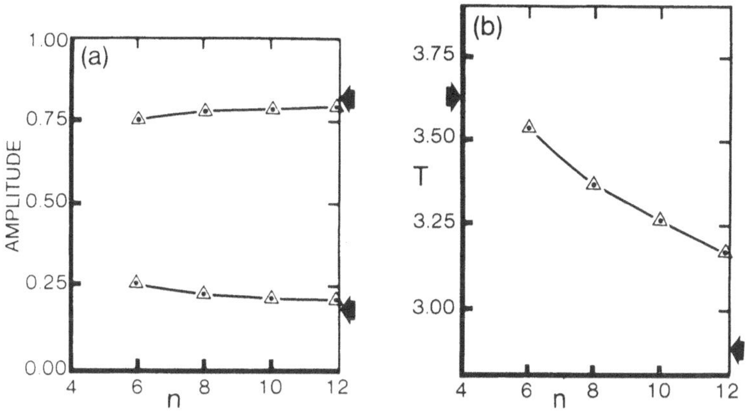

Fig. 7. (a) The amplitudes and (b) the period T of the oscillations found from integrating Eq. (27), $N = 3$, for varying values of n. Except for the value $n = 4$, global limit cycle attractors were found. The arrows on the right-hand side of the diagrams represent the theoretical values for the piecewise linear equation in the limit $n \to \infty$. The arrow on the left-hand side of (b) is the period predicted by the Hopf bifurcation theorem.

For finite n, we use a method of proof that has been developed recently to show limit cycle oscillation in the Field–Noyes mechanism for the Belousov–Zhabotinsky reaction[64] and in feedback inhibition.[46] The method is applicable to all reaction kinetics whose state transition diagram is given in Fig. 4. Let us call F the face separating the volumes 000 and 001 (Fig. 8). From the state transition diagram F is mapped into itself after one complete circuit through the six volumes of concentration space. Call the mapping that does this Γ. For Eq. (27) the point (0.5, 0.5, 0.5) is a fixed point of F. Any other fixed point of F will be a periodic solution of the dynamical system. To prove the existence of another fixed point it is sufficient to excise the vertex (0.5, 0.5, 0.5) from the face F by a simple smooth curve δ (Fig. 8) and show that the remaining region in F is then mapped into itself by Γ.[46,64] The existence of another fixed point is then guaranteed by the Brouwer fixed-point theorem (Appendix). For $n > 4$, there is one eigenvalue with a negative real part and two complex conjugates with positive real parts. For the present case, if the singular trajectory associated with the negative eigenvalue lies completely in the volumes designated 010, 101, then the curve δ can be defined as the intersection of F with a narrow cylinder surrounding the singular trajectory. In Eq. (27), the singular trajectory in the neighborhood of (0.5, 0.5, 0.5) lies along the diagonal joining (0.0, 1.0, 0.0) and (1.0, 0.0, 1.0). The fixed point can therefore be excised from F, and there must be a periodic solution in the dynamical system, Eq. (27). An extension of this method to feedback inhibition in higher dimensions has been found.[80] This establishes the existence of a periodic solution, but neither uniqueness or stability.

I have only been able to establish uniqueness and stability of the periodic solution for Eq. (27) for the case where n is infinite. For this case, the

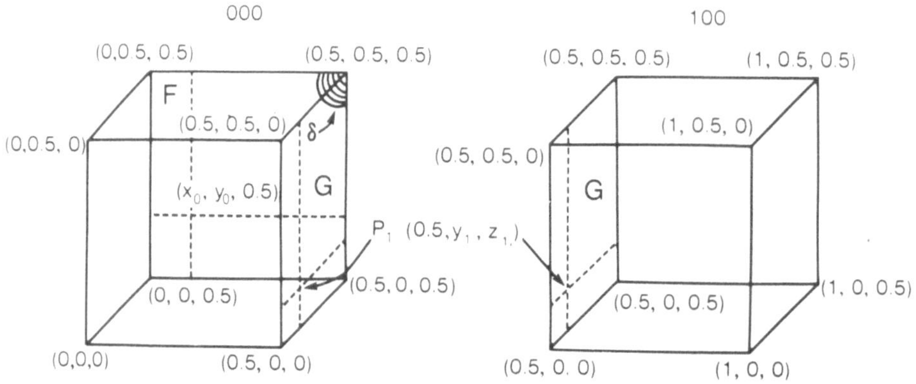

Fig. 8. The volumes represented by 000 and 100 in Fig. 4. In the piecewise linear Eq. (22) with $\theta_i = 0.5$, the trajectories in 000 are straight lines directed toward the vertex (1,0,0). This maps all points on face *F* to face *G*. All trajectories in 100 are in turn straight lines directed toward the vertex (1,1,0) (not shown), and the trajectories in the remaining volumes are again straight lines that eventually lead back to face *F*. From the symmetry of the construction there will be a limit cycle if there is a point for which $y_1 = x_0$, $z_1 = y_0$. If the shaded corner on face *F* is excised from face *F*, the map through the remaining six volumes still maps all points on *F* (excluding the shaded region) back onto *F* (excluding the shaded region).

trajectories in each of the six volumes represented by the vertices on the cycle in Fig. 4 are identical under an appropriate symmetry operation of the subcubes.

Consider a trajectory through any point (X_0, Y_0, Z_0) in the volume 000. From the previous section, the equation of the trajectory through this point in volume 000 is

$$\frac{X-1}{X_0-1} = \frac{Y}{Y_0} = \frac{Z}{Z_0} \tag{30}$$

Consider a trajectory starting at $P_0(X_0, Y_0, 0.5)$ on face *F*, which is mapped to $P_1(0.5, Y_1, Z_1)$ on face *G* (Fig. 8). The coordinates Y_1, Z_1 are computed from Eq. (30) as

$$Y_1 = -\frac{0.5\,Y_0}{X_0-1}, \qquad Z_1 = -\frac{0.25}{X_0-1} \tag{31}$$

Since the trajectories in 000 and 100 can be superimposed under a symmetry operation of the cube, it can be seen that there will be a fixed point provided

$$Y_1 = X_0, \qquad Z_1 = Y_0 \tag{32}$$

Substituting these values in Eq. (31) and eliminating Y_0 from the resulting equation gives a cubic equation in X_0:

$$X_0^3 - 2X_0^2 + X_0 - \tfrac{1}{8} = 0 \tag{33}$$

This can be solved for the three roots $X_0 = 1/2, 3/4 \pm 5^{1/2}/4$. The root $X_0 = 1/2$ is the vertex common to all six volumes on the cycle. The only other root that lies on the face F is $X_0 = 3/4 - 5^{1/2}/4 = 0.191$, for which $Y_0 = -1/4 + 5^{1/2}/4 = 0.309$. It is a simple matter to find the five images of this point on the threshold planes as the cycle is traversed. The cycle passes in turn through the points (0.191, 0.309, 0.5), (0.5, 0.191, 0.309), (0.691, 0.5, 0.191), (0.809, 0.691, 0.5), (0.5, 0.809, 0.691), (0.309, 0.5, 0.809), (0.191, 0.309, 0.5). The period of the oscillation is readily computed as $T = -2\ln(5^{1/2} - 2) = 2.88$. The values for the amplitude and the period for this case are shown in Fig. 7.

A local stability analysis can be performed on the transformation defined in Eq. (31). In the neighborhood of the fixed point the linear stability matrix is

$$\begin{pmatrix} -2 + 5^{1/2} & -\frac{1}{2} + \frac{5^{1/2}}{2} \\ +\frac{3}{2} - \frac{5^{1/2}}{2} & 0 \end{pmatrix} \tag{34}$$

The eigenvalues of this are $p = -1.5 + 5^{1/2}/2, -0.5 + 5^{1/2}/2$. Since the two eigenvalues are in the unit circle, the transformation and in this case also the cycle are locally stable.*

A global analysis of the stability can also be performed. Consider the point $P(\alpha, \beta, 0.5)$ initially on face F. The distance between this point and the fixed point on F is called d. On face G the coordinates of the image of P are given from (31) by

$$P'\left(0.5, -\frac{0.5\beta}{\alpha - 1}, -\frac{0.5}{\alpha - 1}\right)$$

Call d' the distance from P' to (0.5, 0.191, 0.309). If $d' < d$ for $0 \le \alpha, \beta < 0.5$, then the fixed point is globally stable and all trajectories (except for the singular trajectory associated with the stable eigenvalue) tend to it in the limit $t \to \infty$. If $d^2 - d'^2 > 0$, then $d' < d$. After some algebra, we find

$$d^2 - d'^2 = C_1(\alpha)\beta^2 + C_2(\alpha)\beta + C_3(\alpha) \tag{35}$$

where

$$C_1(\alpha) = 1 - \frac{0.25}{(\alpha - 1)^2}, \qquad C_2(\alpha) = -2Y_0 - \frac{X_0}{\alpha - 1}$$
$$C_3(\alpha) = \alpha^2 - 2\alpha X_0 - \frac{0.25^2}{(\alpha - 1)^2} - \frac{0.5Y_0}{\alpha - 1} \tag{36}$$

For $\alpha = \beta = 0$, $d^2 - d'^2 > 0$, and for $\alpha = X_0$, $\beta = Y_0$, $d^2 - d'^2 = 0$. By setting the left-hand side of Eq. (35) equal to zero and numerically solving the resulting

*See, for example, Appendix III of May.[36]

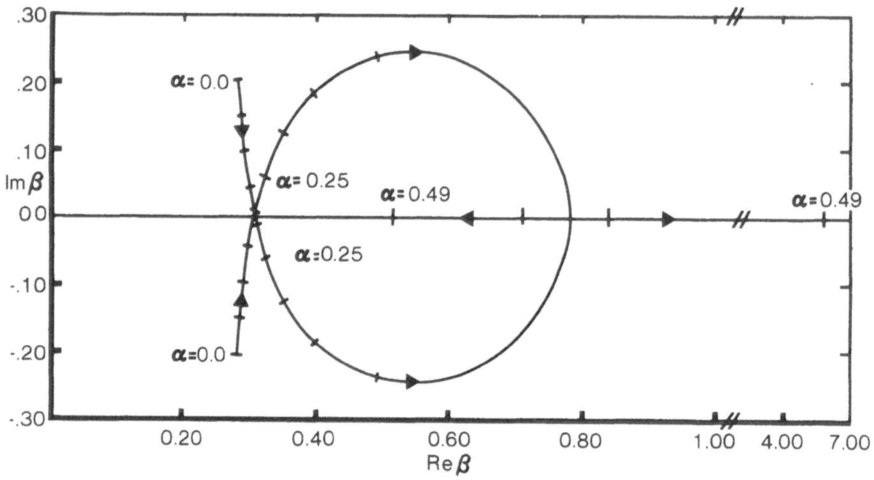

Fig. 9. The solution of Eq. (35) found numerically. As α is varied in the interval $0 \leq \alpha < 0.5$, the roots of β map out of the locus shown. There is only one root for $0 \leq \alpha < 0.5$.

quadratic equation in β for $0 \leq \alpha < 0.5$, we find that the fixed point is the only zero of Eq. (35) in the interval $0 \leq \alpha, \beta < 0.5$ (Fig. 9). Since $d^2 - d'^2 > 0$ at $\alpha = \beta = 0$, it will be greater than zero throughout the interval provided $\alpha = X_0, \beta = Y_0$ represents a minimum of Eq. (35). However, the finding that the limit cycle is locally stable ensures this, so that $d^2 - d'^2$ is positive and the limit cycle is a global attractor.

4.3. Cyclic Attractors and Limit Cycles in Higher Dimensions

In order for Eq. (32) to display cyclic behavior it is necessary that there be a cycle in the state transition diagram. The cycle we have discussed for $N = 3$ (Fig. 4) has a special property. All transitions between a state on the cycle and a neighboring state not on the cycle are oriented toward the states on the cycle. If this is true, we call the cycle a *cyclic attractor.** For $N = 3$, the number of distinct cyclic attractors is 1. Since this case has been of chemical interest, it is of potential interest to investigate the cyclic attractors in higher dimensions.

From the definition, each state on a cyclic attractor specifies the orientations of $N-1$ edges on the state transition diagram. Therefore, the maximum length $L_{\max}$ for a cyclic attractor in N dimensions must satisfy the relation[83]

$$L_{\max} \leq N \cdot 2^{N-1}/(N-1) \tag{37}$$

This upper bound has been sharpened for higher dimensions, $L_{\max} = 8$ for $N = 4$ and $L_{\max} \leq 2^{N-1} - 2$ for $N \geq 5$.[84]

*The simple circuits in the graph of the N-cube defined by the cyclic attractors have been called *snakes* in the combinatorial literature.[82]

For $N = 4$, Gilbert has provided a listing of the distinct cycles on the 4-cube.[85] There are a total of 69 cycles which are unique to the 4-cube and not found in lower dimensions. Of these, six have length 8 and three can be cyclic attractors. These three cyclic attractors are shown in Fig. 10. Using the methods previously described, the piecewise linear equations for each cyclic attractor can be numerically integrated. For $\theta_i = 0.5$, $i = 1, \ldots, 4$ in each case, each cyclic attractor corresponds to a stable limit cycle attractor in Eq. (22). In Fig. 11, we show the asymptotically stable limit cycles found by numerical integration of Eq. (22) for the three cyclic attractors. The cyclic attractor of Figs. 10a and 11a is a higher-dimensional cousin of the cycles for $N = 2, 3$ in

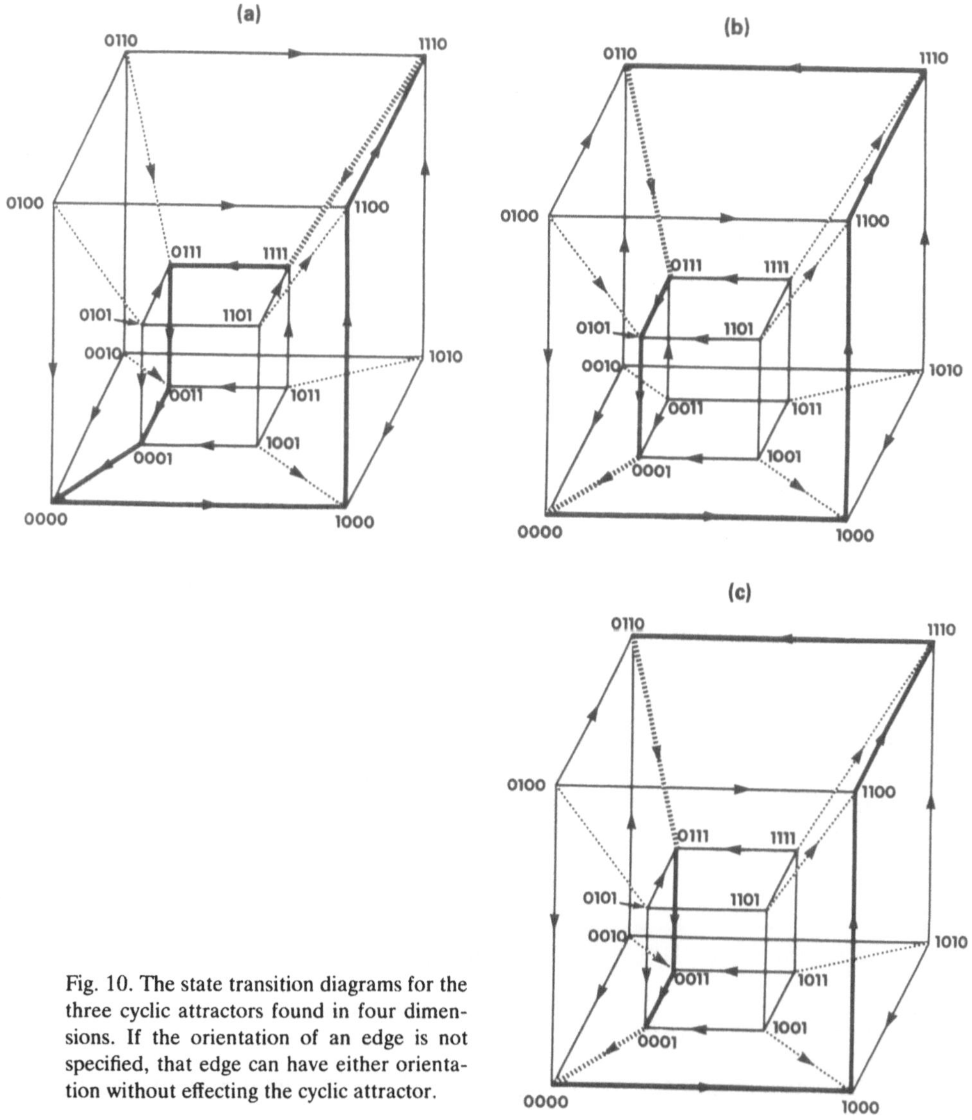

Fig. 10. The state transition diagrams for the three cyclic attractors found in four dimensions. If the orientation of an edge is not specified, that edge can have either orientation without effecting the cyclic attractor.

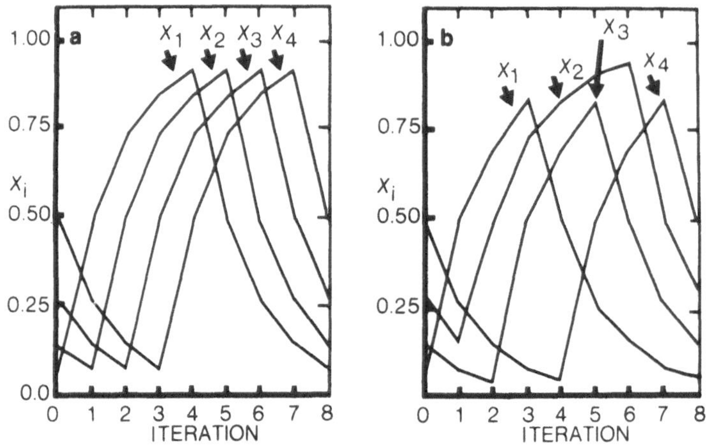

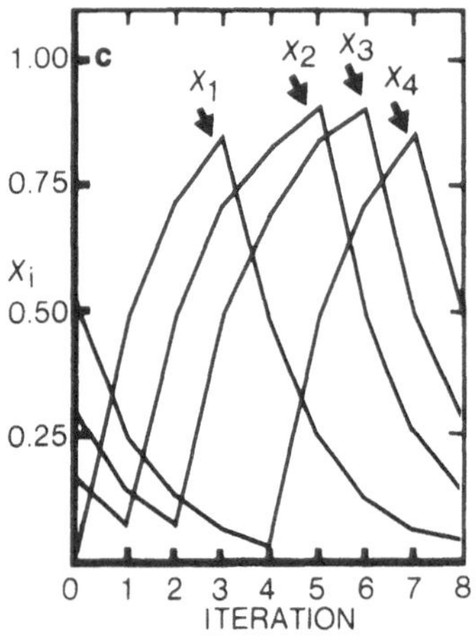

Fig. 11. The three limit cycles found for the cyclic attractors in Fig. 10, by numerically integrating the piecewise linear equation for each cyclic attractor with all thresholds equal to 0.5. Only the values at the thresholds were computed. The concentration of each chemical at each iteration for one cycle are shown. The cycles appear to be stable so that all points in any of the eight volumes on each of the cycles approach the cycle.

Figs. 3d and 4, since it represents the cyclic attractor found in the state transition diagram of Eq. (27), $N = 4$. For any number of dimensions there will always be a cyclic attractor through $2N$ volumes corresponding to the cyclic attractor found for Eq. (27). Numerical integration of Eq. (27) for $n = 8$, $N = 5, 6, 7$ has indicated stable limit cycle attractors for each case where both the period of the oscillation and the amplitude increase as N increases.

These numerical examples and the analysis in Section 4.2 suggest the following conjecture. Cyclic attractors in the state transition diagram of Equation (22) imply stable limit cycle oscillations in Equation (22) provided $N \geq 3$.

5. Reaction and Diffusion in a Ring of Cells

5.1. Eigenvalues of Turing's Equations

In a paper published in 1952, Turing analyzed the stability of two reacting chemicals diffusing in a ring of cells and in a continuous ring of tissue.[11] By performing a stability analysis of linear reaction–diffusion equations, Turing was able to show that spatially homogeneous systems can be unstable with respect to periodic concentration fluctuations of infinitesimal amplitude but finite wavelength and can evolve to spatially periodic concentrations. Prigogine and co-workers have termed the resulting structures "dissipative structures," since they can only be maintained in open systems through which there is a constant flow of energy. They have also done extensive numerical simulation of one kinetic scheme that displays dissipative structures.[13,14] Othmer and Scriven have extended Turing's analysis to study other geometries than a ring of cells.[86] To emphasize the effects of coupling, I shall assume that there are M cells in which there are identical reaction kinetics obeying the boundary conditions in Eqs. (12) and (13) and that if the cells are isolated there is a unique asymptotically stable steady state in each cell. For the case of two reacting chemicals a complete analysis of the linear stability of the reaction system in the neighbourhood of the homogeneous steady state has been obtained.[11] Calling x_i and y_i the concentration differences from the steady state in cell i, the linear equation of motion for M cells in a ring is

$$\begin{aligned} \dot{x}_r &= ax_r + by_r + D_x(x_{r+1} - 2x_r + x_{r-1}) \\ \dot{y}_r &= cx_r + dy_r + D_y(y_{r+1} - 2y_r + y_{r-1}) \end{aligned} \tag{38}$$

for $r = 1, \ldots, M, M+1$, where D_x and D_y give an indication of the strength of diffusive coupling between the cells. The eigenvalues of Eq. (38), p_s and p_s', are solutions of the quadratic equation

$$(p_s - a + 4D_x\varphi_s)(p_s - d + 4D_y\varphi_s) = bc, \qquad s = 0, \ldots, M-1 \tag{39}$$

where

$$\varphi_s = \sin^2 \pi s/M \tag{40}$$

Applying the quadratic formula to Eq. (39) the eigenvalues are computed as

$$\begin{aligned} p_s, p_s' = \tfrac{1}{2}[&a + d - 4\varphi_s(D_x + D_y) \\ &\pm \tfrac{1}{2}\{[a - d - 4\varphi_s(D_x - D_y)]^2 + 4bc\}^{1/2} \end{aligned} \tag{41}$$

The eigenvalues for a single isolated cell displaying identical reaction kinetics are just p_0, p_0'. From the assumption that the steady state in an isolated cell is stable, we find that

$$a + d < 0 \tag{42}$$

We shall examine whether the stable steady state of the isolated cell is ever destabilized by coupling such that (1) a Hopf bifurcation can occur, (2) additional critical points of the original nonlinear equation can be predicted.

5.1.1. Hopf Bifurcation

In a single isolated cell we assume there are stable dynamics, and that Eq. (42) is satisfied. Since D_x, D_y are positive definite, for the cases in which p_s, p_s' are complex conjugates, the real parts will always be negative, and a Hopf bifurcation is impossible.

5.2.2. Topological Theorem Applied to a Ring of Cells

Reaction–diffusion equations for two chemicals in M cells can be displayed in a $2M$-dimensional concentration space where concentrations of each chemical in each cell are represented. Assuming the boundary conditions in Eqs. (12) and (13) for the isolated cells, then in the coupled system it is again impossible for extinction or explosion of any chemical species to occur. Consequently, the dynamics are given in a $2M$-dimensional ball for which the topological theorem must be satisfied. If the number of positive eigenvalues of the homogeneous steady state in the cellular system is an odd number, then from Eq. (14) there must be at least two additional steady states with an even number of positive eigenvalues. Further, since any spatially homogeneous steady state would also be a steady state of the single isolated cell, the additional steady states *must be* spatially inhomogeneous. In Eq. (41) the eigenvalues occur in pairs so that the roots p_k, p_k' are solutions of the same equation as p_{M-k}, $p_{M-k'}$. Therefore, for M an odd number, it is impossible to find a situation in which there are an odd number of positive eigenvalues. However, for M even, for the case $k = M/2$, there are only two eigenvalues, which are

$$p_{M/2}, p'_{M/2} = \tfrac{1}{2}(a + d - 4(D_x + D_y) \pm \{[a - d - 4(D_x + D_y)]^2 + 4bc\}^{1/2}) \tag{43}$$

The real parts of *only one* of the roots in Eq. (43) will be larger than zero, provided

$$\{[a - d - 4(D_x + D_y)]^2 + 4bc\}^{1/2} > |a + d - 4(D_x + D_y)| \tag{44}$$

Consequently, if the linearized equations of a reaction–diffusion equation in the neighborhood of a homogeneous steady state are given by Eq. (38), provided Eq. (44) also holds, there must be in addition at least two inhomogeneous steady states with an even number of positive eigenvalues in the original nonlinear equation.

Although the analysis is able to predict the existence of additional spatially inhomogeneous steady state solutions of nonlinear equations for one special case, the linear analysis of the local stability is quite limited and may be misleading. First, the absence of a Hopf bifurcation does not mean that with two reacting chemicals stable oscillations are impossible. On the contrary,

Winfree has given a numerical computation in which travelling stable concentration waves could be found in a reaction–diffusion equation in which there was a single homogeneous steady state.[87] Further, the computations that are given are only valid for reaction systems with two chemicals and do not generalize to systems with more than two chemicals. For example, Smale has discussed dynamics in systems with two cells but four chemicals (eight dimensions). Here, if the reactions are the same in both cells, it is possible to find dynamics so that in the isolated cells there is a single unique stable steady state, but in the coupled cells there is a globally stable limit cycle attractor.[88]

5.2. Global Analysis of a Nonlinear Cellular Oscillation

In Section 5.1 the assumption was made that the dynamics were identical in each of the M cells prior to coupling. However, it is generally the case, both in biological and chemical systems, that there are spatial heterogeneities present. for example, reaction rates may differ at boundaries and surfaces of reaction vessels. There may also be impurities that act as localized catalysts. One case in which such impurities can act to generate traveling waves has been analyzed theoretically,[89] and there are also indications that impurities can act as pacemakers in the Belousov–Zhabotinsky reaction.[8] Further, if there is more than one catalyst, the qualitative dynamics of the system can depend on the spatial arrangement of the catalysts, so that changes in the arrangements of the catalysts can cause bifurcations in the dynamics. For example, if there are two catalytic sites where the products from the catalysts mutually inhibit the synthesis of each other, then bistability may only be found over a limited range of separations of the localized catalysts.[23]

To demonstrate one of the effects of spatial heterogeneity, a feedback inhibition loop will be considered, where the catalytic sites are localized in two compartments between which diffusion takes place.* The equations describing this system are[74,75,81]

$$\begin{aligned}
\dot{x}_1 &= \lambda[1-S(y_1)]-\gamma x_1+D(x_2-x_1)\\
\dot{x}_2 &= -\gamma x_2+D(x_1-x_2)\\
\dot{y}_1 &= -\gamma y_1+D(y_2-y_1)\\
\dot{y}_2 &= \lambda S(x_2)-\gamma y_2+D(y_1-y_2)
\end{aligned} \tag{45}$$

where we have

$$S(w)=w^n/(\theta^n+w^n) \tag{46}$$

*There are striking similarities in the qualitative dynamics between cellular chemical systems such as the one described in Eq. (45) and chemical systems with localized catalysts which interact by diffusive transport.[23,90–92]

and we have assumed that the parameters characterizing the dynamics are the same for x and y. By redefining variables, Eq. (45) can be cast in more compact form.[75] Defining

$$\bar{x}_+ = \frac{x_1+x_2}{\lambda/\gamma}, \qquad \bar{y}_+ = \frac{y_1+y_2}{\lambda/\gamma}$$

$$\bar{x}_- = \frac{x_1-x_2}{\lambda/\gamma}, \qquad \bar{y}_- = \frac{y_2-y_1}{\lambda/\gamma} \tag{47}$$

$$\tau = \gamma t$$

we find

$$\frac{d\bar{x}_i}{d\tau} = 1 - S\left[\frac{\lambda}{\alpha}\left(\frac{\bar{y}_+ - \bar{y}_-}{2}\right)\right] - \frac{\bar{x}_i}{\tau_i}$$

$$\frac{d\bar{y}_i}{d\tau} = S\left[\frac{\lambda}{\gamma}\left(\frac{\bar{x}_+ - \bar{x}_-}{2}\right)\right] - \frac{\bar{y}_i}{\tau_i} \tag{48}$$

$$i = +, -, \quad \tau_+ = 1, \quad \tau_- = \frac{1}{1+2D/\gamma}$$

A local analysis of the stability properties in the neighborhood of the steady state indicates that Hopf bifurcations can be found in this system.[74,75,93] Numerical integration of Eq. (48) in the unstable regime indicates a globally stable limit cycle oscillation. In Fig. 12, the dynamics are given in the $\bar{x}_+$, $\bar{y}_+$ plane, starting from different initial conditions. In all cases, the limit cycle is a stable attractor. In Fig. 12, although it is theoretically possible that trajectories cross, for the range of initial conditions considered such crossing was not observed. This suggests that there may be an attracting two-dimensional manifold in the four-dimensional space so that when trajectories are projected on the $\bar{x}_+$, $\bar{y}_+$ plane from the attracting manifold, trajectories do not cross.

A good geometrical picture of the flow in the four-dimensional system can be obtained by considering the piecewise linear equations that are found for Eq. (48) in the limit when $n \to \infty$.[75] Defining the Heaviside step function

$$H(w-\theta) = \begin{cases} 1, & w \geq \theta \\ 0, & w < \theta \end{cases} \tag{49}$$

in the limit $n \to \infty$ Eq. (18) can be integrated to give

$$\bar{x}_i(\tau) = \bar{x}_i(0)e^{-\tau/\tau_i} + \tau_i\left(1 - H\left\{\frac{\lambda}{\gamma}\left[\frac{\bar{y}_+(\tau) - \bar{y}_-(\tau)}{2}\right] - \theta\right\}\right)(1 - e^{-\tau/\tau_i})$$

$$\bar{y}_i(\tau) = \bar{y}_i(0)e^{-\tau/\tau_i} + \tau_i H\left\{\frac{\lambda}{\gamma}\left[\frac{\bar{x}_+(\tau) - x_-(\tau)}{2}\right] - \theta\right\}(1 - e^{-\tau/\tau_i}) \tag{50}$$

Once the initial values are assumed, the concentration of each variable is given until the threshold for production of one of the variables is crossed. This

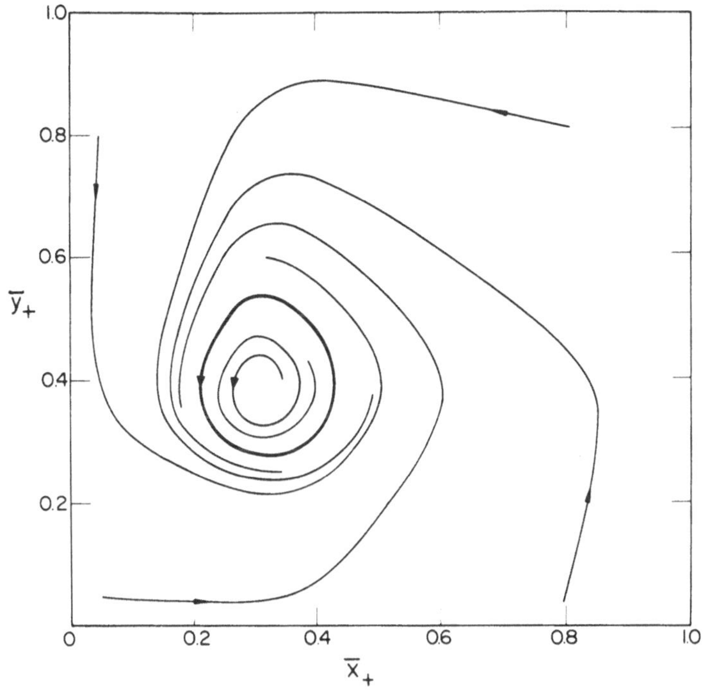

Fig. 12. A two-dimensional representation of the dynamics found by numerically integrating Eq. (48) with the parameters $\lambda/\gamma = 5$, $D/\gamma = 4.5$, $\theta = 0.75$. Each trajectory starts from an initially homogeneous state. (Reprinted from *J. Chem. Phys*, Reference 75, with permission.)

determines a new set of initial values, and the equations can be iterated until a steady state of cycle is reached. I have not found a proof for limit cycle oscillations even in this piecewise linear approximation. However, for $D/\gamma \gg 1$, $\tau_+ \gg \tau_-$, there are two time scales in the problem so that the differences in concentration between the two cells reach a steady state value rapidly compared to changes in $\bar{x}_+$, $\bar{y}_+$. If $x(y)$ are being produced, $\bar{x}_-(\bar{y}_-)$ rapidly reach the limiting values $1/(1+2D/\gamma)$. Similarly if x and y are not being produced, the concentrations reach the limiting values zero. This suggests a decomposition of the $\bar{x}_+$, $\bar{y}_+$ phase space into four regions where the trajectories in each region are given by

$$\frac{\bar{x}_+(\tau)-\bar{x}_+(\infty)}{\bar{x}_+(0)-\bar{x}_+(\infty)} = \frac{\bar{y}_+(\tau)-\bar{y}_+(\infty)}{\bar{y}_+(0)-\bar{y}_+(\infty)} \tag{51}$$

where the asymptotic values $\bar{x}_+(\infty)$, $\bar{y}_+(\infty)$ differ in each region. Calling $\bar{\theta} = \theta/(\lambda/\gamma)$ and $\delta = 1/(1+2D/\gamma)$, the four regions are (Fig. 13)

Region I:

$$\bar{x}_- = 0, \quad \bar{y}_- = \delta, \quad 1 \geq \bar{x}_+ \geq 2\bar{\theta}, \quad 1 \geq \bar{y}_+ \geq 2\bar{\theta} + \delta, \quad \bar{X}_+(\infty) = 0, \quad \bar{Y}_+(\infty) = 1$$

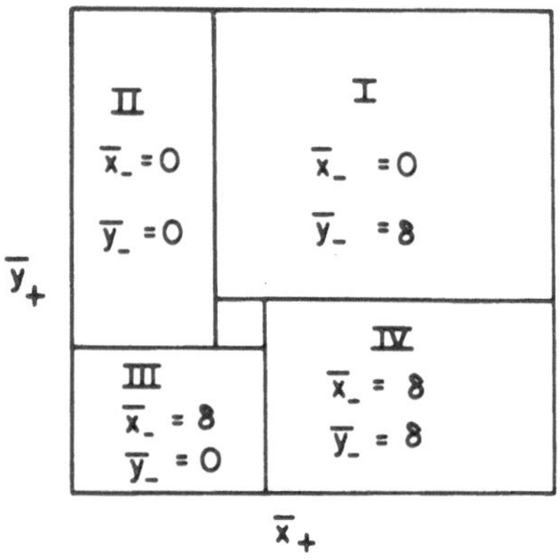

Fig. 13. The four regions of phase space found when $D/\gamma \gg 1$ and there is discontinuous Heaviside step function modulation of production (see text). (Reprinted from *J. Chem. Phys.*, Reference 75, with permission.)

Region II:

$$\bar{x}_- = 0, \quad \bar{y}_- = 0, \quad 2\bar{\theta} > \bar{x}_+ > 0, \quad 1 > \bar{y}_+ \geq 2\bar{\theta}, \quad \bar{x}_+(\infty) = 0, \quad \bar{y}_+(\infty) = 0$$

Region III:

$$\bar{x}_- = \delta, \quad \bar{y}_- = 0, \quad 2\bar{\theta} + \delta > \bar{x}_+ > 0, \quad 2\bar{\theta} > \bar{y}_+ > 0, \quad \bar{x}_+(\infty) = 1, \quad \bar{y}_+(\infty) = 0$$

Region IV:

$$\bar{x}_- = \delta, \quad \bar{y}_- = \delta, \quad 1 \geq \bar{x}_+ \geq 2\bar{\theta} + \delta, \quad 2\bar{\theta} + \delta > \bar{y}_+ > 0, \quad \bar{x}_+(\infty) = 1, \quad \bar{y}_+(\infty) = 1$$

For the trajectories defined in Equation (51) it is possible to prove that there is a unique limit cycle attractor.[75] If p, q are two points on the boundary between regions I and II, and the distance between the two points $d[p, q] = l$, it is possible to show that the distance between the images of these points on the boundary between regions II and III is smaller than $4\bar{\theta}^2 l/(2\bar{\theta} + \delta)^2$. If $f(p)$ is the map that takes point p on the border between regions I and II back to this border after one complete circuit through the four regions, we compute

$$d[f(p), f(q)] < \frac{(2\bar{\theta})^4 (1 - 2\bar{\theta} - \delta)^4}{(2\bar{\theta} + \delta)^4 (1 - 2\bar{\theta})^4} l \tag{52}$$

Since $\delta > 0$, $d[f(p), f(q)] < l$, f is a contraction mapping (see Appendix) and there must be a unique stable limit cycle in the four regions. In Fig. 14, we give this construction for the parameters used to compute Fig. 12. Although there is good agreement between the dynamics in the piecewise linear and the continuous equations, no proof of stable limit cycle oscillations has been found for Eq. (48) or (50). However, there has been a recent proof for the existence of nonlocal periodic solutions of Eq. (45) using fixed-point methods.[81]

6. Discussion

In the preceding sections, several methods for the analysis of nonlinear chemical kinetics have been discussed. The methods share the common feature that they emphasize qualitative features of the dynamics, for example, the number and types of critical points and the phases of oscillation, rather than quantitative features, such as the detailed concentration changes of chemicals as a function of time. Several complementary methods for classification of the qualitative dynamics have been discussed: (i) the stability matrix in the neighborhood of a single critical point (Section 1), (ii) the entire set of critical points (Section 2), (iii) the flows between contiguous volumes in concentration space (Section 3), (iv) the stable attractors of the dynamics outside the neighborhood of the critical points (Section 4).

Whichever qualitative features are examined, the apparently rich chemical literature reduces to a few simple classes. It is difficult to judge the reason why so few dynamic classes have appeared so far. Perhaps there are certain (still unanalyzed) features of kinetic equations that lead to simple dynamics. Another possibility is that chemists have tended to study only a small subclass of chemical kinetics and have ignored the rest. For example, a type of dynamical behavior that it is hard to imagine *not existing* is one in which there are two stable attractors. A limit cycle oscillation and a stable steady state where transitions can occur between the two as a result of large perturbations of concentrations. An example has been previously given in which this type of

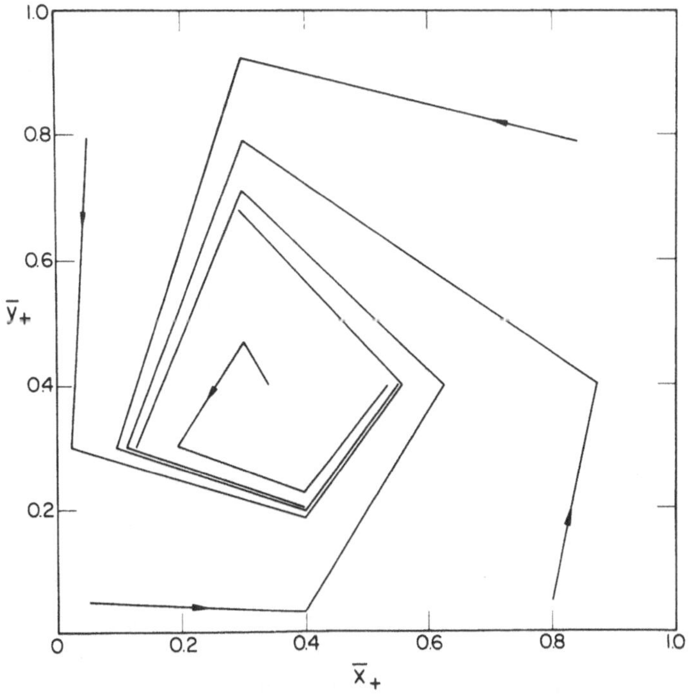

Fig. 14. The construction of the trajectories in the $\bar{x}_+$, $\bar{y}_+$ plane for the piecewise linear approximation to Eq. (18). The same parameters used in Fig. 13 are used here. (Reprinted from *J. Chem. Phys.*, Reference 75, with permission.)

dynamics occurs in a system of heterogeneous catalysts of three chemicals in four cells (12 dimensions), but no global analysis was attempted.[74] Another kinetic system that can display similar dynamics is the limit cycle oscillation shown in Fig. 10b. Here it is also possible to have an extremal steady state at vertex 1011.

I have concentrated on the analysis of the qualitative dynamics of comparatively simple kinetic equations. One reason why I think this approach might be useful is that it may be possible to extend these methods to much more complicated naturally occurring biological systems. In their 1961 paper, Monod and Jacob suggested that growth and differentiation in biological systems might be controlled by networks of genes, in which the end products of reaction sequences catalyzed by one set of enzymes modulated synthetic rates of other enzymes.[12] In this formulation, different stable attractors correspond to different cell types. The kinetic equations with Boolean terms, Eqs. (20) and (22), were developed in an effort to represent mathematically the gene networks envisioned by Monod and Jacob. It therefore becomes of great interest to study the qualitative properties of kinetic equation of this sort, when the number of interacting chemicals becomes very large, and to determine the correspondence, if any, between these networks and the control systems of cells. One important step in this direction has already been taken. Kauffman proposed that genes may be considered Boolean switches, and that gene networks might be represented as switching networks in which the states of all genes change synchronously at subsequent iterations.[70] The startling discovery was made that for switching networks in which each "gene" realized a randomly chosen Boolean function on the states of two randomly chosen "genes" the number of stable asymptotic behaviors increased as $N^{1/2}$, even though the total number of accessible Boolean states is 2^N. Kauffman drew an analogy between this behavior and the statistical properties of gene networks in plants and animals where the number of cell types appears to increase roughly as the square root of the DNA content. Although in the Kauffman formulation all genes synchronously change their states, no evidence has been obtained that such synchronous operation is found in normal plant and animal cells. The piecewise linear equations (19) and (22) are asynchronous since only one "gene" can change its state at one time. In these equations does the number of stable attractors also increase as $N^{1/2}$ if the B_i depend on two other elements of the reaction network? How does the number of stable attractors depend on restrictions on the B_i and also on the geometrical arrangement of the catalytic sites? Is it ever possible to find "strange attractors" displaying chaotic dynamics?

In the past few years increasing interest has been directed toward the study of complex systems, in which there are several interacting species and dynamics can display rich dynamical behavior in both time and space. Even if the specific mathematical techniques discussed here prove to be of limited interest, the global topological viewpoint I have tried to express is just taking hold and is almost certain to have a rich and exciting future.

Appendix

In the Appendix, I give a listing of some of the principal mathematical terms and results that have been used in this article. This should facilitate reading the text for those who are not already expert in topology. For those who wish to pursue the mathematics, a number of excellent sources are available for ordinary differential equations,[29–32] for dynamical systems,[94] and for differential topology.[33,34] I have personally found the text by Guillemin and Pollack[34] very useful, and have taken many examples and definitions in the Appendix from this source. Although it seems to me that there *should be* a single source that could be consulted to find out about the various topics discussed, if such a source exists, I am unaware of it.

"A manifold is a space in which the environment of each point is 'just like' a small piece of Euclidean space."[34] Some of the more famous manifolds are the N-sphere (S^N), the set of points for which $\sum_{i=1}^{N+1} y_i^2 = \text{const}$, and surfaces of genus γ (2-spheres to which γ handles have been added to form a γ-holed donut).

A wonderful theorem that almost everyone knows was proved by Euler for polyhedra whose surfaces are *homeomorphic* (topologically equivalent) to surfaces of genus γ. Let F, E, and V be the numbers of faces, edges, and vertices, respectively, of a polyhedron. Then

$$F - E + V = 2 - 2\gamma. \tag{53}$$

For example, for a cube for which $\gamma = 0$, $F = 6$, $E = 12$, $F = 8$. For a second example, imagine M cubes pasted together to form a donut for which $\gamma = 1$, so that $F = 4M$, $E = 8M$, $V = 4M$.

In Poincaré's earliest work a study was made of vector fields defined on surfaces of genus γ. A zero of the vector field is called a *critical point*. If the real parts of the eigenvalues of the critical point are nonzero (*hyperbolic*) then there are only three different kinds of critical points in two dimensions: *nodes* (the two eigenvalues are real and are of the same sign), *foci* (the two eigenvalues are complex conjugate), and *saddle points* (the two eigenvalues are real but of different sign). If $\mathcal{N}$, $\mathcal{F}$, and $\mathcal{S}$ designate the numbers of nodes, foci, and saddle points, respectively, then for a vector field with only hyperbolic isolated critical points on a surface of genus γ, the *Poincaré index theorem* gives[29]

$$\mathcal{N} + \mathcal{F} - \mathcal{S} = 2 - 2\gamma \tag{54}$$

For example, imagine the flow of hot fudge on a chocolate-coated donut standing on its end and prepared in the manner of Guillemin and Pollack (p.

125).[34] The flow will define two nodes (one at the top and one at the bottom) and two saddle points (one at the top and one at the bottom of the hole). The *Euler–Poincaré characteristic*, designated $\chi(M)$ of the surfaces of genus γ is the number $2-2\gamma$.

I have found a theorem that is closely related to (53) and (54).[95] Consider a polyhedron of genus γ satisfying (53) but where each edge is directed. An *adjacent pair of edges* bounds a common face and terminates at a common vertex. The *reversals* of a vertex R are equal to the number of adjacent pairs of edges bounding the face in which one member is directed toward and the other away from the vertex. The *reversals of a face* are equal to the number of adjacent pairs of edges bounding the face in which one member is oriented clockwise and the second counterclockwise. The index I of a face or vertex is then

$$I = 1 - R/2 \tag{55}$$

and the combinatorial index theorem is

$$\sum_{F,V} I = 2 - 2\gamma \tag{56}$$

This was used to derive Eq. (15). A special case of equation (56) was previously used to classify the ways to build a box![96] The Poincaré index theorem was extended by Hopf to vector fields on arbitrary manifolds. For vector fields with m isolated hyperbolic critical points, the Poincaré–Hopf index theorem is[34]

$$\sum_{i=1}^{m} (-1)^{\mu_i} = \chi(M) \tag{57}$$

where $\chi(M)$ is the Euler–Poincaré characteristic of the manifold and μ_i the number of eigenvalues of the ith critical point for which the real part is negative. For example, imagine coating an N-sphere with hot fudge. There will be two critical points, one at the top for which $\mu_i = 0$, and one at the bottom for which $\mu_i = N$. This is one way to compute $\chi(M)$ for N-spheres, Eq. (11).

There have been many results concerning manifolds with boundaries. An example of a manifold with a boundary is the ball B^N that is defined as the set of points for which $\text{const} - \sum_{i=1}^{eN} y_i^2 \geq 0$. The boundary of B^N is just S^{N-1}. One early result is the *Poincaré–Bendixson theorem*.[29–32] Assume all the trajectories on the boundary of B^2 point inward, and that in the interior there is a single hyperbolic critical point with two positive real parts. Then there must be a closed cyclic path in B^2. The Poincaré–Bendixson theorem is not true in higher dimensions.[30] However, the Hopf bifurcation theorem, Section 4.1, is closely related. The conditions of the theorem ensure that in the neighborhood of the bifurcation point there is an attracting 2-manifold for which the Poincaré–Bendixson theorem can be applied. The theorem we give here for B^N in which the vector field points inward along the boundary, Eq. (14) should be compared with a similar expression if the vector field points outward along the boundary (see Milnor,[33] p. 36).

A famous result has been given by Brouwer.[33,34] Consider a smooth map f that takes B^N into itself $f: B^N \to B^N$. The *Brouwer fixed-point theorem* states that f must have a fixed point, that is, $f(P) = P$ for some $P \in B^N$. An early use of the Brouwer fixed-point theorem in chemistry was a demonstration by Wei that closed chemical reactions must have an equilibrium.[97] A recent discussion of the application of topological and fixed-point methods to study the existence and uniqueness of thermodynamic equilibrium states has been given by Wallwork and Perelson.[98] It is sometimes possible to prove that P is unique and stable. Consider any two points $q, r \in B^N$. The distance between q and r is designated $d[q, r]$. If there exists a real number $k, 0 \leq k < 1$, such that $d[f(q), f(r)] \leq kd[q, r]$, then f is called a *contraction map.*[99] If f is a contraction map then

(1) There is a unique fixed point P_0.

(2) If q_0 is any point of B^N and $q_1 = f(q_0)$, $q_2 = f(q_1), \ldots, q_n = f(q_{n-1})$, then $\lim_{n \to \infty} q_n = P_0$.

ACKNOWLEDGMENTS

This manuscript was partially written while I was a participant at a Biophysics Workshop at the Aspen Center for Physics. Numerous conversations with and talks by the other participants, notably G. Oster, A. S. Perelson, S. Smale, and A. T. Winfree were of great benefit. Thanks also to J. Harper and J. Pasternack for helpful conversations. This work has been partially supported by Grant DID71-04010-A02 from the NSF, a grant from the Faculty of Graduate Studies and Research of McGill University, and grant A0091 from the National Research Council of Canada.

References

1. E. S. Hedges and J. E. Meyers, *The Problem of Physicochemical Periodicity*, Arnold, London (1926).
2. W. C. Bray, A periodic reaction in homogeneous solution and its relation to catalysis, *J. Am. Chem. Soc.* **43**, 1262–1267 (1921).
3. R. J. S. Rayleigh, A study of the glow of phosphorus. Periodic luminosity and action of inhibiting substances, *Proc. Roy. Soc. (London)* **A99**, 372–384 (1921).
4. *Bibliography of Liesegang Rings* (2nd ed.), U.S. Dept. of Commerce, National Bureau of Standards, Miscellaneous Publication 292 (1967).
5. A. J. Lotka, Undamped oscillations derived from the law of mass action, *J. Am. Chem. Soc.* **42**, 1595–1599 (1920).
6. E. K. Pye and B. Chance (eds.), *Biochemical Oscillations, Proceedings of the* 1968 *Prague Symposium*, Academic Press, New York (1973).

7. A. N. Zaikin and A. M. Zhabotinsky, Concentration wave propagation in two-dimensional liquid-phase self-oscillating system, *Nature* (*London*) **225**, 535–537 (1970).
8. A. T. Winfree, Spiral waves of chemical activity, *Science* **175**, 634–636 (1972).
9. A. T. Winfree, Scroll-shaped waves of chemical activity in three dimensions, *Science* **181**, 937–939 (1973).
10. J. A. De Simone, D. L. Beil, and L. E. Scriven, Ferroin–collodion membranes; dynamic concentration patterns in planar membranes, *Science* **180**, 946–948 (1973).
11. A. M. Turing, The chemical basis of morphogenesis, *Phil. Trans. Ray. Soc. London* **B237**, 37–72 (1952).
12. J. Monod and F. Jacob, General conclusions: teleonomic mechanisms in cellular metabolism, growth and differentiation, *Cold Spring Harbor Symp. Quant. Biol.* **25**, 389–401 (1961).
13. P. Glansdorff and I. Prigogine, *Thermodynamics of Structure, Stability and Fluctuations*, Wiley (Interscience), New York (1971).
14. G. Nicolis and J. Portnow, Chemical oscillations, *Chem. Rev.* **73**, 365–384 (1973).
15. P. Ortoleva and J. Ross, Theory of propagation of discontinuities in kinetic systems with multiple time scales: front multiplicity, and pulses, *J. Chem. Phys.* **63**, 3398–3408 (1976).
16. E. E. Sel'kov, Self-oscillations in glycolysis. 1. A simple kientic model, *Eur. J. Biochem.* **4**, 79–86 (1968).
17. J. Higgins, The theory of oscillating reactions, *Ind. Eng. Chem.* 59, 18–69 (1967).
18. I. Prigogine and R. Lefever, Symmetry breaking instabilities in dissipative systems. II, *J. Chem. Phys.* **48**, 1695–1700 (1968).
19. J. J. Tyson and J. C. Light, Properties of two-component bimolecular and trimolecular chemical reaction systems, *J. Chem. Phys.* **59**, 4164–4173 (1973).
20. J. J. Tyson and S. A. Kauffman, Control of mitosis by a continuous biochemical oscillation: synchronization, inhomogeneous oscillation, *J. Math. Biol.* **1**, 289–310.
21. L. N. Grigorev, M. S. Polyakova, and D. S. Chernavskii, Model investigation of trigger schemes and the differentiation process. *Mol. Biol. USSR* **1**, 349–356 (1967).
22. B. B. Edelstein, The dynamics of cellular differentiation and associated pattern formation, *J. Theor. Biol.* **37**, 221–243 (1972).
23. R. M. Shymko and L. Glass, Spatial switching in chemical reactions with heterogeneous catalysis, *J. Chem. Phys.* **60**, 835–841 (1974).
24. W. Lamb, Theory of an optical maser, *Phys. Rev.* **134A**, 1429–1450 (1964).
25. A. Rescigno and I. W. Richardson, *in*: *Foundations of Mathematical Biology* (R. Rosen, ed.), Vol. 3, pp. 283–360, Academic Press, New York (1973).
26. A. Andronov and L. Pontriagin, Systemes grossiers, *Dokl. Akad. Nauk SSSR* **14**, 247–251 (1937).
27. M. M. Peixoto, *in*: *Differential Equations and Dynamical Systems* (J. K. Hale and J. P. LaSalle, eds.), pp. 469–480, Academic Press, New York (1967).
28. R. Thom, *in*: *Towards a Theoretical Biology* (C. H. Waddington, ed.), Vol. 3, pp. 89–116, Aldine, Chicago (1970).
29. S. Lefshetz, *Differential Equations*: *Geometric Theory*, Wiley (Interscience), New York (1957).
30. W. Hurewicz, *Lectures on Ordinary Differential Equations*, MIT Press, Cambridge, Massachusetts (1958).
31. N. Minorsky, *Nonlinear Oscillations*, Van Nostrand, Princeton, New Jersey (1962).
32. M. W. Hirsch and S. Smale, *Differential Equations, Dynamical Systems and Linear Algebra*, Academic Press, New York (1974).
33. J. W. Milnor, *Topology from the Differentiable Viewpoint*, The University Press of Virginia, Charlottesville (1965).
34. V. Guillemin and A. Pollack, *Differential Topology*, Prentice-Hall, Englewood Cliffs, New Jersey (1974).
35. L. P. Quirk and R. Ruppert, Qualitative economics and the stability of equilibrium, *Rev. Econ. Stud.* **32**, 311–326 (1965).
36. R. M. May, *Stability and Complexity in Model Ecosystems*, Princeton University Press, Princeton, New Jersey (1974).
37. R. Levins, Qualitative analysis of partially specified systems, *Ann. N.Y. Acad. Sci.* **231**, 123–128 (1974).

38. B. L. Clark, Graph theoretic approach to the stability analysis of steady state chemical reaction networks, *J. Chem. Phys.* **60**, 1481–1501 (1974).
39. B. L. Clark, Theorems on chemical network stability, *J. Chem. Phys.* **62**, 773–775 (1975).
40. J. J. Tyson, Classification of instabilities in chemical reaction systems, *J. Chem. Phys.* **62**, 1010–1015 (1975).
41. B. B. Edelstein, Biochemical model with multiple steady states and hysteresis, *J. Theor. Biol.* **29**, 57–62 (1970).
42. B. C. Goodwin, *in*: *Advances in Enzyme Regulation* (G. Weber, ed.), Vol. 3, pp. 425–437, Pergamon, Oxford (1965).
43. M. Morales and D. McKay, Biochemical oscillations in "controlled" systems, *Biophys. J.* **7**, 621–625 (1967).
44. C. Walter, Oscillations in controlled biochemical systems, *Biophys. J.* **9**, 863–872 (1969).
45. A. Hunding, Limit-cycles in enzyme-systems with nonlinear negative feedback, *Biophys. Struct. Mechanism* **1**, 47–54 (1974).
46. J. J. Tyson, On the existence of oscillatory solutions in negative feedback cellular control processes, *J. Math. Biol.* **1**, 311–315 (1975).
47. H. G. Othmer, The qualitative dynamics of a class of biochemical control circuits, *J. Math. Biol.* **3**, 53–78 (1976).
48. B. H. Lavenda, The theory of multi-stationary state transitions and biosynthetic control processes, *Q. Rev. Biophys.* **5**, 429–479 (1972).
49. L. Glass, Classification of biological networks by their qualitative dynamics, *J. Theor. Biol.*, **54**, 85–107 (1975).
50. D. Sattinger, *Topics in Stability and Bifurcation Theory*, Lecture Notes in Mathematics 309, Springer-Verlag, New York (1973).
51. J. Marsden (ed.), *The Hopf Bifurcation*, Springer-Verlag, New York (1976).
52. G. R. Gavalas, *Nonlinear Differential Equations of Chemically Reacting Systems*, Springer-Verlag, New York (1968).
53. L. Glass, Combinatorial and topological methods in nonlinear chemical kinetics, *J. Chem. Phys.* **63**, 1325–1335 (1975).
54. L. Glass, A topological theorem for nonlinear dynamics in chemical and ecological networks, *Proc. Nat. Acad. Sci.*, **72**, 2856–2857 (1975).
55. D. Ruelle and F. Takens, On the nature of turbulence, *Commun. Math. Phys.* **20**, 167–192 (1971).
56. G. Oster and J. Guckenheimer, *in*: *The Hopf Bifurcation* (J. Marsden, ed.), 327–353, Springer-Verlag, New York (1976).
57. R. May, Biological populations with nonoverlapping generations: stable points, stable cycles and chaos, *Science* **186**, 645–647 (1974).
58. D. Ruelle, Comments on chemical oscillations, *Trans. N.Y. Acad. Sci.* **35**, 66–71 (1973).
59. G. F. Oster and A. S. Perelson, Chemical reaction dynamics, Part I: Geometrical structure, *Arch. Rational Mech. Anal.* **55**, 230–274 (1974).
60. A. S. Perelson and G. F. Oster, Chemical reaction dynamics, Part II: Reaction networks, *Arch. Rational Mech. Anal.* **57**, 31–98 (1974).
61. J. Moser, *Stable and Random Motions in Dynamical Systems*, Princeton University Press, Princeton, New Jersey (1973).
62. S. W. Golomb, *in*: *Information Theory*: *Fourth London Symposium* (C. Cherry, ed.), pp. 404–424, Butterworth, London (1961).
63. M. A. Harrison, *Introduction to Switching and Automata Theory*, McGraw-Hill, New York (1965).
64. S. P. Hastings and J. D. Murray, The existence of oscillatory solutions in the Field–Noyes model for the Belousov–Zhabotinskii reaction, *SIAM J. Appl. Math.* **28**, 678–688 (1975).
65. R. J. Field and R. M. Noyes, Oscillations in chemical systems. IV. Limit cycle behavior in a model of a real chemical reaction, *J. Chem. Phys.* **60**, 1877–1884 (1974).
66. V. Volterra. *Leçon sur la Theorie Mathematique de la Lutte pour la Vie*, Gauthier-Villars, Paris (1931).
67. G. Yagil and E. Yagil, On the relation between effector concentration and the rate of induced enzyme synthesis, *Biophys. J.* **11**, 11–27 (1971).

68. J. Monod, J. Wyman, and J. Changeux, On the nature of allosteric transitions: a plausible model, *J. Mol. Biol.* **12**. 88–118 (1965).
69. M. Sugita, Functional analysis of chemical systems *in vivo* using a logical circuit equivalent. II. The idea of a molecular auromaton. *J. Theor. Biol.* **4**, 179–192 (1963).
70. S. A. Kauffman, Metabolic stability and epigenesis in randomly constructed genetic nets, *J. Theor. Biol.* **22**, 437–467 (1969).
71. L. Glass and S. A. Kauffman, The logical analysis of continuous nonlinear biochemical control networks, *J. Theor. Biol.* **39**, 103–129 (1973).
72. R. Thomas, Boolean formalization of genetic control circuits, *J. Theor. Biol.* **42**, 563–575 (1973).
73. O. Rossler, *in*: *Lecture Notes in Biomathematics*, Vol. 4, pp. 546–582, Springer-Verlag, New York (1974).
74. L. Glass and S. A. Kauffman, Co-operative components, spatial localization and oscillatory cellular dynamics, *J. Theor. Biol.* **34**, 219–237 (1972).
75. L. Glass and R. Pérez, Limit cycle oscillations in compartmental chemical systems, *J. Chem. Phys.* **61**, 5242–5249 (1974).
76. E. Hopf, Abzweigung einer periodischen Lösung von einer stationären Lösung eines Differentialsystems, *Acad. Wiss. Leipzig, Math-Phys. K. Ber.* **94**, 3–22 (1942).
77. N. Kopell and L. N. Howard, Plane wave solutions to reaction–diffusion equations, *Stud. Appl. Math.* **52**, 291–328 (1973).
78. I. -D. Hsü and N. D. Kazarinoff, An applicable Hopf bifurcation formula and instability of small periodic solutions of the Field-Noyes model, *J. Math. Anal. Appl.* **55**, 61–89 (1976).
79. A. S. Perelson, A note on the qualitative theory of lumped parameter systems, *Chem. Eng. Sci.*, **31**, 170–173 (1976).
80. S. Hastings, J. Tyson, and D. Webster, Existence of nonlocal periodic solutions for negative feedback cellular control systems, *J. Diff. Eqs.*, in press (1976).
81. I.-D. Hsü, The existence of nonlocal, periodic solutions for the Glass–Kauffman model of cellular dynamics, preprint (1976).
82. L. Danzer and V. Klee, Lengths of snakes in boxes, *J. Comb. Theory* **2**, 258–265 (1967).
83. W. H. Kautz, Unit-distance error-checking codes, *IRE Trans. Electronic Computers* **7**, 179–180 (1958).
84. R. J. Douglas, Upper bounds on the length of circuits of even spread in the *d*-cube, *J. Comb. Theory* **7**, 206–214 (1969).
85. E. N. Gilbert, Gray codes and paths on the *n*-cube, *Bell Syst. Tech. J.* **37**, 815–826 (1958).
86. H. G. Othmer and L. E. Scriven, Instability and dynamic pattern in cellular networks, *J. Theor. Biol.* **32**, 507–537 (1971).
87. A. T. Winfree, Rotating solutions to reaction/diffusion equations in simply-connected media, *SIAM-AMS Proceedings* **8**, 13–31 (1974).
88. S. Smale, A mathematical model of two cells via Turing's equation, *in*: *Lectures on Mathematics in the Life Sciences*, Vol. 6 (J. Cowan, ed.), American Mathematical Society, Providence, R. I. (1974).
89. P. Ortoleva and J. Ross, Phase waves in oscillatory reactions, *J. Chem. Phys.* **58**, 5673–5680 (1973).
90. K. Bimpong-Bota, P. Ortoleva, and J. Ross, Theory of localized chemical instabilities, *J. Chem. Phys.* **60**, 3124 (1974).
91. H. D. Thames and A. D. Elster, Equilibrium states and oscillations for localized two-enzyme kinetics: a model for circadian rhythms, *J. Theor. Biol.* **59**, 415–427 (1976).
92. D. G. Aronson, A comparison method for stability analysis of nonlinear problems, *SIAM Review*, in press (1976).
93. I.-D. Hsü, Oscillatory phenomena for the Glass–Kauffman model of cellular dynamics, preprint (1976).
94. S. Smale, Differentiable dynamical systems, *Bull. AMS* **73**, 747–817 (1967).
95. L. Glass, A combinatorial analog of the Poincaré Index Theorem, *J. Comb. Theory* **B15**, 264–268 (1973).
96. E. N. Gilbert, The ways to build a box, *Mathematics Teacher* **64**, 689–695 (1971).

97. J. Wei, Axiomatic treatment of chemical reaction systems, *J. Chem. Phys.* **36**, 1578–1584 (1962).
98. D. Wallwork and A. S. Perelson, Restriction on chemical kinetic models, *J. Chem. Phys.* **65**, 284–292 (1976).
99. M. Rosenlicht, *Introduction to Analysis*, p. 171, Scott Foresman, Glenview, Illinois (1968).

Author Index

Boldface page numbers indicate a chapter in this volume.

Subject Index